Tsunamis : Their Science and Engineering

Advances in Earth and Planetary Sciences

Tsunamis:
Their Science and
Engineering

Proceedings of the International Tsunami Symposium
1981 IUGG Tsunami Commission
May, 1981, Sendai–Ofunato–Kamaishi, Japan

Edited by

K. Iida
Professor Emeritus, Nagoya University, Japan

and

T. Iwasaki
Department of Civil Engineering, Tohoku University, Japan

Terra Scientific Publishing Company/Tokyo
D. Reidel Publishing Company/Dordrecht, Boston, London

Library of Congress Cataloging in Publication Data

International Tsunami Symposium (1981 : Sendai—shi,
    Miyagi—ken, Japan, etc.)
    Tsunamis : their science and engineering

    (Advances in earth and planetary sciences)
    Includes index.
    1. Tsunamis—Congresses.   I. Iida, Kumizi.   II. Iwasaki, T. (Toshio) III.
International Union of Geodesy and Geophysics. Tsunami Commission.   IV Title.
V. Series.
GC219.I57   1981   551.47'022   83-8661
ISBN-13: 978-94-009-7174-5   e-ISBN-13: 978-94-009-7172-1
DOI: 10.1007/978-94-009-7172-1

Published by Terra Scientific Publishing Company (TERRAPUB),
307 Shibuyadai-haim, 4-17 Sakuragaoka-cho, Shibuya-ku, Tokyo 150, Japan,
in co-publication with D. Reidel Publishing Company, Dordrecht, Holland

Sold and distributed in the U.S.A. and Canada
by Kluwer Boston Inc.,
190 Old Derby Street, Hingham, MA 02043, U.S.A.,
in Japan by Terra Scientific Publishing Company (TERRAPUB),
307 Shibuyadai-haim, 4-17 Sakuragaoka-cho, Shibuya-ku, Tokyo 150, Japan

In all other countries, sold and distributed
by Kluwer Academic Publishers Group,
P. O. Box 322, 3300 AH Dordrecht, Holland

D. Reidel Publishing Company is a member of the Kluwer Academic Publishers Group

Preface

The international tsunami symposium convened by the Tsunami Commission of the International Union of Geodesy and Geophysics was held during May 25–28, 1981 at Sendai, Ofunato and Kamaishi, North East Honshu, Japan. This symposium was organized by the Japanese National Committee for the Organization of International Tsunami Symposium, 1981. The opening and closing ceremonies of the symposium were held at Sendai and Kamaishi, respectively, and eight sessions at Sendai and two sessions at Ofunato were arranged. About 140 scientists and engineers including accompanied persons from ten countries participated to make the symposium a great success. In all, 55 papers were submitted prior to the opening of the symposium, of which 54 papers being orally presented, were arranged in ten sessions:

Tsunami source and earthquake, warning system

Tsunami waves and spectra

Tsunami potential estimation

Theoretical arguments on tsunami waves

Tsunami generation and numerical simulation of historical tsunamis

Harbor oscillations by long waves and tsunamis (1)

Tsunami runup

Mitigation of tsunami hazards and socio-economic effects

Harbor oscillations by long waves and tsunamis (2)

Historical study of tsunamis.

Besides, two special popular lectures were provided for about 800 citizens in the Sanriku coastal area at Ofunato Nokyo Kaikan, where the 1960 Chilean tsunami caused great destructive damages. The title of the first lecture was on the 1960 Chilean Earthquake and Tsunami by Dr. Edgar Kausel, University of Chile, Chile, and that of second lecture was Tsunamis, Computers and Mankind by Dr. W. M. Adams, University of Hawaii, U.S.A.

After the symposium the editorial committee was established for the publication of the Proceedings of this symposium. The committee members were nominated by the Japanese National Committee for the Organization of International Tsunami Symposium 1981, and the committee was organized by Professors K. Iida, T. Iwasaki, K. Kajiura and N. Shuto.

Submitted manuscripts were revised after evaluating comments of several reviewers and finally 39 papers were accepted for publication. These papers were classified and rearranged in seven major topics as represented in this Proceedings:

The Tsunami Impact on Society (2 papers)

Tsunami Source and Earthquake (6 papers)

Historical and Statistical Studies of Tsunamis (9 papers)

Tsunami Generation and Propagation (4 papers)

Topographic Effects on Tsunami Waves (7 papers)

Sea Walls and Breakwaters (4 papers)

Tsunami Runup (7 papers).

It is hoped that the Proceedings will contribute to increase international developement of scientific and technical knowledge of tsunamis and serve to protect mankind from the ravages of destructive tsunamis.

Finally, on behalf of the editorial committee, we express our sincere gratitude to the Japanese National Committee for organization of International Tsunami Symposium 1981 for their financial support of the Proceedings. We also wish to express our heartfelt thanks to the committee secretary, Professor K. Kajiura for his strenuous efforts in compiling the Proceedings.

Kumizi Iida and Toshio Iwasaki
Editors of the Proceedings

Welcome Address

May I begin by greeting the hosts, the Governor of Miyagi Prefecture, the chairman of the Tsunami Commission, Int. Union of Geodesy and Geophysics, distinguished guests, participants to the Symposium, Ladies and Gentlemen.

I sincerely welcome all of you and I would like to call the International Tsunami Symposium 1981 to order, here in Sendai.

The venue is the entrance on route to the Sanriku Coastal Area. Following two days sessions in Sendai, motor-coach transfer from Sendai to Ofunato will give participants opportunity to inspect a lot of tsunami traces and of tsunami defence works such as sea-walls, the Ofunato Breakwater and the Kamaishi Tsunami Breakwater under construction. Half day sessions will be held in Ofunato and the Symposium will be closed in Kamaishi, the central part of Sanriku Coast.

The Sanriku Coast has been attacked by numerous huge tsunamis in history. Among them, the Teikan Tsunami in 869, the Keicho Tsunami in 1611 and the Great Meiji Sanriku Tsunami in 1896 caused tremendous damage to inhabitants and properties. The Showa Hachi-nen Tsunami in 1933 caused also destructive damage. These and other moderate or minor tsunamis were evoked frequently by underwater earthquakes occurring around trenches such as the Japan Sea Trench and the Off-Tokachi Trench. They are called the Near-Tsunamis.

In addition to these, the Sanriku Coast suffered severely by tsunamis which were provoked off the Pacific Coast of South America. For example, 1586, 1906, 1960 and 1966 tsunamis were caused off Peru and two tsunamis in 18th century, four tsunamis in 20th century were generated off Chile and attacked the Sanriku Coast. Among them, the 1960 Chilean Earthquake Tsunami has been still fresh in people's memory. This tsunami attacked Japan on May 24th in 1960, just yesterday of twenty one years ago. May 24 is one of three memorial days in the Sanriku Coast to remind of disasters caused by tsunamis.

The other is March 3 for the 1933 tsunami, by which 3644 lives were lost. The third is June 15 for the 1896 tsunami, in which loss of lives amounted to 27,122. In these memorial days, training for alarming, evacuation and gate operation has been exercised in almost whole region of the Sanriku Coast.

People dwelling in the Sanriku Area is very much nervous for tsunamis. Then to prevent tsunami disaster, they have been striving to construct counter tsunami measures. Now almost whole low land is protected by seawalls built along shorelines.

However in spite of people's effort to mitigate tsunami hazzards, problems are still not yet solved completely. Hydrodynamically, tsunami waves are one kind of Cauchy-Poisson waves. Tsunami waves are translatory wave with dispersive nature both of frequency and of amplitude. Moreover tsunami waves show remarkable non-linearity in shallow regions and in running-up on lands. Usually two locations which are apart only several hundred meters show remarkable discrepancy of run-up heights between each other. That is to say, tsunami waves are affected by local geography and of bathymetry.

Table 1.   Organizations from which generous donations were received.

Iwate Prefecture
Miyagi Prefecture
Shizuoka Prefecture
Kamaishi City
Ofunato City
Kashimadai Town
Japan Association of Promotion of Science
Japan Association of Reclamation and Dreadging
The Federation of Electric Power Companies
Iwate Association for Cooperation in Construction of Fishery Ports
Iwate Association for Cooperation in Construction Engineering
Miyagi Branch of Japan Association of Construction of Fishery Ports
Nippon Steel Corporation, Kamaishi Works
Tohoku Electric Power Co. Ltd.
Association for Promotion of Construction Engineering
C.T.I. Engineering
Goyo Construction Co. Ltd.
Hazamagumi Construction Co. Ltd.
I.N.A. Institute of New Architecture
Japan Consultant of Construction Engineering Inc.
Japan Information Service Ltd.
Japan Mineral Exploration Co. Ltd.
Japan Technology Co. Ltd.
Japan Tetrapod Co. Ltd.
Kashima Construction Co. Ltd.
Kumagai Construction Co. Ltd.
Maeda Construction Co. Ltd.
Meiho Engineering Co. Ltd.
New Japan Technology Consultant Co. Ltd.
NHK Sendai Station
Nishimatsu Construction Co. Ltd.
Ohbayashi Construction Co. Ltd.
Osaka Association of Construction Engineering
Sumitomo Metallurgy Co. Ltd.
Taisei Construction Co. Ltd.
Toa Construction Co. Ltd.
Tokyo Consultant of Construction Engineering, Inc.
Uminokai
UNIC Corporation

International Association of Seismology and Physics Earth's Interior
International Association for the Physical Sciences of the Ocean

Thus how to mitigate tsunami disaster is difficult to solve or to approach to its solution, if scientific proof is necessary. Thus far, theoretical formula or experimental formulation are not satisfactory to meet practical problems. However owing to the development of theory of plate tectonics, vast increase of computer memories and to the progress of hardware technology, knowledge on tsunami waves are now getting to

Welcome Address ix

Table 2. National Committee for Organization of International Tsunami Symposium 1981.

Chairman	: Toshio Iwasaki, Dr. Prof. Fac. Engg. Tohoku Univ., Sendai
Member	: Chiaki Agemori, Dr. Prof. Fac. Agr. Kochi Univ., Nangoku
	Kiyoshi Horikawa, Dr. Prof. Fac. Engg. Univ. Tokyo, Tokyo
	Masanobu Hosoi, Dr. Prof. Fac. Engg. Nagoya Inst. Tech., Nagoya
	Yuichi Iwagaki, Dr. Prof. Fac. Engg. Kyoto Univ., Kyoto
	Kinjiro Kajiura, Dr. Prof. Earth. Res. Inst. Univ. Tokyo, Tokyo
	Hisashi Miyoshi, Dr. Tokyo Univ. Fisheries, Tokyo
	Akira Ozaki, Dr. Prof. Fac. Engg., Hokkaido Univ., Sapporo
	Toru Sawaragi, Dr. Prof. Fac. Engg. Osaka Univ., Osaka
	Nobuo Shuto, Dr. Prof. Fac. Engg. Tohoku Univ., Sendai
	Ziro Suzuki, Dr. Prof. Fac. Sci., Tohoku Univ, Sendai
	Yoshiaki Toba, Dr. Prof. Fac. Sci. Tohoku Univ., Sendai
	Hiroyoshi Togashi, Dr. Prof. Fac. Engg. Nagasaki Univ., Nagasaki
	Hideo Watanabe, Dr. Nagoya Regional Weather Station, Nagoya
Adviser	: Kumizi Iida, Dr. Prof. Aichi Inst. Tech., Toyota, Chairman of Tsunami Commission, IUGG

Table 3. Executive Committee.

Chairman	: Toshio Iwasaki, Prof. Dr.
Secretary, Abstracts and Proceedings	: Kinjiro Kajiura, Prof. Dr.
Secretary, Session Organization and Study Tours	: Nobuo Shuto, Prof. Dr.
Session Organization	: Yoshiaki Toba, Prof. Dr.
	Tomowo Hirasawa, Prof. Dr.
Ladies Program	: Kimiko Iwasaki, Mrs.
	Noriko Shuto, Mrs.
Social Events	: Atsushi Numata, Prof. Dr.
	Tadayasu Uehara, Ass. Prof.
Reception	: Teruko Aikawa, Miss
	Yukio Kono, ME.
Staff of Secretariat	: Chiaki Goto, ME.
	Akira Mano, Dr.
	Masaru Nishizawa, ME.
	Eiji Sato, Mr.
	Hiroshi Sato, Mr.
	Masataka Unohana, Mr.
	Hiroto Yamaji, Mr.
	Jyunko Shima, Miss
	Chiho Sugawara, Miss
Staff of Ladies Program	: Reiko Mugikura, Mrs.
	Keiko Numata, Mrs.
	Toshiko Uehara, Mrs.
Chairman of the Ofunato Executive Committee	: Katsuzo Usui, Major of the Ofunato City
Auditor	: Yoshiaki Toba, Prof. Dr.

increase. By which, engineering problems are becoming to be solved with profound scientific base.

The principal objective of this Symposium is to bring together scientists and engineers to exchange informations on technical advances and to discuss progress in science. This may be possible because of the state of progress just mentioned above. The objective will meet also people's expectation to mitigate tsunami disasters. This is clear by an example of a reader's voice printed in May 18 edition of Kahoku-Shimpo, a newspaper published in Sendai, which expresses his deep concern and expectation on this Tsunami Symposium.

I sincerely hope this Symposium will attain these objectives.

Before closing my remarks, I would like to say acknowledgement to those who sponsored the Symposium. They are the Tsunami Commission, IUGG, the International Association for Hydraulic Research, the Oceanographic Society of Japan, the Seismological Society of Japan and the Japan Society of Civil Engineers. International Association of Seismology and Physics Earth's Interior (IASPEI) and International Association for the Physical Sciences of the Ocean (IAPSO) provided financial support. Furthermore this Symposium was made possible under the generous financial, administrative and moral auspices of Government Administrations, Science Foundations, Professional Associations, Industry and Government Laboratories in Japan. Especially Japan Association of Promotion of Science, Iwate Prefecture, Miyagi Prefecture, Shizuoka Prefecture, Kamaishi City, Ofunato City and many other organizations listed in Table 1 are to be made a grateful acknowledgement for their financial assistance.

Finally as the chairman of the Japanese National Committee for the Organization of International Tsunami Symposium 1981, I would like to introduce persons who cooperate for organization listed in Table 2. Among them, Dr. K. Kajiura worked as the chief of the secretariat and Dr. N. Shuto presided the executive committee listed in Table 3. He also organized technical sessions with the assistance of ME. C. Goto and symposium and post-symposium study tours with M. Unohana. Dr. A. Mano performed registration duties with E. Sato and social events were arranged by Dr. A. Numata with the assistance of ME. T. Uehara. Mrs. K. Iwasaki organized ladies program with Mrs. N. Shuto, Mrs. R. Mugikura, Mrs. K. Numata and Mrs. T. Uehara. ME. M. Nishizawa financed. Mayor K. Usui chaired the Ofunato Executive Committee which organized popular lectures in Ofunato.

Representing the Japanese National Committee for the Organization of International Tsunami Symposium 1981, I hope all participants to join us to express our sincere gratitude to them all.

Thank you.

T. Iwasaki
Chairman of the Japanese National
Committee for the
Organization of International
Tsunami Symposium 1981

Opening Address

The Governor of Miyagi Prefecture, Distinguished Guests, Honorable Participants in The International Tsunami Symposium, Ladies and Gentlemen.

On behalf of the Tsunami Commission, I have the great privilege of opening the International Tsunami Symposium of the International Union of Geodesy and Geophysics here in Sendai, Japan, and have the great honor of extending to you all a very cordial welcome to this symposium. First of all, as the Chairman of the Tsunami Commission, I wish to express our heartfelt gratitude to the Japanese National Committee for organization of International Tsunami Symposium, 1981 led by Professor T. Iwasaki for their untiring efforts and very attentive consideration in arranging this symposium.

The tsunami symposium convened by the Tsunami Commission is a very important event for the entire tsunami community. The Tsunami Commission of the International Union of Geodesy and Geophysics was created in August 1960 in Helsinki, Finland at the 12th General Assembly of the Union just after the 1960 great Chilean tsunami which caused great damage widely in the Pacific area. The purpose of the Tsunami Commission was determined at the time when the Commission was established: (1) Promotion of tsunami research and exchange of scientific information about tsunamis in cooperation with the International Association of Seismology and Physics of the Earth's Interior (IASPEI) and the International Association for the Physical Sciences of the Ocean (IAPSO), (2) Enlargement of the Tsunami Warning Systems in the Pacific, (3) Collection of tsunami data, (4) Tsunami research in the Atlantic in cooperation with Seismological Association in Europe. Since then, in fulfilment of this purpose of the Commission about every two years tsunami symposia have been convened by the Tsunami Commission, the most representative being in Honolulu 1961 and 1969, Moscow in 1971, Wellington in 1974, and Ensenada, Mexico in 1977, and smaller ones in Berkeley in 1963, Bern in 1967, Lima in 1969, Grenoble in 1975, and Canberra in 1979. Several other symposia relating to tsunami in different parts of the world were organized by other international organizations. In Japan, a symposium on tsunamis and storm surges was held in 1966 as a part of the Eleventh Pacific Science Congress, but an international tsunami symposium has never before been held in Japan. Since the Pacific Science Congress, 15 years have passed and much tsunami data have been accumulated. Progress in tsunami science and engineering has also been made by the advancement of Seismology, Oceanography and Coastal Engineering. It was time to organize a symposium to present the recent outcome of research on tsunamis in various fields.

In recent years many interesting and important work related to tsunamis has been carried out. On the basis of a fault-origin model by using modern seismology, the resulting tsunami is computed numerically. Numerical calculations of tsunami problems have advanced considerably. It is said, however, that systematic measurement of tsunamis in the open ocean for research purposes and for reliable tsunami warning is still not accomplished. Further technical improvements are necessary for fast and efficient tsunami warning, and tsunami risk estimations are required for the prevention of tsunami hazards to coastal inhabitants and important construction in coastal regions of the Pacific and other oceans. To cover these requirements, we need further various tsunami data and their analysis. I am sure that there are, however, no doubts on the general progress of tsunami investigations by the participants in this symposium.

The Sanriku district, North-East Pacific coast of Honshu, Japan has been subjected to numerous destructive tsunamis ever since history has been recorded. In this sense it is very meaningful to hold a tsunami symposium here and to understand the nature of tsunamis through the study tour in the Sanriku tsunami area. Also it would be useful for us to discuss how to prevent disasters caused by tsunamis from the standpoint of civil engineering. I sincerely hope all of you will enjoy your stay and excursions in Japan and that this excellent opportunity for a symposium will be fruitful for us.

Thank you.

Prof. K. IIDA
Chairman of The Tsunami Commission of
The International
Union of Geodesy and Geophysics

Contents

TSUNAMI IMPACT

TSUNAMI SOURCE AND EARTHQUAKE

HISTORICAL AND STATISTICAL STUDIES OF TSUNAMIS

TSUNAMI GENERATION AND PROPAGATION

TSUNAMI IMPACT

FROM THE PAST

Tsunamis—Their Science and Engineering, edited by K. Iida and T. Iwasaki, 3–8.

The Tsunami Impact on Society

George Pararas-Carayannis

*International Tsunami Information Center, UNESCO-IOC,
P. O. Box 50027, Honolulu, Hawaii, U.S.A.*

(Received November 1, 1981)

Although infrequent, tsunamis are among the most terrifying and complex physical phenomena and have been responsible for great loss of life and extensive destruction to property. Because of their destructiveness, tsunamis have important impact on the human, social and economic sectors of our societies. Historical records show that enormous destruction of coastal communities throughout the world has taken place and that the socio economic impact of tsunamis in the past has been enormous. In the Pacific Ocean where the majority of these waves have been generated, the historic record shows tremendous destruction with extensive loss of life and property. In Japan, which has one of the most populated coastal regions in the world and a long history of earthquake activity, tsunamis have destroyed entire coastal populations. There is a history of tsunami destruction also in Alaska, in the Hawaiian Islands, and in South America, although records for these areas are not extensive. The last major Pacific-wide tsunami occurred in 1960. Others also occurred but their effects were localized.

We have witnessed in the last twenty years a great deal of growth and development of the coastal areas in most of the developing or developed Pacific nations. This is the result of a population explosion and of technological and economic developments that have made the use of the coastal zone·more necessary than before. Fortunately, tsunamis are not frequent events and therefore, their effects have not been felt recently in all developing areas of the Pacific. History, however, has proven that although infrequent, destructive tsunamis indeed occur.

A major Pacific-wide tsunami is likely to occur in the near future. A number of Pacific nations are not prepared for such an event. Other Pacific nations have let their guard down. The social and economic impact of future tsunamis, therefore, cannot be overlooked. The purpose of this paper is to address in the form of an overview the social and economic impact of past, recent and future tsunamis, tsunami hazard management, and the need for adequate future planning, at least for the Pacific Ocean, where tsunami frequency is high.

1. Social and Economic Impact

The tsunami impact on societies can be traced back in recorded history. Historical records of destructive tsunamis date back to 1480 B. C., in eastern Mediterranean when the Minoan civilization was wiped out by such waves. Japanese records documenting such catastrophes extend back to A. D. 684 (Iida *et al.*, 1967). North and South

3

American records have dated such events back to 1788 for Alaska and 1562 for Chile. Records of Hawaiian tsunamis go back to 1821.

While most of the destructive tsunamis have occurred in the Pacific Ocean, devastating tsunamis have occurred in the Atlantic and the Indian Oceans, as well as the Mediterranean Sea. A large tsunami accompanied the earthquakes of Lisbon in 1755, that of the Mona Passage off Puerto Rico in 1918, and at the Grand Banks of Canada in 1929.

Most of the people in the Pacific nations live on or quite near the coast since the interior is often mountainous and most of the good flatland is in the form of coastal plains. Many of these nations have populations with a natural maritime orientation. For many of these countries, foreign trade is a necessity and some maintain large fleets of ships and have large extensive ports. Many of the island countries of the Pacific and those with extensive continental coastlines depend also on transport by small coastal ships and many small inter-island ports exist to facilitate inter-island and coastal trade as well. Countries, like Japan for example, maintain many ports and have extensive ship-building facilities, electric plants, refineries and other important structures. Similarly, many of the other developing and developed countries of the Pacific have many harbors, which are bases for large fishing industries. Peru, for example, at the port of Callao near Lima, maintains a large fishing fleet for the fishing of anchovies. Callao is also located near a very seismic and potentially tsunamigenic region. Other coastal sites throughout the Pacific have begun aquacultural industries and canneries. Therefore, there is a combination of factors which make a number of these developed and developing Pacific islands and continental Pacific nations vulnerable socially and economically to the threat of tsunamis. The extensive coastal boundaries, the number of islands, the long coastlines of Pacific nations containing a number of vulnerable engineering structures, the numerous large ports, the productive fishing and aquacultural industries and the large density of population in coastal areas, puts many of these countries in a very vulnerable position. Japan, for example, which has all the factors of vulnerability mentioned, the social and economic impact of a tsunami can be truly devastating. For example, along the Sanriku Coast or in the Tohoku District of northern Honshu there are a number of flatlands with numerous coastal embayments, where large fishing and aquaculture industries have been established. Throughout history, entire settlements in such areas were struck and destroyed by tsunamis often requiring their rebuilding and relocation. A total of 65 destructive tsunamis struck Japan between 684 A. D. and 1960. As early as July 18, 869 the Sanriku coast was hit by a tsunami which resulted in about 1,000 deaths and the destruction of hundreds of villages. On August 3, 1361 a tsunami destroyed 1,700 houses in the same area. On September 20, 1498 a thousand houses were washed away and 500 deaths resulted from a tsunami which struck the Kii peninsula. Kyshu was struck by a destructive tsunami in September 1596. Great loss of life occurred on 31 January 1596 from a tsunami on the island of Shikoku, affecting also a number of regions in Honshu. In recent times, the great Meiji Sanriku tsunami of June 15, 1896 resulted in 27,122 deaths, thousands of injuries, and the loss of thousands of homes. On March 3, 1933 a tsunami in the Sanriku area reached a height of about 90 feet and killed over three thousand people, injured hundreds more and destroyed approximately 9,000 homes and 8,000 boats. On

December 1944 a tsunami in Central Honshu caused almost 1,000 deaths and the destruction of over three thousand houses. The December 21, 1946 Nankaido tsunami resulted in 1,500 deaths and the destruction of 1,151 houses (IIDA *et al.*, 1967).

In the Hawaiian islands tsunamis have struck repeatedly causing great loss of life and immense damage to property. Most noteworthy of the recent Hawaiian tsunamis is that of 1 April 1946 which inundated and destroyed the city of Hilo, killing 159 people. Other recent tsunamis that have hit Hawaii are those of 1952, 1957, 1960, 1964 and 1975, (PARARAS-CARAYANNIS, 1977).

The most destructive Pacific-wide tsunami in recent times was that of May 1960, which killed over 1,000 people in Chile, in Hawaii, the Philippines, Okinawa and Japan, and causing tremendous loss of life and destruction to property.

More recently, on August 16, 1976 a large earthquake in the Moro Gulf in the Philippines, generated a destructive local tsunami which killed over 8,000 persons, leaving 10,000 injured and 90,000 more homeless (ITIC, 1978). Another earthquake on 12 December 1979 centered in the State of Narino in the southwest corner of Colombia generated a tsunami that destroyed completely several fishing villages killing hundreds of people and creating economic chaos in an already economically depressed region of that country (PARARAS-CARAYANNIS, 1980).

Tsunami destruction has not been confined to Japan or to the Pacific Ocean. Destructive tsunamis have occurred also in the Atlantic and Indian Oceans and in the Carribean and Mediterranean Seas.

The above is simply a brief overview of some large historical tsunamis. It is very difficult to comment specifically on the impact each event has had on each stricken area. However, it can be clearly concluded that natural catastrophes, such as tsunamis, have far more important and long-term social and economic impacts that any historical or statistical record can show. Furthermore, the historical record does not reflect the potential damage that can be caused presently from tsunamis, since a great deal of development has taken place in the last twenty years in the coastal areas of many developing or developed coastal nations. The social and economic impact of future tsunamis will be extremely more severe than that of past events. It is therefore, important to plan and prepare for such future events.

2. Tsunami Hazard Management

There is very little that can be done to prevent the occurrence of natural hazards. Floods, droughts, earthquakes, hurricanes, volcanic eruptions, and tsunamis cannot be prevented. But humankind being as adoptable as it is, has learned to live with all these hazards. Our past approach to hazards has been passive. Justifying them as acts of God or nature for which we can do very little about. Perhaps these natural disasters cannot be prevented but their results and effects, such as loss of life and property, can be reduced by proper planning. To plan, however, for the tsunami hazard, we have to have a good understanding not only of the physical nature of the phenomenon and its manifestation in each geographical locality, but we have to know also about the combination of the physical, social or cultural factors in each locality of each nation in order to effectively plan for the mitigation of the tsunami effects. Clearly, there are areas which are more vulnerable to tsunamis than others. Because tsunami frequency in the Pacific Ocean is high, most efforts in hazard management have concentrated in this area of the world. However, in developing coastal zone management and land use, the tsunami impact, however remote, should be considered. While some degree of risk is acceptable, government agencies should regulate new development and population growth into areas of greater safety and less potential risk. These agencies should formulate land use regulation depending on the tsunami risk potential of a given coastal area, particularly, if such area is known to have sustained damage in the past.

3.   Protective and Preventive Measures

Present protective measures involve primarily the use of tsunami warning systems employing advance technological instrumentation for data collection and for warning communications. Nations such as Japan, USSR, Canada, and USA have developed somewhat sophisticated warning systems with the responsibility of sharing warning information with other nations of the Pacific.

In addition to the International Tsunami Warning System, a number of Regional Warning Systems have been established to warn the population in areas where tsunami frequency is high and where immediate response is necessary. Such Regional Tsunami Warning Systems have been established in USSR, Japan, Alaska and Hawaii. However, vast areas exist where tsunamis cannot be adequately detected or monitored in time and the populations warned to prevent extensive loss of life.

4.   Hazard Perception and Credibility

Because of the rarity of large destructive tsunamis, it is difficult to institute successful tsunami prediction schemes for warning the public. However, we can make them aware of the potential hazard. Tsunami warnings are issued to the public for the purpose of convincing people to evacuate endangered areas. Ample time must be allowed for evacuation, which is a rather difficult procedure. Often the public does not understand the meaning of the warning signals and is not aware of the locations of endangered areas. Most people are reluctant to evacuate their homes and businesses, and their response to warnings in general may not be very good, particularly, if a number of false alarms have been issued.

Tsunami hazard perception by the people of a coastal area is based on education and confidence in government agencies responsible for tsunami predication. Over-warning, based on inadequate knowledge of the phenomenon, or inadequate data on which to base the prediction, often leads to false alarms, and lack of compliance with warning and evacuation attempts. Such false alarms result in a loss of faith in the capability of the system and result in reluctance to take action in subsequent tsunami events. Even if a tsunami prediction is based on valid information and data, warning and evacuation may not be sufficient to minimize the impact of tsunamis on coastal population. Hazard perception by the public is based on a technical understanding of the phenomenon, at least at the basic level, and behavioral response based on the understanding of the phenomenon and the confidence of the public for the authorities.

Fortunately, forecasting of tsunamis in recent years has been quite good and the image of the tsunami warning system and its credibility have improved considerably. Forecasting, however, is not an exact science as the phenomenon itself is very complex, and data on which the forecast is based may often be inadequate for certain areas.

5. Educational Effort

A heightened community awareness of the potential threat of tsunamis, can be achieved through a public education program. Civil Defense authorities in each country can initiate such a public education program consisting of seminars and workshops for responsible government officials, can publish informational booklets on the hazards of tsunamis, and can coordinate with the communications media on the announcement of tsunami information. Other government agencies can take action also to mitigate future losses from tsunamis. For example, government agencies can develop sound coastal management policies, which include zoning and planning for tsunami-prone coastal areas. Scientific organizations can undertake research and engineering studies in developing evacuation zones or engineering guidelines for building coastal structures. Audio visual materials can be prepared for educating children in schools and the public in general. Brochures and pamphlets can be printed describing the tsunami warning system and what the public can do in time of tsunami warning. Internally, government agencies can streamline and coordinate their operating procedures and communications so they can perform efficiently when the tsunami threat arises. Procedures related to tsunami warnings should be reviewed frequently to define and determine better respective responsibilities between the different government agencies at all levels.

6. Conclusion and Recommendations

In spite of our technological improvements of the last two decades, we are still unable to provide timely warnings to many areas of the Pacific, and none for other parts of the world. Improvements are necessary in communications to insure that warning information is prompt and accurate. An increased degree of automation is necessary in handling and interpreting the basic data. Research is needed for example in the development of instrumentation such as deep ocean sensors, which could be useful in early tsunami detection. Research is needed also in the real time interpretation of seismic source parameters which in turn may help in tsunami evaluation. Apparently more research is needed in improving our understanding of a tsunami interacting with the coast.

Research can also assist, not only in the improvement of warning systems, but in our land-use management of tsunami-prone coastal areas, or in the development of important engineering guidelines of critical coastal structures.

In conclusion, the long term objective should be for each nation susceptible to the tsunami hazard to build its technical and scientific infrastructures to meet the hazards of a disastrous event. The immediate objectives of each such state should be to assess this hazard in terms of its potential needs and available resources. Preparedness requires several capabilities, such as rapid identification of imminent tsunamis, effective national and regional warning systems to alert coastal population and industries, and Civil Defense and community preparedness to respond to tsunami warnings.

Finally, appropriate improvements in warning capability in the form of improved instrumentation for tsunami monitoring and for communications should be developed, both for effective warning, and for increased knowledge as an aid to long-term protection.

REFERENCES

Iida, K., D. C. Cox, and G. Pararas-Carayannis, Preliminary Catalog of Tsunamis Occurring in the Pacific Ocean, Data Report No. 5, Hawaii Institute of Geophysics, University of Hawaii, Honolulu, 1967.

ITIC, Tsunami Reports, No. 1976–26, 1978.

Pararas-Carayannis, George, Catalog of Tsunamis in Hawaii, World Data Center-A for Solid Earth Geophysics, 24–43, Boulder, Colorado, 1977.

Pararas-Carayannis, George, Earthquake and Tsunami of 12 December 1979 in Colombia, *Tsunami Newsletter*, **13** (1), 1–9, 1980.

Tsunamis—Their Science and Engineering, edited by K. Iida and T. Iwasaki, 9–22.

Tsunami Disasters and Protection Measures in Japan

Kiyoshi Horikawa* and Nobuo Shuto**

*Department of Civil Engineering, University of Tokyo,
Tokyo, Japan
**Department of Civil Engineering, Tohoku University,
Sendai, Japan

(Received August 30, 1981; Revised February 20, 1982)

This paper is composed of three parts. The first part describes actual tsunami disasters caused directly by tsunamis, and also the secondary damage indirectly attributable to tsunami attacks. In order to demonstrate a typical example of disaster criteria, the subject of boat damage was investigated in relation to tsunami cresting height above mean sea level.

The second part is devoted to the evaluation of the recurrence period of disasterous tsunamis based on several approaches. These approaches are principally divided into two categories, the first on the bases of tsunami magnitude records and the second on tsunamigenic earthquake records. As a result of this treatment, it is concluded that the possible occurrence frequencies of the 1896 Sanriku tsunami, the 1933 Sanriku tsunami, and the 1968 Tokachi-oki tsunami are once in 100 years, 70 years, and 40 years, respectively.

The third part of this paper briefly discusses various kinds of tsunami counter-measures which have been adopted in Japan during the last twenty to forty years. As a conclusion, the authors stress the importance of the idea that tsunami attacks cannot be completely prevented by any means, and that a proper combination of various countermeasures is required.

1. Introduction

Twenty-one years have passed since the 1960 Chilean Earthquake tsunami attacked the circum-Pacific Ocean coast and caused tremendous damage on the Japanese islands. Since then, we have had several tsunami attacks in Japan such as the Niigata Earthquake tsunami in 1964, and the Tokachi-oki Earthquake tsunami in 1968. In recent times, numerous tsunami countermeasures have been adopted to prevent future damage at localities where past tsunamis have attacked with serious aftermath.

On the other hand, during the last few years, a great interest has been taken in the possible occurrence of a future serious earthquake in the district of Shizuoka Prefecture. This earthquake is expected to have its epicenter at the trough in Suruga Bay and to generate a tsunami large enough to cause tremendous damage on the coast. Reflecting the above circumstances, the people living along the coastal region are greately worried about the disaster accompanying possible future earthquakes as well as tsunamis.

From the engineering viewpoint, the final target of tsunami researches is to understand tsunami behavior in the coastal region, to predict probable damage caused by tsunami attack, and finally to propose adoptable means for survival of the inhabitants and protection of property in a given area.

"*Saigai wa wasureta koro ni yattekuru.*" This means, "Natural disaster may invade again when the people have forgotten the past one," and is a well-known Japanese Proverb which gives clear warning to the people concerning future natural disasters. It seems to be true that twenty years are long enough for the inhabitants to lose their memories of disasters.

Considering the above situation, the authors have the intention to first review past records of tsunami damage in Japan and to classify disaster types for practical purposes. This classification may be helpful to understand what happens during a tsunami attack, to find out the appropriate tsunami countermeasures, and also to evaluate the functioning of these measures.

In the second part of this paper, the authors have made an effort to evaluate the appropriate occurrence frequency of representative tsunamis in Japan by means of various approaches. This kind of approach might be useful to establish appropriate investment criteria for various tsunami protective measures in combination with the risk probability. In order to proceed with the above, it is needless to say that sociological changes during the past period such as the population growth, living style, and various properties constructed must be taken into account in the analysis of the data.

2.   Tsunami Disasters

As typical examples of tsunami disasters, the records of the Sanriku tsunami in 1933 and of the Chilean tsunami in 1960 will be cited (Horikawa, 1978).

The Sanriku tsunami was generated by a submarine earthquake of magnitude $M = 8.5$, the epicenter of which was located off the Sanriku coast on the north-eastern coast of the main island of Japan (Honshu). The tide record shows that the predominant period of the tsunami was 15 to 20 min. The official records reveal that: (1) the tsunami cresting heights above mean sea level were 10 m at Taro, 23 m at Shirahama, 24 m at Ryori, and 7 m at Tadakoshi, all of these towns located in Iwate Prefecture; (2) the death toll amounted to 3,008, with 1,152 injured, 4,917 houses washed away, 2,346 houses destroyed, 4,329 houses inundated or flooded, 249 houses burned, 7,303 boats washed away, and 901 boats destroyed.

The Chilean tsunami was generated by an earthquake originating in Chile of magnitude $M = 8.25$ to 8.5, with the epicenter located off Valdivia of the southern part of the Chilean coast. The predominant period was reported to be about one hour, and tsunami cresting heights above T.P. (Tokyo Peil) were recorded to be 0.4 m to 5.0 m along the eastern coast of Hokkaido, 0.6 m to 6.4 m along the Sanriku coast, and 1 m to 3 m along the southern coast beyond the Kanto district. In Japan the death toll amounted to 119, with 872 missing, 872 injured, 2,830 houses totally destroyed and washed away, 19,863 houses inundated above the ground level, and a great number of ships and public utilities damaged.

On the basis of the above and other official records, the disaster caused directly by tsunamis can be summarized as falling in the following groups: (1) death and injury, (2) houses destroyed, partly destroyed, inundated or flooded, and burned, (3) property damage and loss, (4) boats washed away, destroyed, and run on to rocks, (5) lumber washed away, (6) marine installations destroyed, and (7) disastrous damage of public utilities such as railroads, roads, electric power supply installations, and water supply installations.

Secondary damages indirectly caused by tsunamis are divided into four categories; namely (1) burning houses, boats, oil tanks and gas stations; (2) drifting matter such as houses, lumber, boats, drums, automobiles, and sea culture nursery rafts. These are sometimes quite dangerous and caused destruction of houses, bridges and so on; (3) environmental pollution caused by drifting materials, oil, polluted sea bed and epidemic prevention, which might be serious for the present high-density living; and (4) traffic obstruction due to the destruction of roads and railroads, which is of primary importance to the first stage of rescue activities.

In order to consider a typical example of disaster criterion, the records of boat damage caused by the 1933 Sanriku tsunami will be used. Figure 1 shows the relationship between the tsunami cresting height above mean sea level and the boat disaster rate. From this diagram, destruction of boats can be caused by tsunamis higher than 2 m, and the total destruction rate of boats increases with increase of tsunami cresting height. Figure 2 gives the disaster rate of boats with or without an engine. From these curves, it can be realized that the disaster rate for small boats without engines is in general tremendously high.

3. Prediction of Disastrous Tsunami Recurrence Period

In order to clarify tsunami damage, IMAMURA (1949) defined a tsunami magni-

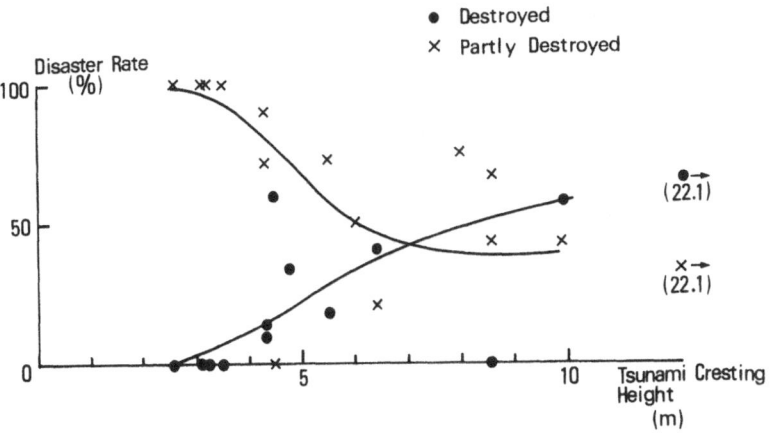

Fig. 1.   Relationship between tsunami cresting height and ship disaster rate.

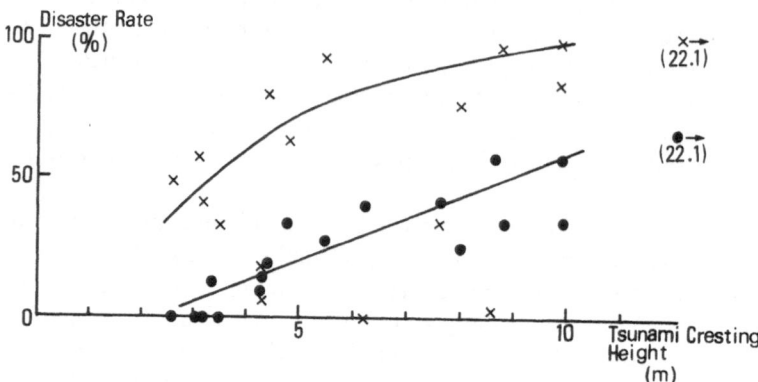

Fig. 2.   Relationship between tsunami cresting height and disaster rate of ships (destroyed and washed away).

tude, m, as similar to the earthquake magnitude, M. Table 1 gives the tsunami height which means the tsunami cresting height along the open coast, the tsunami total energy, and the description of damage in relation to the scale of tsunami magnitude. IIDA (1958) proposed a linear relationship between m and M, and SOLOVIEV (1970) presented a functional relationship between the tsunami magnitude and the tsunami run-up height. These are expressed by the following equations, respectively;

$$m = 2.61M - 18.44 \qquad \text{(Iida)} \qquad\qquad (1)$$

$$m = \log_2 \sqrt{2\bar{h}} \qquad \text{(Soloviev)} \qquad\qquad (2)$$

where $\bar{h}$ denotes mean tsunami run-up height.

The "Chronological Table in Science," issued in 1981 through Iwanami Publishing Co., contains a list of earthquakes which occurred in Japan. From this table, 106 tsunami records can be read and these are divided into six categories based on the

Table 1.   Tsunami magnitude m and related parameters.

m	H (Tsunami height)	Total energy (10^{22}erg)	Damage
4	30 m	64.00	Considerable damage along more than 500 km of coastline
3	10–20	16.00	Considerable damage along more than 400 km of coastline
2	4–6	4.00	Damage and lives lost in certain landward areas
1	2	1.00	Coastal and ship damage
0	1	0.25	Very small damage
−1	50 cm	0.06	None

tsunami magnitude as shown in Table 2. Among these, extreme tsunamis with $m = 4$ are selected and listed in Table 3. From the latter table, it can be realized that extremely large tsunamis have widely attacked the Pacific Ocean coast of Japan, especially frequently on the Sanriku coast.

In the following discussion, the possible occurrence frequencies of the 1896 Sanriku tsunami, the 1933 Sanriku tsunami, and the 1968 Tokachi-oki tsunami will be evaluated as representative cases for Japan.

HATORI (1975, 1977) presented the following two tables. Table 4 is for the historical tsunamis off the Sanriku coast occurring before 1861, and Table 5 is for the recent Sanriku tsunamis occurring in the period from 1896 to 1971. Figure 3 shows the chronology of Sanriku tsunamis with magnitude m larger than or equal to 1, where "a" denotes the 1896 tsunami, "b" the 1933 tsunami, and "c" the 1968 tsunami. Recorded data before 1585 are very scarce, hence if these are not taken into account, tsunamis with the magnitude larger than or equal to 2 occurred once in 50 to 100 years. By using the relationships between m and $\bar{h}$ as well as h_{max} proposed by Soloviev, Table 6 is provided to convert m into $\bar{h}$ or h_{max}.

The historical Sanriku tsunamis during the 1,100 years in the period from 869 to 1971 were rearranged in the order of tsunami magnitude (Table 7). If it is assumed that in the years not appearing in the above list that tsunamis of $m \geq 1$ were not generated, the possible occurrence frequency for each tsunami magnitude can be evaluated as shown in the fourth row of this table. That is to say, the occurrence frequency of the

Table 2. Tsunamis recorded in Japan (416–1978).

m	Number of records	
4	7	(7 %)
3	14	(13 %)
2	25	(24 %)
1	33	(31 %)
0	10	(9 %)
−1	17	(16 %)
Total	106	(100 %)

Table 3. Extreme tsunamis in Japan ($m = 4$).

Date	M	Region
869 (VII 13)	8.6	Sanriku
1611 (XII 2)	8.1	Sanriku ~ Hokkaido
1707 (X 28)	8.4	Tokaido ~ Kyushu
1711 (IV 24)	7.4	Miyako ~ Yaeyama
1854 (XII 24)	8.4	Boso ~ Kyushu
1896 (VI 15)	7.6	Sanriku
1960 (V 23)	8.5	(Chile)

Table 4.   Historical tsunamis off Sanriku (after Hatori).

Date	M	m
869 (VII 13)	8.6	4
1257 (X9)	7.0	1
1585 (VI 21)		1
1611 (XII 2)	8.1	3~4
1616 (IX 9)	7.0	1
1677 (VI 13)	8.1	2.5
1689 (——)		1
1751 (VII 24)		0
1763 (I 29)	8	2.5
1763 (III 15)	7.5	1
1793 (II 17)	7.7	2
1835 (VI 20)	7.5	1.5
1846 (III —)		0
1856 (VIII 23)	7.8	2.5
1861 (X 21)	7.4	1

Table 5.   Sanriku tsunamis in 1896–1971 (after Hatori).

Number	Date	M	m
1.	1896 (VI 15)	7.6	3~4
2.	1897 (II 20)	7.4	0
3.	1897 (VIII 5)	7.7	2
4.	1898 (IV 23)	7.3	−1
5.	1901 (VIII 9)	7.5	0
6.	1901 (VIII 10)	7.4	0
7.	1915 (XI 1)	7.5	0
8.	1927 (VIII 6)	6.9	−1
9.	1928 (V 27)	7.0	−1
10.	1931 (V 9)	7.6	0
11.	1933 (III 3)	8.3	3
12.	1933 (VI 19)	7.1	−0.5
13.	1935 (X 13)	7.2	0
14.	1935 (X 18)	7.1	−1
15.	1936 (XI 3)	7.7	0.5
16.	1943 (VI 13)	7.1	0
17.	1945 (II 10)	7.3	−1
18.	1960 (III 21)	7.5	0.5
19.	1960 (III 23)	6.7	−1
20.	1960 (VI 30)	6.7	−1.5
21.	1962 (IV 12)	6.8	−0.5
22.	1968 (V 16)	7.9	2.5
23.	1968 (V 16)	7.5	(0.5)?
24.	1968 (VI 12)	7.2	0.5
25.	1971 (VIII 2)	7.0	−0.5

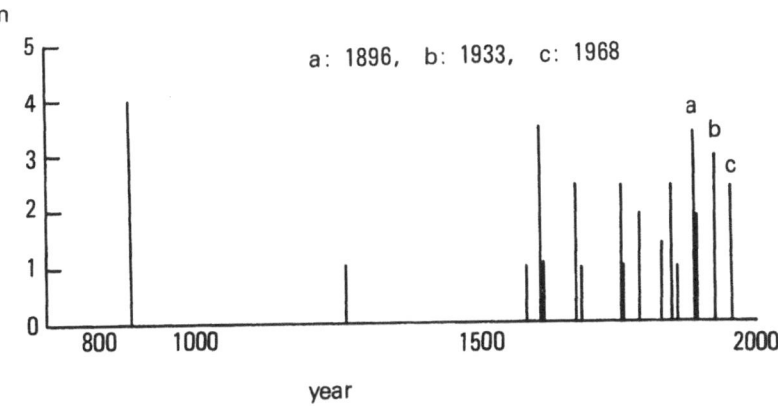

Fig. 3.   Chronology of tsunamis ($m \geq 1$).

Table 6.   Relationship between tsunami magnitude and tsunami run-up height (after Soloviev).

m	$\bar{h}$ (m)	h_{max} (m)
4.5	16	73.9
4	11.3	40.3
3.5	8	22.9
3	5.7	13.4
2.5	4	7.9
2	2.8	4.8
1.5	2	3.1
1	1.5	2.1
0.5	1	1.3
0	0.7	0.9
−0.5	0.5	0.6
−1	0.4	0.4
−1.5	0.25	0.3
−2	0.2	0.2
−2.5	0.125	0.125
−3	0.1	0.1

1896 tsunami is once in 300 years, the 1933 tsunami once in 250 years, and the 1968 once in 140 years.

However the number of tsunamis recorded before 1585 is very small, and if these were neglected, the occurrence frequency of each tsunami is given as shown in the fifth row of Table 7. From this result it can be said that the 1896 tsunami is once in 200 years, the 1933 tsunami once in 150 years, and the 1968 tsunami once in about 90 years.

Figure 4 gives a chronological plot of recent Sanriku tsunamis. From this diagram, it is realized that tsunamis of $m \geq 2.5$ seem to be generated once in 30 to 40 years.

Table 7. Occurrence frequency of tsunamis ($m \geq 1$).

m	N	Date	Occurrence frequency	
			after 869	after 1585
4	1	869 (VII 13)	once in about 1000 years	
3–4	2	1611 (XII 2) 1896 (VI 15)	once in about 300 years	0.0013 0.0039
3	1	1933 (V 3)	once in about 250 years	0.0065
2.5	4	1677 (IV 13) 1763 (I 29) 1856 (VIII 23) 1968 (V 16)	once in about 140 years	0.0091 0.0117 0.0142 0.0168
2	2	1793 (II 17) 1897 (VIII 5)	once in about 110 years	0.0194 0.0220
1.5	1	1835 (VII 20)	once in about 100 years	0.0246
1	6	1257 (X 9) 1585 (VI 21) 1616 (IX 9) 1689 1763 (III 15) 1861 (X 21)	once in about 65 years	0.0272 0.0298 0.0324 0.0350 0.0376

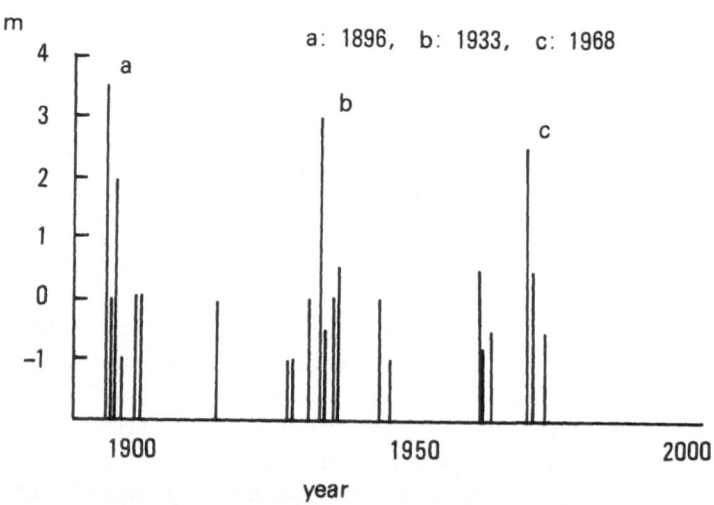

Fig. 4. Chronology of recent tsunamis ($m \geq -1.5$).

By taking the largest record for each year, and by assuming that magnitudes of tsunamis generated in non-recorded years were $m < -1$, the occurrence frequency of each recorded tsunami can be evaluated as shown in Table 8. Then is concluded that the 1896 tsunami is once in 100 years, the 1933 tsunami once in 65 years, and the 1968 tsunami once in 30 years.

As another approach, the magnitude data of earthquakes which generated Sanriku tsunamis were considered. The largest for each year were arranged in order of M. Under the assumption that the earthquake magnitude M reflects the size of tsunamis in a unique way, these data are plotted in Figs. 5 and 6 depending upon whether the assumed occurrence frequency was expressed by a log-normal or a normal distribution, respectively. In the former case, it can be read that an 1896 type tsunami occurs once in 17 years, a 1933 tsunami once in 200 years, and a 1968 tsunami once in 50 years. In the latter case, an 1896 tsunami occurs once in 17 years, a 1933 tsunami once in 125 years, and a 1968 tsunami once in 40 years.

Soloviev presented a relationship between N and M as follows,

$$\log_{10}N = a - bM \tag{3}$$

where N is the number of earthquakes with magnitude M occurring in 100 years. For the east coast of Japan, the values of $a = 9.21$ and $b = 1.08$ were given. On the other hand, the relationship between m and n is expressed by,

$$\log_{10}n = \alpha - 0.31\,m \tag{4}$$

Table 8. Occurrence frequency of tsunamis (1896–1971).

Order	Number	Year	m	Frequency (per year)
1	1	1896	3.4	0.0067
2	11	1933	3	0.0200
3	22	1968	2.5	0.0333
4	3	1897	2	0.047
5	15	1936	0.5	0.060
6	18	1960	0.5	0.073
7	5	1901	0	0.087
8	7	1915	0	0.100
9	10	1931	0	0.113
10	13	1935	0	0.127
11	16	1943	0	0.140
12	21	1962	−0.5	0.153
13	25	1971	−0.5	0.167
14	4	1898	−1	0.180
15	8	1927	−1	0.193
16	9	1928	−1	0.207
17	17	1945	−1	0.220

18 K. HORIKAWA and N. SHUTO

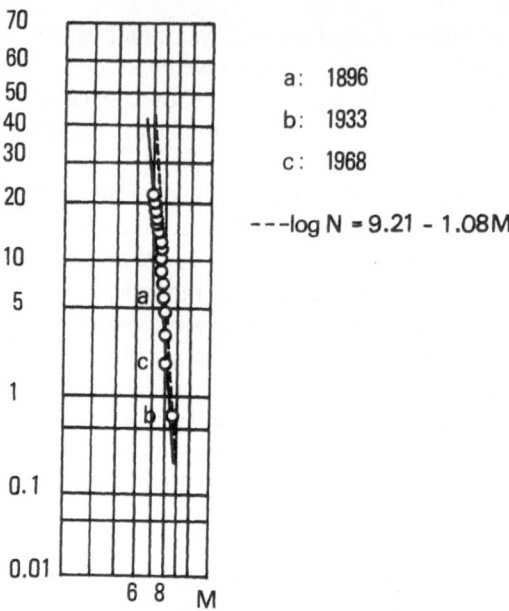

Fig. 5. Occurrence frequency of earthquakes which produced tsunamis (based on records of 1896–1971).

a: 1896, b: 1933, c: 1968

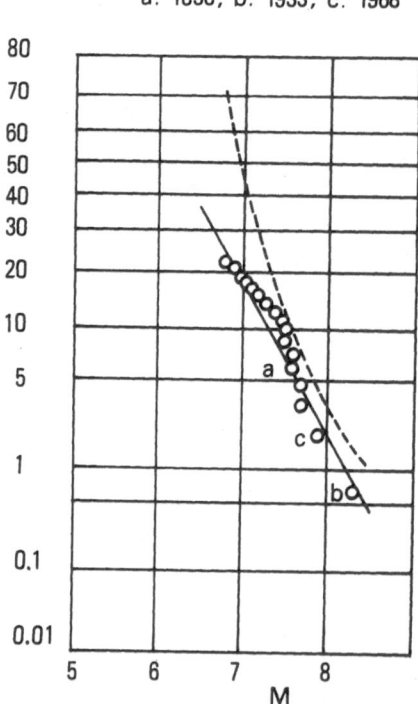

Fig. 6. Occurrence frequency of earthquakes which produced tsunamis (based on records of 1896–1971).

where n is the number of tsunamis with the magnitude m occurring in 100 years. The constant α is different for each region, and is characterized by the ratio between $n(0)$ for $m = 0$ and $N(7.5)$ for $M = 7.5$.

Soloviev obtained the following result for the east coast of Honshu

$$T = \log_{10}\frac{n(0)}{N(7.5)} = -0.4 \tag{5}$$

and gave a value of $N(7.5) = 17.3$, which is larger than the value evaluated by using Eq. (3). Therefore $n(0) = N(7.5)10^{-0.4} = 6.89$. On the other hand, $\alpha = \log_{10}n(0)$ from Eq. (4), hence $\alpha = \log_{10}6.89 = 0.838$. Based on the above relationship, the occurrence frequency of each tsunami can be evaluated. That is to say, the 1896 tsunami occurs once in 200 years, the 1933 tsunami once in 130 years, and the 1968 tsunami once in 90 years.

Finally, the following equation proposed by IIDA (1958) will be applied to the present question. The equation is

$$\log_{10}n = 1.00 - 0.77 \log_{10}H \tag{6}$$

where H is the tsunami cresting height in meters evaluated as the mean for each rank of m in Table 1, and n is the number of tsunamis with height H occurring in 100 years. Then the evaluated results are as follows: the 1896 tsunami is once in 100 years, the 1933 tsunami once in 75 years, and the 1968 tsunami once in 55 years.

In order to check the adaptability of the above various recurrence curves, a comparison was made using the collected data. Figure 7 is for the tsunamis with $m \geq 1$ recorded from 1585 to the present, while Fig. 8 is for the tsunamis recorded in the last 80 years. From these diagrams, it is realized that Eq. (4) agrees well with the data in Fig. 7, but does not agree well with those in Fig. 8, while Eq. (6) gives values which fit rather well with the data for small value of m in Fig. 7, and for large value of m in Fig. 8.

In parallel to the above treatments, records of earthquakes which generated tsunamis in the period from 1896 to 1971 were used in order to investigate the adaptability of Eq. (3). The solid lines in Figs. 5 and 6 are the regression lines on log-normal and normal distribution graphs, while the dashed lines are the curves given from Eq. (3). Based on these comparisons, Eq. (3) seems to give a slightly higher value of occurrence frequency in comparison with the actual data.

The above results are summarized in Table 9. It should be pointed out in all these treatments, except Iida's, only tsunamis generated off the Sanriku coast were taken into consideration, and the region of interest covers the wide area from Tokachi in Hokkaido as the northern boundary, to Miyagi as the southern boundary.

As a conclusion of the present investigation, the occurrence frequencies of the 1896, 1933 and 1968 tsunamis are possibly once in 100 years, 70 years, and 40 years, respectively.

4. Tsunami Countermeasures

It is quite dangerous to believe that the violent attack of tsunamis can be completely prevented by man-made structures. Based on past experience, evacuation

K. HORIKAWA and N. SHUTO

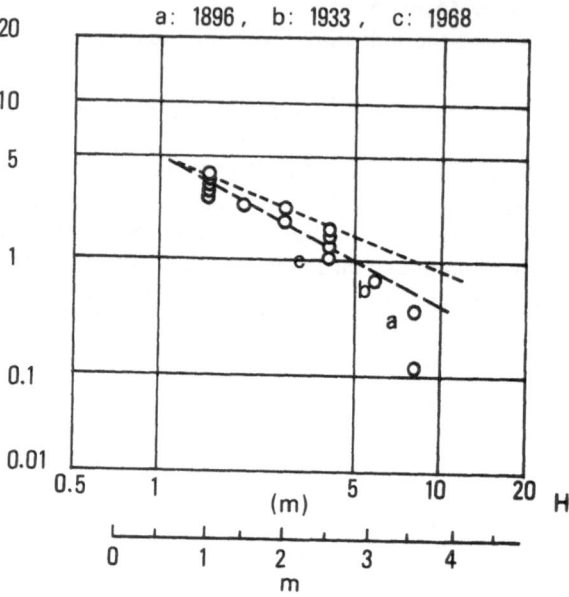

Fig. 7.   Occurrence frequency of tsunamis (based on tsunamis with $m \geq 1$ since 1585).

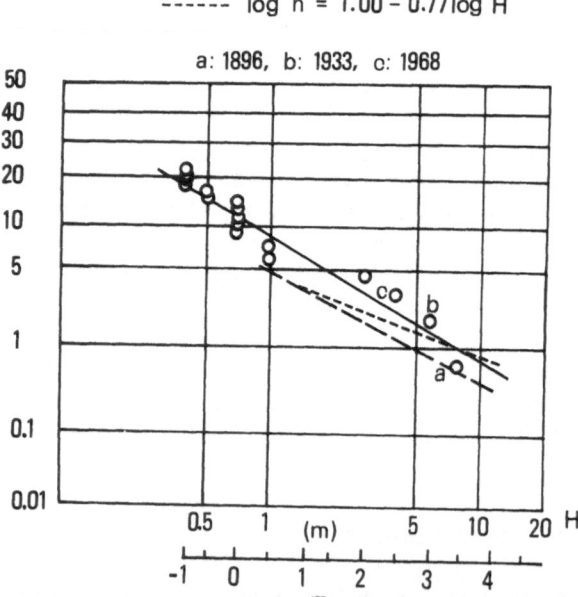

Fig. 8.   Occurrence frequency of tsunamis (based on records of 1896–1971).

Table 9. Summarized results.

	1896	1933	1968
Records after 1585	1/200	1/150	1/90
Records in 1896–1971	1/100	1/65	1/30
Soloviev	1/200	1/130	1/90
Iida	1/100	1/75	1/55
	1/100	1/70	1/40

to a safe area before a tsunami attack is the best recourse for the inhabitants. It is incorrect to depend too much on the functioning of coastal defense structures. The Meteorological Agency of Japan has the responsibility of issuing warnings to the affected people through the appropriate media. To fulfill this important mandate, the Agency and its branch offices collect seismological and oceanographical data. The inhabitants of coastal areas frequently attacked by tsunamis should regularly drill in warning communications and refuge taking.

Standard tsunami countermeasures are: (1) to establish a warning system; (2) to relocate houses from the low-lying land to safe areas; (3) to construct escape routes; (4) to carryout evacuation training; (5) to plant and cultivate forests in order to reduce the tsunami attack; (6) to construct coastal dikes; (7) to leave open spaces for absorbing the tsunami energy; and (8) to construct permanent structures along the waterfront. In addition, tsunami breakwaters have been completed at several locations in Japan since the Chilean tsunami destruction. A new tsunami breakwater is planned for Kamaishi Harbor, and its construction is now in progress. The details of the various construction works carried out in Sanriku District are described in a booklet entitled "An Introduction to Tsunamis and Defense Works in the Sanriku Coastal Area" (Organizing Committee of International Tsunami Symposium, 1981).

The authors believe that the planning of appropriate tsunami countermeasures should be based upon the following terse observation: tsunami can not be completely prevented by any means. The proper combination of various countermeasures is strongly recommended in order to reduce the disaster in the total views.

REFERENCES

HATORI, T., Tsunami magnitude and wave source regions of historical Sanriku tsunamis in northeast Japan, *Bull. Earthq. Res. Inst.*, **50**, 397–414, 1975 (in Japanese).

HATORI, T., Past Sanriku tsunamis and their source areas, *Report on Prediction of Disasters Caused by Future Sanriku Great Tsunamis*, edited by T. Iwasaki, pp. 15–24, Natural Disaster Science Research Group, 1977 (in Japanese).

HORIKAWA, K., Coastal Engineering—An Introduction to Ocean Engineering, pp. 337–346, University of Tokyo Press, 1978.

IIDA, K., Magnitude and energy of earthquakes accompanied by tsunami, and tsunami energy, *J. Earth Sci.*, *Nagoya Univ.*, **6**, 101–112, 1958.

IMAMURA, A., A classification of tsunamis in Japan, *Jisin (Earthquake)*, **2**, No. 2, 23–28, 1949 (in Japanese).

SOLOVIEV, S. L., Recurrence of tsunamis, *Tsunamis in the Pacific Ocean*, edited by W. M. Adams, pp. 149–163, East-West Center Press, Honolulu, 1970.

TSUNAMI SOURCE AND EARTHQUAKE

Tsunamis—Their Science and Engineering, edited by K. Iida and T. Iwasaki, 25–36.

Investigation of Rayleigh Wave Spectra for a Set of Tsunamigenic and Nontsunamigenic Earthquakes

V. K. GUSIAKOV

Computing Center, Siberian Branch, the U.S.S.R. Academy of Sciences, Novosibirsk, U.S.S.R.

(Received September 20, 1981)

The purpose of the present study is the evaluation of prognostic possibilities of Rayleigh waves on the basis of the investigation of their spectral characteristics for the set of tsunamigenic and nontsunamigenic Pacific earthquakes. The long period seismograms of seventeen underwater earthquakes recorded at the WWSS stations located around the Pacific were used as the source data. The parts of seismograms corresponding to the Rayleigh waves were digitized with equispaced samples equal to one second. The amplitude spectra of digitized series obtained is calculated by using discrete Fourier transform, and spectral-time diagrams are obtained by the so-called multiply filter technique.

It is shown that there are distinct differences in the form of amplitude spectra of Rayleigh waves for two groups of earthquakes. The main maximum of spectra for the tsunamigenic earthquakes is much wider than for those nontsunamigenic and its position is shifted to the low-frequency domain. The level of spectral amplitudes within the band of periods from 50 to 150 sec for the first group of the earthquakes is about two times higher than for the earthquakes of the second group. Studying the spectral-time diagrams also shows that there are differences in their structure between both groups of earthquakes, which can be used to recognize tsunamigenic earthquakes.

1. Introduction

The problem of evaluation of tsunamidanger of underwater earthquakes is one of the important practical tasks of seismological investigations of the Pacific region. The probability of tsunami generation is determined mainly by the earthquake magnitude, that is why the existent tsunami warning systems are based on seismic magnitude criterion. According to it, earthquakes are regarded to be able to generate destructive tsunamis when their magnitude exceeds some threshold value. If the threshold magnitude is set correctly, this method does not give missings of potentially dangerous tsunamis, but produces a high percentage of false alarms (SOLOVIEV, 1978a).

The results of many recent theoretical and experimental investigations show that the probability of tsunami generation depends not only on magnitude but also on other parameters of an earthquake, such as depth and mechanism of a source (IIDA, 1970; WATANABE, 1970; BALAKINA, 1972; IVASHENKO and GO, 1973; GUSIAKOV, 1972, 1974), the rupture velocity and the rise time of source deformation (KANAMORI, 1972;

TAKEMURA *et al.*, 1977), therefore these parameters must be taken into account for the improvement of the tsunami warning system. However the elaboration of sufficiently reliable, precise and operational methods for the determination of the above parameters from the observations at one or few stations still presents many problems.

Under existing conditions a number of attempts were undertaken to detect some phenomenological signs of tsunamigenity on records of submarine earthquakes (PISARENKO and POPLAVSKY, 1971; IVASHENKO *et al.*, 1977; ORLOV and POPLAVSKY, 1978). For that purpose, mainly the initial part of seismograms between arrivals of P and S waves was examined.

The present study was carried out for the purpose of the evaluation of prognostic possibilities of Rayleigh waves on the basis of the investigation of their spectral characteristics for the set of strong underwater Pacific earthquakes. The amplitude spectra and the frequency-time diagrams of Rayleigh waves are calculated and their features for two groups (tsunamigenic and nontsunamigenic) earthquakes are considered.

2. Computation of Spectra of Seismic Oscillations

Computation of spectrum of any function $f(t)$ satisfying certain restrictions known as Dirichlet conditions is based on Fourier transform formula

$$F(\omega) = \int_{-\infty}^{\infty} f(t)e^{-i\omega t}dt \tag{1}$$

where $F(\omega)$ is the complex function representing the spectral density or, simply, the spectrum of signal $f(t)$ and ω is angular frequency.

Let $x(t)$ be the displacement of the ground caused by a passing seismic wave and $y(t)$ the y-coordinate of a record of displacement on a seismogram. Using (1) we can find the complex spectrum of a record of displacement.

$$Y(\omega) = \int_{-\infty}^{\infty} y(t)e^{-i\omega t}dt = \int_{-\infty}^{\infty} y(t)\cos \omega t\,dt - i\int_{-\infty}^{\infty} y(t)\sin \omega t\,dt$$
$$= B_1(\omega) - iB_2(\omega). \tag{2}$$

Presenting $Y(\omega)$ as a product $|Y(\omega)|e^{i\psi(\omega)}$ we obtain that the amplitude spectrum of a record is equal to

$$|Y(\omega)| = [B_1^2(\omega) + B_2^2(\omega)]^{1/2} \tag{3}$$

and the phase spectrum of a record is

$$\Psi(\omega) = \text{arctg}\left(-\frac{B_2(\omega)}{B_1(\omega)}\right) + 2\pi n, \qquad n = 0, \pm 1, \pm 2,\dots. \tag{4}$$

Complex spectrum of the displacement of the ground $X(\omega)$ can be represented by a formula similar to (2), but its direct use for calculation of $X(\omega)$ is impossible, because the function $x(t)$ is, in principle, unknown. The determination of $x(t)$ i. e. the restoration of the ground displacement by its record on a seismogram, is one of the basic problems of seismometry and may be approximately carried out by different methods.

In practice, the function $X(\omega)$ can be found from the function $Y(\omega)$ if the complex amplitude-phase characteristic of seismograph $W(\omega)$ and its normal magnification $\bar{V}$ are known

$$X(\omega) = \frac{Y(\omega)}{\bar{V} W(\omega)} = \frac{|Y(\omega)|e^{i\psi(\omega)}}{\bar{V} U(\omega)e^{i\gamma(\omega)}} = |X(\omega)|e^{i\varphi(\omega)}. \tag{5}$$

Then the amplitude spectrum of the ground displacement is

$$X(\omega) = \frac{|Y(\omega)|}{\bar{V} U(\omega)} \tag{6}$$

and the phase spectrum

$$\varphi(\omega) = \Psi(\omega) - \gamma(\omega) \tag{7}$$

where $U(\omega)$ and $\gamma(\omega)$ are the amplitude-frequency and phase-frequency responses of the instrument, respectively. In the following, the amplitude spectrum of the ground displacement will be named, simply, the spectrum and denoted by S; the amplitude spectrum of a record of displacement will be denoted by Z.

Thus for the determination of the spectrum $X(\omega)$ it is necessary to compute the function $Y(\omega)$ by formula (2) and to use formula (6).

A real seismic record is always limited in time, therefore the integration in (2) should be carried out within the finite interval $[T_1, T_2]$ of the length T, outside of which the function $y(t)$ is regarded to be zero. The next step is the replacement of the given continuous signal $y(t)$ by discrete values $y_i, i = 1, \ldots, N$ which are the readings of $y(t)$ at times $t_i, i = 1, \ldots, N$. The quantity $\tau = t_i - t_{i-1}$ is the sampling interval and we consider it to be constant throughout the record.

Therefore the application of transform (1) to the analysis of real seismic records is reduced to the calculation of integrals of the form

$$I(\omega) = \int_{T_1}^{T_2} y(t) \begin{Bmatrix} \cos \omega t \\ \sin \omega t \end{Bmatrix} dt \tag{8}$$

where $y(t)$ is given by its values over equal intervals τ.

The convenient and efficient method of computation of this type of integrals is Filon's method of numerical integration, which is based on the approximation not of the whole integrand in (8) but only the function $y(t)$ (see, for instance, HAMMING, 1962).

As a result of computations we find the approximate spectrum, which tends to the accurate spectrum for every value ω with the decrease of sampling interval τ. The choice of specific value of τ is imposed, mainly, by the type of record used. Numerical tests for model signals, for which analytical formulas for spectra are known, show that the value τ providing 6–8 samples for a minimal visible period of oscillations is sufficient for the computation of spectrum with the accuracy of 3–5%.

The spectral-time transform $R(\omega, t)$ of the signal $y(t)$ is a result of the convolution of $y(t)$ with a narrow-band filter and is defined as follows (DZIEWONSKY et al., 1969)

28

V. K. GUSIAKOV

$$R(\omega, t) = \frac{1}{\pi} \int_0^\infty Y(\lambda) e^{-[\alpha(\frac{\omega-\lambda}{\omega})^2]} e^{it\lambda} d\lambda \qquad (9)$$

where $Y(\lambda) = \int_{T_1}^{T_2} y(t) e^{-i\lambda t} dt$ is the spectrum of the signal $y(t)$ given in the interval $[T_1, T_2]$. The result of the transform $R(\omega, t)$ is the complex function, its module $|R(\omega, t)|$ gives an amplitude of the envelope of a narrow-band signal at the output of the filter with the center frequency ω and arg $R(\omega, t)$ presents the phase of this signal. Usually, the result of transform (9), a so-called frequency-time diagram, is drawn as contours of the value of the module $|R(\omega, t)|$ with the maximum normalized to the level of 100 db.

3. Seismological Data

In this study seventeen underwater earthquakes with magnitudes around 7.0 that occurred mainly in the North-West part of the Pacific, were considered. Nine of them were tsunamigenic, i.e. they generated detectable tsunamis with different intensity. The long period seismograzms of these earthquakes, recorded at the WWSS stations located around the Pacific were used as the source data.

The list of the earthquakes with their basic parameters and the data on intensity of tsunamis taken from (SOLOVIEV, 1978b) in given in Table 1. A large number of records of these earthquakes were examined and there were selected seismograms suitable for digitation, i.e. those having the clear trace not extending the limits of a seismogram.

Table 1. List of earthquakes.

No.	Data	Time			Coordinates		Magnitude M	Depth h km	Intensity of tsunami I
		h	m	s	N	E			
1	25.07.70	22	41	10	32.2	131.7	7.0	34	-0.5
2	02.08.71	07	24	54	41.4	143.5	7.0	50	-1
3	05.09.71	18	35	27	46.5	141.1	7.2	15	1
4	24.11.71	19	35	31	52.7	159.5	7.1	100	-
5	08.01.72	05	27	53	21.2	120.5	6.9	40	-
6	29.02.72	09	22	58	33.6	140.8	7.2	56	-1
7	04.12.72	10	16	12	33.3	140.7	7.4	60	-1
8	26.06.73	22	32	00	43.2	146.7	6.9	50	-
8	26.06.73	22	32	00	43.2	146.7	6.9	50	-
9	28.07.74	11	34	59	46.5	153.2	6.9	45	-
10	27.09.74	05	47	07	43.2	146.7	7.2	46	-2
11	13.06.75	18	08	03	43.5	148.2	7.0	20	-
12	15.06.75	00	19	30	43.2	148.0	7.0	30	-
13	20.07.75	14	37	29	6.5*	154.9	7.3	33	1
14	29.11.75	14	47	37	19.3	155.0**	7.2	47	2
15	06.01.76	21	08	14	51.6	159.4	6.7	0	-
16	21.01.76	10	05	23	44.6	149.3	7.2	33	-2.5
17	19.02.77	22	34	00	51.8	170.6	6.9	33	-

*S, **W.

Since all the earthquakes considered were strong events generating intensive surface waves most of the records made at the stations with instrument magnification more than 1,500 or at those located at short distances from the epicenters did not satisfy these conditions.

In all, 52 seismograms recorded at 12 stations were selected. Among them, for each event we could find several seismograms from different stations, 3–4 on average, but for two events (No. 4 and No. 13) it was possible to find only one record for each of them.

The selected seismograms were digitized with equispaced sampling interval equal approximately to one second which provided from 15 to 20 points for a characteristic period of oscillations. The length of time window selecting the Rayleigh waves from the whole record of an earthquake was, on average, about 500 sec. Thus under indicated values of τ and T the limits of spectra were the frequencies $\omega_0 = 0.0125$ sec^{-1} and $\omega_N = 3.14$ sec^{-1}, corresponding to the periods of $T_{max} = 500$ sec and $T_{min} = 2$ sec.

The amplitude spectra of digitized series obtained were calculated by method described in the previous section. For the sake of convenience of comparison between spectra obtained under different conditions (earthquake magnitudes, epicentral distances) the computed spectra were normalized to the same level of the main maximum. Before being able to analyze the computed spectra, we investigated their stability and significance under the used value of sampling, the length of digitized intervals and the position of the zero-line on records. The results showed that the main features of computed spectra are reliable and significant within the periods from 10 to 150 sec. This band was used for the analysis and calculation of spectral estimates.

4.   Comparison of Spectra and Frequency-Time Diagrams of Tsunamigenic and Nontsunamigenic Earthquakes

Spectra of all the records considered are characterized by the presence of the pronounced peak at the periods 15–20 sec, named, later, the main maximum, the gradual decrease of the spectral level to longer periods, a steeper dip of it to shorter periods and the existence of a number of secondary peaks. The spectra of one earthquake recorded at different stations have, in general, a similar form, although distinctions between them exceed possible errors envolved in digitizing and computing process. The reasons of that are, evidently, the availability of a radiation pattern in the earthquake source, differences in the length and the character of wave paths and also the effect of local geological structures in the neighbourhood of the stations.

The computed spectra were divided into two groups corresponding to tsunamigenic and nontsunamigenic earthquakes. The comparative investigation of the spectra in both groups discovered a number of features of spectra of tsunamigenic earthquakes:

1)   the increased level of spectral components which equals 1/3–1/2 of the level of the main maximum within the band of periods 30–150 sec;

2)   the shift of the main maximum to longer periods;

3)   the noticeable increase of the width of the main maximum, in a number of cases the appearance of secondary peaks around it;

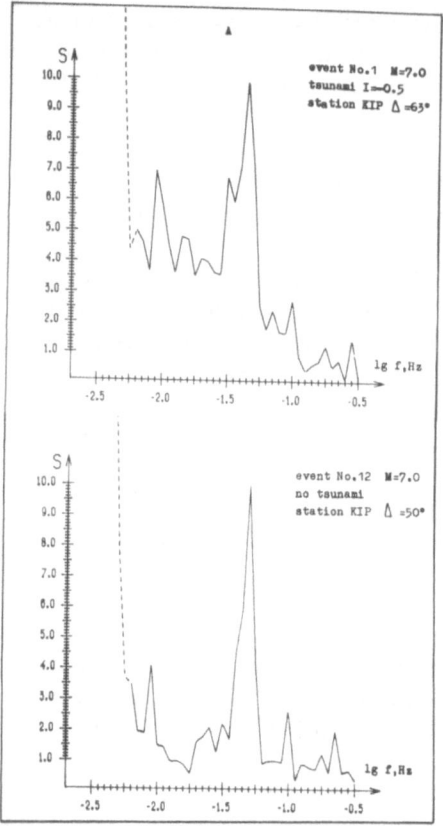
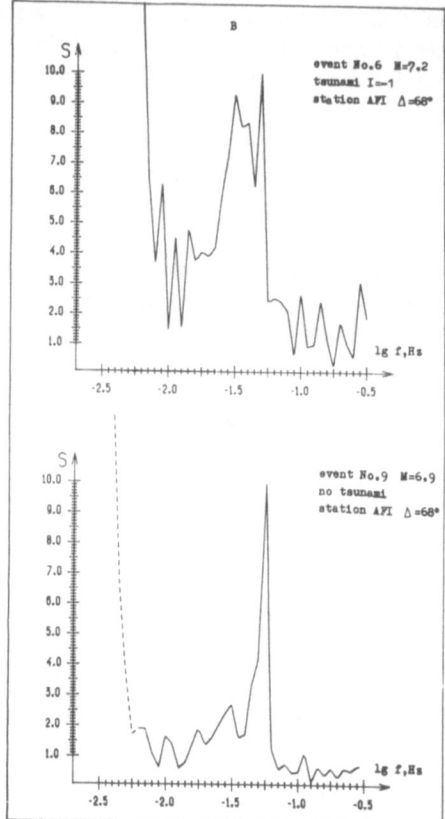

Fig. 1. Comparison of Rayleigh wave amplitude spectra for tsunamigenic and nontsunamigenic earthquakes with relatively close amplitudes.

4) the increase of average level of a spectrum within the whole regarded range of periods as well as the increase of area under a spectral curve.

The above features are distinctly seen in the comparison of spectra for several pairs of tsunamigenic and nontsunamigenic earthquakes with close magnitudes, recorded at the same stations (Fig. 1).

At the same time these features were not discovered for two earthquakes from the first group (No. 10 and No. 16). Their spectra have the form typical for nontsunamigenic earthquakes, that is the narrow main maximum and low level of spectral components equal to 1/5–1/4 of the maximum within the periods 30–150 sec. Appealing to the data of Table 1, we see that these earthquakes generated tsunamis with minimal intensity (-2 and -2.5, respectively), which were detectable only by the analysis of mareograph records. From the point of view of tsunami prognosis both of these earthquakes can be attributed to the group of tsunamigenic events.

On the other hand, in the second group there were two earthquakes (No. 4 and

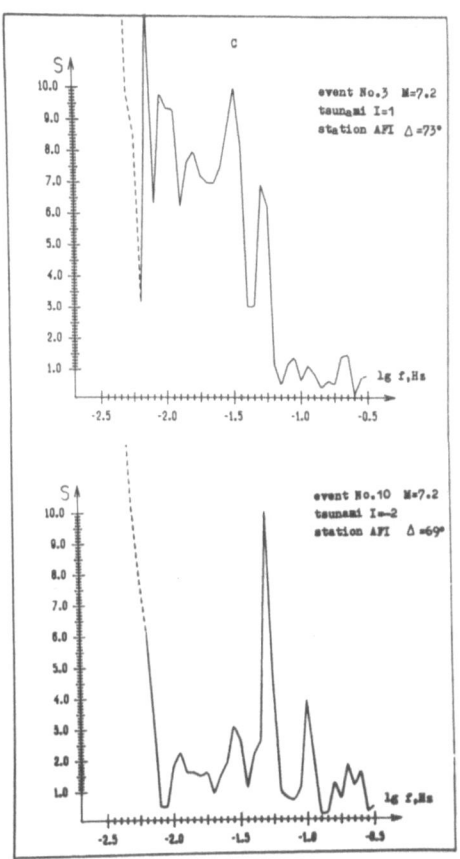

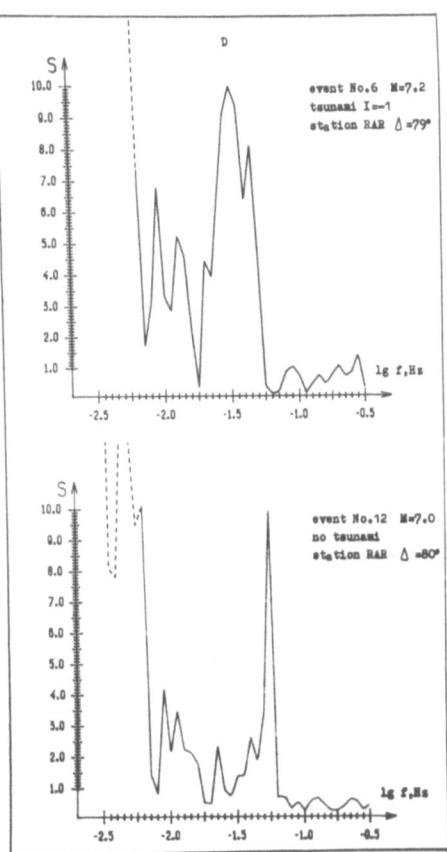

Fig. 1

No. 5) having the spectra typical for tsunamigenic events. One of them (No. 4) occurred at the depth of 100 km and, apparently, therefore it did not generate a tsunami. However, on the basis of data available it is difficult to explain the typical tsunamigenic spectrum of earthquake No. 5.

To obtain some quantitative characteristics of the computed spectra, a number of parameters were calculated for each of them, such as:

area under the spectral curve

$$Q = \int_{\omega_1}^{\omega_2} X(\omega)d\omega \tag{10}$$

spectral energy within the frequency band $[\omega_1, \omega_2]$

$$E = \int_{\omega_1}^{\omega_2} |X(\omega)|^2 d\omega \tag{11}$$

moment about $\omega = 0$

$$m = \int_{\omega_1}^{\omega_2} \omega X(\omega) d\omega \tag{12}$$

coordinates of the spectral centroid (the center of gravity under the curve)

$$\omega_c = \frac{\displaystyle\int_{\omega_1}^{\omega_2} \omega X(\omega) d\omega}{\displaystyle\int_{\omega_1}^{\omega_2} X(\omega) d\omega}, \qquad F_c = \frac{\displaystyle\int_{\omega_1}^{\omega_2} |X(\omega)|^2 d\omega}{2 \displaystyle\int_{\omega_1}^{\omega_2} X(\omega) d\omega}. \tag{13}$$

The distributions of these spectral parameters are shown in Fig. 2 as histograms for tsunamigenic and nontsunamigenic earthquakes. It is seen, that despite of rather a wide dispersion, mathematical expectations in both groups distinctively differ. That difference points out the possibility of the recognition of tsunamigenic earthquakes by a set of spectral signs of Rayleigh waves.

The comparative investigation of frequency-time diagrams of Rayleigh waves also

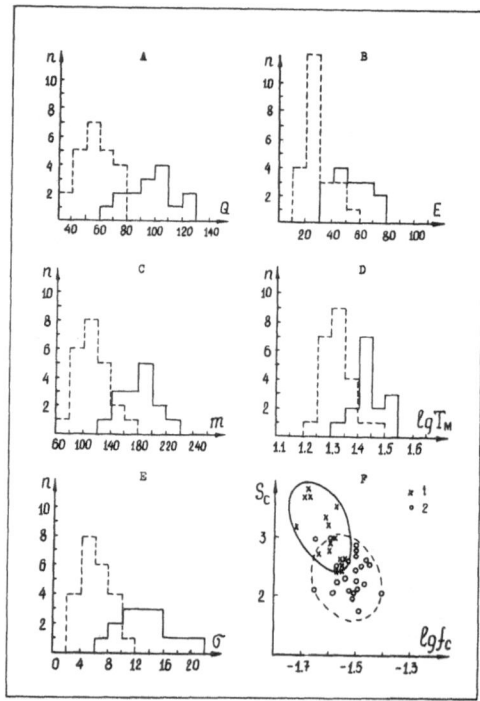

Fig. 2. Histograms of some spectral parameters of Rayleigh waves of tsunamigenic (solid line) and nontsunamigenic (dotted line) earthquakes: A, area under the spectral curve; B, spectral energy; C, moment about $\omega = 0$; D, period of the main maximum of spectrum; E, width of spectrum at the level 1/2 of the maximum; F, distribution of the coordinates of the spectral centroids in the plane of S, $\lg f$ (1, tsunamigenic; 2, nontsunamigenic).

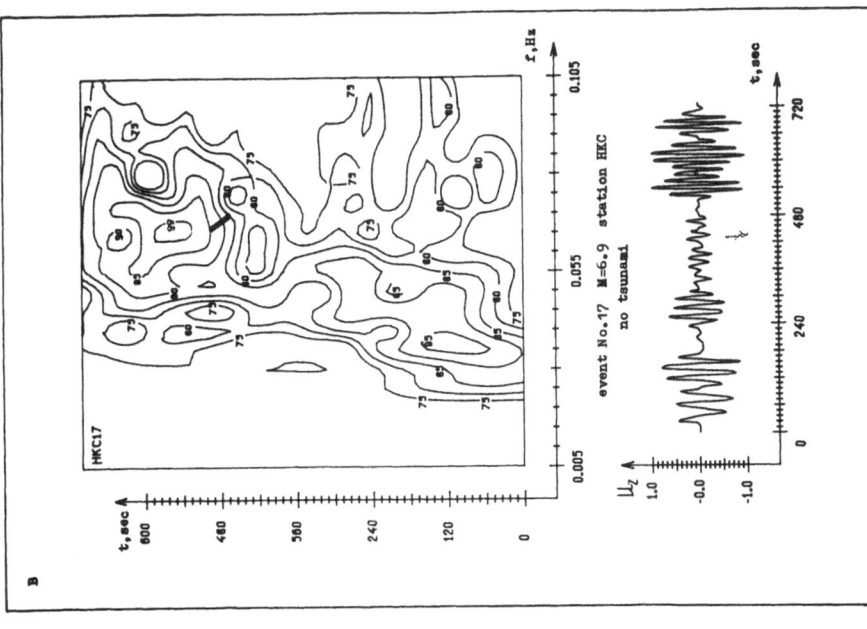

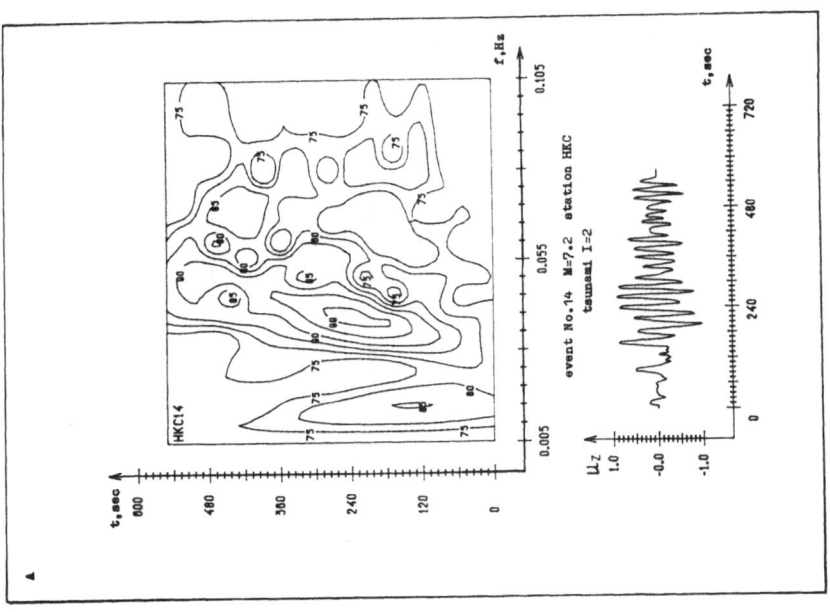

Fig. 3.   Frequency-time diagrams of Rayleigh waves for earthquakes No. 14 (A) and No. 17 (B), recorded at the station HONG KONG.

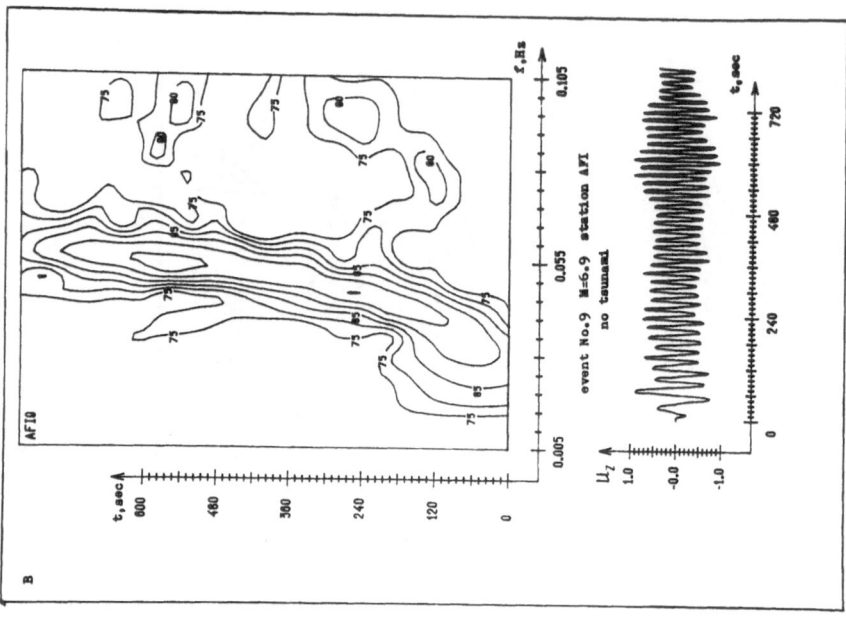

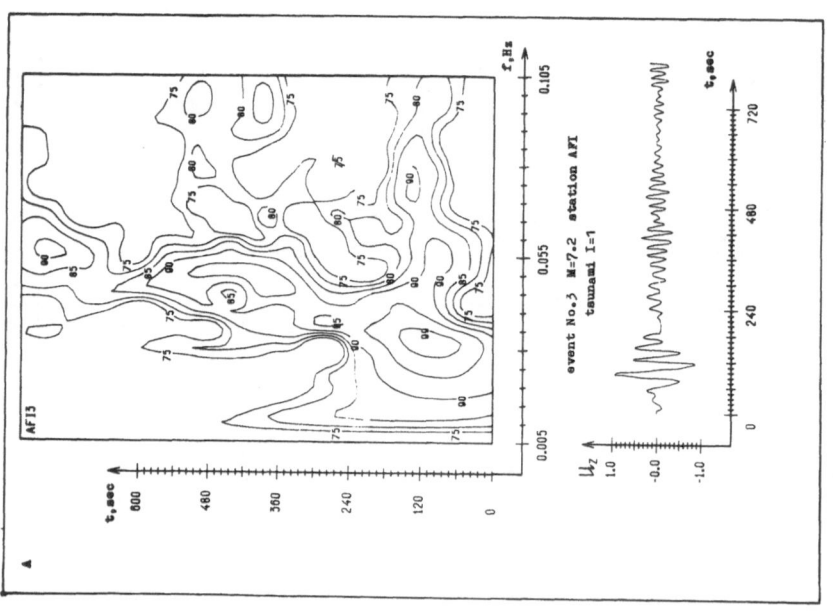

Fig. 4.   Frequency-time diagrams of Rayleigh waves for earthquakes No. 3 (A) and No. 9 (B), recorded at the station AFIAMALU.

discovered a number of features typical for each group of earthquakes. In the diagrams of tsunamigenic events, the contours of maximum level (95–99 db), dispose, mainly, in the left lower quarter of the diagram, while for nontsunamigenic events, they are located in the center of diagrams or, even, they are shifted to the right upper quarter corresponding to the higher frequencies and larger arrival times. The existence of increased levels (80–89 db) of low frequencies (0.01–0.02 Hz) in the initial part of records is also typical for tsunamigenic earthquakes. The diagrams of nontsunamigenic events have a simpler structure with maximum levels forming a single distinct ridge crest (Figs. 3 and 4).

5. Conclusion

The comparison of amplitude spectra and frequency-time diagrams of Rayleigh waves for tsunamigenic and nontsunamigenic earthquakes discovered a number of signs which help to distinguish the earthquakes of one group from another group. The attempt was made to formalize these signs as a set of spectral parameters evaluated quantitatively. Each of these signs has a formal character but as a whole all of them are connected with the predominance of low frequency components in the spectra of tsunamigenic earthquakes. Thus we can speak of a certain correlation between the level of spectral components of Rayleigh waves in the range of periods from 50 to 150 sec and the intensity of tsunami waves. This correlation may be used for the improvement of tsunami warnings by means of the reduction of percentage of false alarms. At present, errors in the prognosis take place, mainly, for earthquakes with magnitudes close to the threshold value. In particular, the use of Rayleigh waves spectral features can increase the reliability of predictions at the cost of exception from the group of tsunamigenic earthquakes of such events whose magnitudes slightly exceed the threshold value used, but spectral parameters are typical for the nontsunamigenic group.

The study presented in this paper was carried out as a part of the joint Soviet-American program of cooperation under US-USSR Environmental Protection Agreement. The preparation of source seismological data was performed at World Data Center-A in Boulder (Colorado), a part of numerical data processing was executed at Joint Institute for Marine and Atmospheric Research in Honolulu (Hawaii). The computation of frequency-time diagrams of Rayleigh waves was done at Novosibirsk Computing Center of the Siberian Branch of the USSR Academy of Sciences.

The author wishes to thank Drs. J. Allen and W. Rinehart of WDC-A for their kind assistance in the selection and digitation of seismograms. The author is also grateful to Dr. H. Loomis of JIMAR for the assistance in computing processing of data and Dr. D. Moore, Director of JIMAR for hospitality and encouragement. I wish to thank also Dr. A.S. Alexeev, Director of Novosibirsk Computing Center for his attention to this study.

REFERENCES

BALAKINA, L. M., Tsunamis and focal mechanism of earthquakes in the North-Western part of the Pacific, Tsunami waves, Proc. of Sakhalin Compl. Res. Inst., No. 29, Yuzhno-Sakhalinsk, 48–72, 1972 (in Russian).

DZIEWONSKY, A., S. BLOCK, and M. LANDISMAN, A technique for the analysis of transient seismic signals, Bull. Seis. Soc. Am., **59**, No. 1, 427–444, 1969.

GUSIAKOV, V. K., The excitation of tsunami and oceanic Rayleigh waves by underwater earthquakes, Mathematical problems in geophysics, No. 3, Novosibirsk, 250–272, 1972 (in Russian).

GUSIAKOV, V. K., On the relationship between tsunami waves and underwater earthquake source parameters, Mathematical problems in geophysics, No. 5, part I, Novosibirsk, 118–140, 1974 (in Russian).

HAMMING, R. W., *Numerical Methods for Scientists and Engineers*, 400 pp., McGraw-Hill, New York-London, 1962.

IIDA, K., The generation of tsunamis and the focal mechanism of earthquakes, Tsunamis in the Pacific Ocean, Honolulu, 3–18, 1970.

IVASHENKO, A. I. and Ch. N. Go, Tsunamigenity and earthquake source depth, Tsunami waves. Proc. of Sakhalin Compl. Res. Inst., No. 32, Yuzhno-Sakhalinsk, 152–155, 1973 (in Russian).

IVASHENKO, A. I., V. A. ORLOV, and A. A. POPLAVSKY, The problem of tsunami operative forecasting as the pattern recognition problem, *Theoretical and Experimental Research on Tsunami Problem*, Moscow, Nauka, 104–113, 1977 (in Russian).

KANAMORI, H., Mechanism of tsunami earthquakes, *Phys. Earth Planet. Inter.*, **6**, 346–359, 1972.

ORLOV, V. A. and A. A. POPLAVSKY, Determination of earthquake source depth by using limited number of dynamic characteristics of record, Seismological observation processing and earthquake prediction on Far East, Vladivostok, 24–33, 1978 (in Russian).

PISARENKO, V. F. and A. A. POPLAVSKY, Statistical method for earthquake source depth recognition by using a single station record, Computational seismology, No. 5, Moscow, Nauka, 1971 (in Russian).

SOLOVIEV, S. L., Tsunamis, Assessment and mitigation of earthquake risk (National Hazards, I), UNESCO, Paris, 118–139, 1978a.

SOLOVIEV, S. L., The basic data on tsunamis near the Pacific coast of the USSR, Investigation of tsunami in open ocean, Moscow, Nauka, 61–136, 1978b (in Russian).

TAKEMURA, M., J. KOYAMA, and Z. Suzuki, Source process of the 1974 and 1975 earthquakes in Kurile Islands in special relation to the difference in excitation of tsunami, *Sci. Rep. Tohoku Univ.*, **24**, No. 4, 113–132, 1977.

WATANABE, H., Statistical studies of tsunami source and tsunamigenic earthquakes occurring in and near Japan, Tsunamis in the Pacific Ocean, Honolulu, 99–117, 1970.

Tsunamis—Their Science and Engineering, edited by K. Iida and T. Iwasaki, 37–49.

Use of Long-Period Seismic Waves for Rapid Evaluation of Tsunami Potential of Large Earthquakes

Hiroo KANAMORI and Jeffrey W. GIVEN

*Seismological Laboratory, California Institute of Technology,
Pasadena, California, U.S.A.*

(Received March 6, 1982)

Two seismological methods for rapid evaluation of tsunami potential of an earthquake are discussed. In the first method, long-period (200 to 300 sec) surface waves are used to determine the seismic moment and the fault geometry. This method allows determination of the source parameters in about 10 minutes after the surface-wave data have been retrieved, and is appropriate for far-field tsunami warning purposes.

In the second method, long-period (100 to 300sec) near-field displacements are used to estimate the seismic moment. Our numerical experiment for a source-station geometry appropriate for Japan demonstrates that a very rapid and robust method can be developed for near-field tsunami warning purposes, if an appropriate (long-period, low gain) recording system is available.

1. Introduction

It is widely known that most, if not all, large tsunamis are caused by sea-bottom deformations associated with large earthquakes. Although many factors such as the magnitude, the depth, the location and the mechanism (rupture mode, dip-slip or strike-slip) of an earthquake determine its tsunami potential, earthquake magnitude is among the most important parameters used by the present tsunami warning system. The magnitude of large earthquakes has been traditionally determined by using 20 sec surface waves (surface-wave magnitude M_s). However, it is known that the surface-wave magnitude M_s saturates beyond a certain limit (about $M_s = 7\frac{3}{4}$) and is not a very reliable measure of the tsunami potential of the earthquake. Hence, it is important to determine the magnitude at long periods (100 to 300 sec).

The next important element is the mechanism of the event. Since tsunamis are primarily excited by vertical motion of the sea bottom, an earthquake with substantial dip-slip motion is more likely to be tsunamigenic than one with a primarily strike-slip mechanism. Therefore, in order to evaluate the tsunami potential of an earthquake, it is desirable to determine the long-period magnitude (e.g. M_w) and, whenever possible, the mechanism.

In order to explore the possibility of using long-period seismic data for tsunami warning purposes, we developed a method for fast retrieval of earthquake source parameters from long-period surface waves (Kanamori and Given, 1981). In this method, Rayleigh-wave and/or Love-wave spectra at a period range from 200 to 300 sec are inverted to obtain either a seismic moment tensor or a fault model. A reasonably accurate estimate of the seismic moment and the fault geometry can be obtained with the data from as few as seven stations within ten minutes after the seismograms have been retrieved. The method is, therefore, adequate for far-field tsunami warning purposes.

For near-field tsunami warning purposes, even faster methods are required. In this case, the method should be designed for a specific geographical area where it is to be applied. In this paper, we discuss two seismological methods, one for far-field and the other for near-field tsunami warning problems.

2.   Method for Far-Field Tsunami Warning

We illustrate how the method we developed (Kanamori and Given, 1981) can be implemented in on-line tsunami warning systems by using the 1979 Colombia-Ecuador earthquake ($M_s = 7.7$, $M_w = 8.2$) as an example. We follow the flow chart shown in Fig. 1.

First, the origin time and the epicenter coordinates of the earthquake are given to the program. Then the Rayleigh-wave train is displayed on a graphic terminal with a group velocity scale superimposed on it (Fig. 2). An appropriate section of the record is then visually windowed. For the Colombia-Ecuador earthquake, the records of R_1 were off-scale so that higher-order Rayleigh waves (e.g., R_2 to R_5) had to be used. However, for the actual tsunami warning purposes, R_1 phases should be used. Spectra are then computed and corrections for the instrument response, phase velocity and Q are applied to obtain the source spectra. The source spectra are inverted to obtain a source moment tensor. However, as described in Kanamori and Given (1981), the inversion is ill-conditioned for a very shallow event so that two out of five moment tensor elements cannot be determined very well. Therefore, the elements M_{zx} and M_{zy} are usually constrained to be 0 during the first inversion. This is equivalent to constraining the mechanism to either 45° dip-slip or a vertical strike-slip. Although the procedure may appear too restrictive, it usually gives a useful first approximation of the source geometry (dip-slip versus strike-slip), the sense of the motion, and the seismic moment.

The result of the constrained inversion for the Colombia earthquake is shown in Fig. 3(a). The mechanism is a reverse dip-slip ($\delta = 45°$, constrained) and the seismic moment is approximately 1×10^{28} dyne-cm. The dip-slip geometry and the large seismic moment suggest a high tsunami potential of this earthquake. Further improvement can be made by applying a correction for the source process time, τ, which can be estimated to be 60 sec from the seismic moment (see Kanamori and Given, 1981). The inversion is repeated with τ included, and the result is shown in Fig. 3(b). The geometry remains the same, but the seismic moment increases slightly. Finally, inversion is attempted with the constraints $M_{zx} = M_{zy} = 0$ removed. When the quality of the data is not very good, the result of this inversion becomes often unstable;

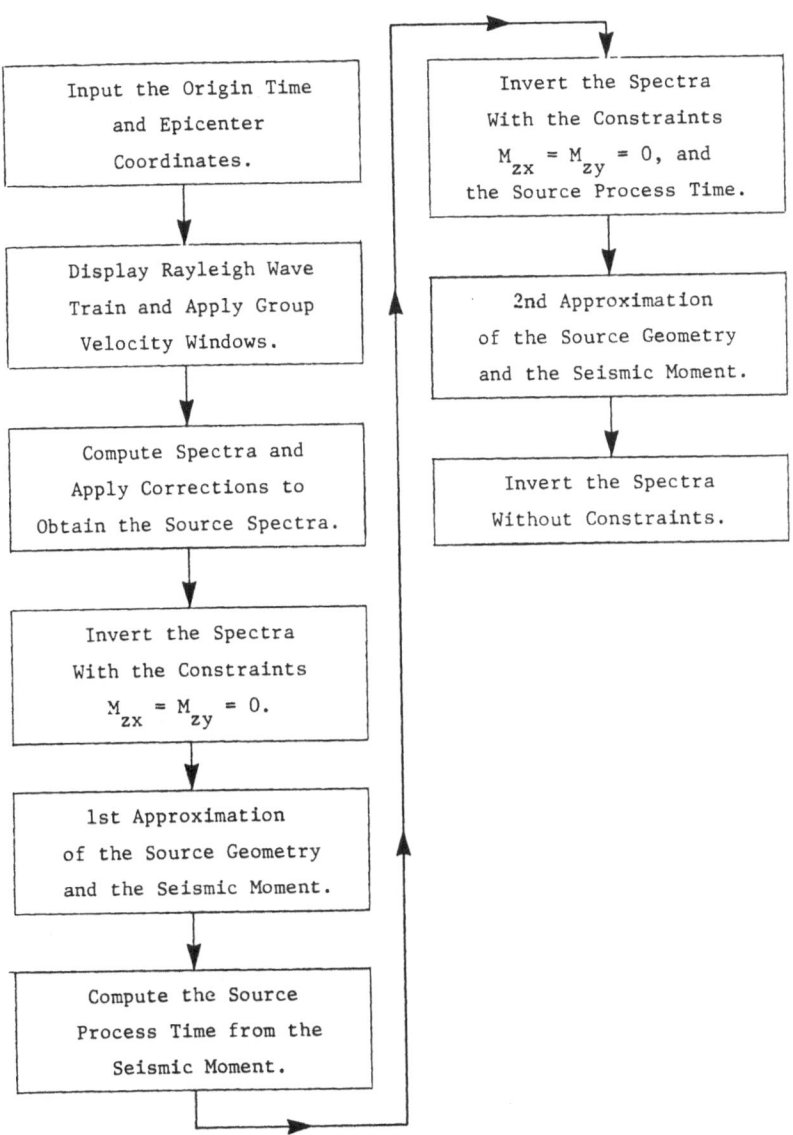

Fig. 1. Flow chart of surface-wave processing procedure for rapid determination of source mechanism and seismic moment.

however, for the Colombia earthquake, a solution is obtained with reasonable stability. The source geometry and the seismic moment for this solution are shown in Fig. 3(c). Although the dip angle and the seismic moment differ significantly from those of the constrained solution, the product $M_0 \sin 2\delta$ is about the same for both solutions. Since the amplitude of Rayleigh waves excited by a shallow dip-slip source is proportional to $M_0 \sin 2\delta$, this quantity can be determined very well even though the

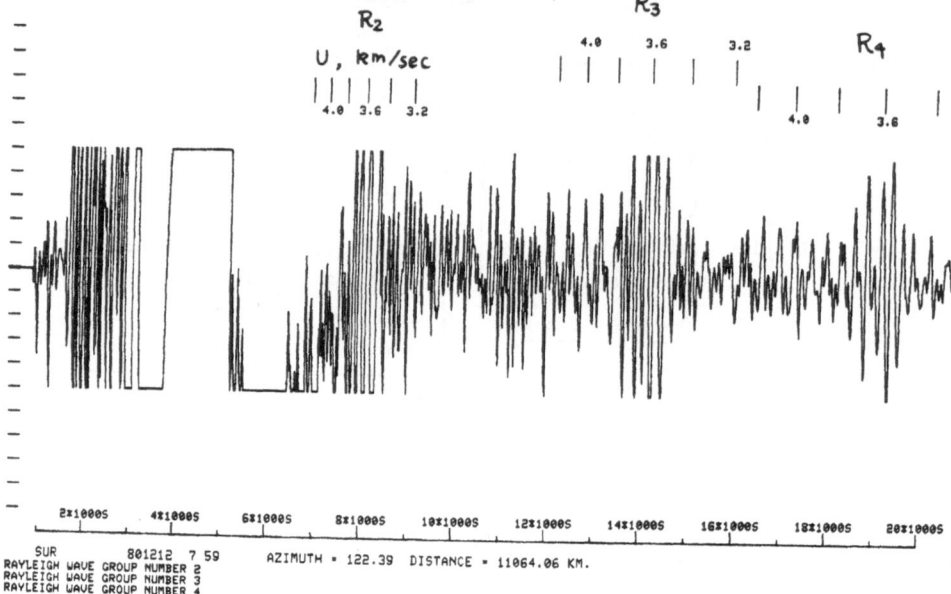

Fig. 2. Rayleigh waves from the 1979 Colombia-Ecuador earthquake recorded at station SUR (Δ = 11,064 km, ϕ_s = 122°) displayed on the screen of a graphic terminal with the group velocity scales. The wave trains within desired group velocity windows can be visually sampled.

Colombia – Ecuador Earthquake, 1979

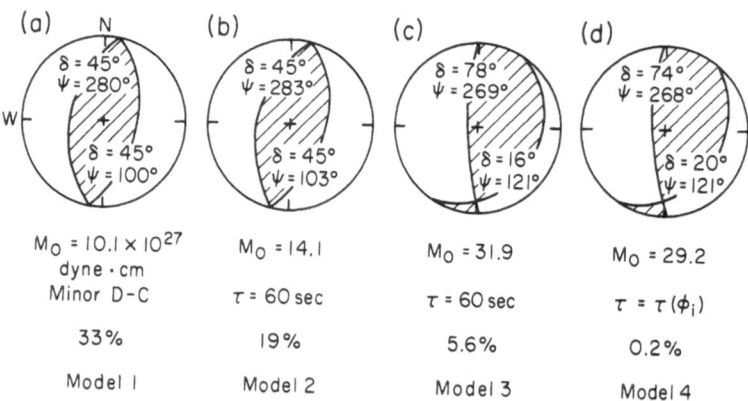

Fig. 3. Mechanisms of the 1979 Colombia-Ecuador earthquake obtained by a moment tensor inversion of Rayleigh waves. (a) constrained ($M_{zx} = M_{zy} = 0$) solution without correction for the source process time; (b) constrained solution with a source process time of 60 sec; (c) unconstrained solution with the same process time; (d) unconstrained solution with azimuthally dependent source process time.

dip angle and the moment cannot be resolved individually. For tsunami warning application, the vertical component of the sea bottom deformation is most important. Hence, $M_0 \sin 2\delta$ is probably a better measure of the tsunami potential of the earthquake than the seismic moment itself.

The solution given by Fig. 3(c) is consistent with the first-motion data and is essentially identical to the solution shown by Fig. 3(d) which is obtained by a more detailed analysis which included the effect of the source directivity. Since solution (c) does not require detailed knowledge of the size of the rupture area and the rupture direction, it can be obtained in a very straightforward manner, and therefore can be used for real-time operations.

Since the entire analysis described above can be performed in about 10 minutes after the surface wave trains have been retrieved, the method described here is promising for far-field tsunami (e.g., tsunamis at Hawaii caused by earthquakes in the Aleutians, Alaska and Chile) warning purposes.

3. Method for Near-Field Tsunami Warning

In the near-field tsunami problem, we usually know the approximate location and the mechanism of the events which are likely to be tsunamigenic. Hence, determination of the mechanism is not very crucial. In this case, it is important to determine the seismic moment at long periods (100 to 300 sec) as quickly as possible (within 15 min after the earthquake origin time) through a relatively simple and robust method such as the one used for the routine magnitude determination. In what follows, we attempt to design a system for a source-station geometry appropriate for Japan.

Figure 4 shows the source areas of some major tsunamis which occurred in the Japanese region. We choose 5 design earthquakes listed in Table 1 and shown in Fig. 4. These earthquakes model the 1952 Tokachi-Oki, 1933 Sanriku, 1896 Sanriku (mechanism assumed), 1923 Kanto, and 1946 Nankaido earthquakes. Also we choose 10 stations shown in Fig. 4 and assume that these stations are equipped with low-gain long-period seismographs similar to the Sprengnether force-balance seismometer (equivalent pendulum and galvanometer periods are 100 and 300 sec respectively) with a gain of 0.1. The stations are chosen to attain a good azimuthal coverage for all the tsunami source areas.

We assume the earthquake mechanism and compute synthetic seismograms (vertical component) at each station. The synthetic seismograms are computed by a simple mode sum using 3271 modes calculated for earth model 1066A (GILBERT and DZIEWONSKI, 1975) by BULAND and GILBERT (1976). These modes are shown in Fig. 5. The cut-off period is 45 sec. For the source, a point double-couple source at a depth of 11 km (at the bottom of the crust of model 1066A) is used. The vertical component of the displacement at the surface U_r is given by (KANAMORI and CIPAR, 1974).

$$U_r = \sum_n \cos \omega_n t (K_0 s_R P_n^0 - K_1 q_R P_n^1 + K_2 p_R P_n^2)$$

where ω_n is the eigen angular frequency, P_n^m, associated Legendre functions, K_i, the excitation functions given by KANAMORI and CIPAR (1974), and s_R, q_R, p_R are the

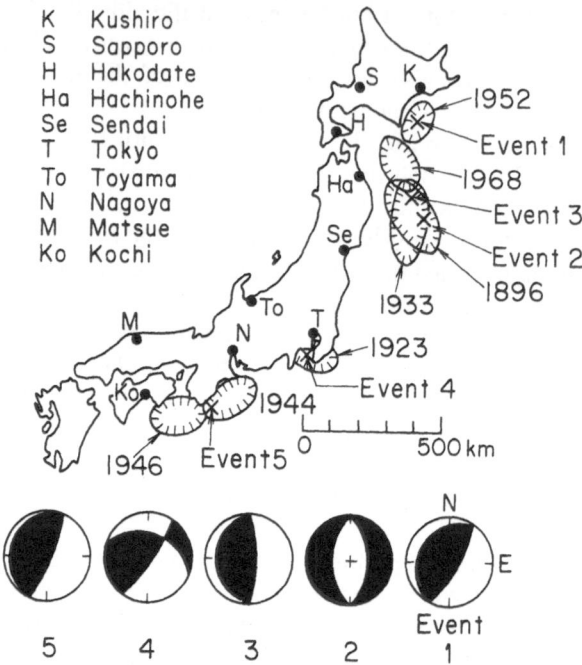

Fig. 4. Locations of design earthquakes No. 1, 2, 3, 4, and 5, and the stations. The source areas of several major tsunamis are shown by elliptic areas. The mechanisms of the design earthquakes are shown at the bottom.

Table 1. Design earthquakes.

Event No.	Lat. (deg. N)	Long. (deg. E)	$\lambda^{(1)}$ (deg.)	$\delta^{(1)}$ (deg.)	$\phi_f^{(1)}$ (deg.)	$\tilde{M}^{(2)}$	Model
1	42.2	143.9	90	20	210	8.5	1952 Tokachi-Oki
2	39.1	144.7	−90	45	0	8.6	1933 Sanriku
3	39.6	144.2	90	20	180	8.5	1896 Sanriku
4	35.2	139.3	162	34	290	8.6	1923 Kanto
5	33.3	135.9	90	10	22	8.3	1946 Nankaido

The seismic moment = 10^{29} dyne-cm.
[1] λ = slip angle, δ = dip angle, ϕ_f = fault strike.
[2] $\tilde{M}$ is the magnitude which corresponds to $\tilde{M}_0 = M_0 \sin 2\delta$ (see text).

constants determined by the fault-station geometry. The summation is taken over the entire 3271 modes. The displacement U_r is then convolved with the instrument response to obtain the synthetic seismogram u_r.

The results for all the events are shown in Fig. 6. The amplitudes of the synthetic seismograms for all the events are listed in Tables 2, 3, 4, 5, and 6. As indicated on the figure, the largest phase arrives following the ray arrival time of the S wave. In general,

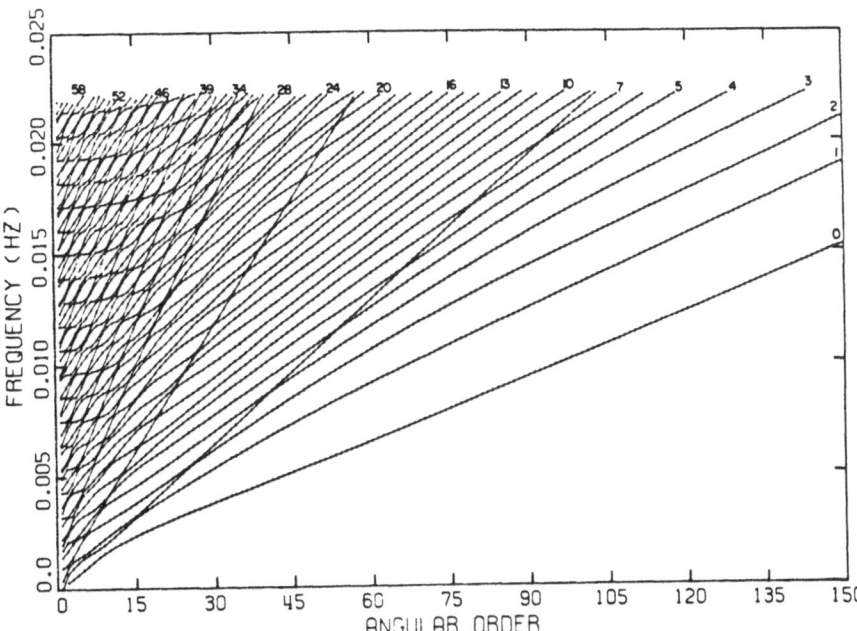

Fig. 5.   The angular order number, *l* vs. frequency (from OKAL 1978). The modes were calculated for earth model 1066A by BULAND and GILBERT (1976).

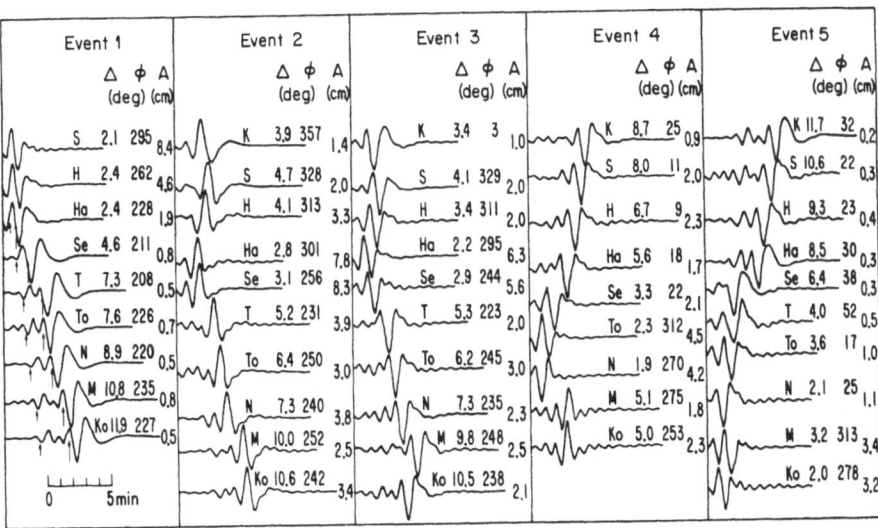

Fig. 6.   Synthetic seismograms computed for the design earthquakes. For station abbreviations, see Fig. 4. Δ is the distance and ϕ is the azimuth. A is the peak-to-peak amplitude of the synthetics. The parameters for the seismograph response are: pendulum period = 100 sec, galvanometer period = 300 sec, damping constant of the pendulum and the galvanometer = 1, coupling constant = 0.05, peak magnification = 0.1. The short and long arrows indicate the ray arrival times of the P and S phases, respectively.

Table 2.   Station data for Event 1.

Event No. 1				$\tilde{M}_0 = 6.4 \times 10^{28}$ dyne cm, $\tilde{M} = 8.5$		
Station	Δ (deg.)	ϕ (deg.)	A (cm)	$\|\sin \Delta\phi\|$	$A\Delta^{0.6}$	$A\Delta^{0.6}/\|\sin \Delta\phi\|$
Kushiro						
Sapporo	2.1	295	8.4	1.0	13.0*	13.0
Hakodate	2.4	262	4.6	0.79	7.8	9.9
Hachinohe	2.4	228	1.9	0.30	3.3	10.8
Sendai	4.6	211	0.8	0.02	1.9	—
Tokyo	7.3	208	0.5	0.04	1.6	—
Toyama	7.6	226	0.7	0.27	2.2	8.3
Nagoya	8.9	220	0.5	0.17	1.8	10.7
Matsue	10.8	235	0.8	0.42	3.1	7.4
Kochi	11.9	227	0.5	0.29	2.3	8.0
				Average	4.1	9.7

Average of $A\Delta^{0.6}$ and $A\Delta^{0.6}/\|\sin \Delta\phi\|$ are computed at the bottom. Asterisk indicates the maximum of $A\Delta^{0.6}$. The values of C_0 computed with the three different methods are shown.
Method 1, $C_0 = 7.8$; Method 2, $C_0 = 8.1$; Method 3, $C_0 = 7.8$.

Table 3.   Station data for Event 2 (for details, see Table 2).

Event No. 2				$\tilde{M}_0 = 10^{29}$ dyne cm, $\tilde{M} = 8.6$		
Station	Δ (deg.)	ϕ (deg.)	A (cm)	$\|\sin \Delta\phi\|$	$A\Delta^{0.6}$	$A\Delta^{0.6}/\|\sin \Delta\phi\|$
Kushiro	3.9	357	1.4	0.06	3.2	—
Sapporo	4.7	328	2.4	0.53	6.0	11.5
Hakodate	4.1	313	4.0	0.73	9.3	12.8
Hachinohe	2.8	301	7.8	0.85	14.5	17.0
Sendai	3.1	256	8.3	0.97	16.3*	16.8
Tokyo	5.2	231	3.9	0.77	10.4	13.5
Toyama	6.4	250	4.8	0.94	14.5	15.4
Nagoya	7.3	240	3.8	0.87	12.5	14.4
Matsue	10.0	252	4.0	0.95	16.0	16.8
Kochi	10.6	242	3.4	0.88	13.9	15.7
				Average	11.7	14.5

Method (1), $C_0 = 7.8$; Method (2), $C_0 = 7.9$; Method (3); $C_0 = 7.8$.

Table 4. Station data for Event 3 (for details, see Table 2).

Event No. 3				$\tilde{M}_0 = 6.4 \times 10^{28}$ dyne-cm, $\tilde{M} = 8.5$		
Station	Δ (deg)	ϕ (deg)	A (cm)	$\lvert \sin \Delta\phi \rvert$	$A\Delta^{0.6}$	$A\Delta^{0.6}/\lvert \sin \Delta\phi \rvert$
Kushiro	3.4	2	1.0	0.04	2.1	—
Sapporo	4.1	329	2.0	0.52	4.6	8.8
Hakodate	3.4	311	2.0	0.75	4.1	5.5
Hachinohe	2.2	295	6.3	0.90	10.3	11.4
Sendai	2.9	244	5.6	0.90	10.7*	11.9
Tokyo	5.3	223	2.0	0.69	5.3	7.7
Toyama	6.2	245	3.0	0.90	8.9	9.8
Nagoya	7.3	235	2.3	0.82	7.6	9.3
Matsue	9.8	248	2.5	0.93	9.7	10.4
Kochi	10.5	238	2.1	0.85	8.5	10.1
				Average	7.2	9.4

Method 1, $C_0 = 7.9$; Method 2, $C_0 = 7.9$; Method 3, $C_0 = 7.8$.

Table 5. Station data for Event 4 (for details, see Table 2).

Event No. 4				$\tilde{M}_0 = 9.3 \times 10^{28}$ dyne-cm, $\tilde{M} = 8.6$		
Station	Δ (deg)	ϕ (deg)	A (cm)	$\lvert \sin \Delta\phi \rvert$	$A\Delta^{0.6}$	$A\Delta^{0.6}/\lvert \sin \Delta\phi \rvert$
Kushiro	8.7	25	0.9	1.0	3.2	3.3
Sapporo	8.0	11	2.0	0.99	6.9	7.0
Hakodate	6.7	9	2.3	0.98	7.3	7.4
Hachinohe	5.6	18	1.7	1.0	4.9	4.9
Sendai	3.3	22	2.1	1.0	4.2	4.2
Tokyo	—					
Toyama	2.3	312	4.5	0.37	7.4*	19.7
Nagoya	1.9	270	4.2	0.35	6.2	17.9
Matsue	5.1	275	1.8	0.27	4.7	17.7
Kochi	5.0	253	2.3	0.61	6.0	9.9
				Average	5.6	10.2

Method 1, $C_0 = 7.9$; Method 2, $C_0 = 8.1$; Method 3, $C_0 = 8.0$.

Table 6. Station data for Event 5 (for details, see Table 2).

Event No. 5				$\tilde{M}_0 = 3.4 \times 10^{28}$ dyne-cm, $\tilde{M} = 8.3$		
Station	Δ (deg)	ϕ (deg)	A (cm)	$|\sin \Delta\phi|$	$A\Delta^{0.6}$	$A\Delta^{0.6}/|\sin \Delta\phi|$
Kushiro	11.7	32	0.2	0.14	0.9	6.9
Sapporo	10.6	22	0.4	0.31	1.4	4.7
Hakodate	9.3	23	0.4	0.29	1.4	4.7
Hachinohe	8.5	30	0.3	0.17	1.0	6.0
Sendai	6.4	38	0.3	0.04	0.9	—
Tokyo	4.0	52	0.5	0.21	1.2	5.7
Toyama	3.6	17	1.0	0.39	2.1	5.5
Nagoya	2.1	25	1.1	0.26	1.7	6.8
Matsue	3.2	313	3.4	1.00	6.8*	6.8
Kochi	2.0	278	3.2	0.85	4.8	5.6
			Average		2.2	5.9

Method 1, $C_0 = 7.8$; Method 2, $C_0 = 8.1$; Method 3, $C_0 = 7.7$.

the amplitude decays as the distance increases, but a clear azimuthal variation due to the radiation pattern is seen. Since the near-field displacement cannot be represented by simple ray arrivals, the standard geometrical spreading and the radiation pattern cannot be used to explain these results. Figure 7 compares the amplitude decay rate of the synthetics computed for a pure dip-slip mechanism with a dip angle of 20°. The decay rate can be approximated by $\Delta^{-0.6}$ where Δ is the distance. This may be compared with the decay rate Δ^{-1} for a homogeneous whole space.

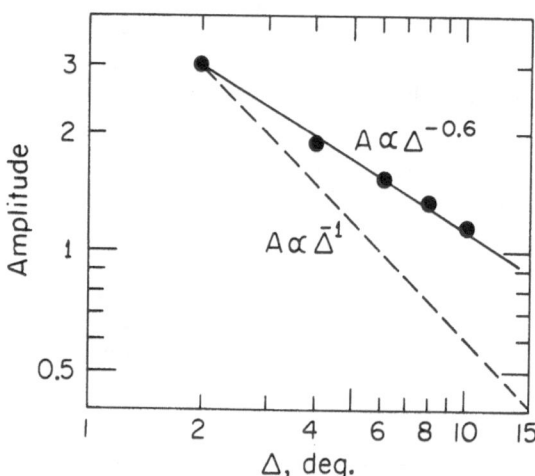

Fig. 7. The amplitude (arbitrary unit) decay curve of the synthetics as a function of distance. The mechanism of the source is pure dip slip dipping 20° to W, and the distance is measured toward west.

Next, the radiation pattern of the synthetics is computed for the dip-slip mechanism and is shown in Fig. 8. The radiation pattern can be roughly approximated by sin ψ where ψ is the azimuth of the station measured from the fault strike. For comparison, the SV-wave radiation patterns in a whole space are shown in Fig. 8 for various take-off angles i_h. The synthetic radiation pattern is similar to that for $i_h = 90°$, the take-off angle appropriate for a homogeneous structure. Since the wavelength is very long with respect to the scale lengths of the structure, this result is reasonable.

Another important result is that the amplitude is larger by a factor of 1.5 for Event 2 (dip angle 45°) than Event 1 (dip angle 20°). This difference arises from the fact that the amplitude from a pure dip-slip source is proportional to $M_0 \sin 2\delta$. This means that, for a dip-slip mechanism, M_0 and δ cannot be determined independently; only the product $\tilde{M}_0 = M_0 \sin 2\delta$ can be determined from the amplitude data. As mentioned earlier, $\tilde{M}_0$ is probably more useful than M_0 for tsunami applications.

Hence, we use M_0 as a parameter which represents tsunami potential of the source. Although this argument does not apply to non dip-slip mechanisms (such as the 1923 Kanto earthquake, Event 4) we use $\tilde{M}_0 = M_0 \sin 2\delta$ as an approximate measure of tsunami potential of an earthquake regardless of the mechanism.

In the following analysis, we assume that the amplitude decay rate is $\Delta^{-0.6}$ and the radiation pattern is approximated by sin ψ. For non dip-slip mechanisms, these assumptions, particularly the sin ψ radiation pattern, may not be valid. However, since most tsunamigenic earthquakes have primarily a dip-slip mechanism, these assumptions are reasonable.

We now determine $\tilde{M}_0$ from the amplitude data obtained at the stations. The amplitude of the seismogram at the i-th station A_i can be written as

$$A_i = f(\Delta_i, \phi_i)\tilde{M}_0 \qquad (1)$$

where $f(\Delta_i, \phi_i)$ is a function of the distance and the azimuth. Strictly speaking, the form of this function depends upon the mechanism. In our simple scheme, however, we can write

$$f(\Delta_i, \phi_i) = \xi \Delta_i^{-0.6} |\sin(\phi_i - \phi_f)| \qquad (2)$$

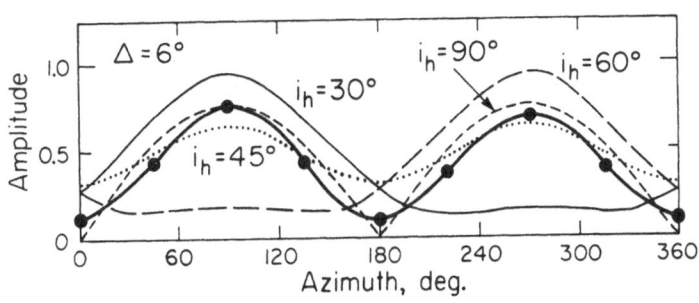

Fig. 8. The radiation pattern of the synthetics computed at $\Delta = 6°$. The mechanism is pure dip slip dipping 20° to W, and the azimuth is measured clockwise from N. SV wave radiation patterns for various take-off angles are shown for comparison.

where ξ is a constant and ϕ_f is the fault strike. From (1) and (2) we have

$$A_i \Delta_i^{0.6} = \xi |\sin (\phi_i - \phi_f)| \tilde{M}_0. \tag{3}$$

The constant ξ includes the effects of instrument response and excitation function, and can be empirically determined using the synthetic seismograms computed for a specific source-station geometry. In order to determine $\tilde{M}_0$ using (3), we try three different methods:

(1)   Calculate $A_i \Delta_i^{0.6}/|\sin (\phi_i - \phi_f)|$ for each station and take the average over i (i.e., all stations). Since the radiation pattern term $|\sin (\phi_i - \phi_f)|$ is very crude, dividing $A_i \Delta_i^{0.6}$ by $|\sin (\phi_i - \phi_f)|$ may introduce a large error for stations where $|\sin (\phi_i - \phi_f)|$ is very small. Hence, we exclude those stations with $|\sin (\phi_i - \phi_f)| \leqslant 0.1$ in taking the average.

(2)   Take the average of $A_i \Delta_i^{0.6}$. Taking the average over i of (3), we have

$$\overline{A_i \Delta_i^{0.6}} = \xi \overline{|\sin (\phi_i - \phi_f)|}\ \tilde{M}_0. \tag{4}$$

Since for a given source-station geometry $|\sin (\phi_i - \phi_f)|$ is a constant, we have $\tilde{M}_0 = \xi' \overline{A_i \Delta_i^{0.6}}$ where ξ' is a constant. Since $|\sin (\phi_i - \phi_f)|$ depends on the location of the source, ξ' is, in general, a function of the source location. However, if the azimuthal coverage of the stations is relatively uniform, it can be treated as a constant.

(3)   Take the maximum of $A_i \Delta_i^{0.6}$. If the azimuthal distribution of the stations is reasonably uniform, we would expect $|\sin (\phi_i - \phi_f)|$ to be close to 1 at least for one station where $A_i \Delta_i^{0.6}$ is maximum. Thus

$$\text{Max} \left(A_i \Delta_i^{0.6} \right) \simeq \xi \tilde{M}_0. \tag{5}$$

Although the seismic moment is physically a more meaningful parameter than the magnitude, use of the magnitude scale is traditional in this type of problem. Hence, we convert $\tilde{M}_0$ to a magnitude $\tilde{M}$ using the standard magnitude-moment relation

$$\log \tilde{M}_0 = 1.5\, \tilde{M} + 16.1. \tag{6}$$

This magnitude $\tilde{M}$ is equivalent to the surface-wave magnitude (or the moment magnitude M_w) for most practical purposes. Since a seismic moment $M_0 = 10^{29}$ dyne-cm is used for the design earthquakes, the magnitude $\tilde{M}$ is 8.5, 8.6, 8.5, 8.6, and 8.3 for Events 1, 2, 3, 4, and 5 respectively after the difference in the dip angle is taken into account.

Combining (6) and (3), (4) or (5), we can write the magnitude $\tilde{M}$ as

$$\tilde{M} = \frac{1}{1.5} \log (\bar{A}_0) + C_0 \tag{7}$$

where

$$\bar{A}_0 = \begin{cases} \overline{A_i \Delta_i^{0.6}/|\sin (\phi_i - \phi_f)|} & \text{(method 1)} \\[2mm] \overline{A_i \Delta_i^{0.6}} & \text{(method 2)} \\[2mm] \text{Max}\,(A_i \Delta_i^{0.6}) & \text{(method 3)} \end{cases} \tag{8}$$

C_0 is a constant which depends upon the method used. We then determine the constant C_0 using the synthetic data set for the five events listed in Table 1. The data and the results are shown in Tables 2, 3, 4, 5, and 6. We average the values of C_0 determined for each event, and obtain $C_0 = 7.8$ for Methods 1 and 3 and $C_0 = 8.1$ for Method 2. In the above, A_i is the peak-to-peak amplitude in cm and Δ_i is in degree. Using these values, we can determine $\tilde{M}$ from the observed amplitude data using (7) and (8).

Among these three methods, Method 3, in which Max $(A_i \Delta_i^{0.6})$ is used, is the easiest to implement. However, when a relatively small number of stations are used, Method 2 would provide a more stable estimate of $\tilde{M}$. Although Method 1 is most straightforward in principle, it would be more cumbersome to use in real-time situations than the other methods.

We believe that these methods, particularly Method 3, are simple and practical enough to be used for a real-time, near-field tsunami warning purposes where a very rapid determination of the source parameters is required. In the numerical experiment described above the cut-off period at the short-period end is 45 sec and the synthetic wave forms are very smooth. However, the seismic waves generated by real events contain energy at periods shorter than 45 sec which would make the long-period amplitude measurements difficult. Some filters would be necessary to remove high frequency energy from the signal. Modification of the present method for recording instruments other than the one used in this numerical study is straightforward.

This research was supported by the U.S. Geological Survey contract No. 14–08–0001–19755. Contribution No. 3756, Division of Geological and Planetary Sciences, California Institute of Technology, Pasadena, California 91125.

REFERENCES

BULAND, R. and J. F. GILBERT, The theoretical basis for the rapid and accurate computation of normal mode eigen frequencies and eigen functions, Unpublished Research News, University of California, San Diego, 1976.

GILBERT, F. and A. M. DZIEWONSKI, An application of normal mode theory to the retrieval of structural parameters and source mechanism from seismic spectra, *Phil. Trans. R. Soc. London, Ser. A*, **278**, 187–209, 1975.

KANAMORI, H. and J. J. CIPAR, Focal process of the great Chilean earthquake May 22, 1960, *Phys. Earth Planet. Inter.*, **9**, 128–136, 1974.

KANAMORI, H. and J. W. GIVEN, Use of long-period surface waves for rapid determination of earthquake source parameters, *Phys. Earth Planet. Inter.*, **27**, 8–31, 1981.

OKAL, E. A., A Physical classification of the earth's spheroidal modes, *J. Phys. Earth*, **26**, 75–103, 1978.

Tsunamis—Their Science and Engineering, edited by K. Iida and T. Iwasaki, 51–60.

A New System for Tsunami Warning in the Japan Meteorological Agency

Masaji Ichikawa* and Hideo Watanabe**

Sendai District Meteorological Observatory, Sendai, Miyagi Pref., Japan
**Meteorological Research Institute, JMA, Tsukuba, Ibaraki Pref., Japan*

(Received October 6, 1981; Revised December 15, 1981)

The Japan Meteorological Agency is responsible for the issuance of tsunami warnings for near or distant tsunami. The average time needed in the issuance of tsunami warnings for local events occurring since 1952 is about 17 minutes. In order to shorten this time and to improve the reliability of the warning, a new system was installed in the Seismological Division of the Agency, one of the regional tsunami warning centers. Test runs of the system demonstrated that the warning service will be improved. The system will be routinely employed from 1982, and the same system will be installed in other regional centers within a few years.

1. Introduction

The tsunami warning service in Japan started on April 1, 1952, as a means to prevent or reduce damage due to tsunamis, and the Japan Meteorological Agency (JMA) is responsible for the issuance of the tsunami warning. A tsunami warning must be given within 20 minutes of the earthquake occurrence at a submarine area of the ocean within 600 km from the Japanese coasts.

From the viewpoint of disaster prevention, prompt issuance of a warning is required, but it is quite difficult to reduce the time limit as long as data processing in the warning service is manually performed. The application of a computer to the data processing, however, will shorten this limit (Ichikawa, 1974). Therefore, a new warning system is planned for installation at each regional tsunami warning center of the JMA. The system was installed in 1979 in the Seismological Division of JMA, one of the regional centers, and will be operational from 1982.

2. Present Status of the Tsunami Warning Service

The present status of tsunami warning in Japan is briefly described here.

2.1 System in JMA

The following are the tsunami warning systems in Japan. 1) tsunami forecast system, 2) dissemination system to transmit forecast messages to coasts, and 3) terminal system to evacuate the general public rapidly on receipt of a warning. Among these, JMA is responsible for the first system.

The warning system is composed of more than 100 seismic monitoring stations, more than 50 tidal monitoring stations, and 6 regional tsunami data analysis centers. The communication between seismic or tidal stations and each regional center is maintained through exclusive JMA lines and public telephone lines. If a land line is broken, a wireless communication system is on standby to operate. The entire coast is divided into eighteen divisions, assigned to the above-mentioned regional data analysis centers as shown in Fig. 1.

Whenever an earthquake is felt, a seismic station promptly measures the arrival times of the P and S waves, and maximum amplitude, and dispatches the results over the exclusive JMA telegraph line to the appropriate regional center within seven minutes after the earthquake occurrence. In addition to the data sent from the seismic monitoring stations, the center can utilize seismograms telemetered from other stations in order to estimate the epicentral area using an analog computer. Furthermore, the tsunami grade magnitude is estimated on the basis of a tsunami forecasting chart, if the

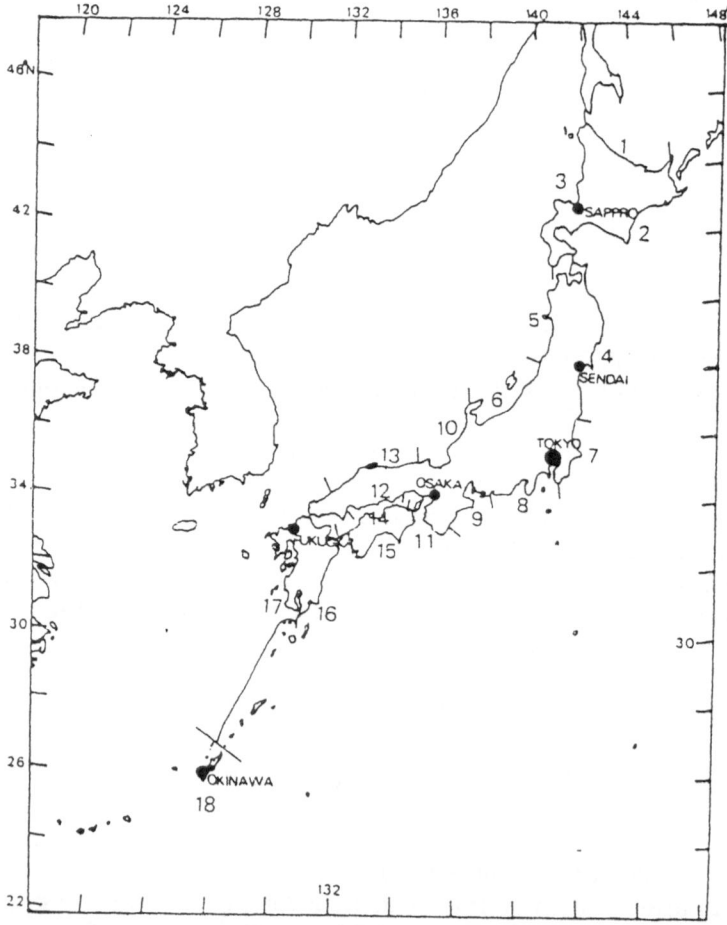

Fig. 1.   Location of regional tsunami analysis centers and coastal divisions used in issuing tsunami warning messages.

epicenter of the earthquake is located under the ocean floor.

The result of the analysis is delivered to relevant sections in the regional center from where the tsunami warning messages are transmitted to the dissemination centers.

The tsunami warning messages are classified into five types: 1) No Tsunami; A tsunami is not predicted. 2) Tsunami Attention; A tsunami may be predicted. Its height may be less than several tens of centimeters. 3) Tsumani; A tsunami is predicted. Its height may be about 2 m at maximum. 4) Major Tsunami; A major tsunami is predicted. Its height may be more than 3 m. 5) Cancellation; No threat is predicted and previous messages are cancelled.

A more detailed description of the JMA tsunami warning system was given by one of the authors (WATANABE, 1977).

2.2 Tsunami warnings issued up to the present

Since 1952, tsunami warnings, including "No tsunami", have been issued for more than 150 events occurring in and near Japan. Figure 2 shows the distribution of earthquakes against which tsunami warnings were issued, and 45 of these events were accompanied by perceptible tsunami.

A statistical study of the time required in the issuance of the tsunami messages, the success rate and the amount of data received at the regional centers will now be presented in order to elucidate the present status of the tsunami warning service in JMA. Figure 3 shows the number of seismo-telegrams received at the regional centers, as a function of time after the occurrence of each event shown in Fig. 2. It is evident from the plot that the installation of a new telecommunication system in 1969 increased in the amount of data received from seism-moniforing stations. At present, we receive, on the average, twenty data messages after the occurrence of an earthquake of magnitude larger than 5.

In view of the time limit requirement of tsunami warning message issuance, the determination of the epicenter and grade magnitude of the predicted tsunami should be finished within fifteen minutes after the earthquake. This suggests that it is difficult to use more than twenty data messages. For the reason, the computerization of the tsunami warning service was planned.

The annual variation in the mean time in the issuance of a warning is calculated and shown in Fig. 4. The average time for the issuance of warning messages is shown in Table 1. It is evident from Table 1 that the issuance time in recent years has shorten remarkably. So far as the time used in the issuance is concerned, the result is quite satisfactory, because most warnings were issued within the time limit.

The success rate for issued tsunami warnings is problematic, too. An evaluation of the issued warnings is conducted on the basis of the following criteria:

Message	Success	Failure
No Tsunami	Tsunami imperceptible	Perceptible tsunami
Tsunami*	Perceptible tsunami	Tsunami imperceptible

*Including Tsunami Attention, Tsunami, and Major Tsunami.

Fig. 2. Distribution of epicenters of earthquakes against which a tsunami warning was issued. Small circle; "No Tsunami" was issued, and tsunami was not generated. Solid small circle; "No Tsunami" was issued, but tsunami was generated. Large circle; tsunami was predicted, and was generated. Solid large circle; tsunami was predicted, but was not generated. Large circle with small solid circle; tsunami was not predicted, but was generated.

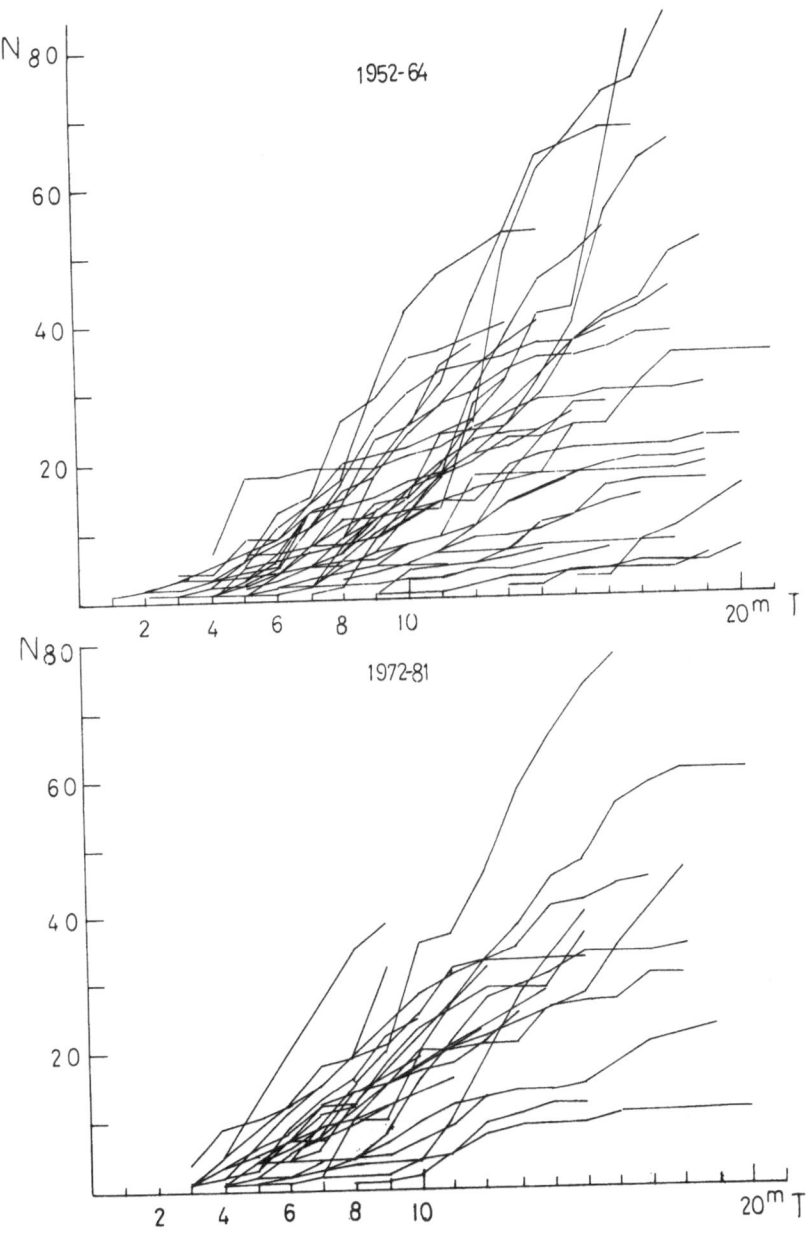

Fig. 3.   Number of seismo-telegrams received at a regional center, as a function of time after the earthquake occurrence (in minutes).

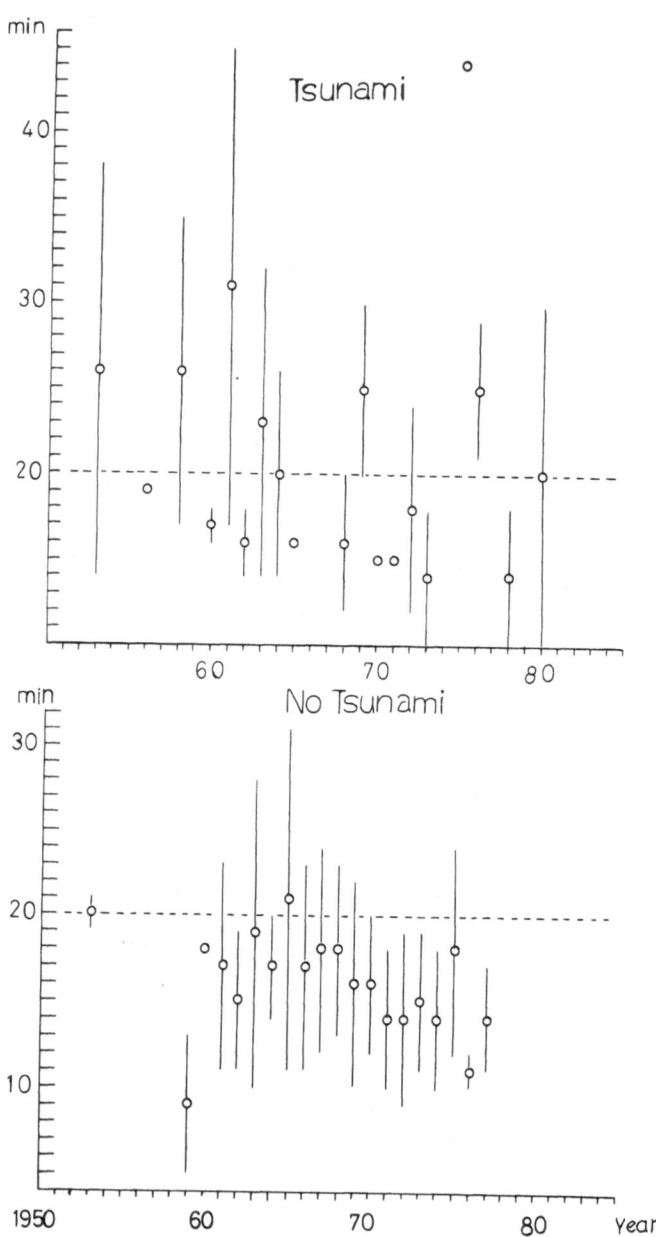

Fig. 4.   Average time and its standard deviation used in the issuance of tsunami warning messages (top; "Tsunami" bottom; "No Tsunami").

Table 1. Average time in minutes needed for the issuance of tsunami warnings.

Message Period	No tsunami	Tsunami
1952–1970	18 ± 6.1	22 ± 10.7
1971–1980	16 ± 9.7	16 ± 6.7
1952–1980	17 ± 8.0	20 ± 9.6

The success rate and the number of warnings issued each year are shown in Fig. 5. In general, the message "No Tsunami" must be issued when a regional tsunami center receives seismo-telegrams, even if the location of the earthquake is not in the ocean bottom. This is the reason for the high success rate in "No Tsunami." It is notable that the figure shows at low success rate for "Tsunami" in recent years, even though the issuance time was remarkably reduced. This suggests that the problem lies in the improvement of the reliability of the tsunami warning, in other words, a more accurate determination of the grade magnitude of the tsunami and its hypocenter. This, however, is quite difficult as long as the data processing is performed manually.

A computer system was applied in the analysis of tsunami warning data by one of the authors (ICHIKAWA, 1974), and the test showed an improvement in the accuracy of the hypocenter determination, as well as in the data processing time. Based on the simulation, a new computer system was installed in the Seismological Division of JMA.

3.   A New Computer System for Tsunami Warning Service

3.1   Hardware

JMA installed a computer system whose block diagram is shown in Fig. 6. The computer is used in common to process seismological data including the tsunami warning data and meteorological telegram data dispatched from weather stations belonging to each regional center (District Meteorological Observatory (DMO)). As shown in Fig. 6, this system is duplex and composed of four central processing units; two front end processors (FEP), and two main processors (one is the host processor and the other is a standby processor).

The meteorological and seismological telegram data are processed by both the master and slave FEPs in order to prevent loss of data due to machine trouble. The FEPs are also in use for gathering earthquake records. Specifically, the processors continually evaluate signals telemetered from weather stations, and transmit seismic signals exceeding present triggering levels to pen-recorders and a magnetic disc. The earthquake records on the pen-recorders are used for the rapid determination of the earthquake parameters, and the digital seismograms stored on the disc are used for more precise interpretation of the earthquake records.

The urgently measured readings of the P and S wave arrival times and the maximum amplitude on the telemetered seismograms are made using an $X–Y$ digitizer under the control of the standby processors. A rapid calculation of earthquake

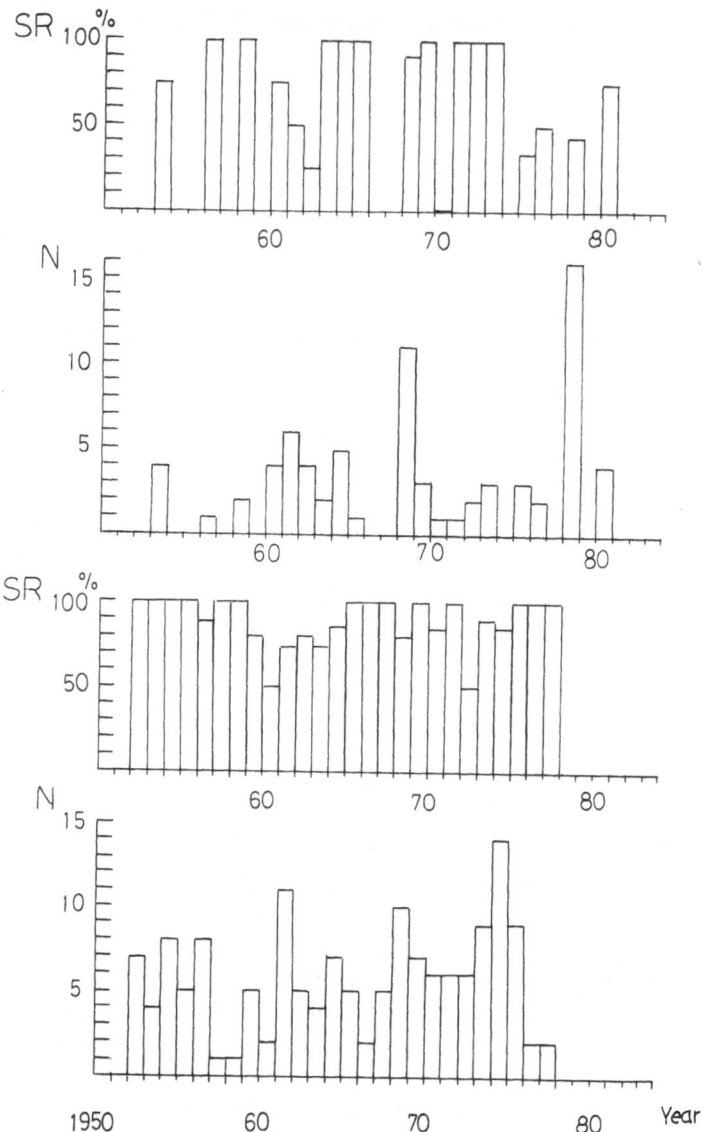

Fig. 5. Number of tsunami warnings issued each year (N) and success rate (SR). The top two plots are for "Tsunami," and the bottom two plots for "No Tsunami."

parameters is also performed by the standby processor. When the standby processor is down, the high priority task of the tsunami warning is performed by the master processor without interruption.

A graphic display in the system plays an important role in the tsunami warning service. On the basis of an operator-computer interactive technique, the rapid determination of earthquake parameters and grade magnitude of the tsunami, and

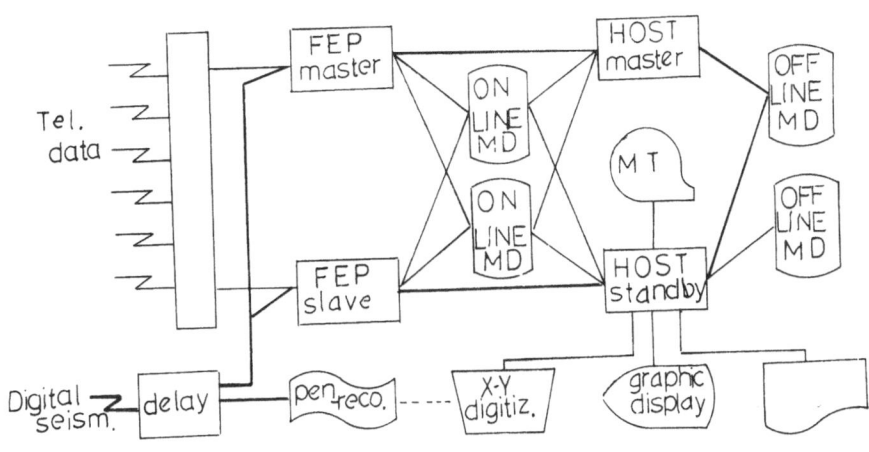

Fig. 6. Block diagram for the new computer system in JMA.

emergent transmission of tsunami messages to relevant divisions in JMA and others are made.

3.2 Procedure of rapid determination of earthquake parameters and tsunami warning

When a regional center receives seismo-telegrams dispatched by weather stations, a rapid determination of earthquake parameters will be done by the system as follows: First, the arrival times of the P and S waves and maximum amplitude will be read on the telemetered seismograms using the $X-Y$ digitizer, and the extracted data are automatically transmitted to other regional centers, if necessary. Second, the earthquake parameters are calculated using both the extracted data and telegram data dispatched from the weather stations.

Three programs using the following data based on the method of least squares are provided for the rapid earthquake determination of a) S–P times, b) P arrival times, and c) both P and S arrival times.

The first method is appropriate for weak local events. When the magnitude of an earthquake is large, it is not always easy to read the S arrival time because of an off scale of the record. The second method is applicable to such large local events, but the hypocenter located, in particular the focal depth, is less reliable when the event occurs outside the seismological network, because of a lack of stations around the epicenter. Use of a few S wave time data in addition to the P wave times makes it possible to determine a more reliable focal depth as well as epicenter. Therefore, the hypocenter should be determined using one of the three methods alternatively selected by an operator in accordance with the quantity and quality of data.

Third, the grade magnitude of the tsunami should be estimated, if necessary. The estimation is automatically done using a table corresponding to the tsunami forecasting chart. The result of the automatic estimation of tsunami magnitude is shown on a map drawn on the graphic display. Tsunami warning messages for forecasting to each coastal division are determined with reference to the tsunami grade map by an operator. The message and other information on the tsunami and earthquake compiled by the computer are automatically transmitted to relevant divisions of JMA, weather stations, and other organizations concerned.

4. Summary

Data processing techniques as well as the seismological network for tsunami warning service have been improved by making effective use of knowledge obtained from earthquakes. As a result of this improvement, the average time required for the issuance of a tsunami warning was reduced to 16 min, but the success rate is not always satisfactory.

A new computer system for tsunami warning service will determine earthquake parameters more accurately in a shorter time, and will make it possible to take the time into consideration of the effect of the focal depth and focal mechanism of the event. Test runs for the system conducted by the staff of the Seismological Division have already demonstrated that the tsunami warning service will be more efficient.

The system will be routinely employed from 1982, and the same system will be installed in other regional centers within a few years.

The present authors wish to express their thanks to the staff of the Seismological Division of JMA for their kind advice in developing the system.

REFERENCES

Ichikawa, M., Analysis of the tsunami warning data by man-machine communication system, *Pap. Met. Geophys.*, **25**, 13–21, 1974.

Watanabe, H., Tsunami warning service and its system in Japan, Technology for Disaster Prevention of National Research Center for Disaster Prevention Science and Technology, The Government of Japan, 313–328, 1977.

Tsunamis- Their Science and Engineering, edited by K. Iida and T. Iwasaki, 61-76.

Some Remarks on the Occurrence of Tsunamigenic Earthquakes around the Pacific

Kumizi IIDA

Aichi Institute of Technology, Toyota, Aichi-ken, Japan

(Received January 16, 1982; Revised February 19, 1982)

The activity of tsunamigenic earthquakes around the Pacific is investigated. During the period from 1900 to 1980, about 370 tsunamis were observed. Of these, 70 tsunamis caused casualities and damage near the source only, and 17 tsunamis caused widespread disasters outside the source region. A great number of tsunamis was repeatedly observed in a certain area related to the activity of earthquakes for short periods of several months.

The activity of earthquakes accompanied by tsunamis is different according to the area. The frequency distributions of these earthquakes are examined. The most tsunami active region is found to be the Japan-Taiwan region where 28.6 percent of the total Pacific tsunamis were generated.

Tsunami runup heights for large tsunamigenic earthquakes are examined in connection with local effects and earthquake magnitudes based on available historical data. The maximum runup height of 525 m was reached by the Alaska earthquake of 1954. It is noticed that these are locally confined effects in the tsunami generation at the time of major earthquake occurrence as seen in bays and straits of Alaska and in the archipelagos of the southwest Pacific.

Regularity of earthquake occurrences in space and time is investigated for regions related to the structure of island arcs and trenches where large earthquakes generate large tsunamis. Some regularities may be seen in the Tokai-Nankai and Sanriku districts of Japan and in the Concepcion and Valparaiso districts of Chile. Some of the other regions examined around the Pacific are also found to have a tendency for regularity of large earthquake occurrences related to the subduction zones where thrust faulting may take place.

1. Introduction

Areas along the boundaries of the Pacific Ocean are known to be the most seismically active regions in the world. Large earthquakes occurring near the coasts of these areas occasionally produce significant tsunamis affecting coastal areas. The most destructive tsunamis have been reported by many investigators, whereas smaller tsunamis have not received the same attention and consequently complete knowledge of the areas of tsunami activity is not yet available. Since there are many dissimilarities in areas of seismic acitivity which could potentially generate tsunamis, regional differences in tsunami activity will be also considered.

In the present paper, the activity of tsunamigenic earthquakes in the Pacific Ocean including the Indian Ocean near Indonesia is investigated together with the effects of tsunamis and the regurality of their occurrences in time and space.

2. Activity of Tsunamigenic Earthquakes

In order to compile and investigate the earthquakes accompanied by tsunamis that occurred in the Pacific, tsunami information was obtained from several literature sources (Cox and Morgan, 1977; Cox, 1981; Iida et al., 1967; Iida, 1981b; Seismol. Soc. Am., 1960–1981; Soloviev and Go, 1969, 1974, 1975; Soloviev, 1975, 1976, 1977; Watanabe, 1968). The present paper covers an investigation of all earthquakes and tsunamis during the 81-year period from 1900 to 1980. The 370 tsunamis which were observed or recorded in the Pacific during this period are classified according to their occurrence region as given in Table 1. The tsunamigenic earthquake activities in the Pacific may be seen in Fig. 1 which indicates their occurrence frequency. A number of tsunamis was frequently observed in areas related to the activity of tsunamigenic earthquakes for a short time period. If the 48 tsunamis generated during two weeks in March of 1953 by the Myojin volcanic eruption, Izu Island are excluded, the most active

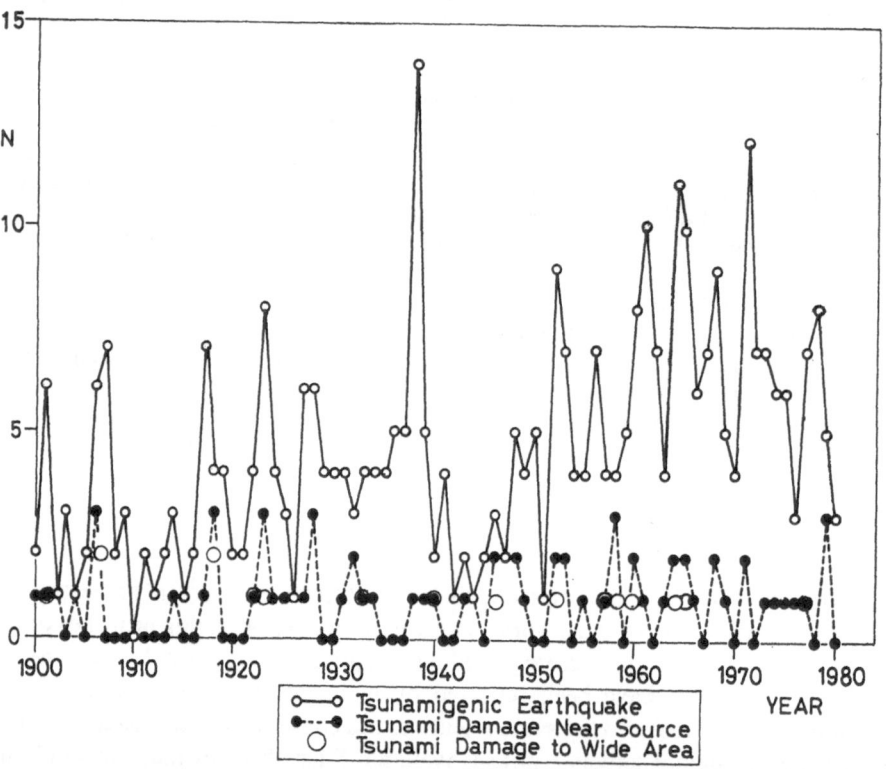

Fig. 1.   Yearly frequency of tsunamis during the 81-year period from 1900 to 1980.

region at the rate of 10 tsunamis per six months was in 1938 off the Pacific coast of Fukushima and Miyagi Prefectures, the northeastern provinces of Japan. None of the tsunamis in this region caused significant damage.

The activity of earthquakes accompanied by tsunamis differs according to the region of the Pacific Ocean. The tsunamigenic earthquake frequency obtained from Table 1 is given in Table 2, in which the tsunami region is divided into eleven and their frequency distributions are shown in Fig. 2. The most active tsunami region is that of Japan-Taiwan where 28.6 percent of the total Pacific tsunamis were generated, though most of these were minor. About two-thirds of the Pacific tsunamis of the last 81 years have emanated from the vicinity of trenches of Japan, Kuril, Aleutian, Peru-Chile. This is related to the structure of island arcs and trenches, as shown in Fig. 2, where large earthquakes frequently generate large tsunamis.

Seventy tsunamis (18.9%) of the total Pacific tsunamis of 370 caused casualities and affected mainly the areas near their sources and 17 tsunamis (4.6%) caused widespread disaster even at places far from their sources, which are as shown in Fig. 3. From Fig. 3 it may be also noticed that most of these tsunami sources are associated with the structure of island arcs and trenches. The earthquake magnitudes of four of these 17 destructive tsunamis are in the ranges of 7.0 to 7.8. These earthquakes are considered as so-called tsunami earthquakes with low frequency. In Table 3 are listed the tsunamis which caused damage at locations far from their sources, with the degree of damage being given.

3. Tsunami Runup Heights of Destructive Tsunamis

The largest value of tsunami runup height around the source area is investigated in relation to local effects and earthquake magnitudes, including the historical data. Tsunami runup heights estimated over 25 m are listed in Table 4 and shown in Fig. 4. These significant tsunamis accompanying earthquakes having a magnitude larger than 8 are attributed to coastal areas of Japan, Kamchatka, Alaska, Peru, and Chile, and those accompanying earthquakes with magnitudes less than 8 are distributed in coastal areas of Alaska, Aleutian, Japan, Indonesia.

It is noted that there are locally confined effects in the tsunami generation at the time of major earthquake occurrence as seen in bays and straits of Alaska and in the archipelagos of the southwest Pacific. Runup values for tsunamis generated in the Aleutian Trench and the Weber Trench area were extraordinarily high as seen in the Alaska region of 51 m in 1964, 60 m in 1899, 115 m in 1788, 525 m in 1954, and in the Banda region of 80 m in 1674 and 90 m in 1629. The runup height is considered to be 85 m in the Ishigaki Island in the Ryukyu area as caused by the 1771 earthquake which occurred in connection with the Ryukyu Trench of Japan.

The configuration of the Alaska fiords and the giant landslide triggered by an earthquake are probably particularly contributing to local tsunami generation as pointed out by Cox (1972). The significant tsunami runup in the Banda Sea areas may have been caused by similar processes.

It may be concluded that we should pay attention to the possibility of local tsunami generation in bays and straits similar to those mentioned above in areas of

Table 1.   Tsunami list in the Pacific during 81-year period from 1900 to 1980.

Year	1	2	3	4	5	6	7	8	9	10	11	12	13	14	15	16	17	total
1900												2						2
1901			1												5			6
1902							1											1
1903		1															2	3
1904										1								1
1905				1						1								2
1906						1				1		2	1		1			6
1907							1	1			5							7
1908													1		1			2
1909							1								1		1	3
1910													1		2			3
1911												2						2
1912															1			1
1913												1			1			2
1914									1					1	1			3
1915															1			1
1916							1					1						2
1917											3		2	1		1		7
1918	2									1		1						4
1919											1	1		1			1	4
1920										1		1						2
1921										1				1				2
1922										1	1		1	1				4
1923		2										1	1	1	3			8
1924											1			3				4
1925				1			1			1								3
1926														1				1
1927		1				1				1		1	2					6
1928							1			1				1	3			6
1929			1		1									1	1			4
1930											1	1	1		1			4

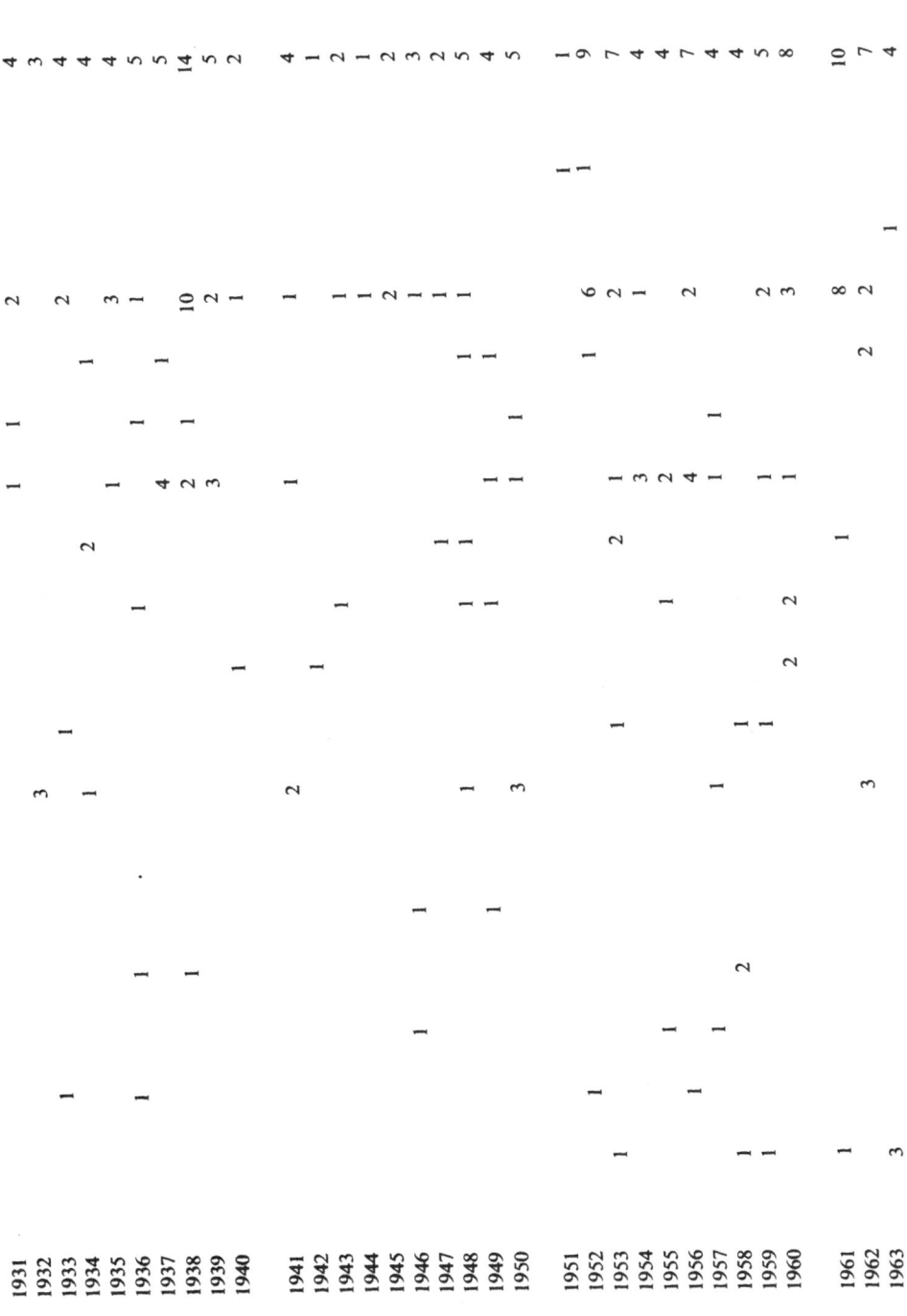

Table 1. (continued)

	1	2	3	4	5	6	7	8	9	10	11	12	13	14	15	16	17	total
1964	1										4	1	1		3			11
1965	1		3								1	1			1			10
1966				5					1	1	2	2				1		6
1967							1		1	1	2	1	1	1				7
1968		1										1	1	1	5			9
1969		1										1	1		2			5
1970									1			1		1	1			4
1971	6		1							1		3		1	1			12
1972				1							1	1		1	3	1		7
1973	2						1			1					2			7
1974									1			2		1	2			5
1975	1										1	1		1	1			5
1976											1			1			1	3
1977											1			1				3
1978	5							1			2	4	1		2	1		7
1979							1					1			1			8
1980				1								1			2			5
Total	20	14	9	13	3	2	24	6	9	20	31	61	21	24	101	5	7	370

1, Kuril Is.; 2, Kamchatka; 3, Aleutian; 4, Alaska; 5, Canada; 6, U.S.A.; 7, Central America; 8, Colombia; 9, Peru; 10, Chile; 11, New Zealand, Kermadec, Fuji, Samoa, New Hebrides; 12, New Ireland, New Britain, Bismark, New Guinea, Solomon; 13, Celebes, Java, Sumatra, Timor; 14, Palau, Philippine; 15, Japan; 16, Taiwan; 17, Hawaii.

Table 2.   Tsunami activity around the Pacific Ocean.

Region	Tsunami generation number	Percent
New Zealand-Kermadec-Tonga-Fiji-Samoa- New Caledonia-New Hebrides-Solomon	31	8.4
New Britain-New Guinea	61	16.5
Celebes-Java-Sumatra	21	5.7
Palau-Philippine	24	6.5
Taiwan-Japan	106	28.6
Kuril-Kamchatka	34	9.2
Aleutian-Alaska	22	5.9
Canada-U.S.A.	5	1.3
Mexico-Guatemala-Panama	24	6.5
Colombia-Peru-Chile	35	9.5
Hawaii	7	1.9
Total	370	100.0

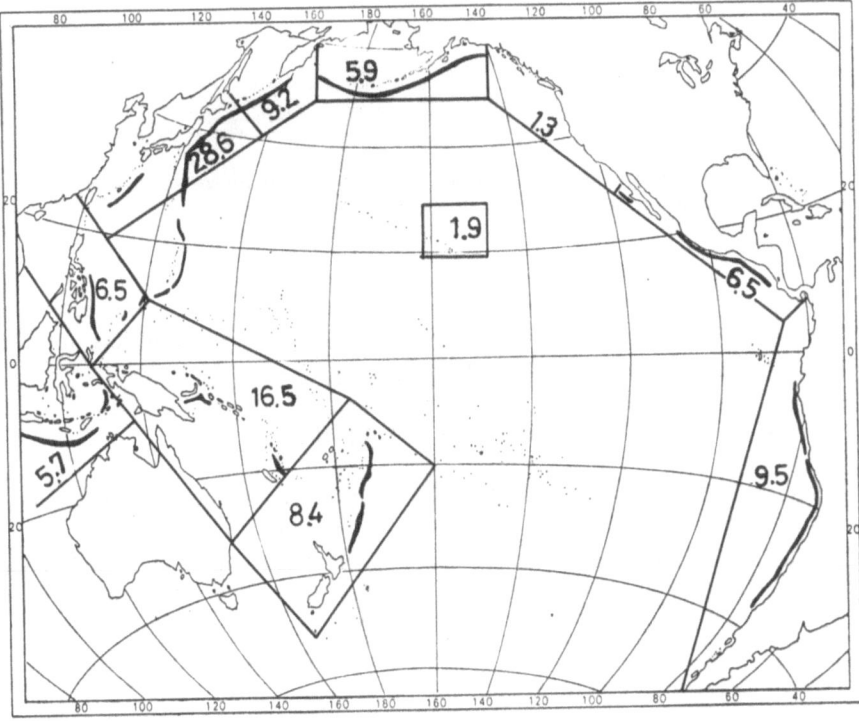

Fig. 2.   Regional tsunami activities in the Pacific Ocean expressed in percent. Thick lines denote the ocean trenches.

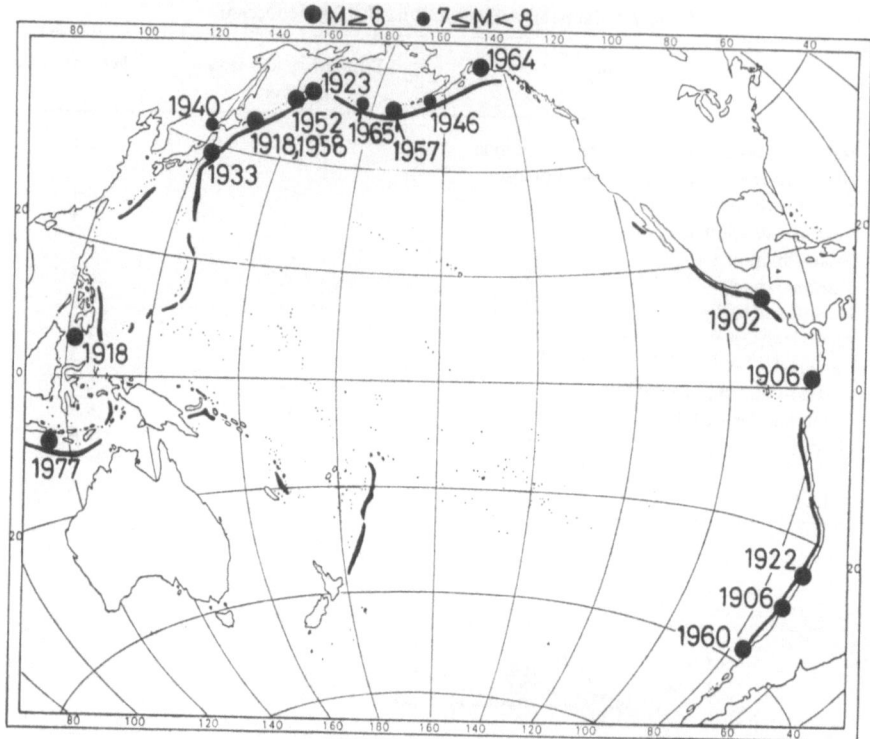

Fig. 3.   Distribution of epicenters of tsunamigenic earthquakes with tsunami-caused damage at locations far from their source. Thick lines denote the ocean trenches.

other earthquake active zones associated with the structure of island arcs and trenches. This is true even though the generation of local significant tsunami has not been reported as accompaniments of large earthquakes.

4.   Occurrences of Large Tsunamigenic Earthquakes

Japan and its adjacent area are known to be the most tsunamic active region in the world and Japan has been attacked by destructive tsunamis since 684. Consequently, the investigation of time and space characteristics of large tsunamigenic earthquakes has been executed based on the available historical data. It is important to know whether these earthquakes took place or not in a certain region with regularities of occurrences for the prediction of an earthquake occurrence and the mitigation of disasters due to earthquakes and tsunamis.

As is known among Japanese seismologists, there are some regions in Japan showing time regularities of destructive earthquake occurrences and these areas may be related to the structure of island arcs and trenches.

During the period from 684 to 1946 there took place 7 destructive earthquakes in

Table 3.   Tsunami damaged area except near taunami source.

Date	Earthquake M	Tsunami source (nearest coast)	Damaged area, damage excluded near source
1902. 2. 26	8.2	El Salvador	San Diego: Small
1906. 1. 31	8.6	Ecuador-Colombia	California, Hawaii: Samll
1906. 8. 17	8.4	Chile	Hawaii: Samll
1918. 8. 15	8¼	Celeves Sea	Indonesia: Small. Philippine: Moderate
1918. 9. 7	7¾	S. Kuril	Chichijima: Small. Hawaii: Small
1922. 11. 11	8.3	N. Chile	Hawaii, Samoa, Sanriku (Japan): Small
1923. 2. 3	8.3	E. Kamchatka	Hawaii: Moderate
1933. 3. 2	8.3	Sanriku	Hawaii: Small
1940. 8. 1	7.0	W. Hokkaido	Primorskiy, Korea: Moderate
1946. 4. 1	7.4	E. Aleutian	Hawaii, California: Severe
1952. 11. 4	8¼	E. Kamchatka	Hawaii: Severe. Sanriku, Samoa: Small. Peru: Small. Chile: Small
1957. 3. 9	8.3	Aleutian	Hokkaido: Moderate. California: Small. Hawaii: Severe. El Salvador: Small
1958. 11. 6	8¼	S. Kuril	Hokkaido, Sanriku: Moderate
1960. 5. 22	8.5	Chile	California: Small. Hawaii: Moderate. Japan: Severe. Samoa: Moderate
1964. 3. 28	8.4	Alaska	California: Severe. Hawaii: Moderate. Japan: Small. Canada: Severe
1965. 2. 4	7¾	Aleutian	Japan: Small.Hawaii: Moderate
1977. 8. 19	8.0	S. Sambawa	Australia: Small

the Tokai district and 9 earthquakes in the Nankai district as shown in Fig. 5, by taking into consideration of the 1605 earthquake and the 1498 earthquake for each of the Tokai and Nankai area, respectively (IIDA, 1981a). Within the time intervals of 1–2 hours to 2–3 years after the occurrence of a large earthquake in the Tokai area, a similar shock occurred in the Nankai area with a high probability after 1096. Further, the intervals between large earthquakes in each of the Tokai and Nankai areas are in the ranges of 90 to 263 years and 92 to 262 years, respectively. Therefore, though this each area does not reveal a regularity of eathquake occurrence, mean intervals are 141.3 years and 157.8 years in the areas of Tokai and Nankai, respectively.

There may be seen another regularity for destructive earthquakes with similar characteristics in the Sanriku district as shown in the top of Fig. 5. The destructive tsunamigenic earthquakes in the south Sanriku district may be classified into two groups according to their mechanism of occurrence. The earthquakes of 1793 and 1896 belonging to one group are of the type of so-called tsunami earthquake with a magnitude less than 8 and low frequency, and with intervals of about 100 years being caused by thrust faulting. The earthquakes of 869, 1257, 1611, and 1933 are of another group of ordinary tsunamigenic earthquakes with a magnitude of 8. These common tsunamigenic earthquakes took place in the similar region with intervals of about 350 years, the epicenter successively migrating from south to northeast as shown in Fig. 5. These earthquakes are considered to be caused by the predominance of normal type faulting.

Table 4. Large tsunami runup height over 25 m.

Earthquake Date	M	Tsunami source	Runup height (m) (Place)	
1586. 7. 9	8.5	Peru	25.6	(Coast of Lima)[3]
1611. 12. 2	8.1	Sanriku, Japan	25	(Taro, Yamada, Iwate Pref.)[3]
1629. 8. 1	7	Banda Is., Indonesia	90	(Neira)[5]
1674. 2. 17	7	Banda Sea, Indonesia	80–100	(Ceyt)[5]
1707. 10. 28	8.4	Nankai, Japan	25.7	(Kure, Kochi Pref.)[3]
1737. 10. 17	8.5	SE. Kamchatka	30	(Avacha)[6]
1746. 10. 28	8	Peru	24–26	(Callao)[2]
1771. 4. 24	7.4	Ryukyu Is.	85.4	(Makinaka, Miyara village, Ishigaki Is.)[3]
1788. 7. 22.	8	Alaska Pen.	115	(Sonak, Unga Is.)[5]
1820. 12. 29	7.5	Flores Sea, Indonesia	25	(Boelekomba)[5]
1871. 3. 3*		Sangihe Is., Indonesia	27	(Buhias)[2]
1883. 8. 26*		Sunda Strait, Indonesia	35	(Merak, Java)[2]
1896. 6. 15	7.6	Sanriku, Japan	38.2	(Shirahama, Ryori Bay)[3]
1899. 9. 10	8.6	Alaska	60	(Lituya Bay)[4]
1905. 7. 4	7.2	Alaska	35	(Yakutat Bay)[4]
1923. 4. 14	7.2	E. Kamchatka	20–30	(Ust' Kamchatsk)[6]
1933. 3. 3	8.3	Sanriku, Japan	29.3	(Shirahama, Ryori Bay)[3]
1946. 4. 1	7.4	E. Aleutian	30.5	(Unimak Is.)[2]
1958. 7. 9	7.5	Alaska	525	(Lituya Bay)[1]
1960. 5. 22	8.5	S. Chile	25	(Isla Mocha)[2]
1964. 3. 27	8.4	Alaska	51.8	(Port Valdez)[1]

*Volcanic eruption.

1. Cox (1972), 2. IIDA *et al.* (1967), 3. IIDA (1981b), 4. SOLOVIEV (1975), 5. SOLOVIEV (1976), 6. SOLOVIEV (1977).

The relation between the large tsunamigenic earthquakes of 1677, 1763, 1856, and 1968 in the north Sanriku area is another example of such a regular interval as also shown in Fig. 5. These earthquakes with a magnitude of approximately 8 are probably caused by a thrust-type mechanism because of the similar characteristics of tsunami generation, and took place with time intervals of 86, 93 and 112 years, respectively. Therefore, the mean time interval between these earthquakes is about 100 years (HATORI, 1975; USAMI, 1979). It is interesting that the migration of the epicentral areas of these earthquakes appears to take place successively northward.

To examine regularities of occurrences of great tsunamigenic earthquakes, large tsunamigenic earthquakes in other regions around the Pacific similar to Japan were investigated in connection with subduction zones where thrust faulting in the characteristic deep trench accompanies the underthrusting of an oceanic plate beneath a continental plate. As is generally known, the Chilean region is located in a subduction zone where tectonic plates converge.

As shown in Fig. 6, the relation between the large tsunamigenic earthquakes of 1570, 1657, 1751, 1835 and 1928 in the Concepcion area gives an interval of about 90 years. All of these earthquakes have a magnitude of 8 and the accompaning tsunamis have similar characteristics. The location of the epicenters of these earthquakes

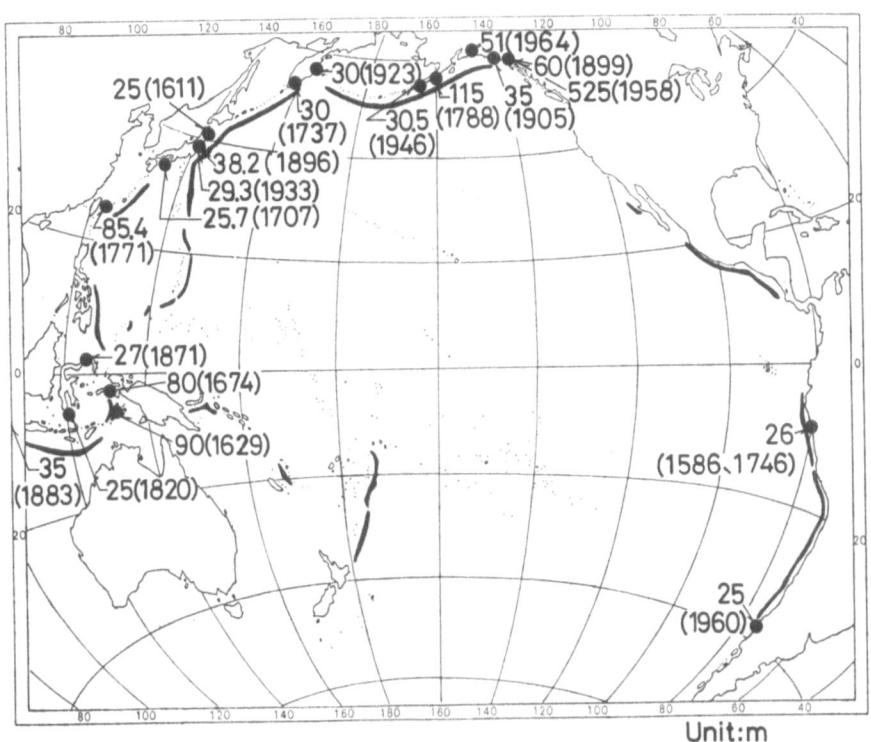

Fig. 4.   Distribution of epicenters of tsunamigenic earthquakes with estimated tsunami runup height exceeding 25 m, based on historical data. Thick lines denote the ocean trenches.

appears to have migrated successively northward. Another example of the regular interval may be seen in the Valparaiso area where the occurrence intervals of destructive earthquakes of 1562, 1647, 1730, 1822 and 1906 with a magnitude of more than 8 is also about 90 years. The locations of their epicenters during the 260-year period from 1562 to 1822 successively migrated from south to north and the epicenter of the 1906 earthquake returned back to almost the original location of the 1562 earthquake epicenter. The intervals between destructive earthquakes of 1575, 1737, 1837 and 1960 in the Valdivia area seem to be longer than those of both the Concepcion and the Valparaiso area, that is, more than 100 years as shown in Fig. 6. If the same process of earthquake occurrence in these areas is assumed, it suggests the possibility of future great tsunamigenic earthquakes.

As discussed above, some regularities are found in the sequence of occurrence of great tsunamigenic earthquakes in each of the Concepcion and the Valparaiso areas. However, this regular pattern in the generation of earthquakes is not apparent for earthquakes which alternately migrated from the Valparaiso area to the Concepcion area or from the Concepcion area to the Valdivia area, though the migration feature from the Valparaiso area to the Valdivia area through the Concepcion area may be recognized.

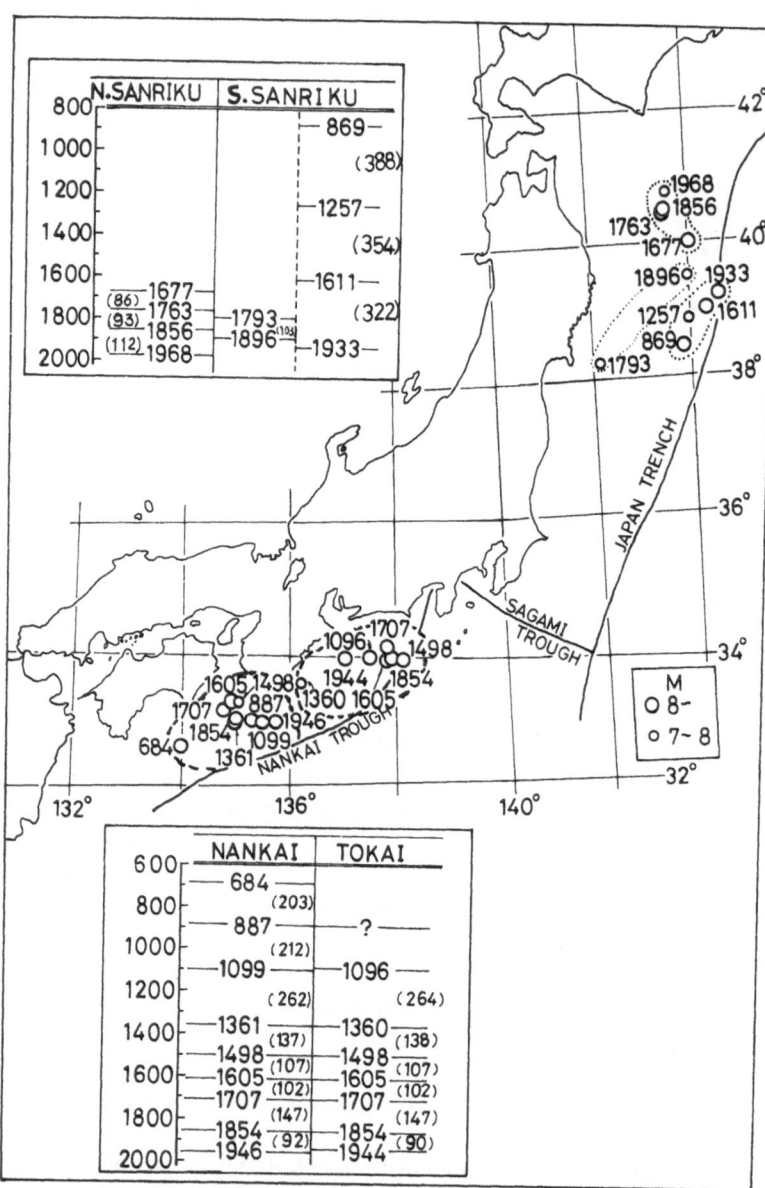

Fig. 5.   Epicentral distribution and time intervals of sequential occurrences of great tsunamigenic earthquakes in the Tokai-Nankai region and both the south Sanriku and north Sanriku areas in Japan. Bracket numeral indicates time interval.

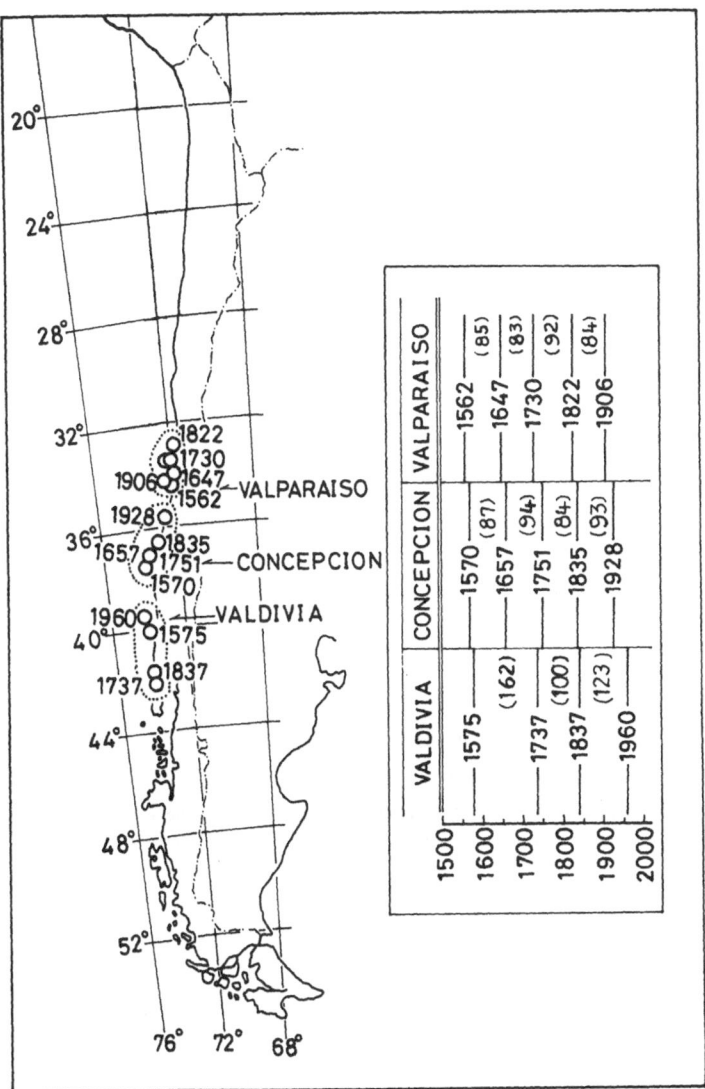

Fig. 6. Epicentral distribution and time intervals of sequential occurrences of great tsunamigenic earthquakes in the Valparaiso, Concepcion, Valdivia regions in Chile. Bracket numeral indicates time interval.

Somewhat regularities in earthquake occurrences may be found in several regions around the Pacific as shown in Fig. 7. These regions are also given in Table 5 together with the time interval in round numbers, earthquake and tsunami magnitude. The time intervals among great earthquakes which occurred in the same area are different from the regions in the range of about 30 years to 300 years. This suggests a difference in the fracture strength and the structural state of regions around the Pacific. For regions

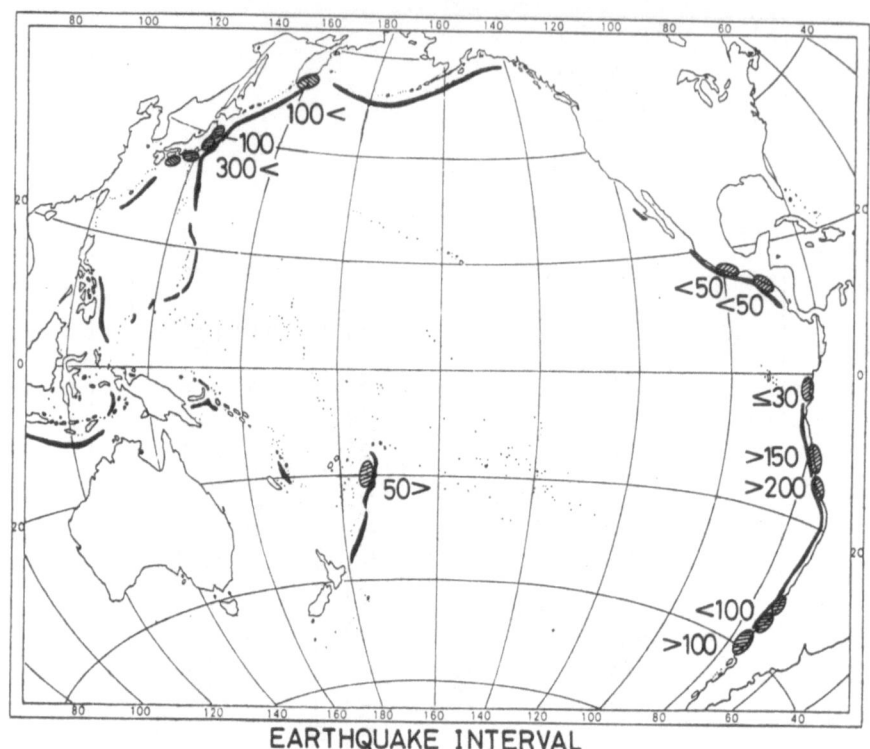

EARTHQUAKE INTERVAL

Fig. 7.   Time intervals of sequential occurrences of tsunamigenic earthquakes with similar characteristics in the same region around the Pacific. Thick lines denote the ocean trenches.

which were not presented in Table 5, an interpretation was not possible for lack of historical data.

5.   Summary and Concluding Remarks

The activity of tsunamigenic earthquakes in the Circum-Pacific zone was investigated for the period from 1900 to 1980 and a compilation was made of those earthquakes and tsunamis which caused damage in coastal areas. During this period the number of tsunamis accompanying submarine earthquakes totaled 370, of which approximately 19 percent caused damage near their sources and about 5 percent caused damage at locations far from their source.

Tsunami runup heights for large tsunamigenic earthquakes were examined in connection with local conditions and earthquake characteristics based on historical data. It is noticed that there are some regions where an extraordinary runup height is generated, and are locally confined effects in the tsunami generation triggered by a major earthquake occurrence associated with the subduction zone of plate tectonics. It seems possible that such tsunamis of local significance may be generated in bays and

Table 5. Time intervals of sequential occurrences of great tsunamigenic earthquakes with similar
characteristics in the same region.

Region	Time interval (year)	Earthquake	Earthquake magnitude M	Imamura-Iida tsunami scale m
SE. Kamchatka	100<	1737, 1841, 1952	8.3–8.5	2–4
Sanriku, Japan	300<	869, 1257, 1611, 1933	8.1–8.6	3–4
N. Sanriku, Japan	100±	1677, 1763, 1856, 1968	7.8–8.1	2–3
Tonga	50>	1865, 1919, 1948	7.5–8.3	1–2
Valdivia, Chile	100<	1575, 1737, 1837, 1960	8–8.5	2–4
Concepcion, Chile	100>	1570, 1657, 1751, 1835, 1928	8–8.5	3–4
Valparaiso, Chile	100>	1562, 1647, 1730, 1822, 1906	8.3–8.6	1–3
Arequipa, Peru	200<	1604, 1868	8.5	3–4
Callao, Peru	150<	1586, 1746, 1966	7.5–8.5	2–4
Colombia-Ecuador	30≥	1906, 1933, 1958, 1979	7.5–8.5	1–2
Guatemala-Nicaragua	50>	1859, 1902, 1950	7.2–7.5	1–2
S. Mexico	50>	1732, 1754, 1787, 1820, 1845, 1868, 1907, 1928, 1950	7.5–8.5	1–3

straits similar to those of Alaska or in similar archipelagos to those of the Southwest
Pacific.

Regularities of occurrence of large tsunamigenic earthquakes in time and space
was investigated for regions related to the structure of island arcs and trenches where
large earthquakes generate large tsunamis. Some regularities for time interval of
similar large earthquake occurrence may be found in the Tokai-Nankai and Sanriku
districts of Japan and in the Concepcion and Valparaiso districts of Chile.

There took place seven pairs of large tsunamigenic earthquakes in the Nankai and
Tokai areas in Japan. Within the intervals of 1–2 hours to 2–3 years after the
occurrence of a large earthquake in the Tokai area, a similar shock occurred in the
Nankai area with a high probability after 1096. This shows the regular migration of
the epicentral area of these earthquakes from east to west. The regular interval of large
tsunamigenic earthquake occurrences in the north Sanriku area is about 100 years,
while that in the south Sanriku area is about 350 years. The epicentral area of these
earthquakes in both areas migrated successively northward.

The above-mentioned regular pattern found for the large tsunamigenic
earthquakes occurred in each area of Concepcion and Valparaiso reveals about
90–year interval with northward migration of the epicentral area, while the time
interval among similar large earthquakes in the Valdivia area seems to be more than
100 years. The time interval of repeated occurrence of similar large tsunamigenic
earthquakes in some other regions in the Pacific appears to be in the range from about
30 years to more than 300 years.

If the same process of large tsunamigenic earthquake occurrence around the Pacific is assumed, from the present results the possibility is suggested of future large tsunami-genic earthquakes with the above-mentioned regular intervals.

REFERENCES

Cox, D. C., Marine effects and hazards of earthquakes, The great Alaska Earthquake of 1964, Oceanography and Coastal Engineering, NAS, Pub. 1605, Washington, National Academy of Sciences, 337–349, 1972.

Cox, D. C., Notes on reported tsunamis in New Hebrides, Solomon Island and Solomon Sea, Bismark Arch and Sea, and north coast of New Guinea, 1–18, 1981 (personal communication).

Cox, D. C. and P. G. Morgan, Local tsunamis and possible local tsunamis in Hawaii, HIG–77–14, Hawaii Inst. of Geophysics, Univ. of Hawaii, 1–118, 1977.

Hatori, T., Tsunami magnitude and wave source regions of historical Sanriku tsunamis in northeast Japan, *Bull. Earthq. Res. Inst., Univ. of Tokyo*, **50**, 397–414, 1975 (in Japanese).

Iida, K., Destructive historical tsunamis in Aichi Prefecture of Japan, Report of Earthquake Disaster Prevention Committee of Aichi Prefecture, Japan, 1–119, 1981a (in Japanese).

Iida, K., List of tsunamis in and near Japan, Data Report, Aichi Institute of Technology, 1–46, 1981b.

Iida, K., D. C. Cox and G. Pararas-Carayannis, Preliminary catalog of tsunamis occurring in the Pacific Ocean, Data Report No. 5, HIG–67–10, Hawaii Inst. of Geophysics, Univ. of Hawaii, August, 1–270, 1967.

Seismological Society of America, Seismological Notes, *Bull. Seismol. Soc. Am.*, **50–71**, 1960–1981.

Soloviev, S. L., Additions and corrections to the catalog of Pacific tsunamis, 1975; 1976; 1977 (personal communication).

Soloviev, S. L. and Ch. N. Go, Catalog of tsunamis in the Pacific, Academy Nauka, USSR, Moscow, 1–83, 1969 (in Russian).

Soloviev, S. L. and Ch. N. Go, Catalog of tsunamis on western coasts of the Pacific, Academy Nauka, USSR, 1–310, 1974 (in Russian).

Soloviev, S. L. and Ch. N. Go, Catalog of tsunamis on eastern coasts of the Pacific, Academy Nauka, USSR, 1–204, 1975 (in Russian).

Usami, T., Study of historical earthquake in Japan, *Bull. Earthq. Res. Inst., Univ. of Tokyo*, **54**, 399–439, 1979.

Watanabe, H., Descriptive table of tsunamis in and near Japan, *Zisin 2nd series*, **21**, 293–313, 1968 (in Japanese).

Tsunamis—Their Science and Engineering, edited by K. Iida and T. Iwasaki, 77–89.

Tsunami Earthquakes and Undersea Deformation

Robert P. COMER

*Department of Earth and Planetary Sciences, Massachusetts Institute
of Technology, Cambridge, Massachusetts 02139, U.S.A.*

(Received August 24, 1981)

The tsunami earthquakes of 1963 October 20 and 1975 June 10 differed from neighboring tsunamigenic events in the Kurile Trench–Japan Trench region in their patterns of accompanying vertical deformation of the sea floor, as indicated by the initial motions of tsunamis at tide stations, as well as in their greater efficiency of tsunami generation (relative to the 20-sec surface wave magnitude, M_s). Tsunami records indicate that the trenchward portions of the tsunami generating areas of these two earthquakes underwent subsidence and that the landward portions were elevated. This is in strong contrast to the deformation typically accompanying earthquakes of more average tsunamigenic efficiency, which appear to involve mostly sea floor uplift except for subsidence in a small landward portion of the generating area. Therefore any proposed tsunami generating mechanism for the earthquakes of 1963 October 20 and 1975 June 10 must account for two first-order observations: (1) an exceptionally high tsunamigenic efficiency relative to M_s and (2) the unusual pattern of seafloor deformation. The assumption of long process times for these earthquakes can explain (1), but an additional explanation is needed for (2). Steep secondary in faulting in the sedimentary wedge landward of the trench is a possibility.

1. Introduction

A tsunami earthquake is defined (KANAMORI, 1972) as an earthquake which generates a tsunami with unusually large amplitudes with respect to the earthquake's magnitude. Magnitude in this case refers to M_s, determined from the amplitudes of seismic surface waves of 20 second period. KANAMORI (1972) investigated the 1896 Sanriku and 1946 Aleutian tsunami earthquakes and concluded that they involved unusually slow rupture and thus radiated substantially more strongly at the long periods important for tsunami generation than at 20 seconds. Thus the tsunami amplitudes were large relative to M_s because M_s gave an inadequate measure of the earthquake size.

Two additional tsunami earthquakes have been defined in the recent literature. TAKEMURA *et al.* (1977) and GELLER and SHIMAZAKI (1978) studied a tsunamigenic earthquake that occurred on 1975 June 10 east of Hokkaido and generated a surprisingly large tsunami for its magnitude $M_s = 7.0$, and ABE (1979) identified the earthquake of 1963 October 20 near the Kurile Trench (an aftershock of a larger event on 1963 October 13) as a tsunami earthquake. HATORI (1975, 1979) has also taken note of the abnormally large tsunami amplitudes produced by these earthquakes. Both recent tsunami earthquakes have been studied, largely through seismic means, by FUKAO (1979), and they also constitute the major topic of the present work. However, the emphasis here will be on tsunami rather than seismic data. The next two sections will treat two kinds of first-order tsunami information: (1) tsunami amplitudes, and their comparison with earthquake magnitudes, and (2) the directions of first motions (up or down) of tsunamis recorded by tide gauges and their relation to the pattern of seafloor deformation accompanying the earthquake.

2. Amplitudes and Magnitudes

ABE (1979) noted that a straightforward way to identify a tsunami earthquake is to compare its seismic magnitude to an estimate of seismic magnitude based on tsunami observations. ABE (1979) defined an earthquake magnitude scale for large shallow submarine earthquakes that is based on tsunami amplitudes:

$$M_t = \log H + B \tag{1}$$

where H is the average of far-field maximum tsunami amplitudes observed in a given region and B is roughly constant, with a weak dependence on both the source region and the observation region. This scale, which has recently been extended to incorporate near-field data and therefore smaller events (ABE, 1981) is naturally useful for comparing the relative tsunamigenic efficiency of earthquakes. It is calibrated by choosing appropriate values of B for each source region, observation region pair so that M_t is a good estimate of M_w for a set of calibration earthquakes. M_w denotes KANAMORI's (1977) moment-magnitude scale which is based on an estimate of the seismic energy from the seismic moment, M_0. As shown by COMER (1980), the relation which follows from substituting M_w for M_t in Eq. (1)

$$M_w = \log H + B \tag{2}$$

may be derived approximately from scaling arguments, hydrodynamics, and some empirical relations in seismology. Both M_w and M_t have an advantage over the 20-sec surface wave magnitude M_s in indicating the true size of an earthquake, because they are based on waves of much longer periods. For very large earthquakes the evergy radiated at a period of 20 seconds reaches a limit as the seismic moment M_0 increases and this saturation places an upper bound on M_s of about 8.2 to 8.4 (see GELLER, 1976, for example). In general, $M_s = M_w$ for $M_w \geq 8.2$, but M_w can be substantially larger than M_s for larger M_w.

Thus, to try to identify tsunami earthquakes, which may violate Eq. (2) and the scaling relations from which it follows, one may compare M_w or M_s to M_t. In Table 1 values of these three magnitudes are given for six recent tsunamigenic earthquakes in the Kuril Trench–Japan Trench region. The epicenters are plotted in Fig. 1. These earthquakes include the tsunami earthquakes of 1963 October 20 and 1975 June 10 plus four nearby events to which, for the sake of comparison, we shall refer as "ordinary tsunamigenic earthquakes." The latter four consist of the 1963 October 13 Kurile earthquake, the 1968 Tokachi-oki earthquake, the 1969 east Hokkaido earthquake,

Table 1.   Magnitudes and patterns of seafloor uplift and subsidence for ordinary tsunamigenic earthquakes and tsunami earthquakes.

	M_s	M_w	M_t	Deformation
Ordinary Tsunamigenic Earthquakes				
1963 Oct. 13	8.1[a]	8.5[b]	8.4[c]	uplift
1968 May 16	8.1[a]	8.2[b]	8.3[c]	uplift, except for a small region of subsidence in the
1969 Aug. 11	7.8[a]	8.2[b]	8.2[c]	landward portion of the source region
1973 June 17	7.7[a]	7.8[d]	8.0[c]	uplift
Tsunami Earthquakes				
1963 Oct. 20	7.2[c]	8.0[e]	7.9[c]	mostly uplift in the landward portion of
1975 June 10	7.0[f]	7.6[g]	7.8[h]	the source region, subsidence in the trenchward portion

Sources for the magnitudes: a. ABE and KANOMARI (1980), b. KANAMORI (1977), c. ABE (1979), d. calculated from $M_0 = 6.7 \times 10^{27}$ dyne cm (SHIMAZAKI, 1974), e. calculated from $M_0 = 11 \times 10^{27}$ dyne cm (HANKS, 1971), f. GELLER and SHIMAZAKI (1978), g. calculated from $M_0 = 3 \times 10^{27}$ dyne cm (TAKEMURA et al., 1977), h. estimated by comparing tsunami amplitudes to those from the 1969 Aug. 11 earthquake at three tide stations. An identical value has been obtained by Abe (personal communication, 1981).

and the 1973 Nemuro-oki earthquake. For each of the four ordinary tsunamigenic earthquakes the three magnitudes are in close agreement with one another, within a range of 0.4 in all cases. For the 1963 October 20 and 1975 June 10 events, $M_t - M_s$ = 0.7 or 0.8, which justifies their classification as tsunami earthquakes. However, in both cases M_w is quite close to M_t (differing by 0.1 or 0.2) and M_s is substantially smaller than M_w. although both are below the usual saturation level (of about 8.2).

Thus the first basic observation to be made regarding the two tsunami earthquakes is that in both cases M_t is significantly larger than M_s and is comparable to M_w. Before considering its interpretation, we examine a second kind of first-order tsunami information regarding these events.

3.   Sea Floor Deformation

Tsunami are indicators of the static, co-seismic vertical deformation accompanying shallow submarine earthquakes. This is evident from recent works by ABE (1973), ANDO (1975, 1982), and AIDA (1978), in which fault models of earthquakes derived from seismic or geodetic data have been shown to be in reasonable accord with the directions and magnitudes of the initial motions of the tsunamis as recorded by tide gauges, or even with the initial parts of the wave forms. The pattern of uplift and subsidence in a tsunami source area, as indicated by tide gauge records when refraction

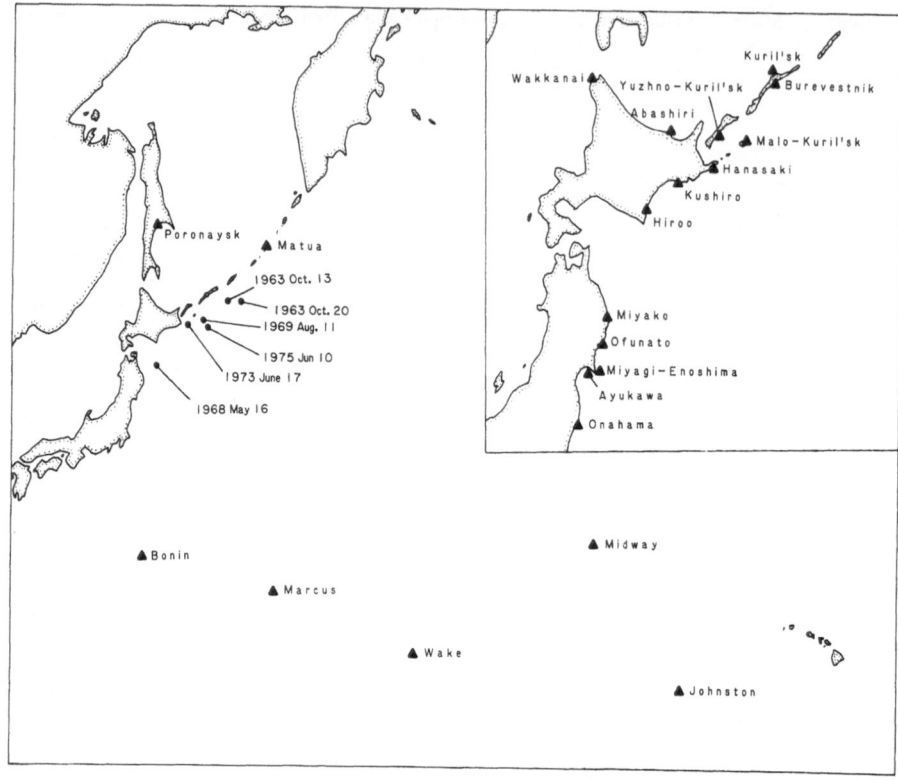

Fig. 1.   Epicentral locations of the six earthquakes of Table 1 and locations of the tide stations listed in Table 2. Inset shows the stations in Japan and the southern Kurile Islands.

and other propagation effects are taken into account, seems to agree in general with the pattern of static, vertical sea floor uplift and subsidence predicted from the fault model of the generating earthquake. (Such deformation may be calculated from a fault model by using the results of MANSINHA and SMYLIE, 1971, for example.) It is important to note that although the works cited above involved analysis of near-field tsunami data, far-field data may also be used to gain insight into the sea floor deformation in a tsunami source region. Dispersion can strongly modify tsunami waveforms over long propagation paths, but because the nature of the tsunami dispersion is such that the leading wave of a tsunami is an Airy phase with a stationary phase velocity, as well as group velocity (KAJIURA, 1963; STONELEY, 1963; Fig. 2), the polarity of the first motion should be preserved even to far field distances. We will use this fact to incorporate data from Pacific island tide stations into an analysis of the deformations in tsunami source regions.

We now review the evidence regarding the patterns of vertical deformation accompanying the six earthquakes of the previous section, beginning with the four ordinary tsunamigenic earthquakes and then considering the tsunami earthquakes:

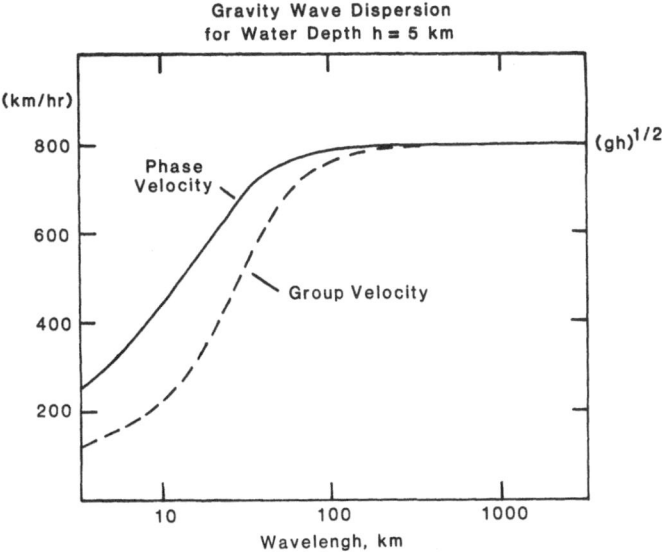

Fig. 2. Theoretical dispersion relation for a tsunami in an idealized ocean (incompressible, inviscid, rigid bottom) of 5 km depth. At long wavelengths both group velocity and phase velocity are stationary and at a maximum.

3.1 1963 Oct. 13

SOLOV'EV (1965a,b) presents an inverse refraction diagram indicating the tsunami source region and tide gauge records from stations in the U.S.S.R. At all the tide stations for which the initial motion was clearly above the noise, initial uplift is apparent. Upward motions were also observed at Miyagi-Enoshima (HATORI, 1979), other stations in Japan (Hatori, unpublished data, 1981), and at the U.S. Midway station (see Figs. 1 and 3). These observations suggest an essentially monopole uplift of the sea floor in the tsunami source region.

3.2 1968 May 16

The inverse refraction diagram constructed by KAJIURA et al. (1968) and their comments regarding the tsunami source indicate that initial uplift occurred at most tide stations in Japan, and consequently over most of the source region, but that because of initial downward motions at a few stations there was probably a small region of subsidence in the northeastern part of the source region. The displacements calculated from fault models by ABE (1973) and AIDA (1978) substantiate this pattern of uplift subsidence. It is also supported by the initial uplift of the tsunami at Midway (Fig. 3). Note that because the great circle path from the source is roughly normal to the trend of the Kurile and Japan Trenches, There should be few refraction effects to alter the actual propagation path from the direct one, and the sense of the tsunami initial motion at Midway should be a good indication of the polarity of the sea floor deformation at the southeastern edge of the source region.

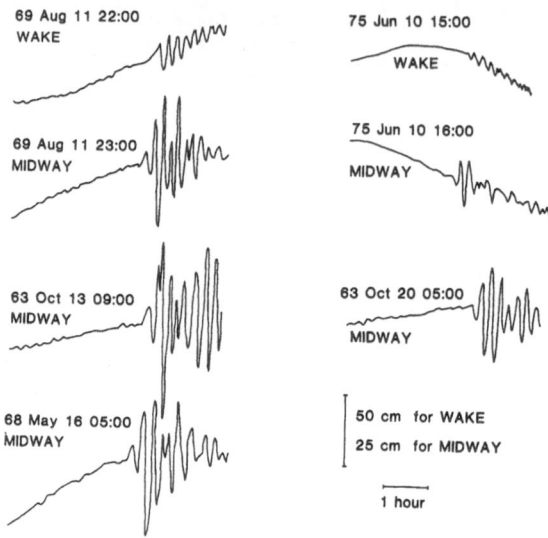

Fig. 3. Tide gauge records from Wake and Midway for tsunamis from three ordinary tsunamigenic earthquakes (left) and two tsunami earthquakes (right). The beginning time of each trace is given (U. T.).

3.3 1969 Aug. 11

A similar pattern is evident for this event: Predominant uplift, with a small area of subsidence along the northwestern edge of the tsunami source region, according to HATORI's (1970) inverse refraction diagram and discussion and an improved diagram (HATORI, 1976) which incorporates additional data showing subsidence at Hanasaki and subsidence and uplift, respectively, at the Soviet stations Yuzhno-Kuril'sk and Burevestnik. This pattern is also in accord with observations of initial uplift and Wake and Midway (Fig. 3) and with the modelling of ABE (1973) and AIDA (1978).

3.4 1973 June 17

HATORI (1976) constructed an inverse refraction diagram for the tsunami source, based on observations at tide stations in Japan. All stations recorded upward initial motion of the tsunami, suggesting monopole uplift as in the case of the 1963 October 13 tsunami. The static sea floor deformations calculated from fault models by SHIMAZAKI (1974) and AIDA (1978) show fairly large amounts of uplift (40–60 cm at maximum) in Hatori's tsunami source area, and also a much smaller (5 cm) amount of subsidence in an area northwest of the tsunami source which includes part of eastern Hokkaido. There is geodetic evidence for such subsidence, also (SHIMAZAKI, 1974). That there is no indication of the subsidence in the tide gauge recordings is probably a consequence of the relatively small amount of the subsidence and the fact that the closest tide stations, in eastern Hokkaido, underwent subsidence themselves.

3.5 1963 Oct. 20

An inverse refraction diagram has been constructed by SOLOV'EV (1965a,b) for the tsunami source. Although the sense of initial motion is unclear in many of the tsunami records from the Soviet tide stations used by SOLOV'EV (1965b), all those where it was clear showed upward motion. It was also upward at Miyagi-Enoshima (HATORI, 1979) and was downward at Midway (Fig. 2). Noting the locations of the tide stations in Fig. 1, Table 2, which summarizes these observations, suggests that there was uplift of the sea floor in the landward portion of the source and subsidence in the trenchward (and oceanward) portion.

3.6   1975 June 10

As in the case of the 1963 Oct. 20 event, the initial motion of the tsunami at Midway was downward, and it was also downward at Wake Island (Fig. 3). HATORI (1975, 1979) has constructed inverse refraction diagrams using data from Japanese and Soviet tide stations, all which show upward initial motion, with the possible exceptions of two Soviet stations, Malo-Kuril'sk and Yuzhno-Kuril'sk (HATORI, 1979). The records at these two stations are rather difficult to interpret, though they seem to show initial downward motion. Table 2 summarizes the observations. Like the 1963 October 20 tsunami earthquake, this event seems to have involved subsidence of the oceanward portion of the source region and uplift of the landward portion, although this pattern may be complicated by the presence of an area of subsidence in the landward portion as suggested by the records from Malo-Kuril'sk and Yuzhno-Kuril'sk. Source complexity is also suggested by the tide gauge record from Miyagi-Enoshima, which, as HATORI (1975) shows, has much shorter periods than records of tsunamis from neighboring sources made by the same recorder.

The results above are summarized in the final column of Table 1 and in Fig. 4. They constitute a fundamental observation: there is a major qualitative difference between the patterns of sea floor deformation accompanying the tsunami earthquakes and the ordinary tsunamigenic earthquakes. Figure 3 shows this clearly, contrasting

Table 2.   Tsunami earthquakes: Directions of tsunami initial motion at tide gauges.

1963 Oct. 20		1975 June 10	
up	down	up	down
Matua[a]	Midway I.[b]	Bonin I.[c]	Midway I.[b]
Poronaysk[a]		Hanasaki[c]	Wake I.[b]
Kuril'sk[a]		Wakkanai[d]	Marcus I.[f]
Miyagi-Enoshima[e]		Abashiri[d]	Johnston I.[f]
		Kushiro[d]	Malo-Kuril'sk?[e]
		Hiroo[d]	Yuzhno-Kuril'sk?[e]
		Miyako[d]	
		Ofunato[d]	
		Ayukawa[d]	
		Onahama[d]	
		Burevestnik[e]	

Sources: a. SOLOV'EV (1965b), b. NOAA microfiche records, c. Hatori (unpublished data, 1981), d. HATORI (1975), e. HATORI (1979), f. OLSEN (1979).

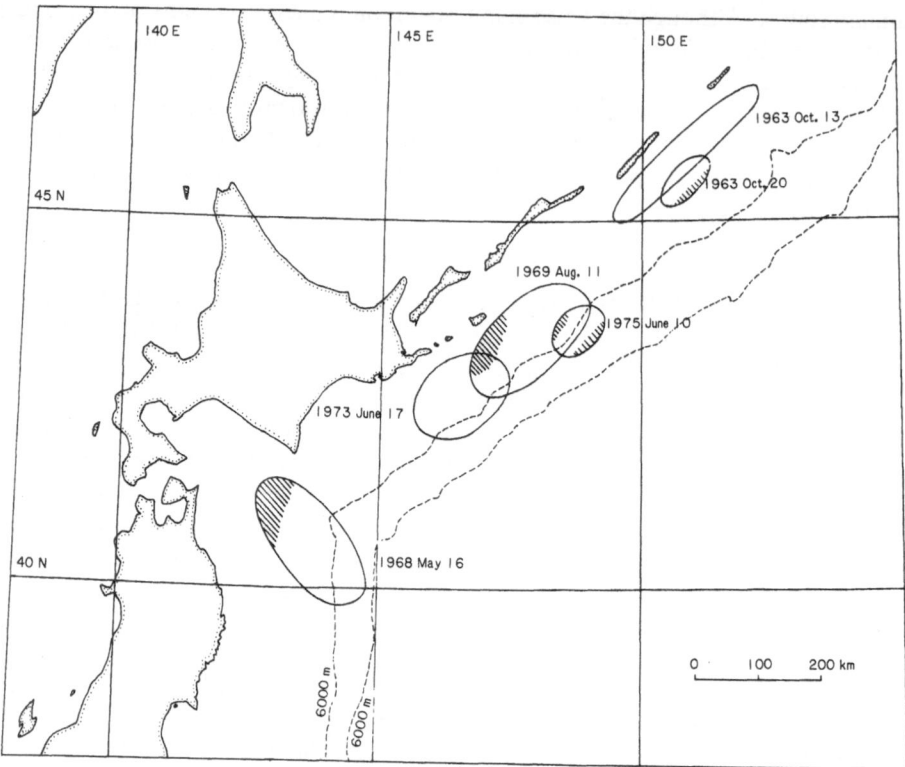

Fig. 4.   Source areas for six tsunamis, based on published inverse refraction diagrams (see text for sources). The hatching denotes regions of inferred subsidence; elsewhere in the source areas uplift is inferred. The Japan-Kurile trench axis is outlined by the 6000 m bathymetric contours.

the upward initial motions of tsunamis from three of the four ordinary tsunamigenic earthquakes at Wake and Midway to the downward initial tsunami motion resulting from the two tsunami earthquakes.

4.   Relationship to Focal Mechanism

The patterns of vertical sea floor deformation indicated by the tsunamis accompanying the ordinary tsunamigenic earthquakes are not surprising. Published focal mechanisms for the earthquakes of 1963 October 13 (STAUDER and BOLLINGER, 1966), 1968 May 16 (KANAMORI, 1971), 1969 August 11 (ABE, 1973; STAUDER and MUALCHIN, 1976), and 1973 June 17 (SHIMAZAKI, 1974) all indicate low-angle thrusting to the northwest. This is in accord with the plate tectonic idea that these earthquakes involve the overriding of the Pacific plate by the Japan and Kurile island arcs, since solutions for global present-day plate motions (CHASE, 1978; MINSTER and JORDAN, 1978) indicate that the northwestern portion of the Pacific plate is moving in a direction almost normal to the island arcs and their associated trenches. As shown by ABE

(1973) and AIDA (1978), fault models corresponding to such low-angle thrusting predict static elevation of the sea floor above the trenchward portion of the fault and subsidence above the landward portion. Such a pattern of sea surface initial deformation is indicated by the initial motions of recorded tsunamis from the 1968 and 1969 events. In the case of the 1963 October 13 and 1973 tsunamis, all clearly observed initial motions were upward, but, as noted in the previous section, the sea floor subsidence calculated for the 1973 earthquake was small and occurred in a region which included the nearest tide gauges so that a sort of cancellation may have occurred. This probably accounts for the lack of evidence for any subsidence in the tide gauge records, and a similar configuration may well account for the 1963 October 13 observations. Uplift of the oceanward portion and subsidence of the landward portion of the source region have also been apparent in the case of other tsunamigenic earthquakes near Japan or associated with other subduction-zone plate boundaries around the Pacific ocean. Examples include the 1952 Tokachi-oki earthquake (AIDA, 1978), the 1964 Alaska earthquake (PLAFKER, 1965) and the 1975 Tumaco, Columbia earthquake (HEARD et al., 1981).

In contrast to the ordinary tsunamigenic earthquakes, the tsunami first motions resulting from the tsunami earthquakes and the static sea floor deformation which they suggest are not as one would expect from the published fault plane orientations. Published focal mechanisms for the earthquakes of 1963 October 20 (STAUDER and BOLLINGER, 1966) and 1975 June 10 (TAKEMURA et al., 1977; GELLER and SHIMAZAKI, 1978) are very similar to those for the ordinary tsunamigenic earthquakes, although, as noted in the previous section, the tsunami initial motions are very different. Clearly, it is important to resolve this discrepancy.

5. Mechanism of Tsunami Earthquakes

We have demonstrated two fundamental observations regarding the kinds of tsunamis generated by the tsunami earthquakes of 1963 October 13 and 1975 June 10: (1) they have large amplitudes relative to the surface wave magnitude M_s, but normal amplitudes relative to the moment magnitude M_w, and (2) they have a pattern of initial tsunami motions which suggests static sea floor deformation very different from the deformation which evidently accompanies ordinary tsunamigenic earthquakes (Table 1, Fig. 4), although there is little difference in focal mechanism. To be fully adequate any proposed mechanism for the tsunami earthquakes must satisfy both observations.

The first observation amounts to noting the M_s gives an underestimate of M_t and M_w. Both GELLER and SHIMAZAKI (1978) and TAKEMURA et al. (1977) find that the effective moment of the 1975 June 10 earthquake increases considerably with increasing period and both conclude that a long rise time (or process time), due perhaps to a combination of long dislocation rise time and low rupture velocity, is the main cause of this effect. This is quite similar to KANAMORI's (1972) conclusion regarding the great tsunami earthquakes of 1896 and 1946. TAKEMURA and KOYAMA (1981) note that tsunami earthquakes may be essentially low-frequency earthquakes, whose M_s values underestimate their seismic moments because their corner frequencies are unusually low. This is consistent with the observation that the earthquakes of 1963 October 20 and 1975 June 10 were both very shallow (FUKAO, 1979; GELLER and SHIMAZAKI, 1978; respectively) and are located closer to the trench axis than the four ordinary tsunamigenic earthquakes (Figs. 1 and 4). Many smaller low frequency earthquakes have been found to occur in a shallow zone just landward of the Kurile and Japan trenches by UTSU (1980) and FUKAO and KANJO (1980).

Characterizing the tsunami earthquakes of 1963 October 20 and 1975 June 10 as low frequency events with long processes times may explain their large tsunami amplitudes, but it does not help to explain the patterns of accompanying tsunami first motions and sea floor deformation, nor their discrepancy with the seismically determined shallow fault dip. However, a very different hypothesis advanced by FUKAO (1979) is promising. The idea is that rupture in the tsunami earthquakes originated on a shallow dipping fault plane but branched upward into the wedge of sediments immediately landward of the trench, perhaps in a complex way involving a series of steep, imbricated thrust faults. FUKAO's (1979) support for this hypothesis comes mainly from seismological and geomorphological observations, and the reasoning that a steep, shallow faulting in highly deformable sediments would enhance tsunami generation. However, such faulting would also produce the kind of tsunami first motions that we have cited. This is explained in Fig. 5 and 6.

Figure 5 schematically illustrates two kinds of faulting which might occur on the landward side of a trench: shallow-dipping thrusting along the plate interface, which is very likely the case for most ordinary tsunamigenic earthquakes, and steep thrusting in the sediments at the leading edge of the overriding plate. The latter type of faulting may occur during tsunami earthquakes, although it probably originates as shallow thrusting along the plate interface, which would explain the observed focal mechanisms, before branching upward. Cross sections of the static vertical deformation expected from the two kinds of faulting are shown in Fig. 6. Gently dipping thrusting produces landward subsidence and trenchward uplift, which is also indicated by the tsunami evidence from the ordinary tsunamigenic earthquakes, whereas steep, very shallow thrust faulting should produce trenchward subsidence and landward uplift, like that indicated by the initial motions of tsunamis generated by tsunami earthquakes. Probably the actual faulting accompanying tsunami earthquakes is more

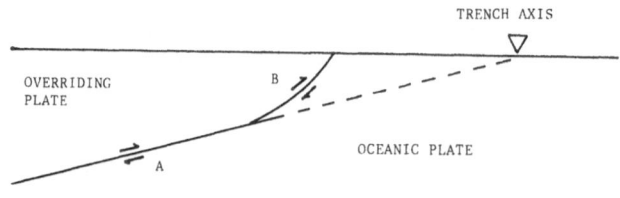

A. Slip along the plate interface

B. Faulting in the sedimentary wedge at the edge of
    the overriding plate

Fig. 5.   Schematic showing two types of faulting in a subduction zone.

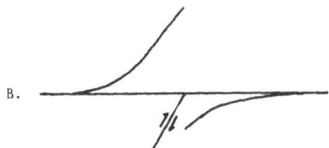

A. Vertical surface displacement due to slip on a
    buried, shallow-dipping fault

B. Vertical surface displacement due to slip on a
    steeply-dipping fault which breaks the surface

Fig. 6. Profiles of theoretical vertical displacement due to two kinds of thrust faulting (after SAVAGE and HASTIE, 1966).

complex than that shown in Figs. 5 and 6; as we have noted, there are indications of source complexity in the tsunami data for the 1975 June 10 event and complexity in the waveforms of the seismic body waves has been reported (TAKEMURA *et al.*, 1977; GELLER and SHIMAZAKI, 1978).

6. Conclusion

We have found that the tsunami earthquakes of 1963 October 20 and 1975 June 10 differ from ordinary tsunamigenic earthquakes not only in their large tsunami amplitudes relative to M_s but also in the accompanying static sea floor deformation as indicated by tide gauge records. This places an additional constraint on proposed mechanisms for these earthquakes. FUKAO's (1979) hypothesis of imbricate thrusts in the sedimentary wedge seems to fit this constraint. However, since faulting in sediments of low rigidity should probably involve long process times, they may also play a role in the generation of tsunamis by tsunami earthquakes.

This study was supported by the National Science Foundation grant EAR8018351. The author is grateful to T. Hatori for providing data which he collected, to K. Abe, K. Aki, D. Davis, K. Kajiura, H. Kanamori and J. Koyama for helpful discussion and comments, and to I. Knudson for assistance in figure preparation.

REFERENCES

ABE, K., Tsunami and mechanism of great earthquakes, *Phys. Earth Planet. Inter.*, 7, 143–153, 1973.
ABE, K., Size of great earthquakes 1837–1974 inferred from tsunami data, *J. Geophys. Res.*, 84, 1561–1568, 1979.

ABE, K., Physical size of tsunamigenic earthquakes of the northwestern Pacific, *Phys. Earth Planet. Inter.*, **27**, 194–205, 1981.

ABE, K. and H. KANAMORI, Magnitudes of great shallow earthquakes from 1953 to 1977, *Tectonophysics*, **62**, 191–203, 1980.

AIDA, I., Reliability of a tsunami source model derived from fault parameters, *J. Phys. Earth*, **26**, 57–73, 1978.

ANDO, M., Source mechanisms and tectonic significance of historical earthquakes along the Nankai Trough, Japan, *Tectonophysics*, **27**, 119–140, 1975.

ANDO, M., A fault model of the 1946 Nankaido earthquake derived from tsunami data, *Phys. Earth Planet. Inter.*, **28**, 320–336, 1982.

CHASE, C. G., Plate kinematics: The Americas, East Africa, and the rest of the world, *Earth Planet. Sci. Lett.*, **37**, 355–368, 1978.

COMER, R. P., Tsunami height and earthquake magnitude: Theoretical basis of an empirical relation, *Geophys. Res. Lett.*, **7**, 445–448, 1980.

FUKAO, Y., Tsunami earthquakes and subduction processes near deep-sea trenches, *J. Geophys. Res.*, **84**, 2303–2314, 1979.

FUKAO, Y. and K. KANJO, A zone of low-frequency earthquakes beneath the inner wall of the Japan trench, *Tectonophysics*, **67**, 153–162, 1980.

GELLER, R. J., Scaling relations for earthquake source parameters and magnitudes, *Bull. Seismol. Soc. Am.*, **66**, 1501–1523, 1976.

GELLER, R. J. and K. SHIMAZAKI, The June 10, 1975 Kurile Islands tsunami earthquake: an extended abstract, in *Proceedings of Conference III, Fault Mechanics and its Relation to Earthquake Prediction*, National Earthquake Hazards Reduction Program, U. S. G. S. Open File Report 78–380, pp. 213–225, 1978.

HANKS, T. C., The Kurile trench–Hokkaido rise system: large shallow earthquakes and simple models of deformation, *Geophys. J. R. Astr. Soc.*, **23**, 173–189, 1971.

HATORI, T., An investigation of the tsunami generated by the east Hokkaido earthquake of August, 1969, *Bull. Earthq. Res. Inst. Tokyo Univ.*, **48**, 399–412, 1970.

HATORI, T., Tsunami activity in eastern Hokkaido after the off-Nemuro Peninsula earthquake in 1973, *Jishin*, **28**, 461–471, 1975 (in Japanese).

HATORI, T., Tsunami off the Nemuro Peninsula in June 1973, and the tsunami generation in east Hokkaido, Tsunami Research Symposium 1974, edited by R. A. Heath and M. M. Crosswell, *Royal Society of New Zealand Bulletin*, **15**, 61–70, 1976.

HATORI, T., Tsunami activity in the Hokkaido and Kurile regions, *Bull. Earthq. Res. Inst. Tokyo Univ.*, **54**, 543–557, 1979 (in Japanese).

HERD, D. G., T. L. YOUD, H. MEYER, J. L. ARANGO C., W. J. PERSON, and C. MENDOZA, The great Tumaco, Columbia earthquake of 12 December 1979, *Science*, **211**, 441–445, 1981.

KAJIURA, K., The leading wave of a tsunami, *Bull. Earthq. Res. Inst. Tokyo Univ.*, **41**, 535–571, 1963.

KAJIURA, K., T. HATORI, I. AIDA, and M. KOYAMA, A survey of a tsunami accompanying the Tokachi-oki earthquake of May 1968, *Bull. Earthq. Res. Inst. Tokyo Univ.*, **46**, 1369–1396, 1968 (in Japanese).

KANAMORI, H., Focal mechanism of Tokachi-oki earthquake of May 16, 1968, *Tectonophysics*, **12**, 1–13, 1971.

KANAMORI, H., Mechanism of tsunami earthquakes, *Phys. Earth Planet. Inter.*, **6**, 346–359, 1972.

KANAMORI, H., The energy release in great earthquakes, *J. Geophys. Res.*, **82**, 2981–2987, 1977.

MANSINHA, L. and D. E. SMYLIE, The displacement fields of inclined faults, *Bull. Seismol. Soc. Am.*, **61**, 1433–1440, 1971.

MINSTER, J. B. and T. H. JORDAN, Present-day plate motions, *J. Geophys. Res.*, **83**, 5331–5354, 1978.

OLSEN, K. H., Midocean "microtsunami" stations on three Pacific atolls, in *Tsunamis, Proceedings of the National Science Foundation Workshop*, pp. 283–293, Tetra Tech Inc., Pasadena, 1979.

PLAFKER, G., Tectonic deformation associated with the 1964 Alaska earthquake, *Science*, **148**, 1675–1687, 1965.

SAVAGE, J. C. and L. M. HASTIE, Surface deformation associated with dip-slip faulting, *J. Geophys. Res.*, **71**, 4897–4904, 1966.

SHIMAZAKI, K., Nemuro-oki earthquake of June 17, 1973: a lithospheric rebound at the upper half of the interface, *Phys. Earth Planet. Inter.*, **9**, 314–327, 1974.

SOLOV'EV, S. L., The Urup earthquake and associated tsunami of 1963, *Bull. Earthq. Res. Inst. Tokyo Univ.*, **43**, 103–109, 1965a.

SOLOV'EV, S. L., *The Earthquakes and Tsunamis of October 13 and 20, 1963, on the Kurile Islands*, Siberian Division of Academy of Science, USSR, Yuzhno-Sakhalinsk, Sakhalin, 1965b (in Russian).

STAUDER, W. and G. A. BOLLINGER, The S-wave project for focal mechanism studies, earthquakes of 1963, *Bull. Seismol. Soc. Am.*, **56**, 1363–1371, 1966.

STAUDER, W. and L. MUALCHIN, Fault motion in the larger earthquakes of the Kurile-Kamchatka arc and the Kurile-Hokkaido corner, *J. Geophys. Res.*, **81**, 297–308, 1976.

STONELEY, R., The propagation of tsunamis, *Geophys. J.*, **8**, 64–81, 1963.

TAKEMURA, M., J KOYAMA, and Z. SUZUKI, Source processes of the 1974 and 1975 earthquakes in Kurile Islands in special relation to the difference in exitation of tsunami, *Sci. Rep. Tohoku Univ., Ser. 5*, **24**, 113–132, 1977.

TAKEMURA, M. and J. KOYAMA, Seismic source of tsunami and ordinary earthquake (abstract), *Abstracts of papers, IUGG International Tsunami Symposium*, pp. 1–4, 1981.

UTSU, T., Spatial and temporal distribution of low-frequency earthquakes in Japan, *J. Phys. Earth*, **28**, 361–384, 1980.

Tsunamis—Their Science and Engineering, edited by K. Iida and T. Iwasaki, 91–101.

A New Scale of Tsunami Magnitude, M_t

Katsuyuki Abe

Department of Geophysics, Faculty of Science,
Hokkaido University, Sapporo, Japan

(Received August 27, 1981; Revised December 21, 1981)

A new scale of tsunami magnitude, M_t, was recently established on the basis of the logarithm of the maximum amplitude of tsunami waves that were recorded by tide gauges. Because this scale was calibrated for the moment magnitude of an earthquake, it represents not only the overall physical size of a tsunami source, but also seismic moment of a tsunamigenic earthquake. The background of the definition of the M_t scale is summarized, and its significance in the study of tsunami generation and earthquake mecahnism is discussed. Because of the remarkable usefulness of the M_t scale, it is attempted to extend it to historical events for which instrumental records are nonexistent. The results of numerical experiments and inverse refraction drawings in the tsunami study are used. This method is applied to large events that have occurred along the Pacific coast of Japan over the past 400 years. The largest M_t of these events reaches $8\frac{1}{2}$, which is much smaller than $M_t = 9.4$ of the great Chilean event of 1960.

1. Introduction

A new scale of tsunami magnitude, designated by M_t, was recently defined in terms of instrumental tsunami-wave amplitudes (ABE, 1979, 1981b). Among several magnitude scales in current use, this M_t scale has the most direct relevance to the study of the earthquake machanism as well as the measurement of the overall physical size of a tsunami source. The results have an important bearing on the reliability of estimates of the seismic moment of a tsunamigenic earthquake for which seismic data are inadequate or nonexistent, and on the prediction of the maximum tsunami amplitude at a particular site. The latter application is very important for providing the existing tsunami warning system with more quantitative information than is now available. The present paper briefly reviews the definition of the M_t scale, and attempts to determine M_t for historical events for which instrumental records are nonexistent.

2. Tsunami Magnitude, M_t

One of the most reliable seismological parameters representing the physical size of an earthquake is the seismic moment, M_o, which is the product of the surface area of the fault with the average displacement of the fault plane and the rigidity of the material of the fault. Recently, KANAMORI (1977) defined a magnitude scale, called the moment magnitude, M_w, in terms of M_o by

$$M_w = (\log M_o - 16.1)/1.5 \qquad (1)$$

where M_o is in dyn $\cdot$ cm. The quantity M_w is suitable for the study of large earthquakes because, unlike the surface-wave magnitude, M_s, it does not saturate for very large events, and is a convenient extension of the M_s scale to great earthquakes. Since tsunamis are primarily caused by seismic sea-bottom deformations with time scales up to several hundred seconds, it is reasonable to define their overall size in relation to M_w.

The fundamental definition of M_t is given by

$$M_t = \log H + B \qquad (2)$$

where H is the maximum tsunami-wave amplitude measured by tide gauges in meters and B is a correction term, defined by requiring $M_t = M_w$ on the average for the calibration data set (Abe, 1979). This equation has been based on the observed good correlation between M_w and $\log H$. According to Comer (1980), this observation is in close accord with the results of simple theory and scaling assumptions of earthquakes. There are two methods of defining B, depending on the epicentral distance. For far-field tsunami data, that is, for distances of not less than 1,000 km, B is expressed by

$$B = C + \Delta C \qquad (3)$$

where C is 9.1 and ΔC is a small correction term which depends on each pair of source region and observation region. Table 1 lists ΔC values, determined from the observations at 20 tide stations in the Pacific (Fig. 1). The ΔC values for Honolulu were miscalculated in Abe (1979), and the corrected values are given in Table 1; fortunately the effect of this correction to the previous results is negligibly small. Using far-field amplitude data obtained at California, the Hawaiian Islands, the Aleutian Islands and Japan, Abe (1979) determined M_t for 65 important tsunamigenic earthquakes that occurred since 1837 in the Pacific. Table 2 lists the 15 super-great events having M_t = 8.6 or over. The Chilean event of 1960 has the largest M_t of 9.4. Most of the super-great events originated in Chile, the Aleutians, Alaska, and Kamchatka.

The method of determining M_t from the regional data was developed on the basis of a number of tide-gauge data obtained in Japan (Abe, 1981b). The 30 tide stations used are shown in Fig. 2. At moderate distances, the amplitudes H decay with distance; here the distance Δ (in km) is defined as the one from the earthquake epicenter to tide station along the shortest oceanic path (Fig. 2). The relation between $\log H$ and $\log \Delta$ is

Table 1.   Values of ΔC.

Source Region	ΔC				
	Honolulu	Hilo	California	Japan	Aleutian
A: Peru, Chile	+0.2	−0.6	+0.2	0.0	+0.2
B: Alaska, Aleutian	+0.5[2]	0.0	+0.2	+0.3	—
C: Kamchatka, Kurile, Japan	0.0	−0.4	+0.1	−0.2[1]	−0.2
Whole Region	+0.2[2]	−0.3	+0.2	0.0	0.0

(1) Except for Kurile Islands and Japan region. (2) Original values are revised.

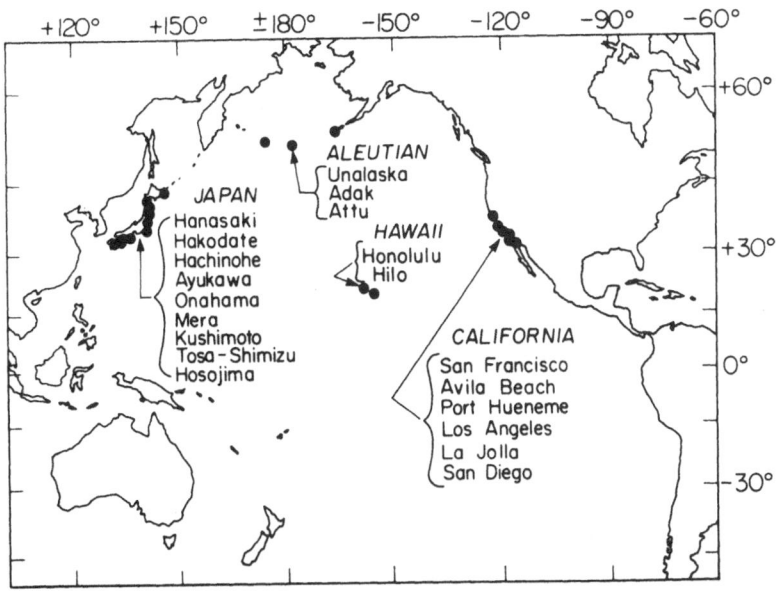

Fig. 1. Location of tide stations in the Pacific.

approximately linear, with a slope of almost unity, as is shown in Fig. 3. This linearity is also supported by numerical experiments (YAMAKI and ABE, 1981). Thus the quantity, B, in Eq. (2) can be modified into

$$B = \log \varDelta + D \tag{4}$$

Table 2. Super-great tsunamigenic earthquakes ($M_t = 8.6$ or over).

Date	Region	M_t	M_s	M_w
May 22, 1960	Chile	9.4	8.5	9.5
April 1, 1946	Aleutian	9.3	7.3	—
November 7, 1837	Chile	9¼	—	—
March 28, 1964	Alaska	9.1	8.4	9.2
May 17, 1841	Kamchatka	9	—	—
August 13, 1868	Chile	9.0	—	—
May 10, 1877	Chile	9.0	—	—
November 4, 1952	Kamchatka	9.0	8.2	9.0
March 9, 1957	Aleutian	9.0	8.1	—
February 3, 1923	Kamchatka	8.8	8.3	—
January 31, 1906	Ecuador-Colombia	8.7	8.7	—
September 7, 1918	Kurile	8.7	8:2	—
November 11, 1922	Chile	8.7	8.3	—
June 15, 1896	Japan	8.6	—	—
February 4, 1965	Aleutian	8.6	8.2	8.7

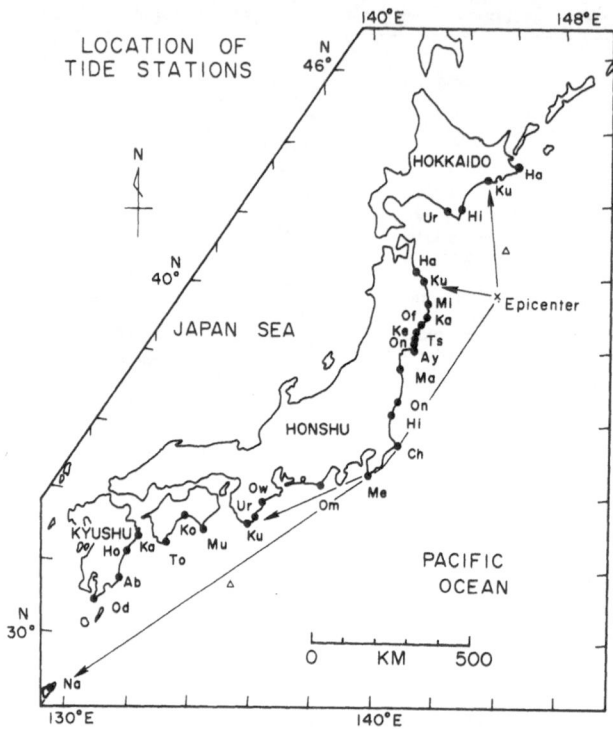

Fig. 2.   Location of tide stations along the Pacific coast of Japan. Example of measuring distance
Δ is shown.

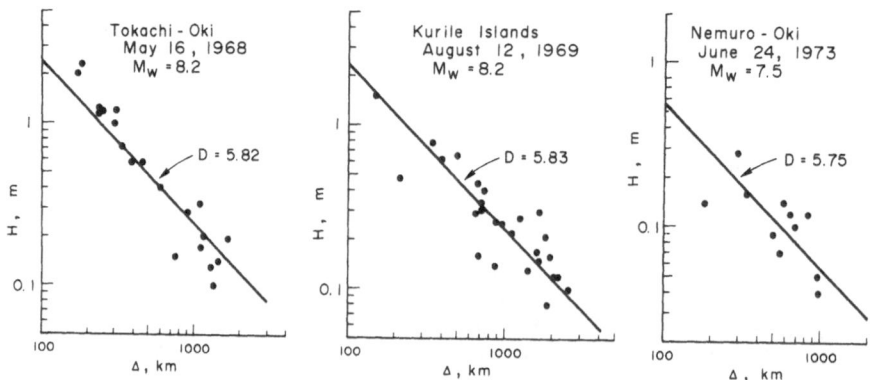

Fig. 3.   Relation between $\log H$ and $\log \Delta$ for three calibration events. H is the maximum amplitude
measured by tide gauges and Δ is the distance from earthquake epicenter to stations. Stations used are
shown in Fig. 2. The straight line is a least-squares fit of $\log H = M_w - D - \log \Delta$ to the data, where
M_w is the earthquake moment magnitude.

where D is constant. By requiring $M_t = M_w$ on the average for the calibration data set, a least-squares solution gives $D = 5.80$. If the maximum crest-to-trough amplitude is used instead of H, the value of 5.55 should be used for D. Equation (4) can be applied to the regional data over the distance from 100 to about 3,500 km and to the earthquakes that occurred offshore from the east coast of Kamchatka, the Kurile Islands and Japan. Using many tide-gauge data obtained in Japan, ABE (1981b) assigned M_t to 80 major tsunamigenic earthquakes that occurred in the northwestern Pacific, mostly in Japan, during the period from 1894 to 1981.

Throughout these studies of M_t, it is not forgotten that the tsunami heights are disturbed by the local topography near the observation site, nevertheless, the maximum amplitudes on the tide-gauge records closely reflect the physical size of a tsunami source. Actually, the M_t values differ from M_w by only 0.2 or less for many earthquakes with known M_w. This much of an uncertainty is tolerable, because an uncertainty of 0.2 in an M_t value results in an uncertainty of only a factor of two in M_0. The good correlation between M_t and M_w is partly due to the fact that most earthquakes offshore from the circum-Pacific have dip-slip faulting in common. If M_t is determined for strike-slip earthquakes, it is only a measure of the tsunami size, not a measure of M_w.

The traditional scale which has long been used in Japan was intended to grade tsunamis in terms of both the maximum local height and the geographical extent of the tsunami hazard. This scale, designated by m, was originally set up by IMAMURA (1942) with five grades. Though the practice of gradation has been modified in various ways (e.g., IIDA, 1958), the physical meaning of m is not clear (KAJIURA, 1981); it is, if anything, analogous to the maximum earthquake intensity rather than the magnitude (SOLOVIEV, 1970; ABE, 1979). Figure 4 shows a comparison between m and M_t for Japanese events. The scatter is considerably large and there appears to be a poor correlation between M_t and m, as is similar to the scatter on the comparison between the maximum earthquake intensity and M_s.

Among many tsunami events, there are unusual ones, called tsunami earthquakes. A tsunami earthquake refers to an earthquake which generates disproportionately large tsunamis for relatively weak seismic waves, indicating that a very slow process or a large vertical deformation is involved (KANAMORI, 1972; FUKAO, 1979). The tsunami earthquake has been identified so far in a qualitative sense, but it is now possible to identify it on the basis of the M_t versus M_s diagram. Here we tentatively define the earthquake of which M_t is greater than M_s by 0.5 or above, as a very unusual one, although a certain consideration is necessary for great earthquakes owing to the saturation of M_s (ABE, 1979). Table 3 lists significant tsunami earthquakes. The notable tsunami earthquakes are the Sanriku (June 15, 1896), the Aleutian Islands (April 1, 1946), the Kurile Islands (October 20, 1963), and the Nemuro-Oki (June 10, 1975) events.

Using the simple theory and scaling relations of earthquake faulting, AIDA (1977) and KAJIURA (1981) recently studied the tsunami energy, E_t, in terms of earthquake fault parameters. This approach permits a more rigorous estimate of E_t than various empirical methods. Their results can be applied to the present study. The relation between E_t and M_t is given, under the assumption of $M_t = M_w$, by

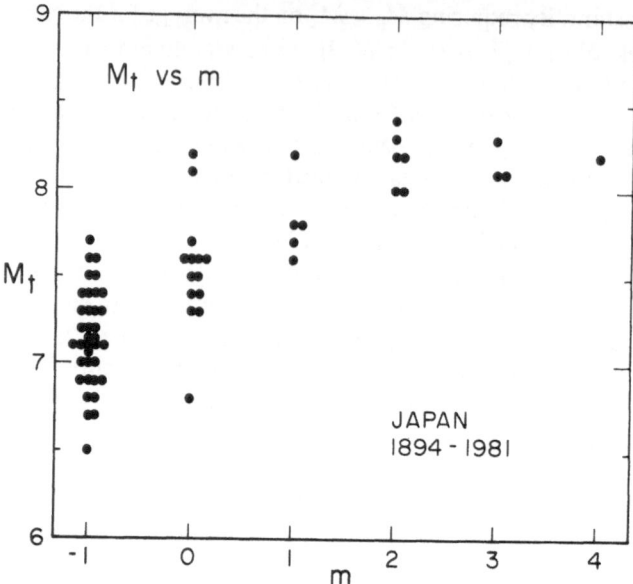

Fig. 4. M_t versus m for tsunami events of Japan. Data are based on the compilation by Abe (1981b).

Table 3.   List of major tsunami earthquakes.

Date[1]	Region	M_s[2]	M_t from world data[3]	M_t from regional data[4]
June 15, 1896	Off NE of Japan	—[5]	8.6	8.2
April 13, 1923	Kamchatka	7.2	8.2	8.2
August 18, 1927	Off central Japan	6.9	—	7.4
April 1, 1946	Aleutian Islands	7.3	9.3	—
November 20, 1960	Peru	7.0	7¾	—
January 16, 1961	Off central Japan	6.5	—	7.1
October 20, 1963	Kurile Islands	7.2	7.9	8.0
August 1, 1968	Philippines	7.2	8.1	—
January 30, 1973	Mexico	7.3	8	—
September 27, 1974	Off NE of Japan	6.5	—	7.0
October 3, 1974	Peru	7.6	8.1	—
June 10, 1975	Off NE of Japan	6.8	7.8	7.9

(1) Date is given in GMT. (2) Surface-wave magnitudes are from Abe (1981a, b). (3) Data are from Abe (1979). (4) Data are from Abe (1981b). (5) M_s is unknown owing to the lack of surface-wave data, but local seismic data suggest a magnitude near 7 (Utsu, 1979).

$$\log E_t = 2 M_t + G \tag{5}$$

where E_t is in ergs and G is a constant plus correction term which depends on the fault geometry. KAJIURA (1981) performed numerical calculations for various fault geometries and obtained $G = 4.54$ as the gross upper limit of E_t. Considering the geometrical similarity of most tsunamigenic earthquakes at island arcs in the circum-Pacific, the variable range of G is very small in actuality. For a typical geometry of faulting (dip angle of the fault $= 30°$, slip angle of the displacement $= 90°$), we have 4.45, which is very close to the above value. Equations (2) and (5) are consistent with the idea that the tsunami energy is proportional to the square of the maximum tsunami-wave amplitude.

From a different point of view, the present results will have obvious importance in predicting the maximum tsunami amplitude at a particular site (ABE, 1981b). If M_w and other fundamental parameters such as location are determined very quickly after a tsunamigenic earthquake (KANAMORI and GIVEN, 1981), the maximum tsunami amplitude can be predicted from Eqs. (2), (3) and (4) by replacing M_t with M_w. Unfortunately no method for the quick determination of the relevant physical parameters is currently available, but recent developments in seismometry will make it possible in the near future.

3. M_t of Historical Events of Japan

Numerical experiments and inverse refraction diagrams are familiar in tsunami studies. Utilizing such techniques, we consider methods of assigning M_t to historical events for which instrumental data are nonexistent.

On the basis of seismic and tsunami data, ABE (1973) confirmed that tsunami waves carry quantitative information about the details of earthquake-generated displacements of the sea bottom in the source region. This discovery has provided a logical basis for numerical experiments. The sea-bottom deformation can now be predicted from the seismic fault parameters or the concept of plate tectonics. Such a deformation is used as an input to shallow water equations to numerically calculate tsunami waves. The input parameters are revised until the optimal combination that best explains the observed coastal tsunami heights and arrival times is obtained. AIDA (1977, 1978a, b) and YAMAKI (1981) applied this technique to several recent events near Japan. The seismic moment derived from their calculations can be converted directly into M_t by Eq. (1), by assuming $M_t = M_w$. A comparison of these M_t values with M_t derived directly from the tide-gauge amplitudes is shown in Fig. 5 by open circles. The correlation is fairly good.

A second method is based on inverse refraction diagrams. The area of tsunami generation can be determined from the arrival time of the tsunami. The wave fronts of the first waves are traced backwards from the coast to the source over a distance equal to ct by means of inverse refraction diagrams (MIYABE, 1934). Here c is the long-wave velocity $\sqrt{gh}$ (g = acceleration of gravity, h = water depth) and t is the arrival time of the tsunami minus the earthquake origin time. The tsunami source area is represented by the region circumscribed by an envelope which touches each of the imaginary wave fronts. It

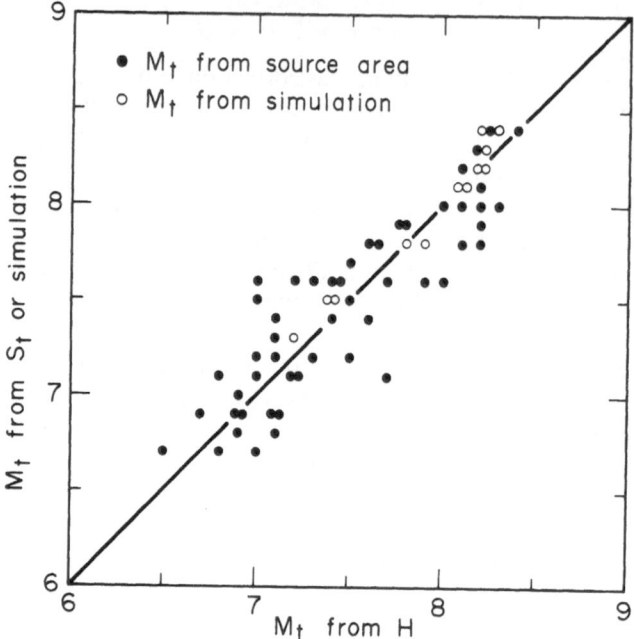

Fig. 5.   Comparison of M_t derived from instrumental tsunami amplitudes, and M_t from results of numerical simulation and consideration of tsunami source area S_t for events that occurred off the Pacific coast of Japan for the period from 1894 to 1981.

has been empirically shown that the tsunami source area, S_t, thus calculated is closely related to the surface area of the fault, S, by

$$S = 0.8 \, S_t \tag{6}$$

for Japanese events (ABE, 1975). On the other hand, a remarkable linearity between log S and log M_0 has been noted (AKI, 1972; ABE, 1975). This linearity is interpreted in terms of a constant average stress drop in tsunamigenic earthquakes. ABE (1975) suggests the relation

$$M_0 = 1.23 \times 10^{22} \, S^{3/2} \tag{7}$$

where M_0 is in dyn · cm and S is in km^2, to represent the overall relation between S and M_0. From Eqs. (1), (6) and (7), we have

$$M_t = \log S_t + 3.9 \tag{8}$$

where S_t is in km^2. The determination of S_t for recent events can be accurately made from tide-gauge records (e.g. HATORI, 1969, 1974c, 1979). A comparison of M_t derived from these S_t with M_t from the amplitude data is shown in Fig. 5 by closed circles. The difference is $- 0.01 \pm 0.25$ on the average, and the correlation is fairly good over a wide range of magnitudes.

AIDA (1977) and HATORI (1975a) extensively applied their techniques to important

Table 4.   Large tsunamigenic earthquakes for 1600–1893 in the Pacific coast of Japan.

Local date	Lat.[1] (°N)	Long.[1] (°E)	$M^{[1]}$	$m^{[1]}$	Region	M_t	References and remarks
February 3, 1605	34.3	140.4 ⎫ 7.9		⎫ 3	Boso-Oki	8	(5), (9)
February 3, 1605	33.0	134.9 ⎭		⎭	Nankai	$8\frac{1}{4}$	(5), (9)
December 2, 1611	39.0	144.5	8.1	4	Sanriku	$8\frac{1}{2}$	(4)
October 31, 1662	31.7	132.0	7.6	2	Hiuganada	—	No details.
April 13, 1677	40.0	144.0	8.1	2	Tokachi-Oki	$8\frac{1}{4}$	(3)
November 4, 1677	35	142 [2]	7.4	3	Boso-Oki	8	(6)
December 31, 1703	34.7	139.8	8.2	3	Boso-Oki	8	(6)
October 28, 1707	33.2	135.9	8.4	4	Tokai and Nankai	$8\frac{1}{2}$	(7), (9)
January 29, 1763	41.0	142.5	7.4	1	Tokachi-Oki	$8\frac{1}{4}$	(3)
April 24, 1771	24.0	124.3	7.4	4	Near Ishigakijima	—	No details.
February 17, 1793	38.5	143.5 [2]	7.1	2	Miyagi-Oki	$7\frac{3}{4}$	(3), (4)
July 20, 1835	37.9	141.9	7.6	2	Miyagi-Oki	$7\frac{1}{2}$	(3)
April 25, 1843	42.0	146.0	8.4	2	Nemuro-Oki?	—	No details, around 8.
December 23, 1854	34.0	137.8	8.4	3	Tokai	$8\frac{1}{4}$	(8), (9), (11)
December 24, 1854	33.0	135.0	8.4	4	Nankai	$8\frac{1}{4}$	(7), (11)
August 23, 1856	40.5	143.5	$7\frac{3}{4}$	2	Tokachi-Oki	$8\frac{1}{4}$	(3), (4)
October 21, 1861	38.5	142 [2]	6.4	2	Miyagi-Oki	$7\frac{1}{2}$	(3)
June 4, 1893	43.5	148	7	1	S. Kurile Is.	—	(10), around 8.

(1) From USAMI (1979). M is local magnitude, and m is Imamura-Iida's tsunami magnitude. (2) Estimated from (3). (3) HATORI (1975a). (4) AIDA (1977). (5) HATORI (1975b). (6) HATORI (1975c). (7) HATORI (1974a). (8) HATORI (1976). (9) AIDA (1981). (10) HATORI (1974b). (11) ABE (1979).

historical events of Japan. The basic data were obtained from historical documents. The values of M_t obtained from their results are given in Table 4, where all the events with $m \geqq 2$ are listed. Values of M_t are given to the large events for the period from 1600 to 1893, and only to the nearest quarter unit; for the events after 1894, values of M_t from instrumental data are available (ABE, 1981b). The present results are admittedly subject to a large uncertainty because of the inevitably poor quality of historical data, but it is to be noted that the present estimates provide the physical size of the tsunamigenic earthquakes, independently of seismic data such as intensity.

The M_t values listed in Table 4 can be compared directly with those of recent events. Two large events of 1611 and 1707 are comparable to or somewhat larger than the Sanriku events of 1896 and 1933, which are the largest events since instrumental data were available. The largest M_t of the events in Japan since 1600 is as large as $8\frac{1}{2}$, which is much smaller than $M_t = 9.4$ of the 1960 Chilean event; in other words, Eq. (5) indicates that the difference of E_t between the largest Japanese and Chilean events is one-sixtieth.

4. Conclusion

Among tsunami magnitude scales in current use, the recently proposed scale, M_t, holds several advantages: (1) it is originally defined in terms of instrumental amplitudes, (2) it scales the overall physical size of the tsunami source, (3) it is a reliable measure of the seismic moment of tsunamigenic earthquakes, particularly historical ones, for which tsunami observations were made but seismic data are inadequate or nonexistent, (4) it permits a direct comparison with the moment magnitude for most events, (5) it gives a reliable estimate of the tsunami energy, (6) it is useful for an easy identification of unusual tsunami earthquakes, and (7) it provides a simple and quantitative means for predicting the maximum tsunami amplitude at a particular site. To make the M_t scale more useful, the method of determining M_t must be examined at various regions worldwide.

I benefited from helpful discussions with Drs. Hiroo Kanamori and Kinjiro Kajiura.

REFERENCES

ABE, K., Tsunami and mechanism of great earthquakes, *Phys. Earth Planet. Inter.*, 7, 143–153, 1973.

ABE, K., Reliable estimation of the seismic moment of large earthquakes, *J. Phys. Earth*, 23, 381–390, 1975.

ABE, K., Size of great earthquakes of 1837–1974 inferred from tsunami data, *J. Geophys. Res.*, 84, 1561–1568, 1979.

ABE, K., Magnitudes of large shallow earthquakes from 1904 to 1980, *Phys. Earth Planet. Inter.*, 27, 72–92, 1981a.

ABE, K., Physical size of tsunamigenic earthquakes of the northwestern Pacific, *Phys. Earth Planet. Inter.*, 27, 194–205, 1981b.

AIDA, I., Simulation of large tsunamis occurring in the past off the coast of the Sanriku district, *Bull. Earthq. Res. Inst., Tokyo Univ.*, 52, 71–101, 1977 (in Japanese).

AIDA, I., Numerical experiments for the tsunami accompanying the Miyagiken-oki earthquake of 1978, *Bull. Earthq. Res. Inst., Tokyo Univ.*, 53, 1167–1175, 1978a (in Japanese).

AIDA, I., Reliability of a tsunami source model derived from fault parameters, *J. Phys. Earth*, 26, 57–73, 1978b.

AIDA, I., Numerical experiments of historical tsunamis generated off the coast of the Tokaido district, *Bull. Earthq. Res. Inst., Tokyo Univ.*, 56, 367–390, 1981.

AKI, K., Scaling law of earthquake source time-function, *Geophys. J. R. Astr. Soc.*, 31, 3–25, 1972.

COMER, R. P., Tsunami height and earthquake magnitude: Theoretical basis of an empirical relation, *Geophys. Res. Lett.*, 7, 445–448, 1980.

FUKAO, Y., Tsunami earthquakes and subduction processes near deep-sea trenches, *J. Geophys. Res.*, 84, 2303–2314, 1979.

HATORI, T., Dimensions and geographic distribution of tsunami sources near Japan, *Bull. Earthq. Res. Inst., Tokyo Univ.*, 47, 185–214, 1969.

HATORI, T., Sources of large tsunamis in southwest Japan, *J. Seismol. Soc. Japan, Ser. 2*, 27, 10–24, 1974a (in Japanese).

HATORI, T., Source area of the tsunami off the Nemuro Peninsula in 1973 and its comparison with the tsunami in 1894, *Spec. Bull. Earthq. Res. Inst., Tokyo Univ.*, 13, 67–76, 1974b (in Japanese).

HATORI, T., Tsunami sources on the Pacific sides in northeast Japan, *J. Seismol. Soc. Japan, Ser. 2*, 27, 321–337, 1974c (in Japanese).

HATORI, T., Tsunami magnitude and wave source regions of historical Sanriku tsunamis in northeast Japan, *Bull. Earthq. Res. Inst., Tokyo Univ.*, 50, 397–414, 1975a (in Japanese).

HATORI, T., Sources of large tsunamis generated in the Boso, Tokai, and Nankai regions in 1498 and 1605, *Bull. Earthq. Res. Inst., Tokyo Univ.*, 50, 171–185, 1975b (in Japanese).

HATORI, T., Sources of tsunamis generated off Boso Peninsula, *Bull. Earthq. Res. Inst., Tokyo Univ.*, 50, 83–91, 1975c (in Japanese).

HATORI, T., Documents of tsunami and crustal deformation in Tokai district associated with the Ansei earthquake of Dec. 23, 1954, *Bull. Earthq. Res. Inst., Tokyo Univ.*, 51, 13–28, 1976 (in Japanese).

HATORI, T., Tsunami activity in the Hokkaido and Kurile regions (1893–1978), *Bull. Earthq. Res. Inst.*,

Tokyo Univ., **54**, 543–557, 1979.

IIDA, K., Magnitude and energy of earthquakes accompanied by tsunami, and tsunami energy, *J. Earth Sci. Nagoya Univ.*, **6**, 101–112, 1958.

IMAMURA, A., History of Japanese tsunamis (translated from the Japanese title), *Kaiyo-no-kagaku*, **2**, 74–80, 1942 (in Japanese).

KAJIURA, K., Tsunami energy in relation to parameters of the earthquake fault model, *Bull. Earthq. Res. Inst., Tokyo Univ.*, **56**, 415–440, 1981.

KANAMORI, H., Mechanism of tsunami earthquakes, *Phys. Earth Planet. Inter.*, **6**, 346–359, 1972.

KANAMORI, H., The energy/release in great earthquakes, *J. Geophys. Res.*, **82**, 2981–2987, 1977.

KANAMORI, H. and J. W. GIVEN, Use of long-period surface waves for fast determination of earthquake source parameters, *Phys. Earth Planet. Inter.*, **27**, 8–31, 1981.

MIYABE, N., An investigation of the Sanriku tsunami based on mareogram data, *Bull. Earthq. Res. Inst., Tokyo Univ.*, Suppl. 1, 112–126, 1934.

SOLOVIEV, S. L., Recurrence of tsunamis in the Pacific, in *Tsunamis in the Pacific Ocean*, (edited by W. M. Adams, pp. 149–163, East West Center Press, Honolulu, 1970.

USAMI, T., Study of historical earthquakes in Japan, *Bull. Earthq. Res. Inst., Tokyo Univ.*, **54**, 399–439, 1979.

UTSU, T., Seismicity of Japan from 1885 through 1925, *Bull. Earthq. Res. Inst., Tokyo Univ.*, **54**, 253–308, 1979 (in Japanese).

YAMAKI, S., Numerical simulation of the 10 June 1975 Nemuro-Oki tsunami, Abstr. Fall Meet. Seism. Soc. Japan, p. 100, 1981 (in Japanese).

YAMAKI, S. and K. ABE, The decay of tsunami amplitudes along the coast as inferred from numerical experiments, Abstr. Fall Meet. Seism. Soc. Japan, p. 175, 1981 (in Japanese).

HISTORICAL AND STATISTICAL STUDIES OF TSUNAMIS

Tsunamis—Their Science and Engineering, edited by K. Iida and T. Iwasaki, 105–119.

Historical Study of Tsunamis at Tofino, Canada

Sydney O. WIGEN

Institute of Ocean Sciences, Sidney, B. C., Canada

(Received October 27, 1981; Revised December 12, 1981)

Procedures proposed by the International Tsunami Information Center have been applied to the historical study of tsunamis at Tofino, Canada. Using the list of 1500 known and possible tsunamigenic events compiled by the Center, the marigrams from the Tofino tide station have been systematically searched for each possible event. Forty-three tsunamis have been identified in a 75-year period of records.

Systematic data have been extracted for these events, including initial and maximum wave heights, periods and travel times from sources. Tidal contributions to the recorded wave heights have been removed, to allow valid comparison of the events. Smallest tsunamis identified show a maximum wave, trough to crest, of 6 cm; the largest, a wave height of 240 cm. All tsunamis were identified with distant generating areas within the Pacific region, and none with local submarine seismic sources.

A logarithmic plot of maximum wave heights appears to be linear and this is applied to establish a magnitude-frequency relationship.

1. Introduction

In 1977 the International Tsunami Information Center initiated a program, the Historical Study of Tsunamis, to access and make available for researchers the tsunami information lying dormant in tide records. The study is designed to present in a standardized form the characteristics of each and every tsunami registered on the analogue records of many tide stations. It also identifies the possible tsunamis that did not appear at each station.

Using the procedures set forth in the Outline (WIGEN, 1978) the analogue tide records for Tofino, Canada from 1906 to 1980 have been systematically searched for possible wave action associated with 1451 events listed in WIGEN (1977), and in TSUNAMI NEWSLETTERS (1977–1980).

2. Tofino and Approaches

Tofino is a fishing community located on the Pacific shore of Vancouver Island. It fronts on a system of channels in Clayoquot Sound, about 5 km from the open sea. A complex system of inlets penetrate 30 km farther inland (Fig. 1). Mid channel depths of these inlets range from about 60 to 150 m, but seaward of Tofino the channel has a least depth of 5.5 m at lowest normal tide.

Offshore the continental shelf reaches a depth of 200 m at 50 km from the coast.

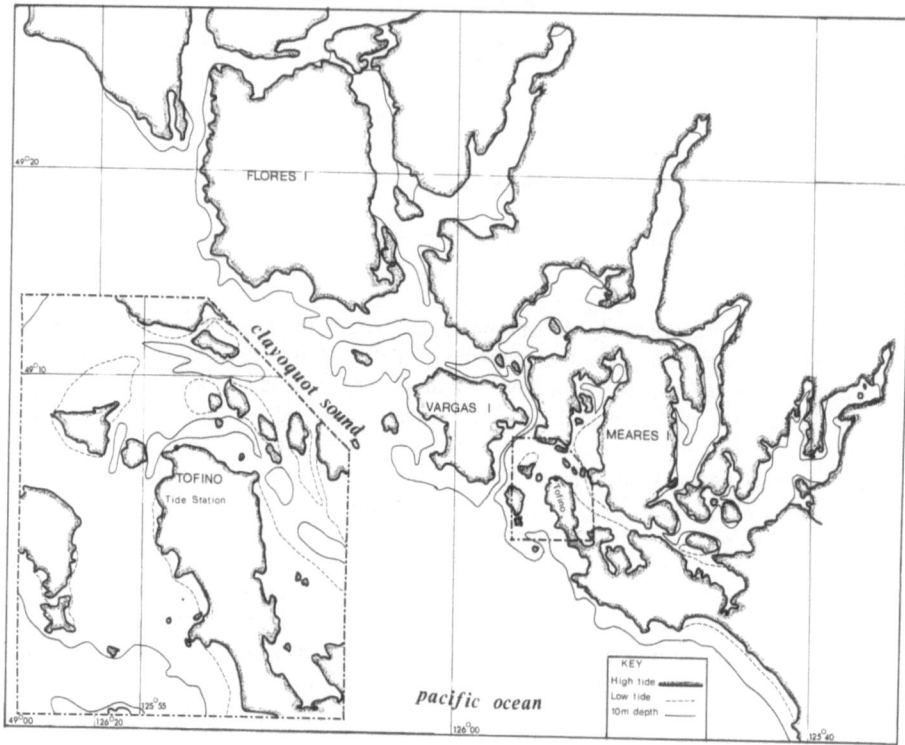

Fig. 1.   Tofino, Clayoquot Sound, and system of inlets.

The continental slope drops to 2,400 m in another 60 km, a 4% gradient (Fig. 2). Beyond lies an irregular collection of seamounts and submarine ridges, in a seismologically active zone of seafloor spreading, terminating in the abyssal plain about 700 km from the coast.

The tide at Tofino is mixed, mainly semidiurnal, with a mean tide range of 2.8 m, and a large tide range of 4.1 m. The highest tide recorded at the gauge station reached 4.8 m, the lowest, −0.3 m (CANADIAN HYDROGRAPHIC SERVICE, 1981).

3.   Tofino Tide Station

The tide station was constructed as a conventional float actuated gauge in 1905 (Fig. 3) (DAWSON, 1906). Although referred to as Clayoquot for the first 60 years of its operation the station has never been moved more than 250 m from its original site in Tofino. Many details about its early operation and performance are available in the files of the Tidal and Current Section, Institute of Ocean Sciences. The gauge has served as the primary tidal reference station for the Pacific Coast of Canada. As such it has been generally maintained to high standards, and has been serviced almost daily by a guage attendant. Occasionally it has experienced problems normal to operations in, until recently, a remote community.

Fig. 2.   Pacific Ocean seafloor and continental shelf off Tofino.

The station was erected on a small wharf which had been extended out to deep water for that purpose. The original stilling well was a rectangular wooden tube, open-ended on the bottom and resting on gravel. In addition to the normal orifice, water will have entered and left the well through the gravel. Tide was recorded originally as a pencil trace on a 24-hour graph on a cylindrical clock. The pencil traces were later inked by hand, with a distinguishing colour for each day of the week. Sometimes small oscillations were smoothed over in the inking, and evidence of some of the smallest tsunamis may have been obscured or lost. A change in 1962 to ink recording on a strip-chart diagram eliminated this problem. Changes were made in the type of stilling well, and in 1971 the well was placed in a vertical shaft in a rocky promontory, with an intake pipe 12 m in length leading from the sea into the well.

4.   Tide Station Response Characteristics

The intake orifice of a stilling well is designed to allow true measurement of tide, but to almost eliminate waves of a period of a few seconds. Ability of the gauge station to record accurately the waves of intermediate periods, tsunamis or seiches, is critically dependent on the stilling well response characteristics. NOYE (1974) correlates the head

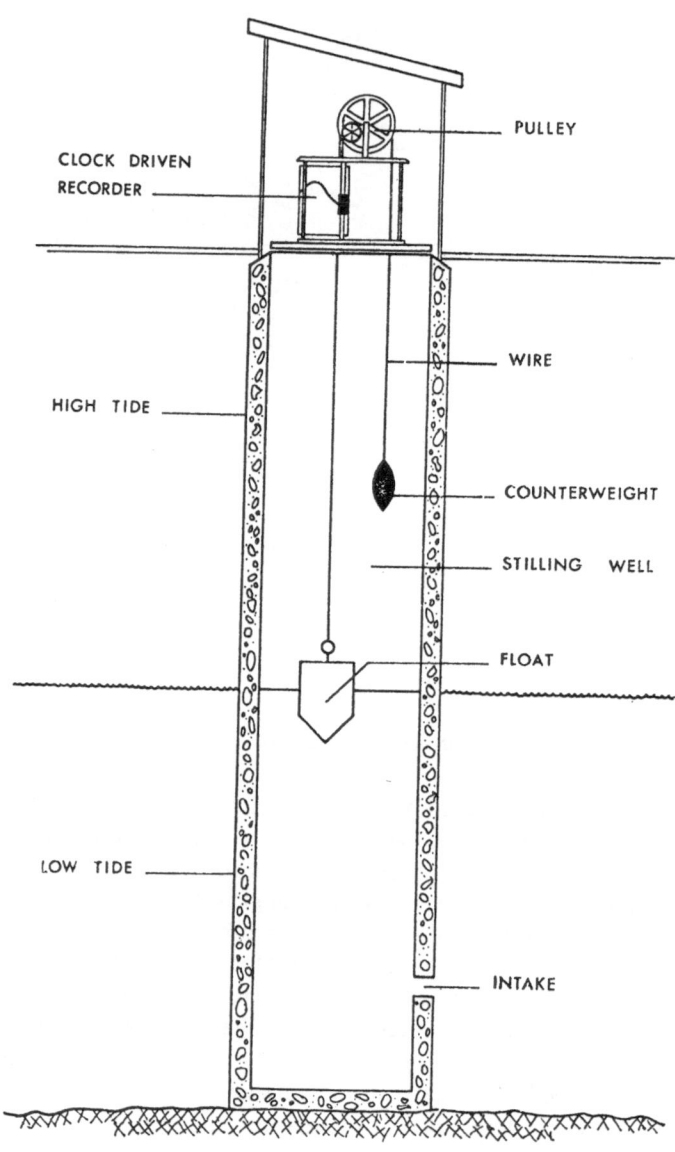

TIDE   GAUGE   STATION

Fig. 3.   Typical tide well and orifice intake.

differences, inside and outside the well with the ratio of cross sectional areas, stilling well and orifice. He also indicates procedures for recreating the actual tsunami wave heights from the gauge diagrams, where the response characteristics are known.

In modern gauging practise the time constant of a stilling well is measured by timing the outflow of a head of water in the well. Only in recent years has the Tofino gauge been so tested during annual gauge inspections. However the validity of historical tsunami records can be estimated from two sources, stilling well to orifice area ratios during the station's operation, and comparative studies of the high frequency oscillations visible on the analogue records.

From annual gauge inspection reports, and construction records available in the files of the Tidal and Current Section, Institute of Ocean Sciences, the following summarized history of Tofino stilling wells and orifices has been compiled.

Period	Stilling well diameter	Orifice	Area ratio, well to orifice
1906–1942	10″ × 12″ wooden well and later 12″ wood or iron pipe, open at the bottom and resting on seafloor	1¼, plus flow throgh gravel	<90:1
1942–1964	6″ cast iron sewer pipe	9/16″	115:1
1964–1971	15″ iron culvert pipe, asphalted	1½″	100:1
1971–1981	18″ concrete pipe	1.7″ intake pipe restricted by sediment in later years	>110:1

The comparative study was made by inspecting one winter month of analogue records for each year to compare how the higher frequency wave oscillations were penetrating into the stilling well.

The winter period is a time of the year when high frequency sea waves and some seiching could be expected.

Six categories of response were identified:

Active—Wave heights of 6 to 15 cm (0.2 to 0.5 feet). Period 1 to 2 minutes or less, so that individual crests could not be distinguished on the alanlgue trace.

Moderately active—Wave heights of 3 to 6 cm, and period of 2 to 3 minutes, individual crests identifiable.

Slightly active—Wave heights of 3 to 6 cm, and periods of 3 to 5 minutes. Shorter period waves registering only slightly.

Very slight—Seiche waves of 3 to 6 cm, registering at 5 to 20 minute periods. Virtually no visible shorter period waves.

Smooth—No response to 5 to 20 minute oscillations, but still appearing to register tidal changes.

Stepping—Gauge record showing irregular rates of rise or fall, with sudden surges. At this stage sediments are blocking flow through the intake until a head difference develops to push the water through.

Active	Moderately active	Slightly active	Very slight	Smooth or Stepping
1906–1923	1924–1926			
1927–1935	1936–1937			
1938–1941	1942–1943			
1944–1946	1947–1948			
1949	1950			
1951–1959	1960–1961			
1962–1965				Aug. 1965–May 1966
	June 1966–1969			
1970–1971	1972–1974	1975–1976	1977–Sept. 1978	Oct. 1978–1980

During the period 1906 to 1974 when the gauge was active or moderately active, it is probable that the Tofino analogue records give a close approximation of actual tsunami waves. From 1975 to 1978 response to tsunamis may have been diminished. From 1979 through 1980 the gauge may at times have been unable to respond to a smaller tsunami, or register accurately a larger one.

5. Tsunami Identification and Data Extraction

As a working hypothesis it was assumed that any possibly tsunamigenic event in the Pacific, Indian, or southern Atlantic Oceans was a conceivable source of tsunami at Tofino. For the period of tide records, 1906 to 1980, each event listed in the Chronological List (WIGEN, 1977) and in Tsunami Newsletters from these regions was systematically investigated. Wave travel time from each epicenter was calculated using the Tofino travel time chart (U.S. DEPT. OF COMMERCE 1969), and projected rates of travel outside the limit of this chart. Estimated wave arrival times were calculated from the listed time of the event, plus travel times. For each event the analogue tide trace, original or microfilm copy, was carefully inspected for any detectable anomaly close to the assumed arrival time. In more than 80% of the cases, no abnormality was visible. In some cases a clearly defined tsunami wave pattern commenced close to the calculated arrival time. The remainder had some seiche action already in progress that made a decision not obvious. Each of these was given further careful scrutiny, but unless significant change in the wave pattern was evident, it was designated as No Tsunami.

A summary of the results for each of the 19 designated areas of the Historical Study of Tsunamis is tabulated in Appendix A. A complete listing of the events investigated is available from the author, of from the International Tsunami Information Center.

Forty-three tsunamis were identified, four of which overlapped on previous events. These tsunamis are listed in Appendix B, along with the following standardized data:

1) Actual arrival time, Universal Time, of the initial wave, point A on Fig. 4.

2) First motion, the height of the initial wave, measured as centimetres of rise or fall above an assumed, manually interpolated tide level. V–B, not V–A.

3) Second motion, the following fall or rise (V–B) + (C–X).

4) Height of the maximum wave, the greatest sum of two successive displacements from the tide level (Y–D) + (E–Z), not Y–Z. This is the water level change contributed by the tsunami. Elimination of tide provides a basis for intercomparison between tsunamis, or between responses at different tide stations to the same tsunami.

5) Time at Y, the commencement of the maximum wave. If two or more waves had the same maximum height, the time of the later one is listed, as the best indicator of how long after the initial arrival the tsunami could be creating maximum hazard.

6) Period of the tsunami, the average time interval between a series of crests.

7) Interval between the time of initial and maximum wave.

With the analogue tide records available from Tofino it was found in practice that the smallest tsunami displacement that could be clearly identified on a good record was 2 cm. The smallest maximum wave listed is 6 cm.

Probably one or more tsunamis with waves larger than 6 cm were obscured by seiche and omitted. Also one or several of the smaller identified tsunamis may have been aliased by a seiche commencing near the anticipated arrival time. These two sources of error in the 75-year sample will tend to cancel each other out.

Frequency distribution of maximum wave heights suggests that many tsunamis may have reached Tofino, too small to be distinguished on the system of recording.

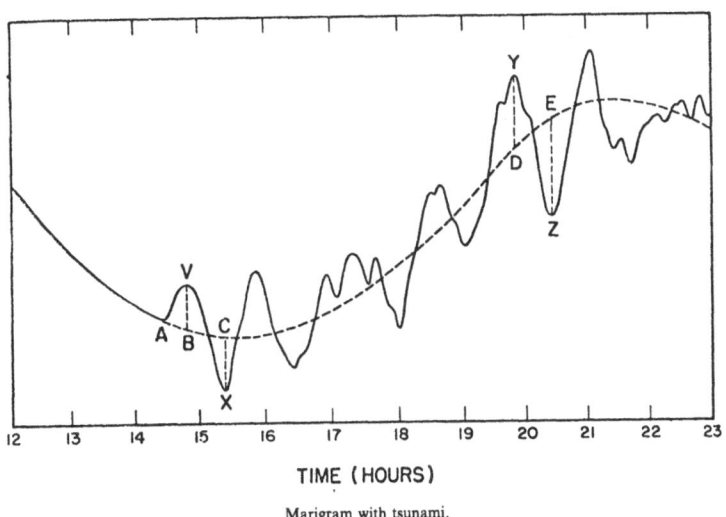

TIME (HOURS)

Marigram with tsunami.

Fig. 4. Tsunamigram, illustrating standardized data extracted.

6. Frequency Analysis

The events, graphed logarithmically and sequentially in order of maximum wave height H_m show an apparently linear relation (Fig. 5). Since the events are placed in order of size, the sequential number, N, on the abscissa represents the number of tsunamis equal to or larger than that event occurring in the 75-year period of data. Regression analysis of the 39 data points gives the equation: $\log H_m = -0.965 \log N + 2.3128$. Correlation coefficient of the analysis is 0.993.

Since N represents the number of events for a 75-year period, the maximum wave height of a once-in-a-century tsunami at Tofino can be calculated by letting $N = 75/100$, and gives a value of 271 cm. The limits to which this frequency-magnitude relationship can be extrapolated are unknown. Validity of the equation itself depends on how accurately a 75-year period of tsunamis typifies the recurrence of these events.

The only historical record uncovered of tsunamis on the Canadian west coast prior to tide gauge operation is associated with the great Meiji Sanriku tsunami of June 15, 1896. The Victoria Daily Colonist of June 21, 1896 reported:

> Coincident with the devastation of the coastal cities of Japan by a tidal wave on Monday last, a similar disturbance of the sea created no little alarm along the entire seaboard of Vancouver Island, sweeping the full length of the coast, but

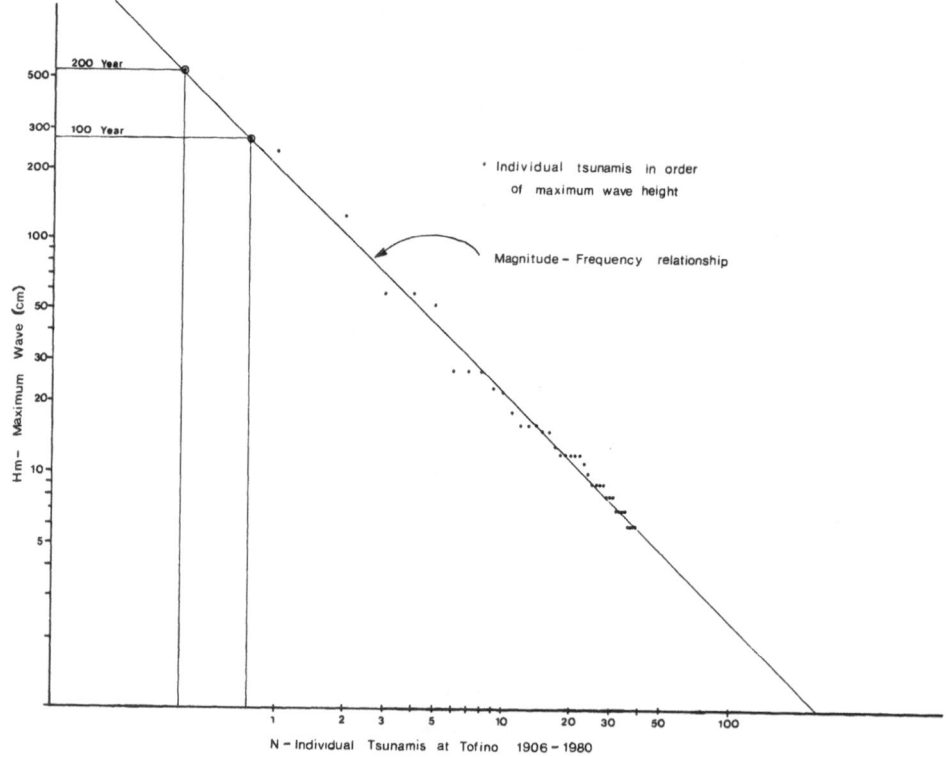

Fig. 5.   Tofino tsunami magnitude-frequency relationship.

fortunately doing no serious damage to any of the settlements .... For upwards of three hours the waters rushed inland for miles, moving strangely to and fro with a velocity of nearly eight miles per hour. The steamer MAUDE which returned yesterday from west coast ports, was at Kyuquot at the time and her officers, crew and passengers watched with curious attention the novel movement of the waters, estimating that at Kyuquot during the few hours of inundation all the sea-skirting land was submerged to a depth of four to five feet.

No specific information for Tofino relative to this tsunami is on record.

7. Sources of Tsunamis Reaching Tofino

Superimposed on the U.S. Dept. of Commerce World Seismicity Map, Figs. 6–9 (pp. 114–117), are the locations of epicenters from which identifiable tsunamis reached Tofino. Most consistent source areas for the larger events are Alaska-Aleutians, Kamchatka-Kuril Islands, and northern Japan. However the second largest event came from Chile, in 1960. No tsunamis were registered from earthquakes in the seismically active zone west of Vancouver Island.

8. Conclusion

The foregoing work has been undertaken in accordance with the Historical Study of Tsunamis program of the International Tsunami Information Center. The collation of similar data from Tofino and many other tide stations will provide a body of knowledge from which the directions of propagation of tsunamis from each source region may be identified.

REFERENCES

CANADIAN HYDROGRAPHIC SERVICE, Canadian tide and current tables, Vol. 6, 1981.
DAWSON, W. B., Tide levels and datum planes on the Pacific coast of Canada, Department of Marine and Fisheries, Ottawa, 1906.
NOYE, B. J., Tide-well systems III: improved interpretation of tide-well records, *J. Marine Res.*, **32**, 183–194, 1974.
TSUNAMI NEWSLETTER, International Tsunami Information Center, Honolulu, 1977–1980.
U. S. DEPT. OF COMMERCE, *Coast and Geodetic Survey*, Tsunami travel time to Tofino, 1969.
WIGEN, S. O., Historical study of tsunamis, chronological and area lists, International Tsunami Information Center, Honolulu, 1977.
WIGEN, S. O., Historical study of tsunamis—an outline. Institute of Ocean Sciences, Sidney, B. C., Pacific Marine Science Report 78–5, 1978.

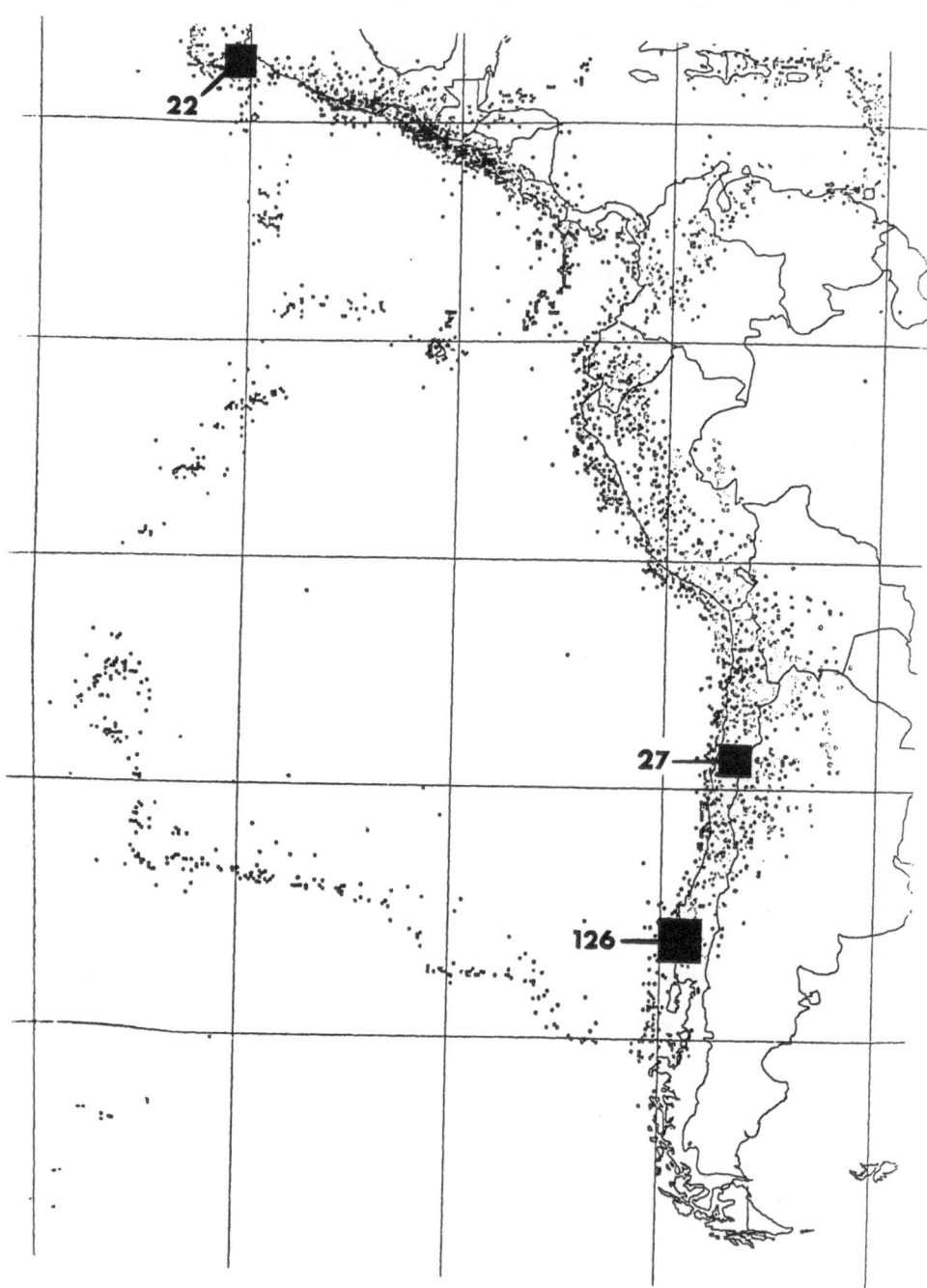

Fig. 6.   Source of tsunamis reaching Tofino from Chile to Mexico.

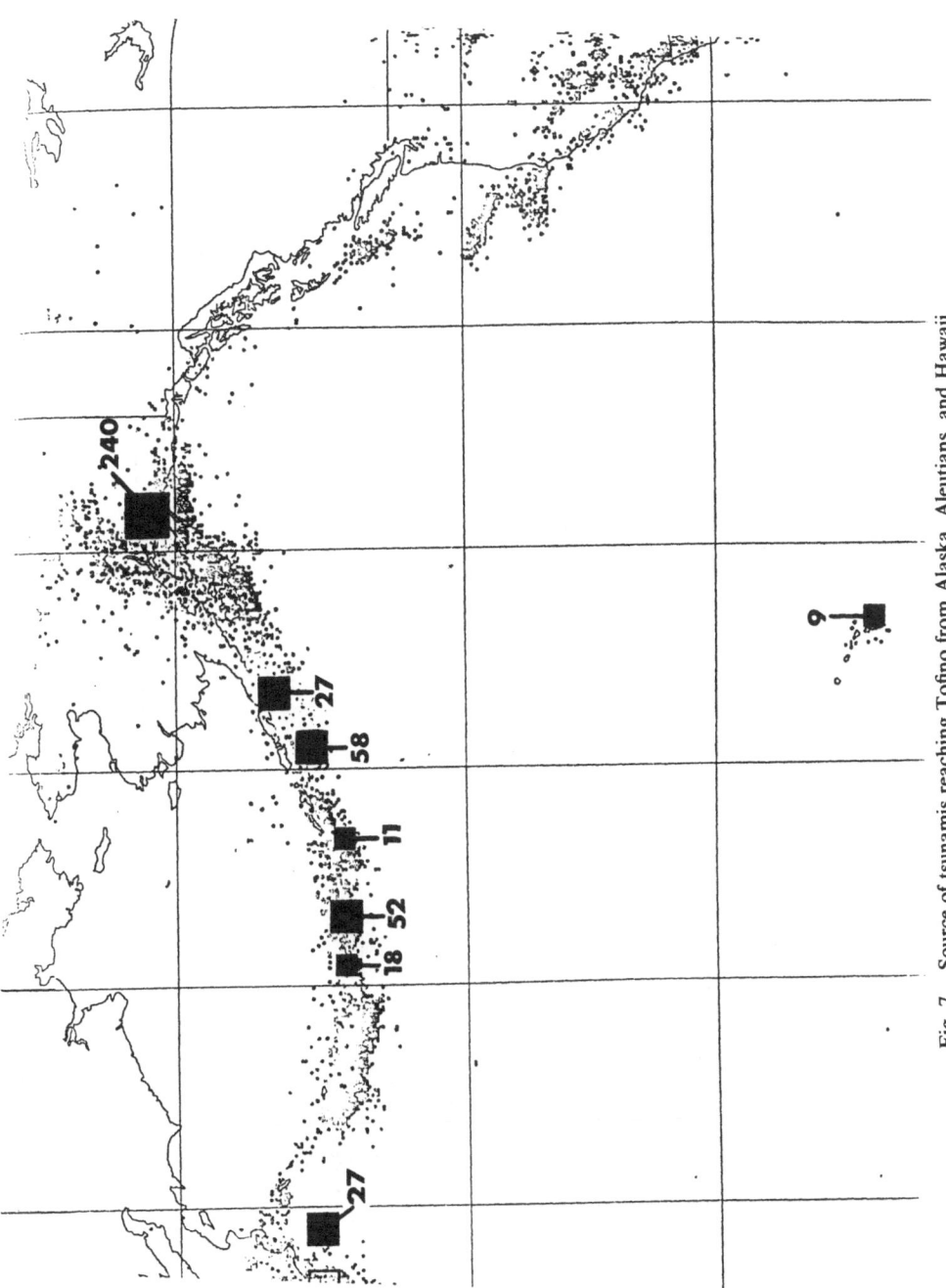

Fig. 7. Source of tsunamis reaching Tofino from Alaska, Aleutians, and Hawaii.

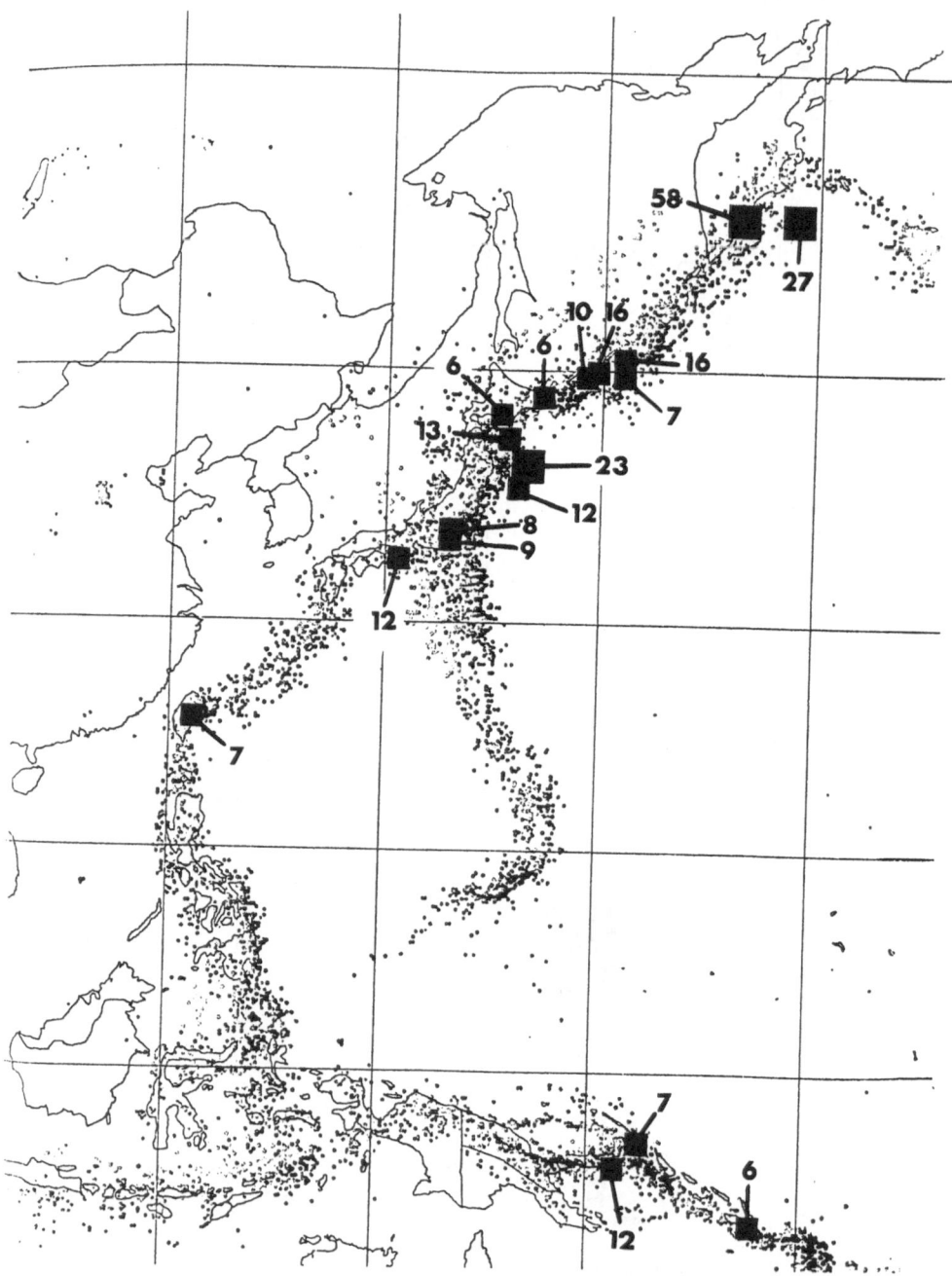

Fig. 8.   Source of tsunamis reaching Tofino from Kamchatka, Kurils, and Japan.

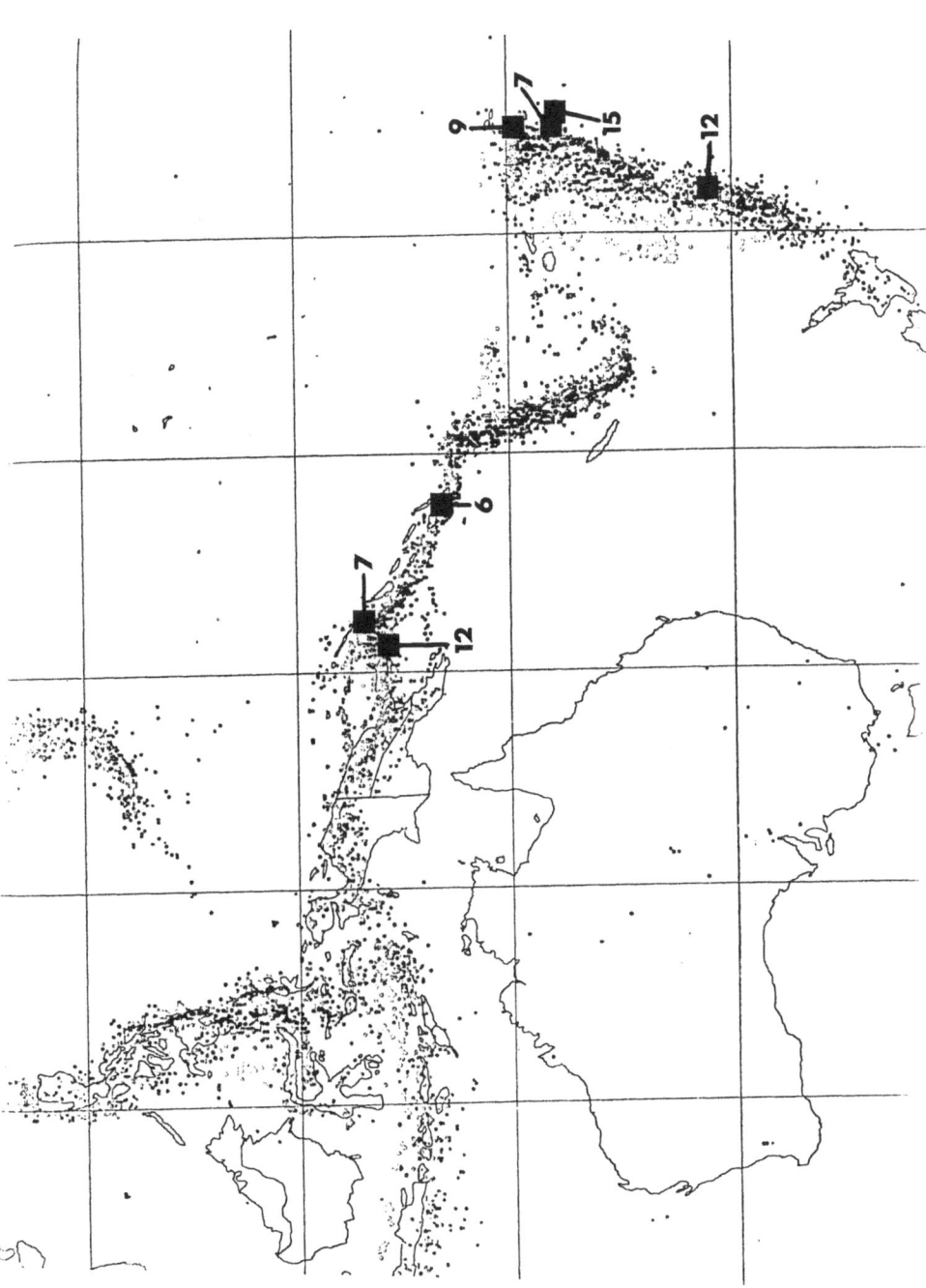

Fig. 9. Source of tsunamis reaching Tofino from the Southwest Pacific.

S. O. Wigen

Appendix A

The effect of tsunamis of Tofino by area of origin.

1906 to 1980 Area of origin	Number of inspected earthquakes	No evidence of tsunamis	No tide record	tsunami recorded	Indefinite evidence
1 Chile-Peru	98	96	0	2	0
2 Caribbean	0	0	0	0	0
3 Ecuador-Mexico	91	89	0	2	0
4 Baja California Alaska Panhandle	42	42	0	0	0
5 Gulf of Alaska Aleutians	112	102	2	8	0
6 Kamchatka Kuril Is.	96	84	1	9	2
7 Mariana Japan Trench	120	111	3	5	1
8 Sea of Japan	17	17	0	0	0
9 Ryukyu Trench	83	78	0	4	1
10 Philippine Trench	60	60	0	0	0
11 S. China-Sulu Celebes Seas	46	45	1	0	0
12 Java Trench Banda Sea	150	147	1	2	0
13 Bismark Arch. New Hebrides	300	291	2	6	1
14 Tasman Sea	46	46	0	0	0
15 Tonga Kermadec	71	64	0	4	3
16 Hawaii	16	15	0	1	0
17 Atlantic	70	68	2	0	0
18 Mediterranean	0	0	0	0	0
19 Indian Ocean	33	32	0	0	1
Total	1451	1387	12	43	9

Appendix B

Tsunamis—recorded at Tofino.

Year	Mo.	Day	Hour (UT)	Lat.	Long.	Mag.	Actual arrival	First motion	Second motion	Maximum arrival	Wave & HT	Period	Time initial to max. wave
Area of origin: 1. Chile-Peru													
1922	11	11	0433	28.5S	070.0W	8.4	2030	3R	5F	12/0333	27F	20	7 hr 03 min
1960	5	22	1911	39.5S	074.5W	8.3	1221	22R	22F	1730	126F	24	5 hr 09 min
Area of origin: 3. Ecuador-Mexico													
1928	6	17	0319	16.3N	098.0W	7.9	2020	2F	3R	2334	8F	15	3 hr 14 min
1934	11	30	0205	18.5N	105.5W	7.0	1008	15R	22F	1022	22F	20	0 hr 14 min
Area of origin: 5. Gulf of Alaska-Aleutians													
1929	3	7	0135	51.0N	170.0W	8.6	0639	4F	8R	0800	11F	13	1 hr 21 min
1938	11	10	2019	55.5N	158.0W	8.7	11/0011	12R	20F	0125	27F	63	1 hr 14 min
1946	4	1	1229	52.8N	163.5W	7.4	1650	21R	48F	2052	58F	15	4 hr 02 min
1957	3	9	1422	51.3N	175.8W	8.3	1907	14R	21F	10/0332	52R	22	
1957	3	9	2039	52.5N	169.5W	7.1	10/0137	49R					
1957	3	11	1455	51.5N	178.5W	7.2	2019	7F	13R	2120	18F	21	1 hr 01 min
1964	3	28	0336	61.1N	147.6W	8.5	0700	99R	138F	0850	240F	28	1 hr 50 min
1970	3	11	2239	57.5N	153.9W	6.1	12/0233	2F	3R	0358	6F	20	1 hr 25 min
Area of origin: 06. Kamchatka-Kuril Is.													
1918	9	7	1716	45.5N	151.5E	8.3	8/0148	9R	16F	0156	16F	16	0 hr 08 min
1918	11	8	0438	44.5N	151.5E	7.9	1347	2R	4F	1410	7F	15	0 hr 23 min
1923	2	3	1602	54.0N	161.0E	8.4	2354	6R	11F	4/0350	27F	18	3 hr 56 min
1923	4	13	1531	56.5N	162.5E	7.2	14/0027	9R	15F	0806	15F	40	7 hr 39 min
1952	11	4	1658	52.8N	159.5E	8.4	5/0042	21R	47F	0347	58R	21	3 hr 05 min
1958	11	6	2258	44.5N	148.5E	8.7	7/0803	5F	9R	1655	10F	9	8 hr 52 min
1963	10	13	0518	44.8N	149.5E	8.2	1403	3R	4F	1857	16F	22	4 hr 54 min
1969	8	11	2127	43.6N	147.8E	7.1	0625	3F	4R	1413	6F	20	7 hr 48 min
1973	6	17	0355	43.2N	145.8E	6.5	1345	4R	5F	1457	6R	14	1 hr 12 min
Area of origin: 7. Mariana-Japan trench													
1915	11	1	0724	38.0N	144.0E	7.8	1658	2F	3R	1912	12F	12	2 hr 14 min
1933	3	2	1731	39.3N	144.5E	8.9	3/0312	10R	18F	0457	23F	7	1 hr 45 min
1952	3	4	0123	42.5N	143.0E	8.6	1150	3R	5F	2112	12F	25	9 hr 22 min
1968	5	16	0049	40.8N	143.2E	6.1	1021	4R	8F	1714	12F	20	6 hr 53 min
1968	5	16	1039	41.5N	142.7E	6.4	—	—	—	6/0100	13R	20	—
Area of origin: 9. Ryukyu Trench													
1923	9	1	0259	35.3N	139.5E	8.3	1616	2F	3R	1705	8R	15	0 hr 49 min
1923	9	2	0247	35.0N	139.5E	7.7	1547	7F	9R	1550	9R	18	0 hr 03 min
1944	12	7	0436	33.8N	136.0E	8.3	1559	5F	12R	1605	12R	14	0 hr 06 min
1951	10	22	0543	24.0N	121.3E	7.1	2145	4R	5F	23/0033	7R	18	2 hr 48 min
Area of origin: 12. Java Trench-Banda Sea													
1918	7	18	0230				1035	4F	6R	1250	16R	20	2 hr 15 min
1931	3	28	1239	07.0S	129.5E	7.3	29/0510	2F	3R	0545	9F	18	0 hr 35 min
Area of origin: 13. Bismark Arch.-New Hebrides													
1931	10	3	1913	10.5S	161.8E	8.1	0952	2R	3F	1313	6F	17	3 hr 21 min
1931	10	3	2155	11.0S	163.0E	7.0							
1931	10	3	2248	11.0S	161.5E	7.3							
1944	12	27	1526	18.0S	168.0E	7.3	28/0712	4F	7R	0743	12F	10	0 hr 31 min
1946	9	29	0302	04.5S	153.5E	7.7				2237	8F	14	
1971	7	26	0123	04.9S	153.2E	6.6	1528	3R	4F	2303	7R	19	7 hr 35 min
Area of origin: 15. Tonga-Kermadec													
1917	5	1	1826	29.0S	177.0W	8.6	0824	12F	12R	0824	12F		0 hr 00 min
1917	6	26	0550	15.5S	173.0W	8.7	1718	2F	4R	1845	9R		1 hr 27 min
1919	4	30	0717	19.0S	172.5W	8.4	2000	2F	4R	2358	15R	10	3 hr 58 min
1921	2	27	1824	18.5S	173.0W	7.2	28/0718	2R	4F	0735	7R		0 hr 17 min
Area of origin: 16. Hawaii													
1975	11	29	1448	19.4N	155.1W	7.2	2056	3R	9F	2202	9F	14	1 hr 06 min

Tsunamis—Their Science and Engineering, edited by K. Iida and T. Iwasaki, 121–130.

Historical Study of Tsunamis at Miyako, Japan

Masami Okada* and Masaomi Tada**

Japan Meteorological Agency, Tokyo, Japan
**Miyako Weather Station, Miyako, Japan*

(Received January 16, 1982; Revised February 24, 1982)

In order to study the characteristics of small tsunamis, all tidal records on strip chart kept by the Miyako Weather Station were inspected around the estimated time of tsunami arrival expected from seismic and volcanic events compiled in the catalog by Wigen (1977a). Fifty-eight events since 1937 were regarded as records of "tsunami" or "possible tsunami." Using the present data, we revealed the fact that the relationship between amplitude and frequency of tsunamis at Miyako can be denoted by

$$n(A)\mathrm{d}A = kA^{-p}\mathrm{d}A, \quad \text{or} \quad N(A > a) = \frac{k}{p-1}a^{-p+1},$$

as found for Tofino in Canada by Wigen (1977b). Here $n(A)\mathrm{d}A$ and $N(A > a)$ indicate those frequencies of tsunamis whose maximum amplitude are from A through $A + \mathrm{d}A$ and larger than a, respectively; k and p are constants. p was estimated as 2.1 for Miyako, being a little larger than for Tofino.

1. Introduction

Various efforts have been focussed on the preparation of catalogs listing the arrival of tsunamis at the coast of Japan (e.g., Watanabe (1968); Sendai District Observatory (1979, 1980)). Iida et al. (1967) also compiled tsunami records for occurrences in the Pacific Ocean. Those catalogs usually contain not only a description of the tsunamigenic event but also the data on fairly large waves at some tide stations. However, data on smaller waves were disregarded at other stations, even if tsunami waves were noticed on the tidal records. Further, we believe that some small events might possibly have been omitted from those catalogs. It is therefore necessary to inspect the original tidal records whenever we study in detail the characteristics of small tsunamis arriving at a certain station.

According to the list of potentially tsunamigenic events given by Wigen (1977a, 1978), the authors attempted to compile all the tsunamis which arrived at the tide station at Miyako into a single table, by examining all marigrams since 1937 kept at the Miyako Weather Station. This kind of work will surely contribute forward a more thorough catalog of tsunamis in the Pacific Ocean. Using the present data, we found that the relationship between maximum amplitude and frequency of tsunamis at Miyako can be expressed by the formula, that is found for Tofino by Wigen (1977b).

2. Tidal Observation at Miyako

Miyako is located along the Sanriku Coast of Japan where large tsunamis frequently attacked in the past. The tide station was established in 1919 by the Iwate prefectural government as a part of the Miyako Weather Station, but it was destroyed completely by the Sanriku Tsunami in 1933. In 1937 it was reconstructed at the point shown in Fig. 1. This station is in the central part of Miyako Bay, and the tsunami waves observed there are generally smaller in amplitude than those at neighboring stations such as Hachinohe. This fact, pointed out by HATORI (1979), is not essential to the present study, because the disturbance of a tsunami wave by seiche action is also small at this station. The tidal records on strip charts kept at the Miyako Weather Station comprise one of the longest records in Japan, even though the records prior 1934 have been lost.

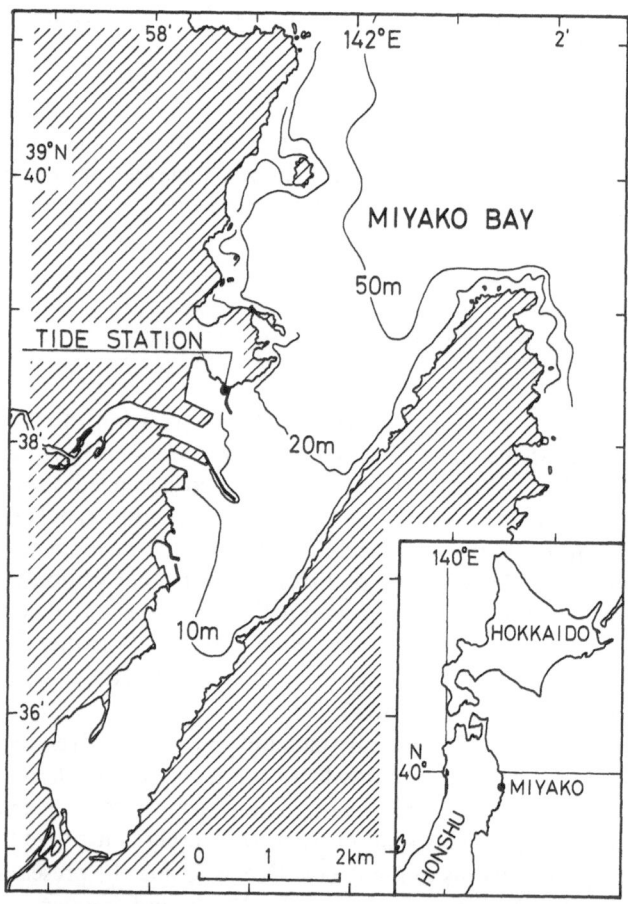

Fig. 1. Location of tide station at Miyako.

3. Method

All tidal records on strip charts were inspected around the estimated time of the tsunami arrival expected from seismic and volcanic events compiled in the catalog by WIGEN (1977a). This catalog contains about 1,000 events from 1937 through 1976. In addition to this catalog, other tsunami catalogs were also used. Figure 2 shows the travel time chart to Miyako, which was used to estimate arrival times in the study. A record of a small tsunami generated in an area far from the station appears similar to the record of a seiche caused by a meteorological disturbance. Therefore, the beginning time of the undulation is an important factor in distinguishing a tsunami from seiches, which is compared with the estimated arrival time expected from a potentially tsunamigenic event.

The records were classified into six categories, namely, (1) tsunami; definite tsunami on record. (2) possible tsunami; possibly a tsunami rather than a seiche caused by meteorological disturbance. (3) possible seiche; possibly a seiche rather than a tsunami. (4) no tsunami; no tsunami on record, or an unrecognizable tsunami due to large disturbances of other causes. (5) unknown; indistinguishable from larger tsunami caused by a main shock. (6) no observation; malfunction of tide gage.

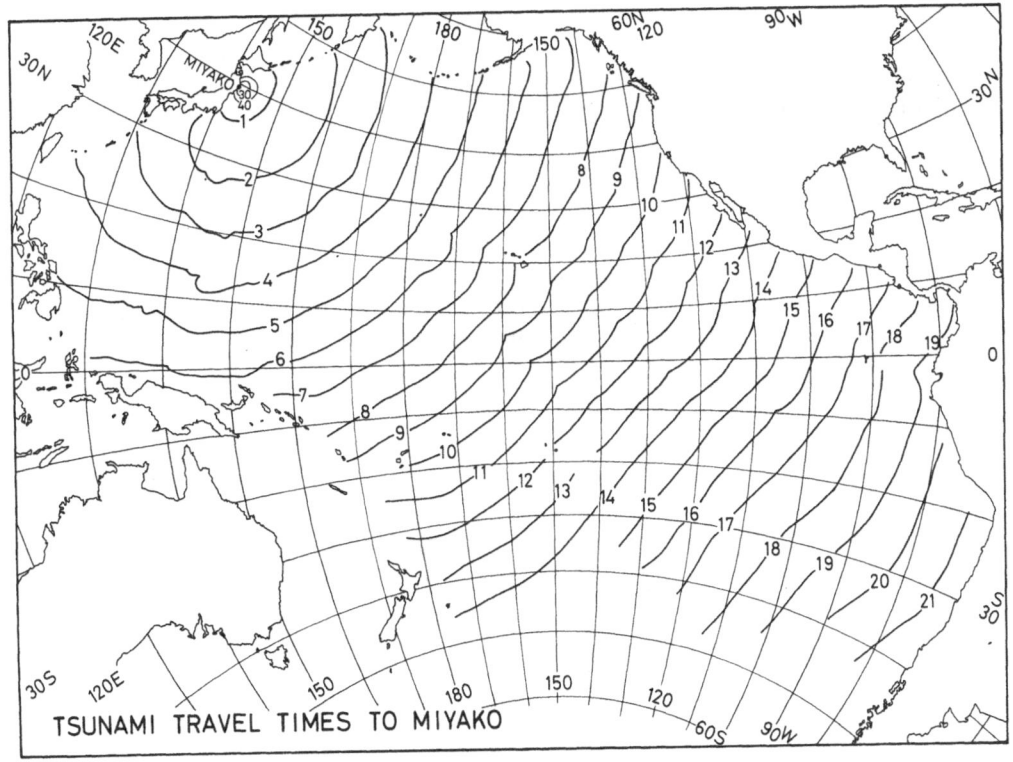

Fig. 2.   Travel time chart in hour to Miyako.

On the records of "tsunami" and "possible tsunami," some characteristics, as shown in Fig. 3, were studied, that is, (1) arrival wave; the time of A, and motions (BC and DE). (2) maximum motion; the successive peak and trough shown by FG and IJ whose sum is the largest in a series of undulations. The time at mean level is shown by H. (3) maximum wave; the time of I, the period shown by FK, and the maximum difference in height from trough to peak shown by IL, from a higher peak or deeper trough to the mean value of the troughs or peaks before and after it.

The characteristics found are listed in Table 1. A + (−) in "motion" and P in "C" mean upward (downward) motion and "possible tsunami," respectively. The "maximum motion" is dependent on the smoothing curve of the tidal record; on the other hand, the "maximum wave" is independent of the curve. Earthquake data were collected from various publication. The numbers of "tsunami" and "possible tsunami" were 33 and 26, respectively. Aside from the table, "possible seiche" were determined for 20 cases. As a tsunamigenic event on April 30 in 1956 was not found except for Wigen's list, it was not numbered.

The largest tsunami was observed on May 16, 1968. It was off scale on the ordinary tide gage, and the characteristics for this event were determined for the record sent to the weather station by radio telemetering. Owing to poor resolution of the record, the values of tsunami are a little less accurate.

Seiching caused by meteorological disturbances occurs far more often than tsunami. Moreover, sometimes this seiching is similar to a tsunami on the record. Some records of seiching beginning near the estimated arrival time might have been misunderstood as "possible tsunami." On the other hand, it is probable that some small tsunamis of arrival were not noticed due to seiching.

4. Frequency of Tsunami at Miyako

Fifty-eight events from 1937 through 1981, except for the one on April 30, 1956, were regarded as the record of "tsunami" or "possible tsunami." At Miyako, tsunamis originating in the Pacific Ocean arrive more frequently than once per year. Sixty per cent of them were generated by earthquakes occurring near Japan.

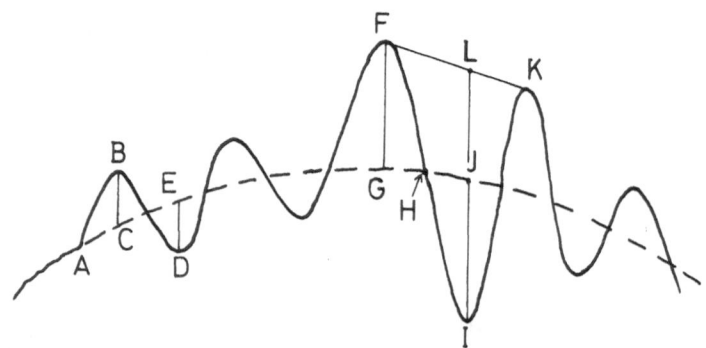

Fig. 3.   Studies of tsunami on tidal record.

Figure 4 shows the frequency distribution of maximum amplitude of tsunamis at Miyako. The abscissa and ordinate are on a logarithmic scale. The linear relation between them is denoted by

$$n(A)dA = kA^{-p}dA, \quad \text{or} \quad N(A > a) = \frac{k}{p-1}a^{-p+1},$$

as found for Tofino in Canada by WIGEN (1977b). Here $n(A)dA$ and $N(A > a)$ indicate those frequencies of tsunamis whose amplitude are from A through $A + dA$, and larger than a, respectively. k and p are constants.

The parameter p was estimated as 2.1 for Miyako, being a little larger than that for Tofino. It should be noted that this relationship is similar to the frequency distribution of the maximum amplitude of earthquake waves at a certain station, namely the formula of ISHIMOTO and IIDA (1939). $p = 2.1$ for tsunami at Miyako is also a little larger than normal value for usual earthquakes, which is from 1.7 through 2.0.

Extrapolating above equation, tsunami larger than 5 m and 10 m in amplitude might attack at the station once per 110 years and 250 years, respectively. At the Sanriku Coast, the greatest tsunamis attacked in 1896 and in 1933. Although traces of these tsunamis are still marked at places higher than 20 m in the heads of bays in this area, the water heights found in the vicinity of the tide station were about 8 m and 3 m, respectively. Historical tsunamis as great as the two above occurred in 869 and in 1611. As disastrous great tsunamis have occurred four times in the past 1,200 years, the estimation based on Wigen's formula seems to be plausible.

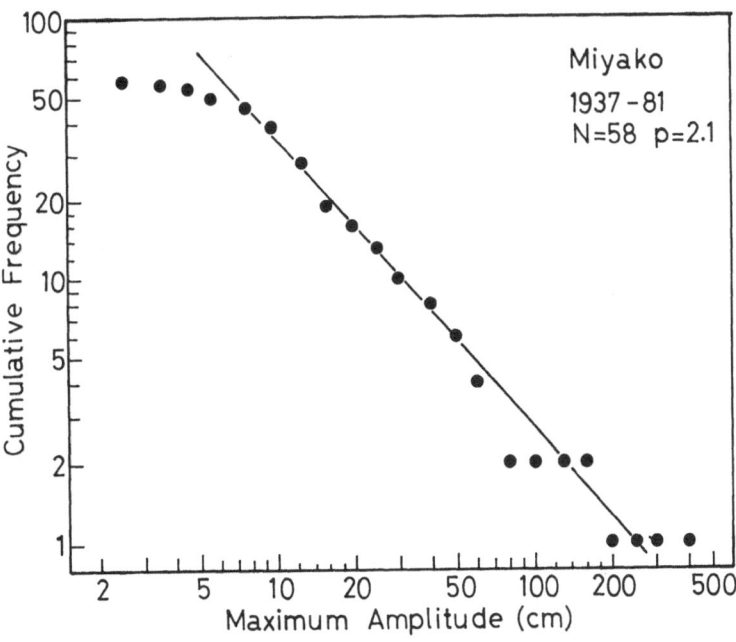

Fig. 4.   Frequency of maximum amplitude of tsunami at Miyako, 1937–1981.

Table 1. Catalog of Tunami Observed at Miyako since 1937

NO	DATE & TIME (Y mo d h m)	EARTHQUAKE DATA — Dep (km)	Mag	lat.	long.	LOCATION	ARRIVAL WAVE time (d h m)	motion	motion	MAX. MOTION time (d h m)	motion	motion	MAX. WAVE time (d h m)	amp (cm)	T	C
1	1938 11 5 0843	20	7.7	37.1N	141.7E	OFF FUKUSHIMA PREF	5 0935	+7	-15	5 1205	-19	+19	5 1210	34	21	P
2	1938 11 6 0854	0	7.5	37.5N	141.8E	OFF FUKUSHIMA PREF	6 0950	-10	+5	6 1055	+10	-8	6 1058	17	15	P
3	1938 11 6 2139	0	7.1	37.0N	141.7E	OFF FUKUSHIMA PREF	-	-	-	7 0145	-5	+3	7 0140	7	16	P
4	1938 11 13 2231	60	6.0	37.0N	141.5E	OFF FUKUSHIMA PREF	-	-	-	14 0240	+6	-5	14 0246	14	17	P
5	1938 11 22 0114	10	6.7	37.0N	141.8E	OFF FUKUSHIMA PREF	-	-	-	22 0515	+2	-4	22 0510	3	15	P
6	1940 5 24 1634	60	8.4	10.5S	77.0W	NEAR COAST OF PERU	-	-	-	25 1833	-7	+7	25 1750	15	22	P
7	1943 3 21 2036	33	7.3	5.8S	152.3E	SOLOMON SEA	-	-	-	22 1710	+6	-6	22 1715	10	25	P
8	1943 4 6 1607	60	7.9	30.8S	72.0W	OFF CENTRAL CHILE	-	-	-	8 0840	-14	+8	8 0849	25	53	P
9	1943 6 13 0512	20	7.1	41.1N	142.7E	OFF SANRIKU DIST	13 0545	+4	-8	13 0555	+4	-8	13 0757	12	15	P
10	1943 12 23 1900	50	7.3	5.5S	153.5E	SOLOMON SEA	24 02--	-	-	24 1030	-10	+9	24 1042	15	43	P
11	1944 12 7 0435	0-30	8.0	33.7N	136.2E	KUMANO-NADA	-	-	-	7 1020	+15	-13	7 1010	27	50	P
12	1945 2 10 0458	30	7.3	40.9N	142.1E	E OFF AOMORI PREF	10 0544	-10	+2	10 0837	+6	-9	10 0842	12	25	P
13	1945 4 14 1735	33	7.0	57.0N	164.0E	E OFF KAMCHATSKA PEN	15 06--	-	-	15 0820	-5	+3	15 0822	6	12	P
14	1946 4 1 1229	33	7.4	52.8N	163.5W	E OFF KAMCHATSKA PEN	-	-	-	2 1020	+17	-14	2 1014	26	23	P
15	1946 12 20 1919	30	8.1	33.0N	135.6E	OFF KII PEN	20 2140	+6	-1	21 0430	-13	+5	21 0423	15	27	P
16	1950 1 30 0056	33	7.0	53.5S	71.5W	S COAST OF CHILE	-	-	-	31 0350	-5	+5	31 0405	8	33	P
17	1951 2 13 2213	33	7.1	56.0N	156.0W	S COAST OF ALASKA PEN	14 0352	+4	-4	14 0710	-3	+6	14 0652	9	52	P
18	1951 3 4 0123	40-50	8.1	42.2N	143.9E	SE OFF HOKKAIDO	4 0209	+63	-37	4 0230	+63	-37	4 0223	73	25	P
19	1952 3 9 1704	0-20	7.0	41.7N	143.5E	SE OFF HOKKAIDO	-	-	-	9 2025	-5	+4	9 2020	8	22	P
20	1952 11 4 1658	30-60	8.25	52.8N	159.5E	SE OFF KAMCHATSKA PEN	4 1910	+27	-24	5 0805	+40	-36	4 2300	73	44	P
21	1953 11 25 1748	40-60	7.5	34.3N	141.8E	OFF CHIBA PREF	25 1859	+2	-3	26 0300	+3	-3	26 0245	5	23	P
22	1955 3 18 0006	33	7.4	54.0N	161.0E	SE COAST OF KAMCHATSKA PEN	18 0345	-3	+3	18 1600	-4	+4	18 1650	8	55	P
23	1956 3 30 0611	-	-	55.0N	160.9E	E COAST OF KAMCHATSKA PEN	30 0915	-4	+2	30 1228	-3	+5	30 1236	8	22	P
‡	1956 4 30 0611	-	-	55.0N	160.9E	E COAST OF KAMCHATSKA PEN	30 0933	+2	-2	30 1307	+3	-4	30 1258	7	43	P
24	1957 3 9 1422	33	8.25	51.3N	175.8W	MID ALEUTIAN IS	9 1850	+4	-5	10 0320	+7	-7	10 0317	13	18	
25	1958 4 7 1804	20	6.5	38.3N	143.8E	OFF MIYAGI PREF	-	-	-	7 1950	+2	-2	8 1913	3	10	P
26	1958 11 6 2258	80	8.25	44.3N	148.5E	NEAR ETOROFU IS	7 0005	+10	-12	7 0250	-12	+8	7 0242	19	23	

‡ ------ Erruption of Mt. Bezymianny ? ------

------ The date of earthquake is doubtfull. ------

No.	Year	Origin (Mo Da Time)	h	M	Lat.	Long.	Region	Obs. 1	Obs. 2	Obs. 3 (max)	
27	1959	1 22 0510	30	6.8	37.6N	142.4E	OFF FUKUSHIMA PREF	—	22 0630 −3 +6	22 0638 8 16	
28	1960	3 20 1707	20	7.5	39.8N	143.5E	OFF SANRIKU DIST	20 1726 +20 −12	20 1739 +20 −12	20 1735 31 22	
29	1960	3 23 0023	20	6.7	39.3N	143.8E	OFF SANRIKU DIST	23 0053 −3 +6	23 0120 −7 +7	23 0142 11 10	
30	1960	5 22 1911	33	8.5	39.5S	74.5W	OFF S CHILE	23 1748 +32(−12)	23 1830 −107 +99	23 1835 173 50	
31	1962	4 12 0053	40	6.8	38.0N	142.8E	OFF FUKUSHIMA PREF	12 0130 +1 −2	12 0735 −2 +2	12 0728 4 26	
32	1963	10 13 0517	20	8.1	43.8N	150.0E	OFF ETOROFU IS	13 0640 +9 −5	13 1030 +7 −5	13 1041 12 55	
33	1964	3 28 0336	20	8.5	61.1N	147.6W	S COAST OF ALASKA	28 1040 +6 −3	29 0030 −9 +9	29 0022 18 50	
34	1965	2 4 0500	40	7.5	51.3N	178.6E	ALEUTIAN IS	4 0748 +7 −3	4 2050 −9 −7	4 2051 15 44	
35	1965	7 2 2059	50	6.7	53.5N	168.0W	E ALEUTIAN IS	3 0210 −2 +1	3 0403 −2 +2	3 0220 4 33	P
36	1965	9 17 1621	40	6.2	36.3N	141.3E	OFF IBARAKI PREF	—	17 1740 +2 −3	17 1743 5 28	P
37	1966	10 17 2142	30	7.5	10.7S	78.6W	OFF PERU	—	19 0333 −7 +7	19 0330 14 20	
38	1968	1 29 1019	30	6.9	43.2N	147.0E	E OFF HOKKAIDO	—	29 1420 −4 +3	29 1415 7 22	
39	1968	5 16 0049	0	7.9	40.7N	143.6E	E OFF N HONSHU	16 0129 +286 −140	16 0348 −282 +215	16 0347 472 25	P

------ Out of scale, maximum values were obtained with teleometering system. ------

No.	Year	Origin (Mo Da Time)	h	M	Lat.	Long.	Region	Obs. 1	Obs. 2	Obs. 3 (max)	
40	1968	5 16 1039	40	7.5	41.4N	142.9E	E OFF AOMORI PREF	—	16 1353 −20 +27	16 1356 48 22	P
41	1968	6 12 1342	0	7.2	39.4N	143.1E	OFF IWATE PREF	12 1405 −3 +35	12 1416 +35 −10	12 1410 44 12	
42	1969	8 11 2128	30	7.8	42.7N	147.6E	E OFF HOKKAIDO	11 2235 +27 −32	11 2247 −32 −32	11 2255 56 22	P
43	1970	5 31 2023	43	7.8	9.2S	78.8W	COAST OF N PERU	1 16-- —	1 1630 +3 −3	1 1635 5 42	P
44	1971	7 14 0619	47	7.9	5.5S	153.9E	NEW IRELAND REGION	—	14 2027 −3 +7	14 2033 10 20	P
45	1971	7 26 0123	48	7.9	4.9S	153.2E	NEW IRELAND REGION	26 09-- —	26 1950 −5 +7	26 2000 12 22	
46	1971	8 2 0725	60	7.0	41.2N	143.7E	S OFF HOKKAIDO	2 0808 −8	2 0818 −8 +8	2 0825 14 23	
47	1972	12 4 1016	50	7.2	33.2N	141.1E	E OFF HACHIJO IS	4 11--	4 1640 +7 −3	4 1638 10 20	P
48	1973	6 17 0355	40	7.4	43.0N	146.0E	OFF NEMURO PEN	17 0448 +31 −13	17 0740 +40 −26	17 0731 64 19	P
49	1973	6 24 0243	30	7.1	43.0N	146.8E	OFF NEMURO PEN	24 0348 +5 −9	24 0359 +5 −9	24 0403 15 25	
50	1975	6 10 1347	0	7.0	42.8N	148.2E	E OFF HOKKAIDO	10 1447 +17 −3	10 1512 +10 −12	10 1523 23 13	
51	1975	11 29 1448	5	7.2	19.3N	155.0W	NEAR HAWAII	29 2330 +2 −4	30 0440 +6 −5	30 0441 12 10	
52	1978	3 24 1948	40	7.3	44.3N	149.8E	NEAR ETOROFU IS	24 2100 +6 −5	24 2110 +6 −5	25 0036 8 25	
53	1978	6 12 0814	40	7.4	38.2N	142.2E	OFF MIYAGI PREF	12 0839 +18 −11	12 1253 −12 +13	12 1245 23 24	
54	1979	2 20 0632	40	6.5	40.5N	143.5E	OFF TOHOKU	20 0720 —	20 0722 +3 −2	20 0725 5 8	P
55	1979	12 12 0759	32	7.9	1.6N	79.4W	NEAR COAST OF EQUADOR	13 0330 −11 +4	13 0947 +13 −8	13 0831 21 15	
56	1980	2 23 0551	30	6.8	43.5N	146.6E	E OFF HOKKAIDO	23 05-- —	23 0727 +3 −4	23 0733 7 25	P
57	1980	7 17 1942	33	7.9	12.5S	165.9E	SANTA CRUZ IS	18 0410 +3 −3	18 0612 +5 −5	18 0648 10 10	P

------ see 'Further Remarks' ------

No.	Year	Origin (Mo Da Time)	h	M	Lat.	Long.	Region	Obs. 1	Obs. 2	Obs. 3 (max)	
58	1981	1 18 1817	0	7.0	38.6N	143.0E	E OFF MID-TOHOKU	18 1850 +9 −3	18 1932 −5 +8	18 1934 12 10	

5. Further Remarks

On July 17, 1980, a large earthquake occurred near the Santa Cruz Islands. About nine hours later, undulations were observed at many stations along the Pacific Coast of Japan. Figures 5 and 6 show the location of several tide stations in Japan and the tidal records at these stations on next day, respectively. Although these undulations look like a tsunami, it was difficult to confirm them in situ. This is because we had neither recieved the information on the tsunami occurrence from the International Tsunami Information Center, nor found a precedent case of tsunami arriving to the coast of Japan from that region. Therefore, we had to carefully study all the available information.

It was concluded that these undulations were tsunami waves generated by the earthquake which occurred near the Santa Cruz Islands, based on following fact:

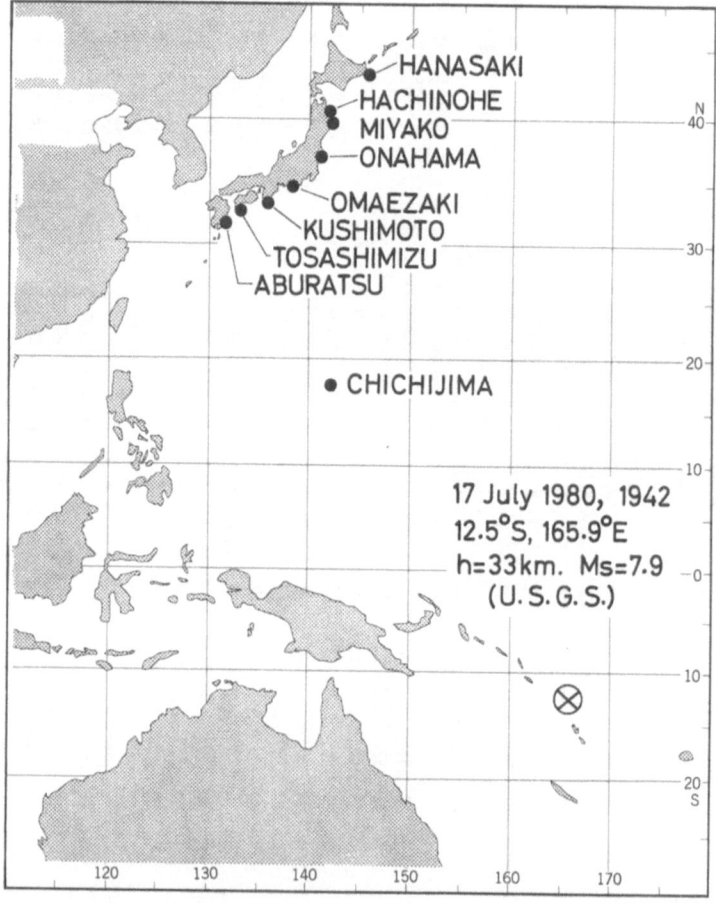

Fig. 5.   Location of tide stations and earthquake epicenter.

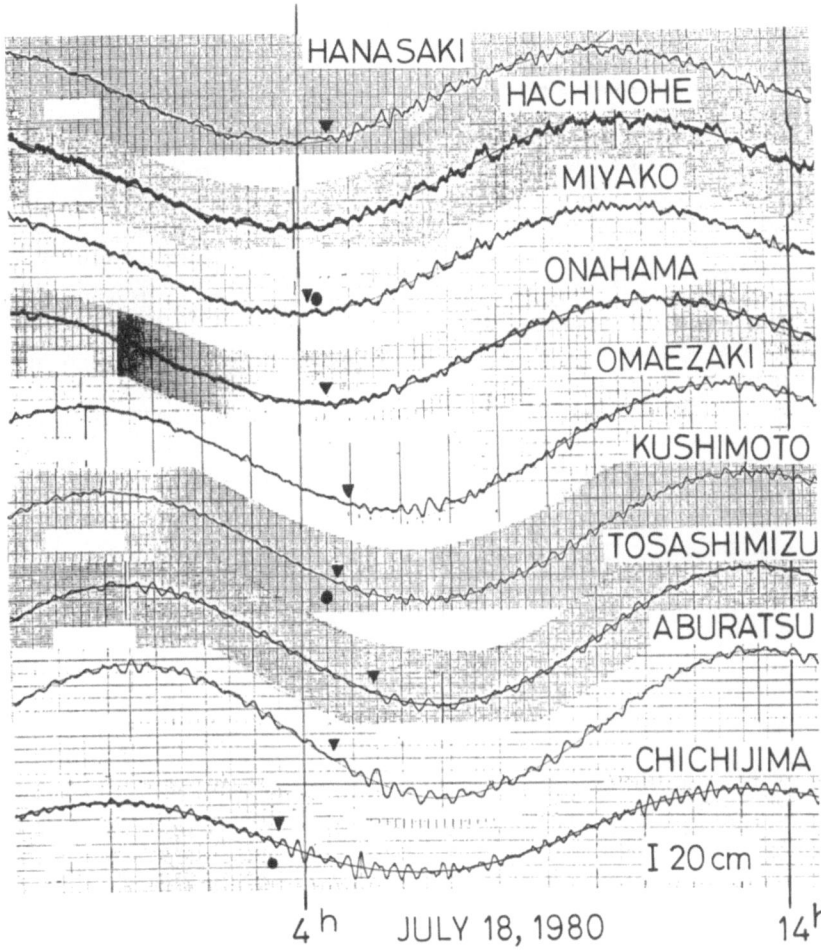

Fig. 6.   Tidal record at the stations shown in Fig. 1, July 18, 1980. Triangles show the beginning time of tsunami, and circles mean the estimated arrival time.

(1)   The closed circles in Fig. 6 show the estimated arrival time of the tsunami. The undulations began almost at the same time as the estimated arrival time.

(2)   According to the "Preliminary Determination of Epicenters" published by the U.S. Geological Survey, small tsunami generated by this event were observed st some stations in Hawaii.

From this experience, it seems possible that some small tsunamis have not been observed by the present operational tsunami network. Similar considerations were taken to make sure of the existence of tsunami for the earthquakes Nos. 44 and 45, as an observation of tsunamis had not been reported in Japan.

REFERENCES

Hatori, T., Relation between tsunami magnitude and wave energy, *Bull. Earthq. Res. Inst.*, **54**, 531–541, 1979 (in Japanese).

Iida, K., D. C. Cox, and G. Pararas-Carayannis, Preliminary catalog of tsunamis occurring in the Pacific ocean, Hawaii Inst. Geophy. Univ., Hawaii, 1967.

Ishimoto, M. and K. Iida, Observations sur les seismes enregistres par le microseismographe construit dernierement (1), *Bull. Earthq. Res. Inst.*, **17**, 443–478, 1939 (in Japanese).

Sendai District Observatory, Catalog of tsunamis arrived at Tohoku District generated by earthquakes occurring in foreign region, Tohoku Tech. Rep., Vol 100, 1979 (in Japanese).

Sendai District Observatory, Catalog of tsunamis arrived at Tohoku District generated by earthquakes occurring near Japan, 1980 (in Japanese).

Watanabe, H., Descriptive table of tsunamis in and near Japan, *Zisin*, Ser. 2, **21**, 293–313, 1968 (in Japanese).

Wigen, S. O., Historical study of tsunamis—Chronological list of events—, ITIC, 1977a.

Wigen, S. O., Tsunami threat to port Alberni, Prepared for consulting engineers Allan M. McCrae, Victoria, B.C., and Reid Crowther and Partners Ltd., Vancouver, B.C., 1977b.

Wigen, S. O., Historical study of tsunamis—an outline, Inst. Ocean Sci., Patricia bay, Sidney, B.C., 1978.

Tsunamis—Their Science and Engineering, edited by K. Iida and T. Iwasaki, 131–145.

Some Statistics Related to Observed Tsunami Heights along the Coast of Japan

Kinjiro KAJIURA

*Earthquake Research Institute,
University of Tokyo, Tokyo, Japan*

(Received October 20, 1981: Revised December 10, 1981)

Some spatial statistics related to tsunami heights observed along the coast of Japan are made on the basis of data collected during field surveys taken after the occurrence of large tsunamis. For a regional frequency histogram of tsunami heights along a coast of about 300 km, the long-normal distribution is found to be a reasonable first approximation, supporting VAN DORN's result (1965). However, the standard deviation seems to vary by a factor of about 1.5 around a central value of 0.24 (logarithmic scale) depending on the characteristic "period" of the tsunami as well as the topography of the coast. The height ratio of the observed maximum, H_{max}, to the mean, $\tilde{H}$, is about 4. On the other hand, for local statistics in the range of about 20 km of the coast, the height ratio of the maximum, $H_{n,\,max}$, to the mean, $\tilde{H}_n$, is about 2. Statistically speaking, the maximum double amplitude of tide gauge records is close to the run-up height observed in the vicinity, although the individual ratio varies by a factor of two.

The maximum local-mean tsunami height, $\tilde{H}_{n,\,max}$, is related to the moment-magnitude of earthquakes, M_w, by $\log \tilde{H}_{n,\,max} = 0.5\,M_w - 3.30$. The uncertainty factor of estimated $\tilde{H}_{n,\,max}$ for individual tsunami is about 1.5 and the formula fails for very great earthquakes ($M_w > 8.5$).

1. Introduction

In Japan, there are a considerable amount of coast tsunami data (inundation or run-up heights) collected by field survey teams after the occurrence of large tsunamis. The observed values of tsunami height for a single event vary not only regionally but also locally. It is noted that the variation of heights is very large even within a short stretch of the coast, say 10 to 20 km. Besides errors involved in measurements, the diversity of tsunami heights observed along the coast is customarily attributed to various factors such as wave refraction, scattering, resonance and non-linear effects due to topographic irregularities of various scales. Instead of trying a deterministic approach to this problem, a simple statistical description of the wave height diversity is attempted here to express the gross features of tsunamis. A similar study was made by VAN DORN (1965), who proposed a log-normal distribution for the occurrence frequency of observed tsunami heights along a coast.

On the basis of this kind of statistical knowledge we may be able to attain a better understanding of the relationship between the "maximum" tsunami height observed along the coast and the earthquake magnitude (the moment-magnitude) and also establish better criteria for designing protective measures against tsunamis. Local tsunami risk analysis, which has been up to the present mainly concerned with probabilistic statements of the maximum tsunami height with respect to time at a specific location, may also be improved by a knowledge of the spatial statistics.

2. Sampling Problems Related to Run-Up Heights

When undertaking a statistical analysis of run-up data, we are immediately faced with the problems of data quality and sampling. In most field surveys of tsunamis, the coastal areas covered by survey teams are limited within the region where tsunami heights are significantly large: the peripheral regions with an average tsunami height less than about 0.5 m are seldom covered unless some special interest exists. Even in the vicinity of the central area of tsunami attack, data sampling is very dense in the region of populated and relatively flat lands of bay coasts, whereas remote locations posing difficult access such as a rocky open sea coast are poorly sampled. The nature of this data sampling causes difficulty in the interpretation of statistics.

Other problems are: 1) different criteria for the height measurements, and 2) accuracy of the measurements. Some survey groups appear to choose water marks which represent more or less the mean height in the locality while rejecting exceptional values. Other groups prefer to choose the highest mark in the locality even if it might be a result of water-splash. Furthermore, because of the difficulty of finding a reference level, say, the position of a nearby tidal level at the time of measurement or a geodetic bench mark, the accuracy of height measurements varies considerably depending on the local situation as well as the survey team. For some old records, even the reference levels relative to which the observed height values were reduced were not clearly stated. For more recent tsunamis, the height values are usually referred to T.P. (the reference level of land maps in Japan which is close to the local mean sea level at Tokyo: Tokyo Peil).

If the data taken by various survey teams are combined, the problem of redundant data arises (the same water marks are likely to be measured repeatedly by different survey teams when water marks are very distinct). There may be other problems to be considered in the interpretation of a statistical description.

Examples of the variation in data sets taken by different survey teams in the same area for the same tsunami are shown in Fig. 1. For the 1896 Sanriku tsunami, the overlapping area of interest is the coast of Iwate Prefecture about 200 km long from Hachinohe to Hirota Bay, where the data listed in the "Records of Tsunami Disasters" (Civil Engineering Section of Iwate Prefecture, 1936; number of samples 84) are compared with those listed in the report by KUNITOMI (1933). Kunitomi used a combination of data by IKI (1897) and those obtained by the local weather stations (number of samples 91). For the 1933 Sanriku tsunami, the data collected by survey teams of the Central Meteorological Observatory and local weather stations (KUNITOMI, 1933; number of samples 201) and those collected by a survey team of the

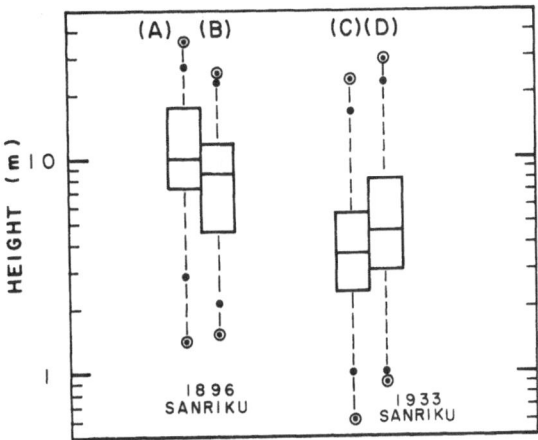

Fig. 1.   Comparison of run-up data in the same region taken by different survey teams. (A) and (B) are the data by Iwate Prefecture (1936) and KUNITOMI (1933), respectively for the 1896 Sanriku tsunami. (C) and (D) are the data by KUNITOMI (1933) and EARTHQ. RES. INST. (1934), respectively for the 1933 Sanriku tsunami. A box represents the height range of 1/2 of all data with the median value indicated by a bar. The maximum and minimum values (◉) and the second maximum or minimum values (●) are also shown.

Earthquake Research Institute (EARTHQ. RES. INST., 1934; number of samples 523) are compared for the whole Sanriku coast of about 300 km length south of Hachinohe.

The presentation is in the form of box plots, where the box gives values between upper and lower quartiles of data with the horizontal bar as a median value. The maximum and minimum values are shown by circles with dots and the second maximum and minimum by black circles. As seen in Fig. 1, the median values differ considerably indicating the bias of different sampling, and this difference can not be attributed to the statistical variability of random sampling from the same population.

3.   Statistics of Run-Up Measurements

VAN DORN (1965) presented a spatial distribution of observed tsunami run-up heights around the Hawaiian Islands. He placed together data of two different tsunamis of 1946 and 1957 observed around the islands of Kauai, Oahu, and Hawaii, by representing the coastal locations in terms of the azimuth measured from the local meridian passing through the center of each island. From the frequency histogram of these data (samples 254), he proposed a log-normal distribution with the standard deviation, σ_*, found equal to 1.8 (log $\sigma_* = 0.255$). A similar analysis for the 1933 Sanriku tsunami also yielded the identical standard deviation.

In the present paper, six tsunamis observed in Japan with sufficient data available are discussed; they are the 1896 Sanriku, 1933 Sanriku, 1946 Nankaido, 1960 Chile, 1964 Niigata, and 1968 Tokachi-oki tsunamis. Except for 1946 and 1964 tsunamis, run-up data determined are all for the Sanriku coast in the north-eastern part of Japan. The 1946 tsunami attacked the region from Kii to Shikoku in the southwestern part of Japan, and the 1964 tsunami occurred off the Niigata coast in the Japan Sea. The

Table 1. List of tsunami data.

Date	Region	Epicenter	M	M_w	N	H	$\hat{H}$	$\hat{H}_{n,max}$	H_{max}	H_{min}	SDT	c_3	c_4	Data source
1896 VI 15	Sanriku	144.2°E 39.6°N	7.7	8.5	248	7.0	6.7	17.0	32.6	0.9	0.318	−0.159	2.26	(1), (2), (3)
1933 III 3	Sanriku	144.7°E 39.1°N	8.3	8.4	724	4.5	4.5	11.0	28.7	0.6	0.284	0.067	2.64	(3), (4)
1946 XII 21	Nankaido	135.6°E 33.0°N	8.1	8.1	164	3.2	3.1	4.5	6.8	0.9	0.179	0.028	2.53	(5), (6), (7) (8), (9)
1960 V 23	Chile	73.5°W 38.0°S	8.3	9.5	669	3.3	3.2	4.6	6.4	0.7	0.142	−0.656	3.56	(10)
1964 VI 16	Niigata	139.2°E 38.4°N	7.5	7.6	111	1.6	1.7	3.5	6.4	0.6	0.257	0.187	2.18	(11)
1968 V 16	Tokachi-oki	143.6°E 40.7°N	7.9	8.2	508	2.6	2.6	4.5	6.8	0.4	0.195	−0.554	3.64	(12)
									Synthetic histogram		0.239	−0.083	3.08	

The format of the table is specified as follows: Date, Japan standard time; M, JMA magnitude of earthquake; M_w, moment magnitude of earthquake defined by KANAMORI (1977); N, number of run-up data; H, median value of tsunami height in m; $\hat{H}$, mean (logarithmic) value of tsunami height in m; $\hat{H}_n$, mean value of the local mean tsunami height $\hat{H}_n$ in m; H_{max}, maximum value of observed tsunami heights in the region under consideration in m; H_{min}, minimum value of observed tsunami heights in the region under consideration in m; SDT, standard deviation in terms of the logarithmic scale in height values; c_3, skewness parameter in terms of the logarithmic scale in height values; c_4, kurtosis parameter in terms of the logarithmic scale in height values.

Data Source: (1) SECTION OF CIVIL ENGINEERING, IWATE PREF. (1936); (2) IKI (1897); (3) KUNTOMI (1933); (4) EARTHQ. RES. INST. (1934); (5) EARTHQ. RES. INST. (1947); (6) HYDRO. DEP'T OF JAPAN (1948); (7) CENTRAL MET. OBS. (1947); (8) YOSHIDA et al. (1947); (9) YOSHIDA (1947); (10) COMMITTEE FOR FIELD INVESTIGATION, CHILEAN TSUNAMI (1961); (11) AIDA et al. (1964); (12) KISHI (1969).

height values for the 1968 tsunamis are corrected by adding 75 cm to give the height from the low tide level, because this tsunami occurred at low tide. For all other tsunamis, no correction is made. The relevant information concerning earthquakes and tsunamis to be discussed is listed in Table 1.

3.1 Regional statistics

For each tsunami, data obtained by different survey teams are lumped together for the total range of about 300 km of the coastal stretch where major tsunamis were experienced: for the 1896, 1933 1960, 1968 tsunamis, the region is from Hachinohe in Aomori Prefecture to Ishinomaki in Miyagi Prefecture, for the 1946 tsunami, the region is from Shingu in Wakayama Prefecture to Sukumo in Kochi Prefecture, and for the 1965 tsunami, the region is from Funakawa in Akita Prefecture to Naoetsu in Niigata Prefecture, including Sado Island.

Frequency histograms of run-up or inundation heights, H, are presented in Fig. 2, where the abscissa is expressed in terms of percentage of occurrence, and the ordinate is in the logarithmic scale of the observed value, H, normalized by the logarithmic mean value, $\bar{H}$. The histograms for the 1896 and 1933 tsunamis are flatter than the normal distribution, for the 1960 and 1968 tsunamis they are skewed to the right considerably, and for the 1964 tsunami a small secondary peak is seen in the histogram. It is quite clear from these frequency histograms that the difference in the sample distributions can not be explained simply as a result of random sampling from the same normal population.

However, if we combine all these frequency distributions with equal weight, the resultant distribution is closely approximated by a log-normal distribution only slightly skewed to the right as shown in Fig. 3, suggesting that the type of frequency

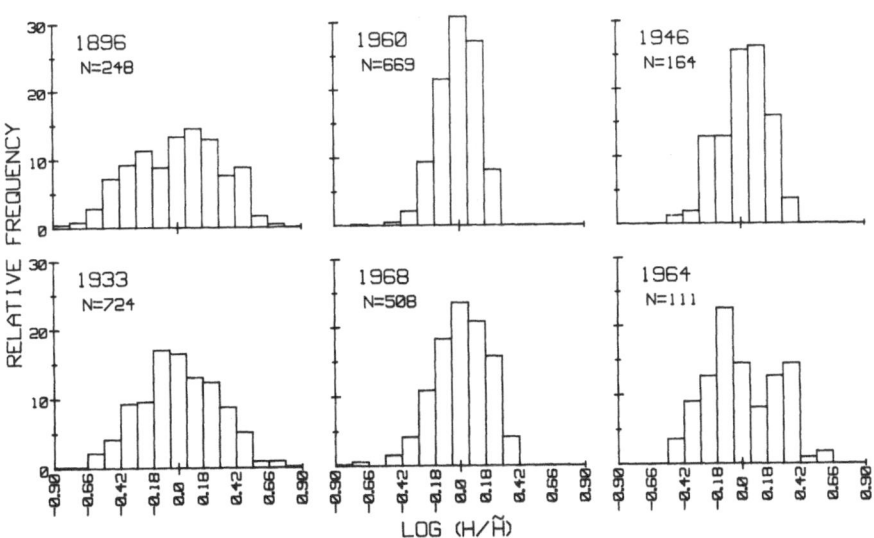

Fig. 2.   Histograms of tsunami run-up heights observed in the range of 300 km of the coast.

distribution for an individual tsunami deviates considerably from the log-normal one, but that there is no systematic deviation of parameters as a whole.

As shown in Table 1, the standard deviations, SDT, (logarithmic scale in height) for the six tsunamis are within a range of 0.14 to 0.32 and the average value, $\langle SDT \rangle$, for all data combined is 0.24. This average value favorably compares with the Van Dorn's estimate ($\simeq 0.26$).

The skewness, c_3, varies between -0.66 to 0.07, and the kurtosis, c_4, between 2.2 to 3.6 (logarithmic scale in height). There appears to be a tendency for c_4 to increase with decrease of SDT. The parameter values for all data combined are $c_3 = -0.08$ and $c_4 = 3.08$. These values are very close to the normal distribution for which $c_3 = 0$ and $c_4 = 3.0$.

Even taking uncertainty of parameter values due to biased sampling into consideration, we can recognize a distinct difference of the values of SDT, c_3 and c_4 between the data for the 1896 and 1933 tsunamis and for the 1960 and 1968 tsunamis, although the data were sampled from the same geographical region of the Sanriku coast. The former group has larger SDT values than the latter. This difference may be attributed to the physical difference of the tsunami characteristics such that the former group had shorter predominant periods of incoming waves (10 to 15 minutes) than those of the latter group (more than 25 minutes), and that the bay responses to incoming tsunamis were quite different.

Regional maximum and minimum values   Since the frequency distribution for an individual tsunami deviates from log-normal, it is difficult to discuss the statistical characteristics of extreme heights such as the maximum, H_{max}, and the minimum, H_{min}. As a tentative attempt, the values of log $(H_{max}/\tilde{H})/SDT$ and log $(H_{max}/H_{min})/(2SDT)$ are plotted in Fig. 4 as a function of the number of total samples. For reference, the theoretical probabilities of standardized normal extremes are also shown in Fig. 4, where the thick line gives the mode and thin lines represent lines below which 10 percent and 90 percent of sample values of the extremes fall respectively (GUMBEL, 1958: Graph 4.2.2.(2)). In the figure, data for the 1923 Kanto, 1944 Tonankai, 1952 Tokachi-oki, and 1968 Hiuganada are also included although the number of sampled data is limited. Relevant data are listed in Table 2.

It is seen that all but one of the maximum values of the tsunami data considered fall

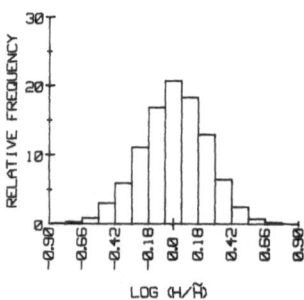

Fig. 3.   A synthetic histogram of tsunami run-up heights for 6 tsunamis in Fig. 2.

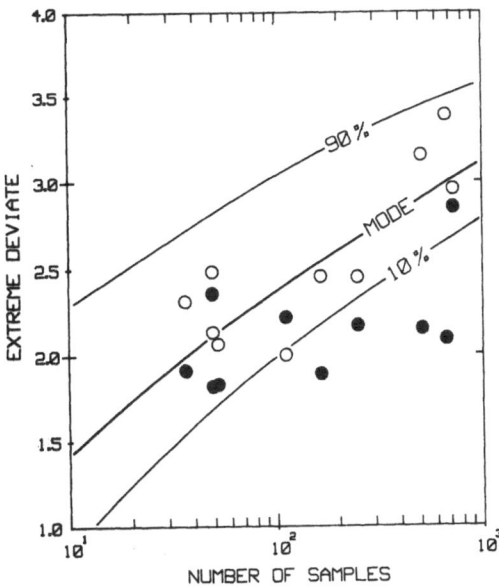

Fig. 4. Dependence of the observed extreme values on the number of samples. The extreme deviates are expressed by log $(H_{max}/\tilde{H})/SDT$ (●) and by log $(H_{max}/H_{min})/2SDT$ (○) with SDT the observed standard deviation (logarithmic scale). A thick line and thin lines are theoretical probabilities of extremes for the normal distribution.

Table 2.   Additional list of tsunami data.

Date	Region	Epicenter	M	M_w	N	SDT	$\tilde{H}$	$\tilde{H}_{n,\,max}$	H_{max}	H_{min}	Data source
1923 IX 1	Kanto	139.3°E 35.2°N	7.9	7.9	49	0.375	2.5	3.9	12.0	0.3	(13), (14)
1944 XII 7	Tonankai	136.2°E 33.7°N	8.0	8.1	52	0.297	2.4	5.4	8.4	0.5	(15), (16), (17)
1952 III 4	Tokachi-oki	143.9°E 42.2°N	8.1	8.1	49	0.224	1.9	3.7	6.5	0.5	(18)
1968 IV 1	Hiuga-nada	132.5°E 32.3°N	7.5	7.4	36	0.362	0.9	1.9	4.6	0.1	(19)

For explanation of the symbols used, see Table 1.
Data source: (13) IKEDA (1925); (14) HATORI *et al.* (1973); (15) CENTRAL MET. OBS. (1945); (16) OMOTE (1946); (17) IIDA (1977); (18) CENTRAL MET. OBS. (1953); (19) KAJIURA *et al.* (1968).

below the median line of the normal extreme because of relatively low values of kurtosis or large values of skewness. If the standard deviation is reduced to about 0.8 to 0.9 of the actual value, all the extreme values might fall in the range expected from the log-normal distribution reasonably well. On the other hand, all the plotted points of log $(H_{max}/H_{min})/(2SDT)$ fall within the 10 and 90 percent lines, although the statistics of this variate obey a slightly different law from the former variate.

If we adopt the normal distribution, the mode of the normalized extreme values for 100 and 500 samples are about 2.4 and 2.9 respectively. Thus, if we assume the standard deviation, σ, to be $0.9 \times SDT$ with $SDT \simeq 0.24$, we have; $H_{max}/\tilde{H} \sim 3.3$ and 4.2 for 100 and 500 samples, respectively. These values may be favorably compared with the range of $H_{max}/\bar{H}$ directly computed from data listed in Tables 1 and 2, where the ratio is from 2.0 to 4.9 with a mean (logarithmic) value of 3.6.

3.2   Local statistics

The height statistics for all observed data of a given tsunami include various physical processes related to the regional as well as the local character. To separate local characteristics from regional ones, we have divided the data for four tsunamis experienced in the Sanriku coast along a stretch of about 300 km from Hachinohe to Ishinomaki into 17 local sets in intervals of about 20 km each as shown in Fig. 5. This smaller length of coast includes roughly one or two units of topographic features such as coastal indentations and embayments. In most segments, the number of height samples for each tsunami varies from about 5 to 70, depending on the topographic features as well as the interest of the survey teams.

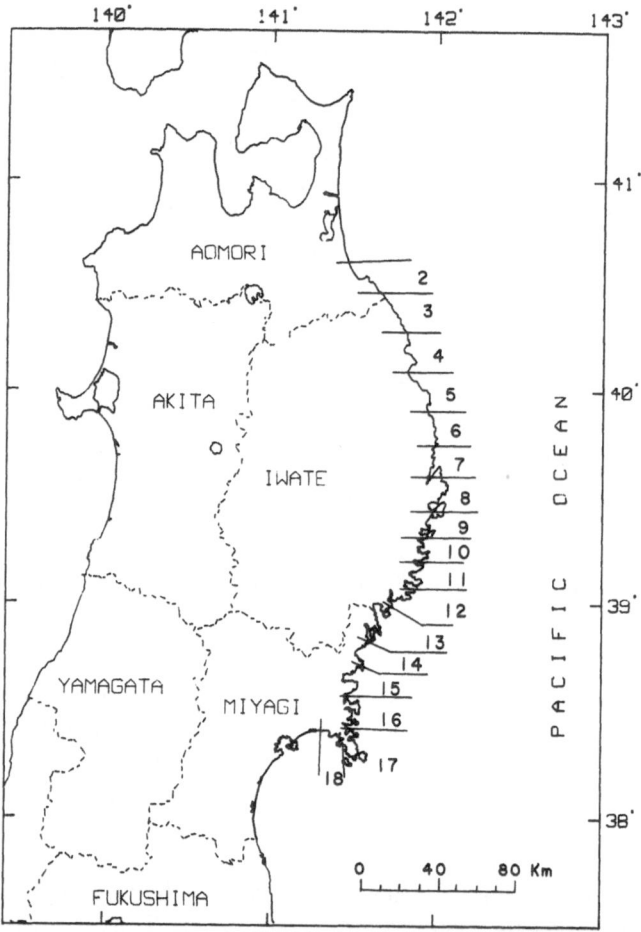

Fig. 5.   Distribution of segments along the Sanriku coast for use in the local statistics of run-up data.

For each tsunami, the distribution of logarithmic mean values, M_n ($=\log \tilde{H}_n$), and the standard deviation, SD_n, therefrom with respect to their locations is shown in Fig. 6. The geographical variations of the mean values, $\tilde{H}_n$, should be better understood by the deterministic approach because they are mainly dictated by the gross coastal topographic features of the location in relation to the tsunami source characteristics (this is evident in view of successful tsunami numerical simulations). The 1896 and 1933 tsunamis show similar variations of $\tilde{H}_n$ in the southern half of the Sanriku coast (segment numbers greater than 5). In the northern region, the mean heights of the 1896 tsunami are larger than those of the 1933 tsunami, suggesting the location of the source area of the 1896 tsunami to be farther north of the source of the 1933 tsunami. In contrast, the 1968 and 1960 tsunamis exhibit quite different distributions. In particular, the geographic variation is small for the 1960 Chilean tsunami because this tsunami propagated across the Pacific Ocean, and the incident waves along the Sanriku coast were believed to be relatively uniform with long periods. These differences of mean height distribution are also reflected in the frequency histograms already shown in Fig. 2.

If we calculate the standard deviations, MSD, of these local mean values, M_n, within the region considered, numerical values fall between 0.08 and 0.26 with the average value, $\langle MSD \rangle$, for all tsunamis being 0.17 as shown in Table 3. Recalling the average standard deviation, $\langle SDT \rangle$, of all data, which was 0.24, it is found that about half of the variance comes from the regional variability of the mean values. The ratios of the maximum, $\tilde{H}_{n, \max}$, of the local mean value $\tilde{H}_n$ to the total mean value $\tilde{H}$, i.e., $\tilde{H}_{n, \max}/\tilde{H}$, are in the range of $1.4 \sim 2.5$ with the mean (logarithmic) value of about 2.0.

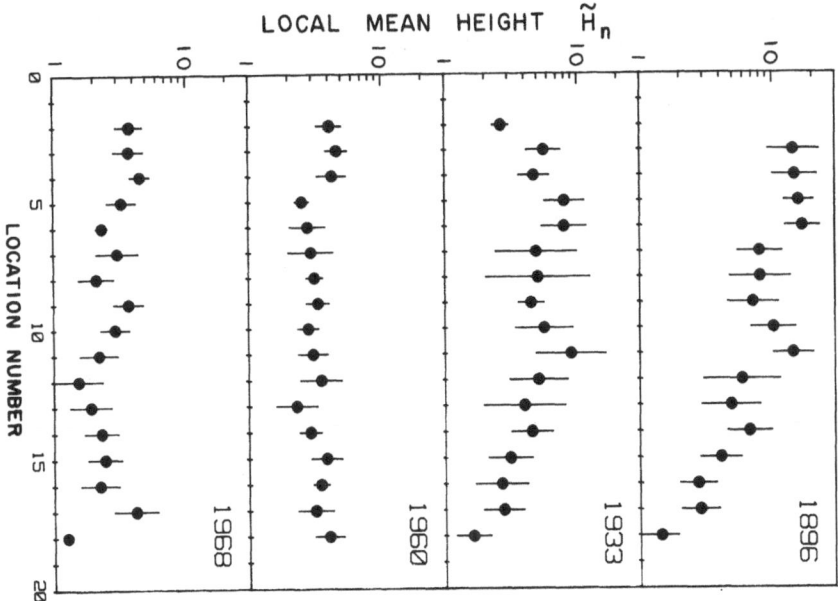

Fig. 6.   Distribution of local mean run-up heights $\tilde{H}_n$ together with their standard deviation SD_n along the Sanriku coast. Absissa is in terms of the segment number shown in Fig. 5.

In Fig. 6, it is seen that the regional variation of the standard deviation, SD_n, is considerable (see also Table 3). The mean value, $\overline{SD}_n$, over the whole region varies in the range of 0.1 and 0.2 depending on the particular tsunami, with an average value over all tsunamis, $\langle \overline{SD}_n \rangle$, of 0.16. Here again $\overline{SD}_n$ is larger for the 1896 and 1933 tsunamis than for the 1960 and 1968 tsunamis, reflecting differences in characteristics between these two groups of tsunamis.

If we normalize individual SD_n by the $\overline{SD}_n$ of each tsunami, $SD_n/\overline{SD}_n$ varies from place to place in the range from 0.6 to 1.5, reflecting the local topographic effects. The variation of $SD_n/\overline{SD}_n$ for various tsunamis at the same location is mostly within 30 percent, but at some specific sites it becomes very large, suggesting different responses of the locations to the incoming tsunamis.

Local maximum and minimum values   The ratio of the maximum, $H_{n,\,max}$, to the mean, $\tilde{H}_n$, of run-up heights (logarithmic scale) within each segment is normalized by the individual standard deviation, SD_n, and is shown in Fig. 7 as a function of a number of samples. The thick and thin lines drawn in the figure are the same as in Fig. 4. It is noted that the plotted points scatter over a wide range, although centered very roughly on the mode of extremes for the normal distribution. On the other hand, the observed ratios of maximum $H_{n.\,max}$, and minimum, $H_{n.\,min}$, (logarithmic scale) normalized by $2SD$ (not shown here) lie within the range of the 10 and 90 percent lines for the normalized extreme deviate based on the log-normal distribution. This character is similar to the case of regional statistics (see Fig. 4).

Although the values of SD_n vary over a wide range, let us tentatively assume the log-normal distribution with the local standard deviation SD_n to be 0.16 as a model of the local run-up height frequency. Then, since the mode of the normalized extreme values for about 30 samples is 2.0, we have, $H_{n,\,max}/\tilde{H}_n \sim 2.0$. On the other hand, since $\tilde{H}_{n,\,max}/\tilde{H} \sim 2.0$ as already mentioned, we have, $H_{max}/\tilde{H} \sim 4.0$, for the total regional data, which is consistent with the result of the regional statistics.

In passing, it is noted that the ratio, $H_{n,\,max}/\tilde{H}_n$ (~ 2.0), here derived is considerably larger than that suggested by SOLOVIEV (1978) who indicated that $H_{max}/H_{av} \sim \sqrt{2}$, with H_{av} representing average values of wave run-up over a certain

Table 3.   Regional and local statistics of run-up heights along the Sanriku coast.

Tsunami name	$\log \tilde{H}$	$\log \tilde{H}_{n,\,max}$	$\log (\tilde{H}_{n,\,max}/\tilde{H})$	MSD	$\overline{SD}_n$	Range of SD_n
1896 Sanriku	0.824	1.23	0.406	0.257	0.176	0.11–0.29
1933 Sanriku	0.649	1.04	0.391	0.153	0.204	0.10–0.40
1960 Chile	0.510	0.664	0.154	0.078	0.106	0.05–0.17
1968 Tokachi-oki	0.411	0.653	0.242	0.134	0.133	0.07–0.18
		Mean value	0.298	0.168*	0.160*	

*Computed from the mean variance.

$\tilde{H}$, regional mean value of run-up height H in m; $\tilde{H}_{n,\,max}$, maximum of the local mean run-up height $\tilde{H}_n$ in m; MSD, standard deviation of $\log \tilde{H}_n$; SD_n, standard deviation of $\log H$ in the nth local segment; $\overline{SD}_n$, regional mean value of SD_n.

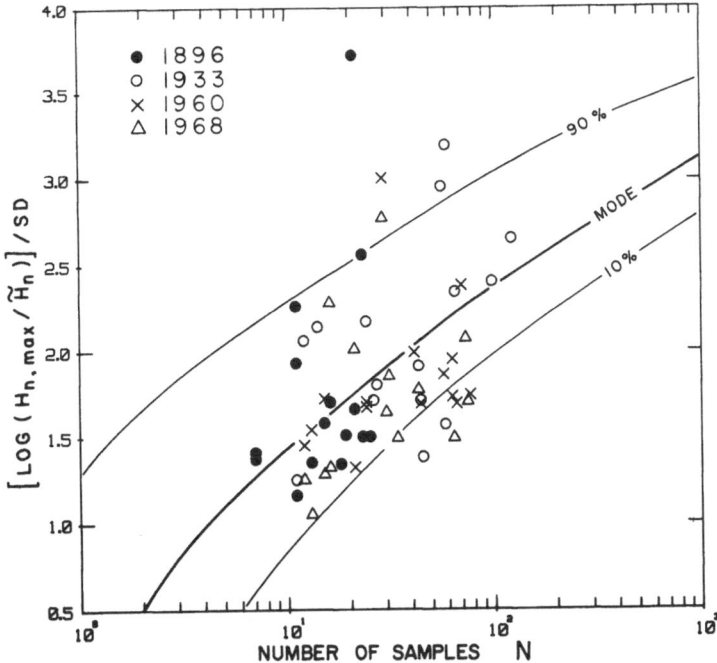

Fig. 7. Dependence of the normalized observed local extreme values on the number of samples. Symbols ●, ○, ×, △ correspond to data for 1896, 1933 1960, and 1968 tsunamis, respectively.

length of the coast (not mentioned explicitly).

4. Relation of Field Data and Tide Gauge Data

In an ideal case when a tide gauge registers the water level outside of the tide-well faithfully without any filtering effects, the maximum water level at the site should be equal to the maximum crest height recorded on the gauge relative to the same reference level. In reality, the maximum crest heights of tide records (very often represented by the maximum value of the crest elevation relative to the ordinary tide level) are almost always smaller than the observed inundation heights measured in the vicinity. In practice, the maximum wave height, A, (crest to trough distance) measured by a tide gauge is often used to represent a tsunami height at the coast. However, the relation between A and the inundation heights, H, measured by field survey has not been studied systematically because data sets are scarce. WILSON (1972) quoted an opinion of Cox that H would be from 1 to 1.5 times the maximum height, A, recorded on a tide gauge in the vicinity, this factor being a function of location, period of waves, and damping characteristics of the tide gauge.

This relation is examined here on the basis of available data for four tsunamis: the 1933, 1952, 1960, and 1968 tsunamis observed on the Hokkaido and Sanriku coasts, although the compared run-up heights were not measured at the exact site of the tide

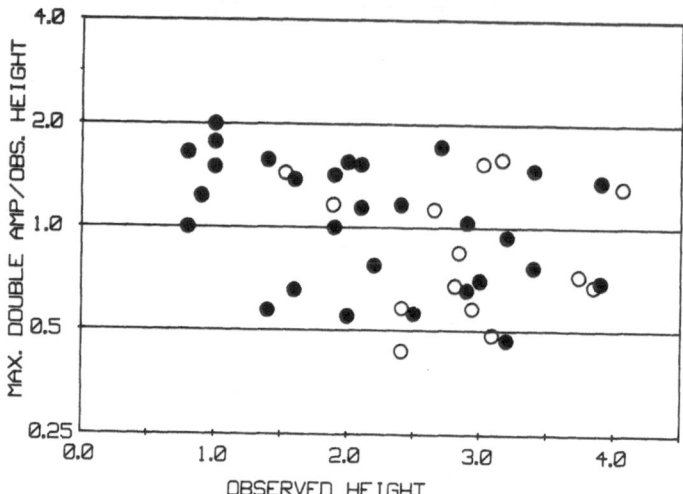

Fig. 8.   The relation between the maximum double amplitude A (crest to trough height) measured by a tide gauge and the run-up height H or the local mean height $\tilde{H}_n$. Ratios A/H and $A/\tilde{H}_n$ are shown by symbols ● and ○, respectively.

gauge. In Fig. 8, we can find that A/H lies between 0.5 and 2 with a mean (logarithmic) value of approximately 1.0. Thus, the wave height, A, measured by a tide gauge can be considered to represent the run-up height, H, measured in the vicinity. Furthermore, the value of A might be used in place of $\tilde{H}_n$ in a rough estimation as shown also in Fig. 8.

5.   Dependence of Coastal Tsunami Heights on the Moment-Magnitude of Earthquakes

The relationship between the "maximum" tsunami height, H'_{max}, (or the tsunami "magnitude") and the earthquake magnitude, M, has been studied by various investigators (see, for example, IIDA, 1963; WILSON, 1972). Since it is believed that the moment-magnitude, M_w, of earthquakes (KANAMORI, 1977) is better suited for correlation with the size of the tsunami than M, the correlations of the observed maximum, H_{max}, maximum local mean, $\tilde{H}_{n, max}$, or regional mean, $\tilde{H}$, of run-up heights to the moment-magnitude, M_w, of earthquakes are shown in Fig. 9. In this figure, some data are added (since 1920) for the observed maximum run-up heights of large tsunamis observed outside of Japan (IIDA et al., 1967).

The full line in the figure is the relationship given by

$$\log \tilde{H}_{n, max} = 0.5\, M_w - 3.30. \qquad (1)$$

It is interesting to notice that the above formula is nothing but the relation assumed by ABE (1981) in his Fig. 8 between the upper limit of the tsunami height (the maximum double amplitude of the tide gauge) and the moment-magnitude M_w. In other words, the limiting tsunami height assumed by ABE (1981) corresponds to $\tilde{H}_{n, max}$. Dashed lines in the figure are drawn by taking the empirical relations such that

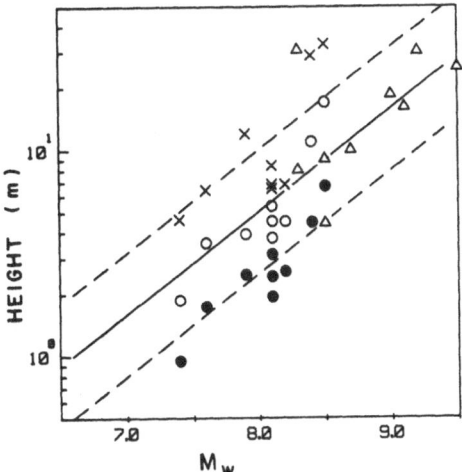

Fig. 9. Relationships between various tsunami height values (the observed maximum ×, maximum local mean ○, regional mean ●) and the moment-magnitude of earthquake M_w. The data of the maximum tsunami height observed outside of Japan are also shown by a symbol △. For thick and dashed lines, see text.

$H_{max}/\tilde{H}_{n, max} \sim 2.0$ and $\tilde{H}_{n, max}/\tilde{H} \sim 2.0$ into account (the constant term in (1) becomes 3.00 or 3.60, respectively).

As already discussed, the observed maximum run-up height, H_{max}, for a given number of samples scatter statistically over a considerable range, so that application of the formula should be made with caution. A rule of thumb may be an uncertainty by a factor of 1.5 for the estimation of maximum run-up height, H_{max}, by this formula. Another notable point is that for great eathquakes with M_w larger than, say, 8.5, the observed maximum values do not follow the trend of the formula and seem to be saturated at about 20 to 30 m. Since the observed maximum heights are rather isolated values and the maximum local mean values are about one half of the observed maximum, it seems that the maximum local mean tsunami height will never exceed 15 m for ordinary tsunamis generated by the tectonic movements due to earthquakes.

It is remarked that the mean slip displacement D of the idealized earthquake fault for a given earthquake moment (ABE, 1975) may be written in terms of M_w as

$$\log D(\text{meter}) \simeq 0.5 \, M_w - 3.63. \tag{2}$$

Compared with (1), it may be said very roughly that $\tilde{H}_{n, max}$ is about two times D, or that D corresponds to the overall mean run-up height $\tilde{H}$ of the tsunami observed along a coastal stretch of about 300 km, at least for data in Japan.

6. Conclusion

Because of the strong dependence of tsunami behavior along the coast on both the characteristics of the incoming tsunami wave train and the topographic conditions of local and regional scales, it is very difficult to draw general conclusions concerning the spatial statistics of coastal wave run-up heights applicable to all regions and tsunamis. However, from the analysis of data in Japan, an idealized statistical model, although very rough, has been constructed. In summary, it was found that: 1) Regional (300 km range) and local (20 km range) frequency distributions of run-up heights are log-normal with the standard deviations 0.24 and 0.16 (logarithmic scale in normalized height), respectively. 2) As a rule of thumb, relations among various height values, such as the observed maximum, H_{max}, the local maximum, $H_{n, max}$, the local mean, $\tilde{H}_n$, the maximum local mean, $\tilde{H}_{n, max}$, and the regional mean $\tilde{H}$ are $H_{n, max}/\tilde{H}_n \sim 2.0$, $\tilde{H}_{n, max}/\tilde{H} \sim 2.0$, and $H_{max}/\tilde{H} \sim 4.0$. 3) The relationship between $\tilde{H}_{n, max}$ and the moment-magnitude M_w, of earthquakes for $M_w \lesssim 8.5$ is roughly represented by Eq. (1). 4) For an individual tsunami, the deviations of observed values from those deduced from this idealized model may lie within a factor of about 1.5.

I wish to thank Miss Kato for her valuable assistance in computation of statistics.

REFERENCES

ABE, K., Reliable estimation of the seismic moment of large earthquakes, *J. Phys. Earth*, **23**, 381–390, 1975.
ABE, K., Physical size of tsunamigenic earthquakes of the northwestern Pacific, *Phys. Earth Planet. Inter.*, **27**, 194–205, 1981.
AIDA, I., K. KAJIURA, T. HATORI, and T. MOMOI, A tsunami accompanying the Niigata earthquake of June 16, 1964, *Bull. Earthq. Res. Inst.*, **42**, 741–780, 1964 (in Japanese).
CENTRAL METEOROLOGICAL OBSERVATORY, General survey of the great Tonankai earthquake on December 7, 1944, Special Pub., CMO, pp. 1–94, 1945 (in Japanese).
CENTRAL METEOROLOGICAL OBSERVATORY, General survey of the great Nankaido earthquake on December 21, 1946, Special Pub., CMO, pp. 1–84, 1947 (in Japanese).
CENTRAL METEOROLOGICAL OBSERVATORY, General survey of the Tokachi-oki earthquake in 1952—Hokkaido, *Quart. J. Seismol.*, **17**, 50–87, 1953 (in Japanese).
COMMITTEE FOR FIELD INVESTIGATION OF THE CHILEAN TSUNAMI OF 1960, *Report on the Chilean Tsunami of May 24, 1960, as Observed along the Coast of Japan*, 397 pp., Maruzen, Japan, 1961.
EARTHQUAKE RESEARCH INSTITUTE, Reports on the Sanriku Tsunami of 1933, Bull. Earthq. Res. Inst. Supplementary volume 1, Figs. 98–209, 1934.
EARTHQUAKE RESEARCH INSTITUTE, Report of the investigation of the Nankaido earthquake on December 7, 1946, *Special Rep.*, Earthq. Res. Inst., **5**, 1–195, 1947 (in Japanese).
GUMBEL, E. J., *Statistics of Extremes*, 375 pp., Columbia Univ. Press, New York, 1958.
HATORI, T., I. AIDA, and K. KAJIURA, Tsunamis in the south Kanto district, *Pub. for the 50th Anniv. Great Kanto Earthq., 1923*, Earthq. Res. Inst., 57–66, 1973 (in Japanese).
HYDROGRAPHIC DEPARTMENT OF JAPAN, Report on tsunamis at the time of the Nankaido earthquake in 1946, Special issue, *Hydrographic Bulletin*, **201**, 1–39, 1948 (in Japanese).
IIDA, K., Magnitude, energy and generation mechanisms of tsunamis and a catologue of earthquakes associated with tsunamis, in *Proc. Tsunami Meetings Associated 10th Pac. Sci. Congr.*, Honolulu, edited by D.C. Cox, I.U.G.G. Monogr. 24, pp. 7–18, 1963.
IIDA, K., Geographical distribution of seismicity and damage caused by the Tonankai earthquake on December 7, 1944, Report prepared for the committee for disaster prevention, Aichi Prefecture, pp. 1–120, 1977 (in Japanese).
IIDA, K., D. C. COX, and G. PARARAS-CARAYANNIS, Preliminary catalog of tsunamis occurring in the Pacific Ocean, Data Rep. No. 5, HIG–67–10, Hawaii Inst. Geophys., Univ. of Hawaii, 1967.
IKEDA, T., Tsunamis in the Izu and Awa districts, *Pub. Earthq. Inv. Committee-B*, **100**, 97–115, 1925 (in Japanese).

IKI, T., Report of the great Sanriku tsunami, *Pub. Earthq. Inv. Committee-B*, **11**, 5–34, 1897 (in Japanese).

KAJIURA, K., I. AIDA, and T. HATORI, An investigation of the tsunami which accompanied the Hiuganada earthquake of April 1, 1968, *Bull. Earthq. Res. Inst.*, **46**, 1149–1168, 1968 (in Japanese).

KANAMORI, H., The energy release in great earthquakes, *J. Geophys. Res.*, **82**, 2981–2987, 1977.

KISHI, T., Tsunami on May 16, 1968, on the coast of Hokkaido and Tohoku, in *General Report on the Tokachi-Oki Earthquake, 1968*, pp. 207–256, edited and published by the Inv. Committee of the Tokachi-oki earthquake 1968, Hokkaido University, 1969 (in Japanese).

KUNITOMI, S., On the Sanriku earthquake and tsunami, *Quart. J. Seis.*, **7**, 1–43, 1933 (in Japanese).

OMOTE, S., The tsunami, earthquake sea waves, that accompanied the great earthquake on December 7, 1944, *Bull. Earthq. Res. Inst.*, **24**, 31–57, 1946 (in Japanese).

SECTION OF CIVIL ENGINEERINGS, IWATE PREFECTURE, *Records of Tsunami Disasters*, pp. 24–30, published by Iwate Pref., 190 pp., 1936 (in Japanese).

SOLOVIEV, S. L., Tsunamis, in *The Assessment and Mitigation of Earthquake Risk*, (*Natural Hazards, I*), pp. 118–139, UNESCO, 1978.

VAN DORN W. G., Tsunamis, in *Advances in Hydroscience*, edited by Ven Te Chow, **2**, 1–48, Acad. Press, New York and London, 1965.

WILSON, B. and A. TØRUM, Run-up heights of the major tsunami on North American coasts, in *The Great Alaska Earthquake of 1964: Oceanogr. and Coast. Engineering*, pp. 158–180, NAS Pub. 1605, Washington, Nat. Acad. Sci., 1972.

YOSHIDA, K., T. YAMAGIWA, K. KAJIURA, and T. SUZUKI, On the tsunami accompanying the Nankai earthquake, Part 1, *Geophysical Note*, Geophys. Inst., Univ. of Tokyo, **1**, 1–20, 1947 (in Japanese).

YOSHIDA, K., On the tsunami accompanying the Nankai earthquake, Part 2, *Geophysical Note*, Geophys. Inst., Univ. of Tokyo, **1**, 1–15, 1947 (in Japanese).

Tsunamis—Their Science and Engineering, edited by K. Iida and T. Iwasaki, 147–159.

Maximum Entropy Spectral Analysis of Tsunamis along the Mexican Coast, 1957–1979

Antonio J. Sanchez* and Salvador F. Farreras**

Estación de Investigación Oceanográfica, Secretaría de Marina, Ensenada, B. C., México
**Centro de Investigación Científica y Educación Superior de Ensenada (CICESE), Ensenada, B. C., México*

(Received August 24, 1981)

Spectral analysis of 45 tidal gauge records of the Pacific Ocean coast of Mexico, from 11 tsunamis as recorded in 12 tidal stations, was performed by means of Fast Fourier Transform (FFT) and Maximum Entropy Method (MEM).

Results for at least 6 stations, show coincidence in one, two, or three spectral peaks excited by different tsunamis in the same place, as an evidence of local resonant modes of oscillation.

Most of the spectra do not show coincidence in peak locations at different stations for the same tsunami. In the case of Ensenada, Acapulco and Manzanillo, some common peaks appear for each tsunami; and even one of them for four different tsunamis in the three places.

A simple analytical model to evaluate the normal modes of oscillation of the continental shelf was used. Model predicted modes agree reasonably with the common spectral peaks for the three former bays.

Spectra for all the stations inside the Gulf of California show most of the energy concentrated in low frequency bands, rather than the higher ones appearing in the open Pacific Ocean coast stations. It is suggested that the low frequencies excited, correspond to transversal modes of oscillation of the Gulf of California.

1. Introduction

Tsunami effects along the Pacific coast of Mexico have been scarcely studied. Tidal gauges were initially installed in the main ports during 1952, but not all of them have been continuously operating since then. Secretaría de Marina (1973, 1974a, b, c) published some of the tidal records containing tsunami evidences. Parameters like arrival time of the first and second waves, and maximum rise or fall, in some of these tidal stations for November 1952, March 1957, May 1960, May 1962, and March 1964 tsunamis, were given by Zerbe (1953), Salsman (1959), Symons and Zetler (1960), Merino y Coronado et al. (1962), and Spaeth and Berkman (1967), respectively.

Energy density spectra for July 1957 tsunami in Acapulco and Salina Cruz bays, and May 1960, March 1964 tsunamis in Ensenada, were obtained using Blackman and Tukey (1958) method, by Munk Cepeda (1961), and Raichlen (1970), respectively. Amplitude spectra for May 1960 and March 1964 in Ensenada bay, and

May 1960 in Acapulco bay, were calculated through the Goertzel method, by Abe and Ishii (1981).

Maximum initial rise recorded since 1952, among all the tidal stations, was 1.43 m at Ensenada, for 1964 tsunami. Only minor damages and inundations have been reported, and for the May 1960 and March 1964 tsunamis solely.

In this work, a search was made, as exhaustive as possible, for tsunami evidences in all available analog tidal records from the Pacific Ocean coast of Mexico, from 1952 to 1979, attempting to identify through spectral analysis the main characteristics and similarities, and intending a reasonable explanation.

2. Raw Data Process

Following Pararas-Carayannis (1977), and Wigen (1977) catalogs as a guide, 65 tidal records from the period 1952–1979 were identified as tsunami event holders. However, only 45 of them, from 11 tsunamis as recorded in 12 different tidal stations, during the period 1957 to 1979 (Fig. 1), were in good enough conditions to perform spectral analysis. Four stations are inside the Gulf of California, and the rest of them in bays along the open ocean coast.

Only two of the recorded tsunamis were from a local source (the Middle American Trench), the rest came from distant areas close to Alaska, the Aleutian Islands, Japan, Hawaii, New Zealand, and Chile. No one tsunami with origin in between South of Japan–North of New Zealand was recorded; it is believed that their energy was dissipated through a trapping wave interaction process with the numerous Oceania

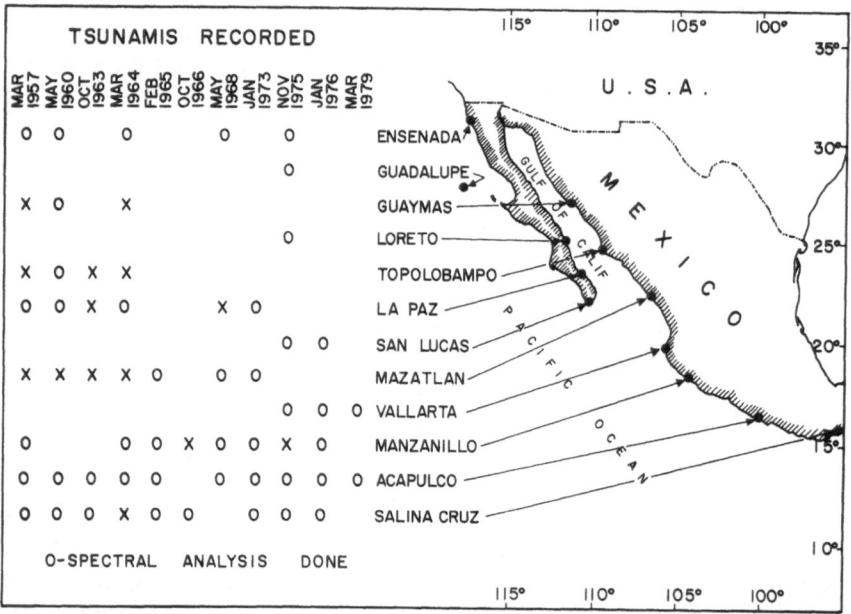

Fig. 1. Tsunami inventory and tidal station locations along Pacific Ocean coast of Mexico.

islands and atolls, a phenomena already mentioned by SHEPARD *et al.* (1950), REID (1979), Bernard, Olsen and Vastano, as quoted in VAN DORN (1979), and LOOMIS (1978).

After tide was filtered out, records were smoothed, and digitized every 5 minute intervals (Nyquist frequency = 0.1 cycles/min) for a total length of 21.33 hours (256 data points).

Spectral analysis, by means of Fast Fourier Transform, was firstly used to locate the regions where the main energy peaks were concentrated. This was accomplished keeping resolution purposely low to get high stability. Afterwards, Maximum Entropy Method, a spectral analysis technique developed by BURG (1967), claimed to have a high resolution for short-length records (ULRYCH, 1972), was applied to determine the peak frequencies in the energy regions previously defined. Sixty bands for Fast Fourier Transform, with a resolution of 0.0017 cycles/min, and 300 bands for Maximum Entropy Method, were used.

3. Maximum Entropy Spectral Computations

The concept of Entropy as a variable proportional to the information contained in a given system, was firstly introduced by SHANNON (1948) in the theory of communications.

In an aleatory process, a "white noise" series has the maximum possible entropy, or maximum information contained. For a given time series, it is always possible to find another series, with the maximum aleatory character consistent with the initial data, and the same entropy as the original series, which will contain the maximum possible information. It is called Maximum Entropy Series, or error series, and can be obtained from the original x_t series through

$$y_t = \sum_k a_k x_{t-k} \tag{1}$$

where a_k are weight coefficients of an error prediction filter. The entropy change between both series is (BURG, 1967)

$$\delta H = \int_{-f_N}^{f_N} \log \left| \sum_{-\infty}^{\infty} a_k e^{-2\pi j f k \delta t} \right|^2 df \tag{2}$$

where f = frequency, δt = sampling interval, and f_N = Nyquist frequency.

To keep the entropy constant, either the above integral can be extremized through a variational formulation; or the difference between a predicted output series for the time $t + 1$, and the original series, can be minimized through the Wiener least square error procedure, making:

$$\frac{\partial I}{\partial a_j} = 0 \tag{3}$$

where

$$I = \sum_t \left(\sum_k a_k x_{t-k} - x_{t+1} \right)^2 \tag{4}$$

or it is obtained

$$\sum_k a_k \theta_{r-k} = \theta_{r+1} \quad r = 0, 1, \ldots, M \tag{5}$$

where M = number of filter elements, to be determined later, and θ is the auto correlation of I. Equation (5) can be written in a Toeplitz matrix form, as

$$\begin{bmatrix} \theta(0)\ \theta(1)\ \ldots\ \theta(M) \\ \theta(1) \\ \vdots \\ \theta(M)\ \ldots\ \theta(0) \end{bmatrix} \begin{bmatrix} a_0 \\ a_1 \\ \vdots \\ a_M \end{bmatrix} = \begin{bmatrix} \theta(1) \\ \theta(2) \\ \vdots \\ \theta(M+1) \end{bmatrix} \tag{6}$$

or, after defining the vector filter λ as

$$\lambda = \begin{bmatrix} 1 \\ \lambda_2 \\ \lambda_3 \\ \vdots \\ \lambda_{M+1} \end{bmatrix} = \begin{bmatrix} 1 \\ -a_0 \\ -a_1 \\ \vdots \\ -a_M \end{bmatrix}. \tag{7}$$

Equation (6) is equivalent to

$$\begin{bmatrix} \theta(0)\ \theta(1)\ \ldots\ \theta(M) \\ \theta(1) \\ \vdots \\ \theta(M)\ \ldots\ \theta(0) \end{bmatrix} \begin{bmatrix} 1 \\ \lambda_2 \\ \vdots \\ \lambda_{M+1} \end{bmatrix} = \begin{bmatrix} P_{M+1} \\ 0 \\ \vdots \\ 0 \end{bmatrix} \tag{8}$$

where P_{M+1} is the mean square error of the output series, or the average energy output of the filter coefficients. Equation (8) can be solved by means of the Levinson algorithm. The energy spectra is defined by BURG (1967) as

$$E(f) = \frac{P_{M+1}}{f_N \left|1 + \sum_{n=1}^{M} \lambda_{n+1} e^{-2\pi j f n \delta t}\right|^2} \tag{9}$$

where P_{M+1} and the λ's are obtained by solving Eq. (8). The adequate number of filter elements M is given by the AKAIKE (1969) criteria.

All the spectra shown in this work were normalized with respect to the mean square of the record heights. They were obtained by the Maximum Entropy Method, and fitted to the Fast Fourier Transform defined energy regions, by adjusting the number of filter elements M.

Results show that, for a given place, some of the energy peaks excited by different tsunamis, are at about the same frequencies. Acapulco bay, the most spectacular example, shows a region of common peaks between 0.03 and 0.04 cycles/min, and another close to 0.089 cycles/min (Fig. 2A). La Paz bay spectra (Fig. 2B) show common peaks at about 0.016 and 0.031 cycles/min for different tsunamis. Manzanillo bay shows a region in between 0.026 and 0.045 cycles/min, and another centered at 0.093 cycles/min (Fig. 3A). Finally, Salina Cruz shows a broad band of energy peaks in

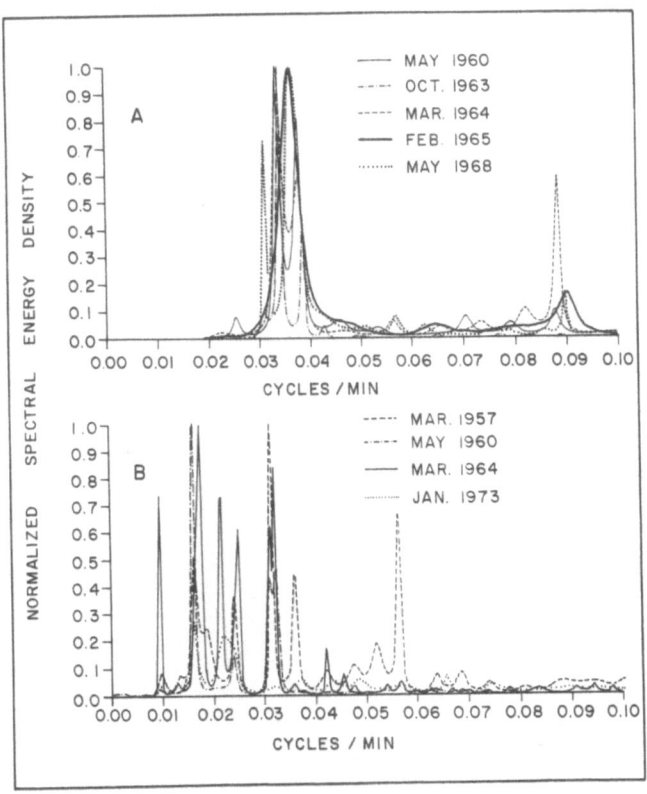

Fig. 2. Tsunami spectra for: A) Acapulco bay, and B) La Paz bay.

between 0.04 and 0.07 cycles/min (Fig. 3B). Spectra for Ensenada and Mazatlan exhibit the same features, and can be seen in SANCHEZ (1980).

The above spectral energy peak distributions, indicate the controlling role of local resonance in defining the tsunami response of the bays; a fact which has also been observed in another places by MUNK *et al.* (1959), RAICHLEN (1976), WATANABE (1964), YAROSHENJA (1974), and FARRERAS (1978), among others. Slight frequency shifts in similar peaks from different tsunamis, at the same place, may be produced by different incident wave directions. Relative height of one excited peak with respect to another may depend on the offshore tsunami energy available for each frequency.

On the other hand, spectra of the same tsunami at different places, do not show joint peak locations; however, a few exceptions can be noted: for March 1964 tsunami in Ensenada, Manzanillo, and Acapulco (Fig. 4A), common peak regions appear about 0.03 to 0.04 cycles/min, 0.06 to 0.07 cycles/min, and 0.09 cycles/min; for January 1973 tsunami in Manzanillo and Acapulco (Fig. 4B), one common peak is present at 0.034 cycles/min; for January 1976 tsunami in Manzanillo and Acapulco (Fig. 5A), one common peak is found at 0.034 cycles/min, and a common excitation region at about 0.075 to 0.090 cycles/min; finally, for March 1979 tsunami in

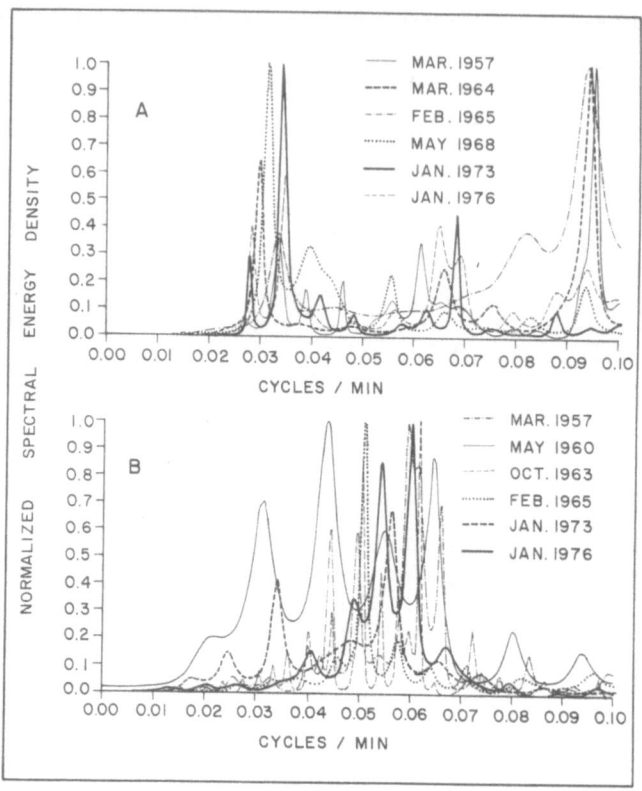

Fig. 3.   Tsunami spectra for: A) Manzanillo bay, and B) Salina Cruz bay.

Manzanillo and Acapulco (Fig. 5B), common peaks are present again at 0.034 and 0.090 cycles/min.

A summary of common peak locations and frequencies for eight different tsunamis, including the four most significant cases shown above, is given in Table 1. The most noticeable region is in between 0.032 to 0.039 cycles/min, frequently excited in Manzanillo and Acapulco, and sometimes in Ensenada, by different tsunamis.

These similar frequencies excited in near by bays may correspond to continental shelf trapped edge wave modes, a mechanism already discussed for the continental shelf off California by MILLER et al. (1962), MUNK et al. (1964), and WILSON (1971). An alternative explanation, particularly suitable for low frequencies, may consider residual long-wave trains reaching the different harbors from the tsunami source area (WILSON and TORUM, 1968).

4.   Shelf Resonance Model

A rectangular flat continental shelf, with a gentle bottom constant slope, and a shallow tsunami wave perturbation incident normal to the rectilinear outer edge and

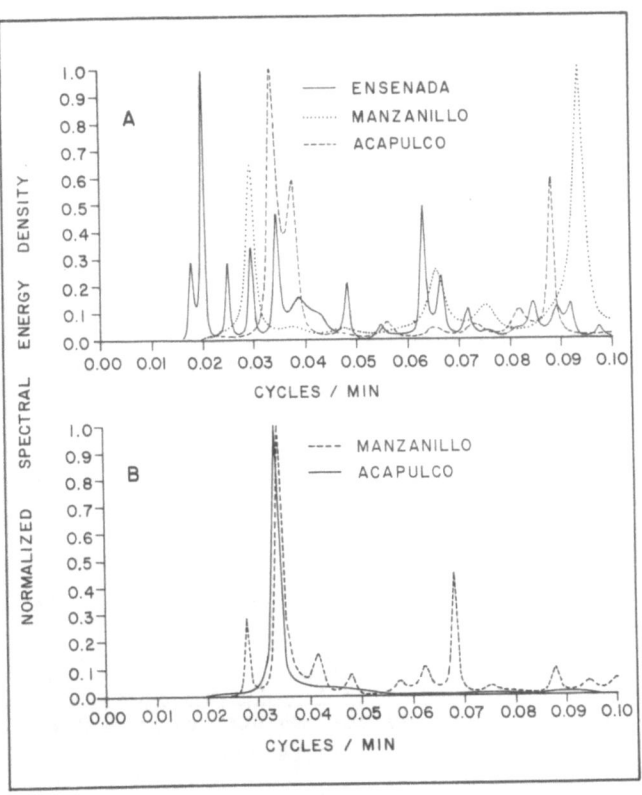

Fig. 4.   A) March 1964 tsunami spectra, and B) January 1973 tsunami spectra.

coast line, are considered. If vertical accelerations and second order horizontal motion terms are neglected, the equation of motion, following LAMB (1932), Art. 169 and 186, can be linearized and written as

$$\frac{\partial}{\partial t}\left(\frac{\partial \mu}{\partial t}\right) = -g\frac{\partial v}{\partial x} \tag{10}$$

and the equation of continuity, for an incompressible fluid

$$v = -\frac{\partial}{\partial x}(h\mu) \tag{11}$$

where v = vertical displacement of the water free surface, g = gravitational acceleration, x = horizontal coordinate parallel to the wave propagation direction, t = time, h = shelf depth, and μ = horizontal water particle displacement. The elimination of μ gives

$$\frac{\partial^2 v}{\partial t^2} = -\frac{\partial}{\partial x}\left(h\frac{\partial^2 v}{\partial t^2}\right) = g\frac{\partial}{\partial x}\left(h\frac{\partial v}{\partial x}\right). \tag{12}$$

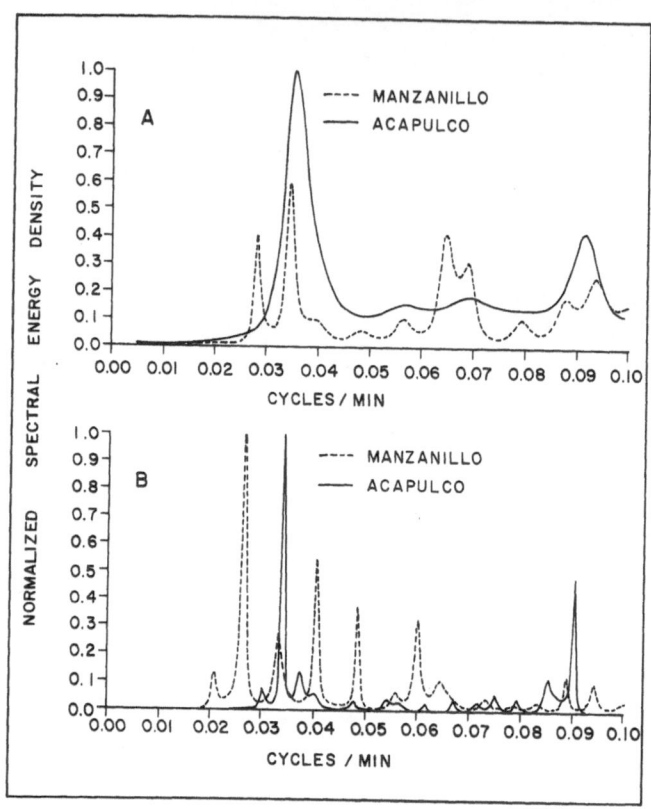

Fig. 5.   A) January 1976 tsunami spectra, and B) March 1979 tsunami spectra.

In the case of simple harmonic motion, and depth increasing uniformly with length, a finite solution for Eq. (12) is

$$v_i(x, t) = S_i J_0(2\sqrt{K_i x}) \cos(\beta_i t + \varepsilon_i) \qquad (13)$$

where S_i = constant amplitude, β_i = angular frequency of the ith mode, ε_i = phase, J_0 = zero order Bessel function, and

$$K_i = \frac{a\beta_i^2}{g h_0} \qquad (14)$$

being a the shelf width, and h_0 the outer edge maximum depth.

Considering a node of the oscillation in the outer edge, as boundary condition, the normal mode frequencies are given by

$$f_i = \frac{R_i \sqrt{g h_0}}{4\pi a} \qquad (15)$$

being R_i the roots of the Bessel function. Frequencies were evaluated for the continental

Table 1.   Locations and frequencies of common spectral peaks for different tsunamis.

Freq. cycl/min	Mar 1957	May 1960	Mar 1964	May 1968	Jan 1973	Nov 1975	Jan 1976	Mar 1979
.016		La Paz Guaymas						
.025			Ensenada La Paz					
.029			Ensenada Manzanillo					
.031		La Paz Guaymas						
.032				Manzanillo Acapulco				
.033		La Paz Salina Cruz						
.034	Manzanillo Acapulco			Manzanillo Mazatlan	Manzanillo Acapulco Salina Cruz		Manzanillo Acapulco	Manzanillo Acapulco
.035		Ensenada Acapulco	Ensenada Acapulco					
.038		Ensenada Acapulco						
.039				Ensenada Manzanillo				
.051		Ensenada Salina Cruz						
.066	Ensenada Acapulco Salina Cruz		Ensenada Manzanillo					
.074	Ensenada Acapulco							
.080						Loreto Acapulco Vallarta		
.085						Guadalupe Vallarta Acapulco		
.086						Guadalupe Vallarta		
.090						San Lucas Vallarta		Manzanillo Acapulco

shelf in front of Ensenada, Acapulco, and Manzanillo, with widths and outer edge depths obtained from MENARD (1960). Results are shown in Table 2. Normal mode frequencies oscillation of the shelf predicted by the model agree reasonable with spectral peak frequencies found in some of the spectra of the three locations. Particularly, the 0.032 to 0.034 cycles/min peak present in Manzanillo and Acapulco, appear to be a resonant mode of oscillation of the continental shelf. Little differences between computed and observed values depend on different incident angles.

5.   Gulf of California Oscillations

Spectra of 2 tidal stations inside, and 3 tidal stations outside the Gulf of California,

Table 2. Continental shelf dimensions and mode frequencies.

Location	Shelf		Model frecuency cyc/min	Spectral Analysis frecuency cyc/min
	Width km	Outside Bound. Depth m		
Ensenada	80	1,500	0.017	0.018
			0.040	0.040
			0.063	0.060,0.065
			0.085	0.079,0.090
Manzanillo	44	1,500	0.032	0.031,0.032
			0.073	0.068,0.077
Acapulco	45	350	0.015	—
			0.034	0.034
			0.054	0.052
			0.073	0.072,0.076
			0.092	0.092,0.097

for May 1960 and March 1964 tsunamis, are shown in Fig. 6. For the stations inside the Gulf, the energy is concentrated in low frequency bands, lower than 0.03 cycles/min; while, for those outside the Gulf it is concentrated in high frequency bands, higher than 0.03 cycles/min.

It is known (WILSON et al., 1968; LOOMIS, 1978; KAJIURA, 1979) that the 1960 and 1964 tsunamis, generated in shallow continental shelves, had main waves with frequencies lower than 0.025 cycles/min. The absence of these spectral peaks in the stations outside the Gulf indicates the predominant role of the high frequency shelf modes amplification, in the resonant phenomena over there. On the other hand, the presence of low frequency peaks, suggests a non linear effect of wave motion with multiple reflections at both sides of the Gulf of California leading to transversal harmonic modes.excitation on the basin, with strong low frequencies amplification. Longitudinal modes of oscillation inside the Gulf of California can not be considered in the case since their frequencies are of the order of 0.0001 cycles/min (MUNK, 1941).

6. Conclusions

Coincidence in one, two, or three spectral peaks excited by different tsunamis in the same place, is an evidence of local bay resonant modes of oscillation.

High frequency common peaks appearing for each tsunami in near by places, can be explained either as trapped edge wave modes of the continental shelf, or residual long wave trains from the tsunami source area. The first possibility is more supported by the shelf resonant analitical model results.

Low frequencies excited inside the Gulf of California, may correspond to transversal modes of oscillation of the basin.

Absence of recorded evidences in the Pacific Ocean coast of Mexico, of tsunamis coming from Oceania, indicates the existence of a dissipation process in the numerous islands of the South Pacific Ocean.

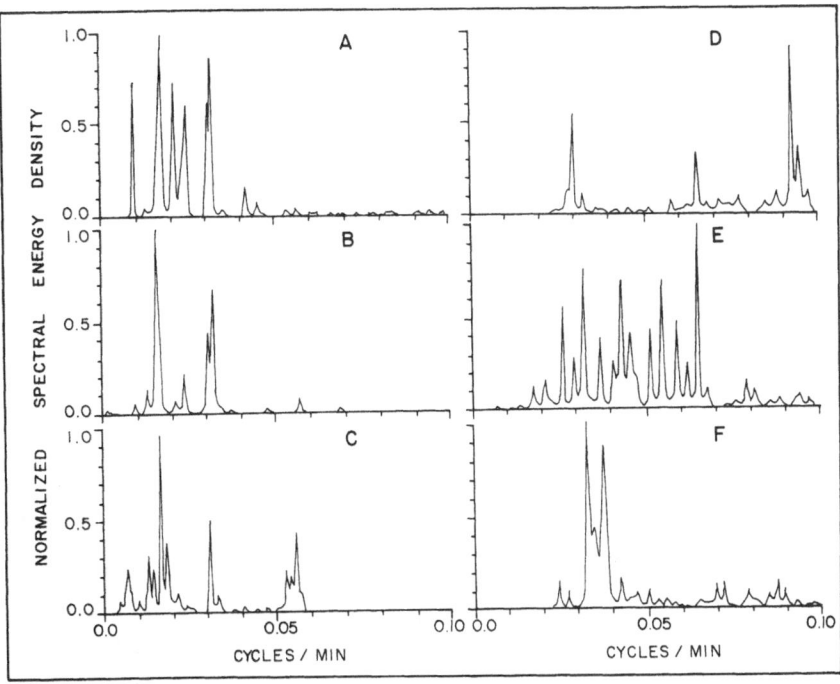

Fig. 6.   Inside Gulf of California tsunami spectra: A) La Paz, March 1964, B) La Paz, May 1960, and C) Guaymas, May 1960; and outside the Gulf: D) Manzanillo, March 1964, E) Salina Cruz, May 1960, and F) Acapulco, May 1960.

The rest of the tsunami spectra obtained in this work, and not shown in this publication, do not considerably modify the results stated above, as can be seen in Sánchez (1980).

To Dr. John Taylor, from the Marine Environmental Data Service, Ottawa, Canada; who gently provided the last computer program version of the Maximum Entropy Method.

REFERENCES

Abe, K. and H. Ishii, Study of shelf effect for tsunami using spectral analysis, in *Abstracts of Symposium Papers*, pp. 23–26, International Tsunami Symposium, I.U.G.G., Japan, 1981.

Akaike, H., Power spectrum estimation through autoregressive model fitting, *Ann. Inst. Statist. Math.*, **21**, 407–419, 1969.

Blackman, R. and J. Tukey, The measurement of power spectra from the point of view of the communications engineer, 190 pp., Dover Pub. Inc., 1958.

Burg, J. P., Maximum entropy spectral analysis, paper presented at the 37th Annual International Soc. Explor. Geophys. Meeting, Oklahoma City, Okla, 1967.

Farreras, S. F., Tsunami resonant conditions of Concepcion bay (Chile), *Marine Geod.*, **1**, 355–360, 1978.

Kajiura, K., Tsunami generation, in *Tsunamis Proceedings of the National Science Foundation Workshop*, pp. 15–40, Tetra Tech Inc., Pasadena, California, 1979.

Lamb, H., *Hydrodynamics*, 6th edition, 738 pp., Cambridge Univ. Press, 1932.

LOOMIS, H. G., Tsunamis, in *Geophysical Predictions, Studies in Geophysics*, pp. 155–165, National Academy of Sciences, Washington, 1978.

MENARD, H. W., Topographic Charts Nos. 1, 5, and 9, U.S. Bureau of Commercial Fisheries and University of California, Institute of Marine Resources, San Diego, California, 1960.

MERINO Y CORONADO, J., E. SALYANO, J. J. ROSALES, and M. MARTINEZ, Los temblores de Acapulco de 1962, *Anales Inst. de Geofísica, Univ. Nac. Autón. de México*, **8**, 23–26, 1962.

MILLER, G. R., W. H. MUNK, and F. E. SNODGRASS, Long period waves over California's continental borderland, Part II, Tsunamis, *J. Marine Res.*, **20**, 31, 1962.

MUNK, W. H., Internal waves in the Gulf of California, *Bull. Scripps Inst. Oceanogr., Contrib.*, #**123**, 81–91, 1941.

MUNK, W. H. and H. CEPEDA, Sobre un pico notablemente agudo en el espectro del nivel del mar en Acapulco, *Geof. Intern*, **1**, 45–54, 1961.

MUNK, W. H., F. E. SNODGRASS, and M. J. TUCKER, Spectra of low-frequency ocean wave, *Bull. Scripps Inst. Oceanog.*, **7**, 283–362, 1959.

MUNK, W. H., F. E. SNODGRASS, and F. GILBERT, Long waves on the continental shelf: an experiment to separate trapped and leaky modes, *J. Fluid Mech*, **20**, 529–554, 1964.

PARARAS-CARAYANNIS, G., *Catalog of Tsunamis in Hawaii*, 78 pp., World Data Center, Solid Earth Geoph. Report, N.O.A.A., 1977.

RAICHLEN, F., Tsunamis: some laboratory and field observations, U.S. Army Coastal Engineering Research Center, Corp of Engineers, pp. 2103–2122, 1970.

RAICHLEN, F., Tsunamis, in *Coastal Wave Hydrodynamics Theory and Engineering Applications*, Chap. 7, pp. 7-1 to 7-62, Ralph M. Parsons Lab. for Water Resources and Hydrodynamics, Dept. of Civil Engineering, Massachusetts Institute of Technology, 1976.

REID, R. O., Island response to Tsunamis, in *Tsunamis Proceedings of the National Science Foundation Workshop*, pp. 182–187, Tetra Tech Inc., Pasadena, California, 1979.

SALSMAN, G. G., The Tsunami of March 9, 1957, as recorded at tide stations, 18 pp., Tech. Bull. No. 6, U. S. Dept. of Comm., Coast and Geod. Survey, 1959.

SÁNCHEZ, A. J., Tsunamis en la costa occidental de Mexico, Tesis de Maestría, 199 pp., Centro de Inv. Cient. y Educ. Sup. de Ensenada, Baja California, Mexico, 1980.

SECRETARIA DE MARINA, Estudio geográfico de la región de Manzanillo, Col., 361 pp., Dir. Gral. de Ocean. y Señalam. Marit., Mexico, 1973.

SECRETARIA DE MARINA, Estudio geográfico de la región de Ensenada, B. C., Dir. Gral. de Ocean. y Señalam. Marit., Mexico, 1974a.

SECRETARIA DE MARINA, Estudio geográfico de la región de Mazatlán, Sin., 353 pp., Dir. Gral. de Ocean. y Señalam. Marit., Mexico, 1974b.

SECRETARIA DE MARINA, Estudio geográfico de la región de Salina Cruz, Oax., 347 pp., Dir. Gral. de Ocean. y Señalam. Marit., Mexico, 1974c.

SHANNON, C. E., A mathematical theory of communications, *Bell System. Tech. J.*, **27**, 379–423, 1948.

SHEPARD, F. P., G. A. MAC DONALD, and D. C. COX, The tsunami of April 1, 1946, *Bull. Scrip. Inst. Oceanog.*, **5**, 391–528, 1950.

SPAETH, M. G. and S. C. BERKMAN, The tsunami of March 28, 1964, as recorded at tide stations, 86 pp., Spec. Pub. 33, U. S. Dept. of Comm., Coast and Geod. Survey, 1967.

SYMONS, J. M. and B. D. ZETLER, The tsunami of May 22, 1960, as recorded at tide stations, 29 pp., Prelim. Rep. U. S. Dept. of Comm., Coast and Geod. Survey, 1960.

ULRYCH, T. J., Maximum entropy power spectrum of truncated sinusoids, *J. Geophys. Res.*, **77**, 1396–1400, 1972.

VAN DORN, Instrumentation and observations, in *Tsunamis Proceedings of the National Science Foundation Workshop*, pp. 281–295, Tetra Tech Inc., Pasadena, California, 1979.

WATANABE, H., Studies of the tsunamis on the Sanriku coast of North-Eastern Honshu in Japan, *Geophys. Mag.*, **32**, 120–127, 1964.

WIGEN, S. O., *Historical Study of Tsunamis: Chronological and Area List*, 78 pp., International Tsunami Information Center, Honolulu, Hawaii, 1977.

WILSON, B. W., Tsunami responses of San Pedro Bay and Shelf, California, J. Waterway, Harbors Coastal Eng. Div., *Proc. Am. Soc. Civ. Eng.*, **97**, 239–258, 1971.

WILSON, B. W. and A. TORUM, The tsunami of the Alaska earthquakes, 1964: engineering evaluation, Tech. Memo No. 25, Coastal Eng. Res. Center, U.S. Army Corp of Engineers, Washington, D. F., 1968.

YAROSHENJA, R. A., A study on natural oscillations in the sea level of Kurile and Kamchatka inlets, in *Tsunami Research Symposium*, Bull. No. 15, pp. 39–49, Intern. Union of Geodesy and Geoph. Tsunami Committee, Royal Society of New Zealand, 1974.

ZERBE, W., The tsunami of November 4, 1952, as recorded at tidal stations, 62 pp., U.S. Coast and Geod. Surv., Spec. Pub. 300, 1953.

Tsunamis—Their Science and Engineering, edited by K. Iida and T. Iwasaki, 161–172.
Copyright © 1983 by Terra Scientific Publishing Company (TERRAPUB), Tokyo.

Study of Shelf Effect for Tsunami Using Spectral Analysis

Kuniaki Abe* and Hiroshi Ishii**

Nippon Dental University, Niigata Branch, Niigata, Japan
**Earthquake Prediction Research Center, Tohoku University, Sendai, Japan*

(Received August 10, 1981; Revised November 14, 1981)

Spectral analysis was made for distant tsunamis observed at a coast facing to the open ocean and compared with the response curve of a linear sloping shelf model between two flat regions for a unit amplitude incidence. Good agreement was found between both spectra for periods longer than a certain critical one. The critical period is about 50 min for almost all cases. Overestimations in the calculation compared with the observation were consistently found for periods shorter than the critical value. These results enable us to assume that a tsunami incident to a shelf possesses a white spectrum and that the main part of the wave train in a marigram observed at the coast is the wave excited on the sloping shelf. Local differences among the spectra can be explained by applying this shelf model to an observation point. Finally, an edge wave arrival was confirmed for the 1952 Kamchatka tsunami by using a method of moving spectrum.

1. Introduction

Although a tsunami is considered to be a simple waveform near its source, it is always observed to be oscillatory at a distant coast. This oscillatory nature can be understood by investigating tsunamis observed at various stations for various origins. Kato *et al.* (1961) attributed the high inunduations from the 1960 Chilean tsunami observed at some bay heads along the Sanriku coast, which is on the Pacific coast of northeast Japan, to resonance of the bays with the tsunami wave. According to this study, the Chilean tsunami consisted of a wave packet with several long period crests and troughs. Kato *et al.* (1961) suggested the possibility of amplification of the long period components by the continental shelf. Ishii and Abe (1980), and Abe and Ishii (1980) discussed the edge wave and reflection for a shelf with a linear slope between two flat regions based on theoretical and observational considerations.

In particular, Abe and Ishii (1980) calculaed the amplitude spectrum for an incident sinusoidal plane wave of unit amplitude, assuming a shelf structure such as that off the Sanriku coast. They showed that the spectrum observed at Onahama for the 1952 Kamchatka tsunami was approximated with the spectrum calculated through applying the shelf model to the observation point. They found the longest resonance period for the shelf to be 70 min. It is interesting that the value of 70 min is nearly equal to that observed at the Sanriku coast for the 1960 Chilean tsunami (Kato *et al.*, 1961).

It is important to consider the amplifying effect of long periods by the shelf. Herein the shelf effect is investigated using spectral analysis for tsunamis observed at various observation points.

2.   Application of Model

The shelf model is shown in Fig. 1. For this model Shaw (1979) theoretically derived the amplification factor at the coast. Ishii and Abe (1980) derived the edge wave dispersion relation and Abe and Ishii (1980) also obtained the amplification factor at the coast to an oblique incidence applying the same expressions as used by Ishii and Abe (1980) to the reflection problem. Since the amplitude ratio at the coast to the incident one at the shelf entrance was given in spectral form, the result leads us analyze the spectrum of the tsunami observed at a coast on the shelf. Although each continental shelf shows a local structure, it generally consists of a shallow flat region and a gentle slope. The shelf model is applied to the real shelf neglecting the trench as shown in Fig. 2. The selection of geometrical parameters was made on the basis of a good approximation of the shelf slope. The vertical scale is enlarged over the horizontal one. Four shelves in the Pacific ocean were selected for the application to Japan. In north America, three shelves were used in this analysis. Each model is designated by the name of an observation point. Table 1 indicates a list of the tsunamis and observation location used. The data employed are the cases of the 1952 Kamchatka, 1958 Itrup, 1960 Chilean, 1963 Itrup, 1964 Alaskan, and 1965 Aleutian tsunamis. The method of spectral correlation given by Abe (1981) is used to determine the incident angle.

Time histories of the sea level were digitized using an x-y reading machine from copies of tide gage records. The digitization was made with various time intervals depending on the time variation of sea level so that the waveform was faithfully reproduced. The average interval is 2–3 min. Time series with a constant time interval of 3 min were obtained using linear interpolation for a coase interval of digitization and neglecting higher frequency component for a dense one. Ordinary tides were eliminated form the records and the spectra were computed using the Goertzel method. The time length of the record was taken from tsunami onset to edge wave arrival. This length is

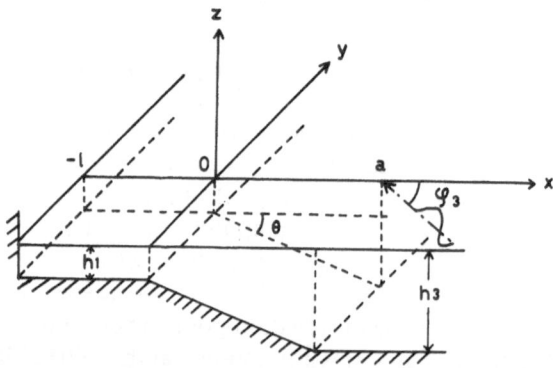

Fig. 1.   Shelf model used in the calculation. The geometrical form is represented with a constant dip angle of θ, two constant depths of h_1, h_3, and a constant width of l for a shallow flat region.

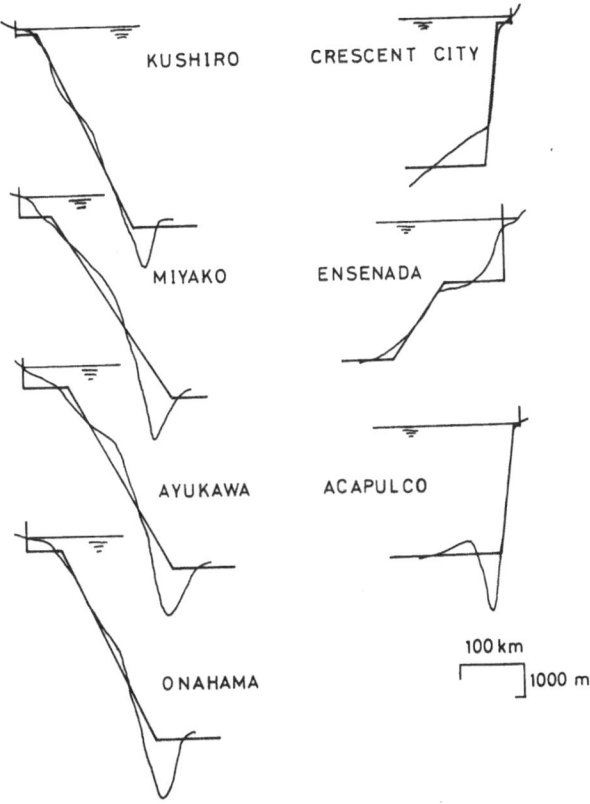

Fig. 2. Application of the model to a local sea bottom topography.

Table 1. Tsunamis and observation points used in the spectral analysis.

Data (G. T)	Origin	Observation point
1952.11.04	Kamchatka	Kushiro, Miyako, Onagawa Onahama, Acapulco, Crescent City
1958.11.06	Iturup	Miyako
1960.05.22	Chile	Miyako, Tsukihama, Onahama, Acapulco, Ensenada, Crescent City
1963.10.13	Iturup	Miyako
1964.03.28	Alaska	Hachinohe, Miyako, Onahama, Acapulco, Ensenada
1965.02.04	Aleutian	Hachinohe, Onahama

about 7 hours for all the tsunamis except the Kurile tsunamis of 1958 and 1963, which are 4 hours.

3. Spectrum Representing a Shelf Effect

3.1 The 1952 Kamchatka tsunami

The first case is the result for the Kamchatka tsunami observed in the northeast of

Japan as shown in Fig. 3. In the figure the solid lines show the observed spectra and the dotted ones show the best fit curves of the amplitude ratio. The best fit curve was determined from the curves calculated for various incident angles using the correlation method (Abe, 1981). Correlation calculation was carried out for some observed spectral peaks having periods longer than 50 min. The incident angles thus obtained are 10° for Kushiro and Miyako, and 20° for Onahama.

Although the individual spectral peak observed is not found in the calculation, the envelopes of the observational spectra agree with the calculation, especially for periods

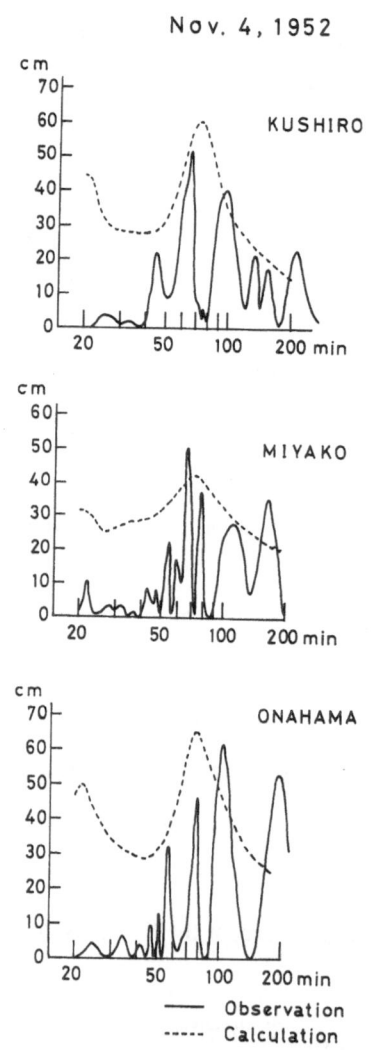

Fig. 3. Spectra observed in the 1952 Kamchatka tsunami (solid lines) and the spectra calculated for the local sea bottom topography (dotted lines). Spectral amplitude in the calculation was normalized using a set of peak values from the spectrum observed for periods longer than 50 min.

longer than 50 min. It is considered that in the observation each spectral peak and trough forming the envelope was generated at the source under the controle of a rise time etc. According to YAMASHITA and SATO (1974) a short rise time contributes to making many spectral peaks. We consider the shelf effect as a receiver of tsunami and do not take into consideration of the source condition. Therefore, the comparison is restricted to the spectral envelope. For periods shorter than 50 min the calculated amplitude is overestimated. The similarlity of observational spectra among these locations is understood to be due to the similarlity of the shelf topographies. The shelf resonances were excited at all the observational locations. Particularly in Kushiro and Miyako the maximum spectral amplitudes were produced by resonance. The agreements between calculated curve and spectral envelope are not so good for Crescent City and Acapulco as for the other three locations, although the figures are not presented.

3.2   The 1960 Chilean tsunami

Spectra obtained for the 1960 Chilean tsunami were compared with the calculated curves. Three cases in North America are shown in Fig. 4. In the figure the dotted lines indicate the calculated amplitude ratios. In applying the shelf model, the incident angles of 20°, 30°, and 10° were utilized as the best fit values for Crescent City, Ensenada, and Acapulco, respectively. The agreement between both spectra is found for periods longer than 50 min. The comparison for each observation point reveals the following facts. In Crescent City three values of spectral amplitudes observed at the periods of 60, 80, and 130 min show good correlations with ones calculated. In Ensenada the maximum amplitude observed for the period of 58 min is well explained by the shelf resonance calculated with the model. In Acapulco the periods observed with large amplitude (29, 41, and 150 min) agree with the ones predicted from calculation. Overestimations are seen at all the observation locations for periods shorter than 50 min in the same way as described for the 1952 Kamchatka tsunami. In particular, the resonance period is short in Crescent City, and relatively large amplitudes are observed in the spectral components shorter than the 50 min period. The calculated spectrum has three peaks for the period range from 25 to 200 min in the case of Acapulco. This complexity results from a very shallow water of the shelf. The observed component of 85 min with relatively large amplitude is inexplicable from an assumption white spectral incidence.

It is considered that various spectra along the North American coast result from a rich variety of shelf structures. For Japanese observation points quite good agreements between calculated responce curve and spectral envelope of the observation are obtained, although not shown here (for Miyako, see Fig. 6).

3.3   The 1964 Alaskan tsunami

Three results of the 1964 Alaskan tsunami are shown in Fig. 5. Incident angles are obtained to be 20° for Onahama, 50° for Ensenada, and 20° for Acapulco and used in the calculation curves. In Ensenada and Acapulco it is understood from the correspondence of the peaks between calculation and observation that the resonances were well excited. The agreement of the resonance periods is not so good for

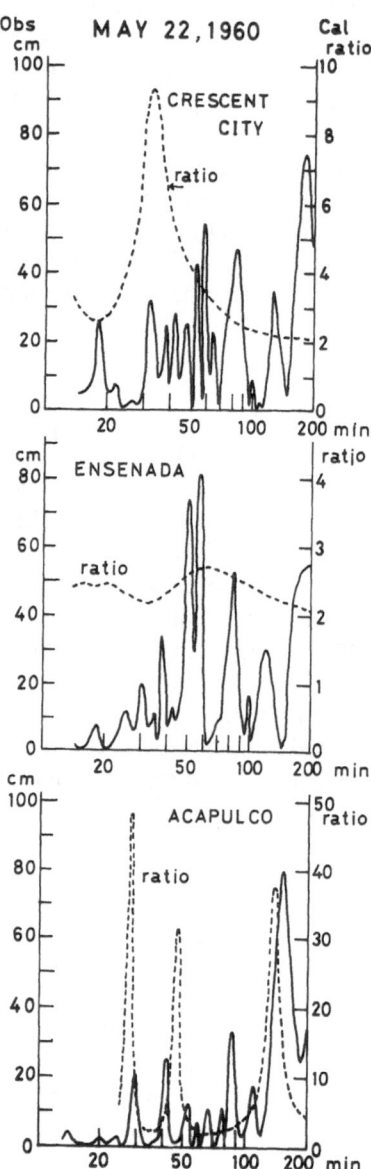

Fig. 4.   Spectra observed in the 1960 Chilean tsunami (solid lines) and the spectral responses calculated with the local sea bottom topography (dotted lines).

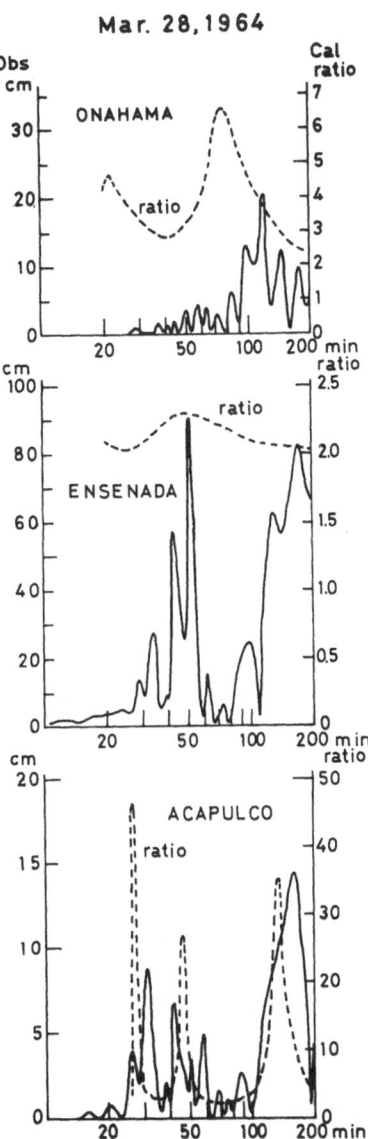

Fig. 5.   Spectra observed in the 1964 Alaskan tsunami (solid lines) and the spectral responses calculated with the local sea bottom topography (dotted lines).

168 Ku. Abe and H. Ishii

MIYAKO

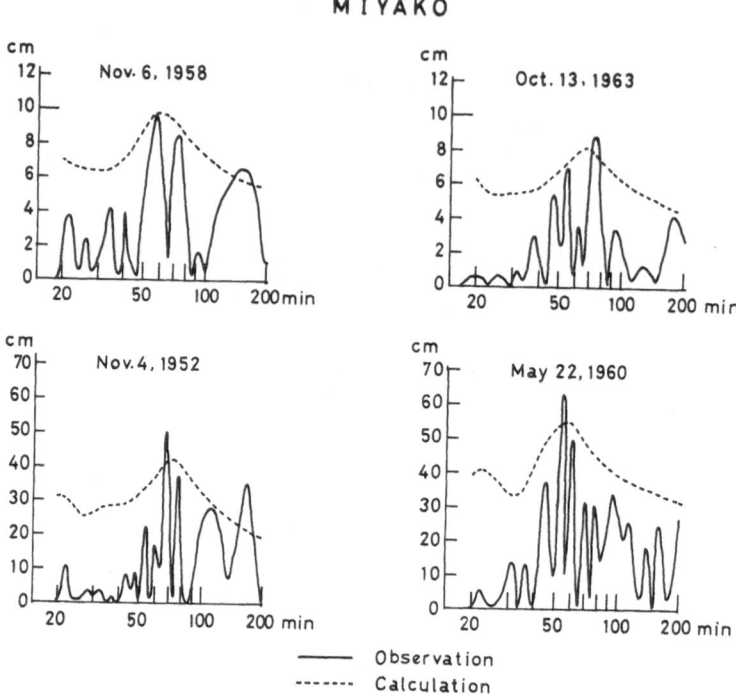

Fig. 6. Four tsunami spectra observed in Miyako (solid lines) and the spectra calculated with the local sea bottom topography (dotted lines). The spectral amplitude ratio in the calculation was normalized using a set of peak values of the spectrum observed for periods longer than 50 min.

Acapulco as for Ensenada. The discrepancy in the case of Acapulco was also shown in the case of the 1960 Chilean tsunami. It is concluded that the three observed resonance periods agree with ones calculated within the accuracy of 15 % in Acapulco. For Onahama it is notable that agreement between two spectra is found for periods longer than 100 min. The critical period of 100 min, an explainable limit of the calculation for the observation, is very long in comparison with the 50 min in the previous two tsunamis. Since the resonance of 70 min at Onahama was excited for the Kamchatka tsunami, as seen in Fig. 3, there is a possibility that the resonance period component was not radiated from the source. It is interesting that this component wave was not observed at either Ensenada or Acapulco.

3.4 Spectra in Miyako

The fact that a kind of similarity in the observed spectra is found among different tsunamis at the same location was discussed by TAKAHASI and AIDA (1963), and RAICHLEN (1979). This tendency was found for Ensenada and Acapulco as described above. We attempted to investigate it for the case of Miyako as shown in Fig. 6. The two tsunamis, which have their origins in the shelf off Iturup Island, were included with the Kamchatka tsunami and the Chilean one. These four tsunamis have different sources. The incident angles are 50° for the 1958 Iturup tsunami, 40° for the 1963 Iturup tsunami, and 60° for the Chilean tsunami. The amplitude ratios calculated were normalized to the coastal amplitude by applying a parallel slide of the envelope curve. It is interesting that Miyako shows the same spectral response for these four tsunamis. One exceptional case is the 1964 Alaskan tsunami, which is not shown here, indicating that the spectra observed at Miyako for the tsunami lacks the period component shorter than 100 min and is very similar to the case of Onahama. Most tsunami spectra of distant origins are explained by the white spectral incidence for a period component longer than 50 min. It is very interesting that the observed spectral envelope is independent of the location and size of the source.

4. Edge Wave Arrival

As described previously, ISHII and ABE (1980) obtained a dispersion equation for edge waves on a shelf with a linear slope between two flat regions, and also found an edge wave in the 1952 Kamchatka tsunami. According to that reference, the smoothed tide gage records at both Ayukawa and Onahama showed a maximum amplitude of arrival corresponding to the minimum group velocity, which is characteristic of an edge wave. The velocity and period of the edge waves were 65 m/sec and 60 min, respectively, according to their results. Using the method of moving spectra, we can also demonstrate the arrival of the edge wave for the 1952 Kamchatka tsunami. The results are presented in Fig. 7. From top to bottom the results are shown for Kushiro, Miyako, Onagawa, and Onahama. These observation locations are distributed along the Pacific coast of northeast Japan. In the figure, the ordinate is the time from origin to start of sampling and each spectrum was calculated using two hours of sampling. Solid lines are the travel times from the epicenter calculated by the dispersion equation assuming the shelf of Miyako extends to all the observation points. The period of 60 min, which is the minimum of the calculation curve, corresponds to the edge wave calculated by ISHII and ABE (1980). In comparing the calculated curve with the moving spectra, the spectral amplitudes at the period of 60 min near the minimum point of the curve are found to be large at all the observation points except Kushiro. Therefore, these results are considered to be proof of an edge wave arrival. This arrival is clearly observed in the time histories of the spectral amplitude for the period of 60 min, as shown in Fig. 8. In this figure, the arrows show the calculated arrival time of the edge wave. The edge wave is small in amplitude at Kushiro and grows with propagation from Kushiro to Onahama. Such a development is also seen in a series of marigrams given by ISHII and ABE (1980). It is indicated that the growth of the edge wave corresponds to the length of distance having a uniform profile of the sea depth.

5. Summary

We investigated the shelf effect in the tsunami spectra observed along the coast based on the response curve calculated for the shelf model with a linear slope between two flat regions. The main results obtained are as follows:
1) A shelf resonance excited by multiple reflection of a direct wave was observed

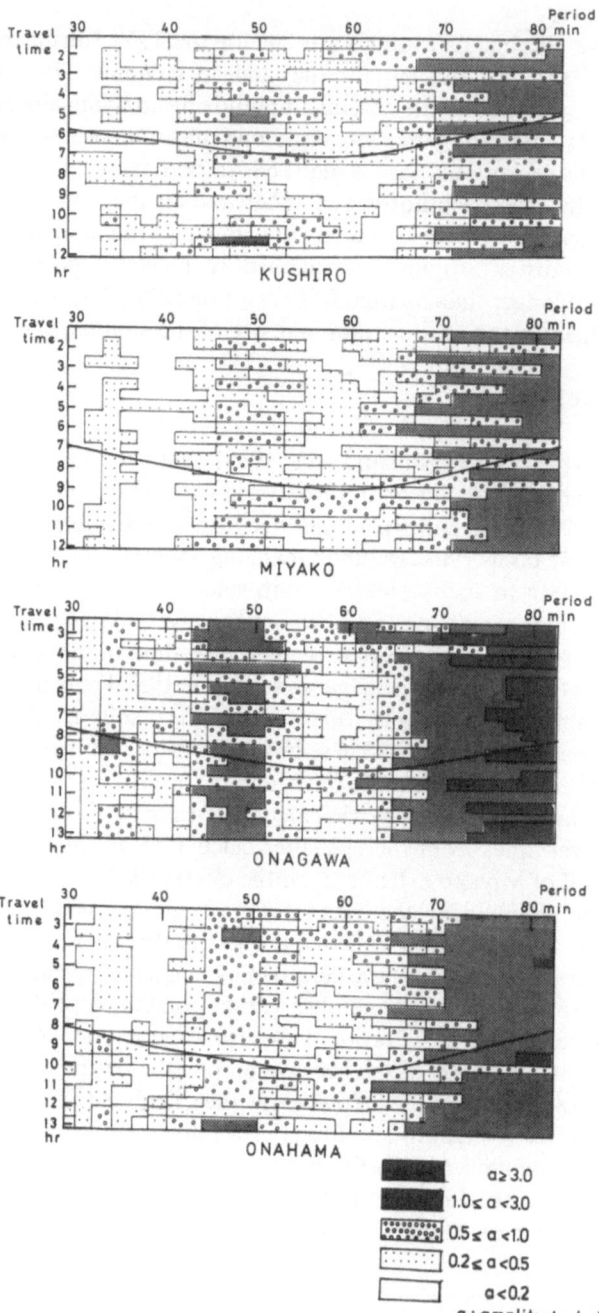

Fig. 7.   Moving spectra observed in the 1952 Kamchatka tsunami. The ordinate is the time from the origin to the onset of sampling. Each spectrum was calculated for the time length of 2 hours with a 3 min sampling. The abscissa is the period in minutes. Amplitude is classified with various patterns described at the bottom of the figure. Solid line curves are travel times calculated from the group velocity dispersion curve using the model.

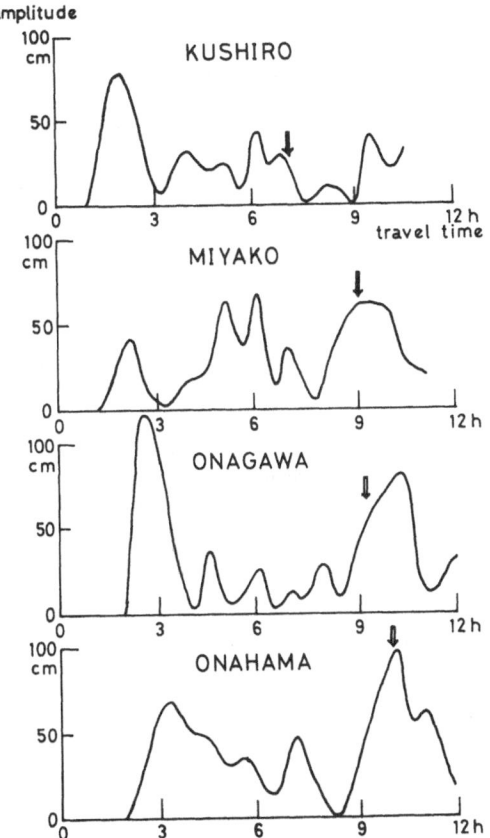

Fig. 8. Spectral amplitude in the period of 60 min versus travel time in hours. Arrows indicate the arrival times of edge wave obtained by ISHII and ABE (1980).

in most distant tsunamis.

2) Good agreement in the spectral forms was obtained between observation and calculation in most tsunamis analyzed, for periods longer than 50 min. This result leads us to the conclusion that incident tsunamis have white spectra for periods of 50–200 min in most cases.

3) Short period components smaller than 50 min were not greatly excited. This is a result of the absence of short period components at the shelf entrance.

4) It was found that the local difference in the spectral behavior among various observation points in the period range longer than 50 min was mainly due to the difference of shelf structures.

5) An edge wave was revealed in the 1952 Kamchatka tsunami propagating along the Pacific coast of north east Japan by the use of a moving spectral analysis.

The conclusions described above indicate the important role of the shelf as a resonator and wave guide for distant tsunami.

REFERENCES

ABE, Ku., Incident angle identification from the spectrum for a tsunami invasion to the shelf, *Bull. Nippon Dental Univ., General Education*, **10**, 87–93, 1981.

ABE, Ku. and H. ISHII, Propagation of tsunami on a linear slope between two flat regions, Part II Reflection and transmission, *J. Phys. Earth*, **28**, 543–552, 1980.

ISHII, H. and Ku. ABE, Propagation of tsunami on a linear slope between two flat regions, Part I Edge wave, *J. Phys. Earth*, **28**, 531–541, 1980.

KATO, Y., Z. SUZUKI, K. NAKAMURA, A. TAKAGI, K. EMURA, M. ITO, and H. ISHIDA, The Chile Tunami of 1960 observed along the Sanriku Coast of Japan, *Sci. Rep. Tohoku Univ., Ser. 5, Geophysics*, **13**, 107–125, 1961.

RAICHLEN, F., Bay and harbor response to tsunamis, in *Proc. National Sci. Found. Workshop*, edited by Li-San Hwang and Y. Keen Lee, pp. 188–221, 1979.

SHAW, R. P., Long waves obliquely incident on a continental slope and shelf with a partially reflecting coastline, Symposium on Tsunami, Manuscript Report Series, No. 48, pp. 122–130, Fisheries and Environment, Canada, 1978.

TAKAHASI, R. and I. AIDA, Spectra of several tsunamis observed on the coast of Japan, *Bull. Earthq. Res. Inst., Tokyo Univ.*, **41**, 299–314, 1963 (in Japanese).

YAMASHITA, T. and R. SATO, Generation of tsunami by a fault model, *J. Phys. Earth*, **22**, 415–440, 1974.

Tsunamis—Their Science and Engineering, edited by K. Iida and T. Iwasaki, 173–183.

Colombia-Peru Tsunamis Observed along the Coast of Japan—Tsunami Magnitude and Source Areas

Tokutaro HATORI

Earthquake Research Institute, University of Tokyo, Tokyo, Japan

(Received August 10, 1981; Revised November 26, 1981)

Based on tide-gauge records of the USCGS, the magnitude and source area of five Colombia-Peru tsunamis generated over the 20 years between 1960 and 1979 were investigated. According to the author's method based on the attenuation of tsunami height with distance, these tsunami magnitudes (Imamura-Iida scale: m) were determined to be $m = 2 - 2.5$. Source areas of tsunamis inferred from an inverse refraction diagram agree well with the aftershock areas. The source areas lay on the continental slope extending 120–270 km along the bathymetric contours.

Along the Japanese coast, the effects of recent Peru tsunamis were evidently have been small, but inundation heights of historical tsunamis in 1586, 1687, 1868 and 1877 locally reached about 2–3 m. The estimated sources for these historical tsunamis (magnitude; $m = 3.5-4$) have a 400–500 km length, lining up along the Peru-Chile Trench. The source areas of recent tsunamis occupied only a part of the historical tsunami source areas. Seismic gaps exist off Chimbote ($8°-10°$S) and Lima ($11°-13°$S). A segment of 1,000 km from the South Peru to North Chile region ($14°-23°$S) should be considered as an area of relatively high tsunami risk.

1. Introduction

Five historical tsunamis (1586–1906) generated in the Colombia-North Chile region were recorded along the coast of Japan (YUMURA, 1961; WATANABE, 1968). The tsunamis of Aug. 13, 1868 and May 9, 1877 hit the Hawaiian Islands (IIDA *et al.*, 1967; PARARAS-CARAYANNIS, 1969). The Ecuador tsunami of Jan. 31, 1906 was observed by tide-gauges in Japan (HONDA *et al.*, 1908). Large historical tsunamis that originated in South America before the 1960 Chile tsunami were studied by the author (HATORI, 1968) with respect to the distributions of tsunami height and travel times in the whole Pacific region.

During the past 20 years (1960–1979), five tsunamis generated in the Colombia-Peru region were also observed at Japanese tidal stations (HATORI, 1981). In the present paper, three aspects of Colombia-Peru tsunamis are treated: 1) Tsunami magnitude and source area of recent tsunamis, 2) Tsunami effects in Japan, and 3) Estimation of the source area of historical tsunamis before 1906. The seismic gap was examined, comparing the source areas of recent and historical tsunamis in the Colombia-North Chile region.

2. Earthquake Data

Five tsunamigenic earthquakes occurred in the Colombia-Peru region during the past 20 years between 1960 and 1979. Earthquake data taken from the PDE reports of the US Coast and Geodetic Survey (USCGS) are listed in Table 1. The earthquake magnitudes, M_s (20-sec surface-wave magnitude), were in the 7.5 to 7.9 range. The Colombia earthquake on Dec. 12, 1979, being $M_s = 7.9$, was the largest among them.

Figure 1 shows the aftershock areas of the five tsunamigenic earthquakes. Aftershock areas except for the earthquake on May 31, 1970 lay on the continental slope, parallel to the bathymetric lines. The length of the aftershock areas was 120–200 km. According to an analysis of long-period surface-waves (ABE, 1972), the 1970 earthquake was caused by a normal fault mechanism and the 1966 earthquake by a low-angle thrust faulting. The 1979 Colombia earthquake was a thrust fault earthquake (H. Kanamori, personal communication, 1981).

3. Source Area and Tsunami Magnitude

According to a field investigation report on the 1979 Colombia tsunami (PARARAS-CARAYANNIS, 1980), many coastal villages in the neighborhood of the Colombia and Ecuador border were destroyed by 2–5 m high waves. Subsidences of 50–60 cm were reported from Rompido and Cascajal Islands near San Juan, Colombia. A tide-gauge record was obtained at Esmeraldas, Ecuador (Fig. 2). The initial wave clearly began with a downward motion and took only 5–6 minutes to reach the tide station after the earthquake. According to eyewitness reports at San Juan, three to four waves observed, the first wave arriving in about 10 minutes. Figure 2 shows the source area of the tsunami inferred from an inverse refraction diagram based on arrival times and aftershock activity. The source length was approximately 270 km, running parallel to bathymetric contours and the area was 25×10^3 km². The source dimension was about the same as tsunamis of similar magnitude (Imamura-Iida scale: $m = 2.5$, see Fig. 7) generated in the vicinity of Japan (HATORI, 1978).

Table 1. Earthquake and tsunami data in the region from Colombia to Peru.

Data	Earthquake					Tsunami
	Epicenter		Location	d	M_s	$m*$
	Lat.	Long.		(km)		
1960 Nov. 20	6.8 S	80.7 W	Peru	–	(6.8)	2.5
1966 Oct. 17	10.7 S	78.7 W	"	38	7.5	2
1970 May 31	9.2 S	78.8 W	"	43	7.8	2
1974 Oct. 3	12.3 S	77.8 W	"	13	7.6	2.5
1979 Dec. 12	1.58 N	79.39 W	Colombia	33	7.9	2.5

*Tsunami magnitude of Imamura-Iida scale.
M_s: Earthquake magnitude (20-sec surface-wave).

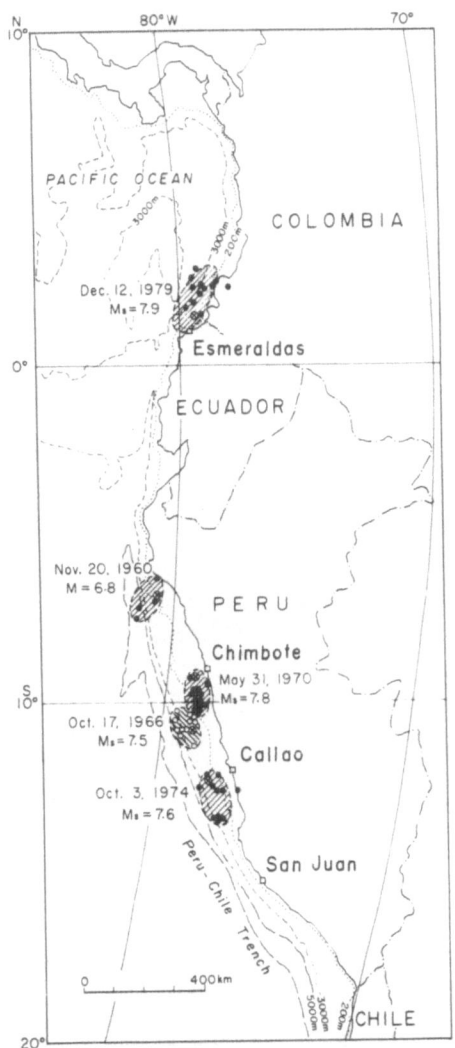

Fig. 1. Distribution of aftershock areas of tsunamigenic earthquakes during the past 20 years (1960–1979).

The Peru tsunami on Oct. 3, 1974 was recorded by tide-gauges at Callao and San Juan (PARARAS-CARAYANNIS, 1975). Two tsunami records and the last wave fronts of the inverse refraction diagram starting from the tide stations are shown in Fig. 3. Taking the aftershock activity into account, the estimated source area of this tsunami was located on the continental slope and extended approximately 200 km in an elongated shape parallel to the bathymetric contours.

Using the author's method (HATORI, 1979), which is based on the attenuation of wave-height with distance from the epicenter, the tsunami magnitude of the Imamura-Iida scale, m, is determined in Fig. 4. The wave-height means in Fig. 4 is the half-

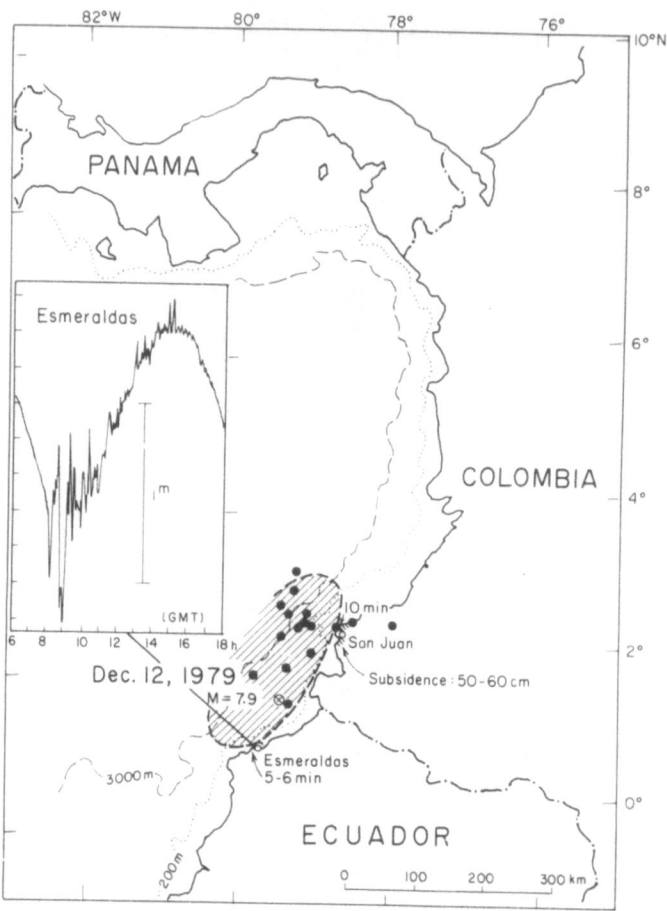

Fig. 2.   Estimated source area of the 1979 Colombia tsunami, distribution of aftershocks (closed circles), and tide-gauge record.

amplitude of the maximum wave above the ordinary tidal level on the records. On the average, the magnitudes of the Peru tsunamis in 1960 and 1974 are determined to be $m = 2.5$. The magnitudes of the 1966 and 1970 tsunamis are $m = 2$. According to the same method, the magnitude of the large Chile tsunami on May 22, 1960 was $m = 4.5$. From the definition of tsunami magnitude (Hatori, 1979), the tsunami energy is reduced by one-fifth when the magnitude decreases by one.

On tide-gauge records observed along the Japanese coast, double amplitudes of the five tsunamis were about 20 cm on the average and increased in northern Japan (Hatori, 1981). The period of the tsunamis observed at most of the Japanese tide stations was 15 minutes, not as long as that of the 1960 Chile tsunami. Figure 5 shows a refraction diagram of the 1979 Colombia tsunami across the Pacific Ocean. The computed travel times agree well with those observed in Japan. The travel time of the 1979 Colombia tsunami to northern Japan was 19.5 hours and that to western Japan

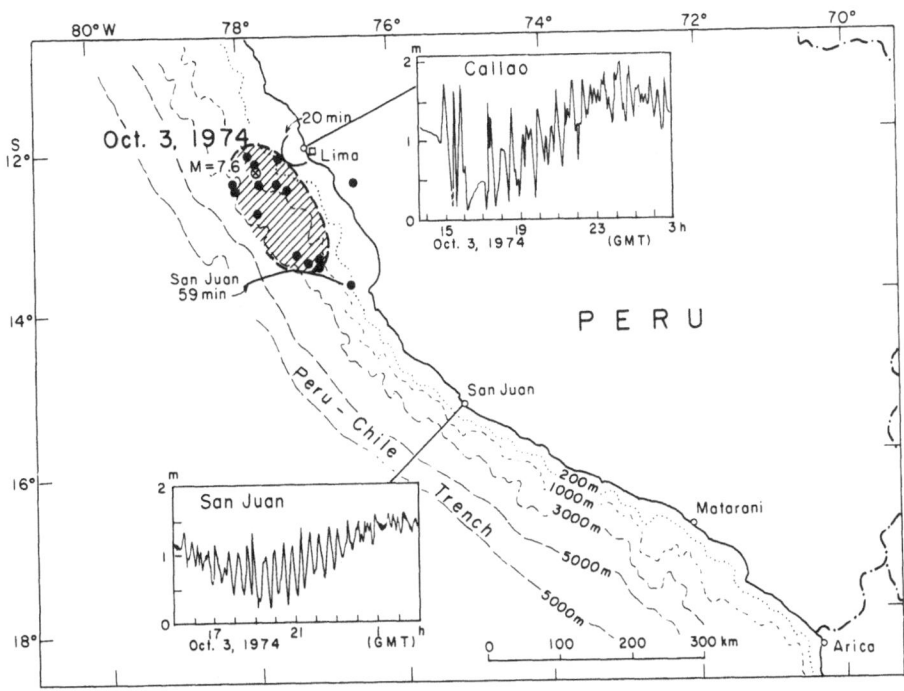

Fig. 3.   Estimated source area of the Peru tsunami on Oct. 3, 1974, distribution of aftershocks (closed circles), and tide-gauge records. The last wave fronts correspond to travel times observed by tide-gauges.

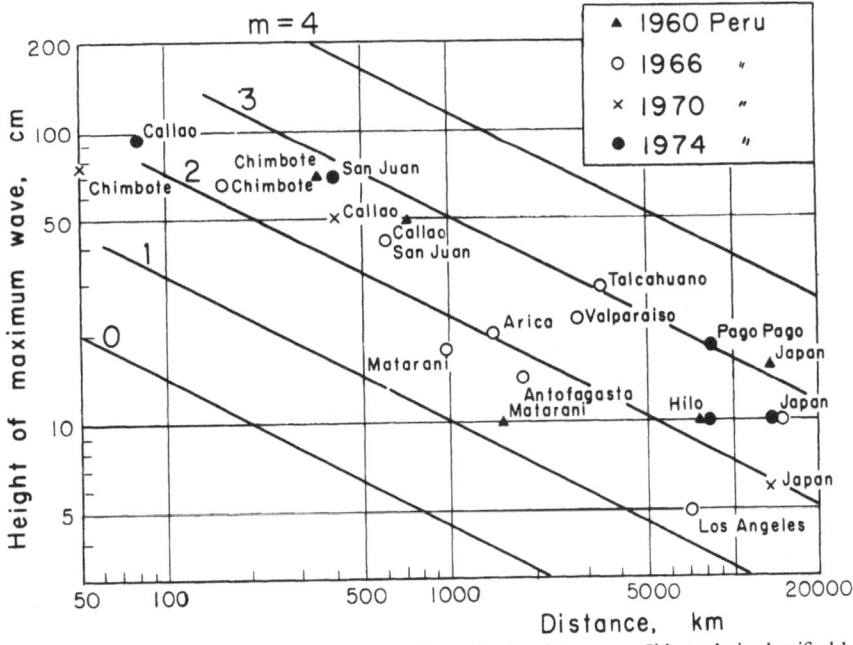

Fig. 4.   Magnitude of the Peru tsunamis. Tsunami magnitude of Imamura-Iida scale is classified by the attenuation of tsunami height with distance from the epicenter.

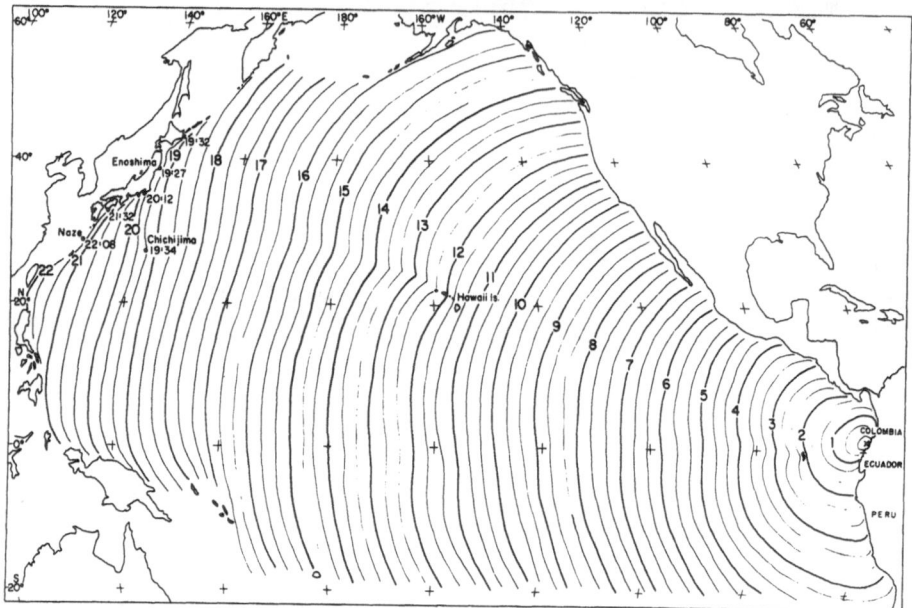

Fig. 5. Refraction diagram of the 1979 Colombia tsunami across the Pacific Ocean (time interval: 20 min). Travel times (h, m) observed at the Japanese tide stations are shown.

was 21 hours. The travel time of the 1974 Peru tsunami to northern Japan was 21.5 hours. The farther south the source is located the longer is the travel time to Japan, and the difference of travel times between northern and western Japan becomes shorter.

4.   Comparison of the Colombia-Ecuador Tsunamis in 1906 and 1979

Figure 6 shows tide-gauge records of the two tsunamis on Jan. 31, 1906 and Dec. 12, 1979 obtained at Ayukawa and Kushimoto (see Fig. 9). Wave-heights of the 1906 tsunami are 40–50 cm which are 2.5 times higher than those of the 1979 tsunami. With additional data obtained at Pacific tide-stations (IIDA et al., 1967; PARARAS-CARAYANNIS, 1980), the magnitude of the 1906 tsunami is compared with that of the 1979 tsunami in Fig. 7. Wave-heights at Hilo for many tsunamis were 3–4 times larger than those at Honolulu. Wave-heights at San Diago always were small in comparison with those at neighboring stations. Thus, the magnitude of the 1906 tsunami is determined to be $m = 3.5$, and that of the 1979 tsunami is $m = 2.5$.

Arrival times of the 1906 tsunami were recorded by a tide-gauge at Panama and observed by eyewitnesses at Guapi and Tumaco. Based on arrival times of the initial motion, the estimated tsunami source area is shown in Fig. 8. It is seen that the source area of the 1906 tsunami lay on the continental slope and extended about 400 km in an elongated shape parallel to bathymetric contours. The source area is 47×10^3 km^2. The source area of the 1979 tsunami lay within that of the 1906 tsunami.

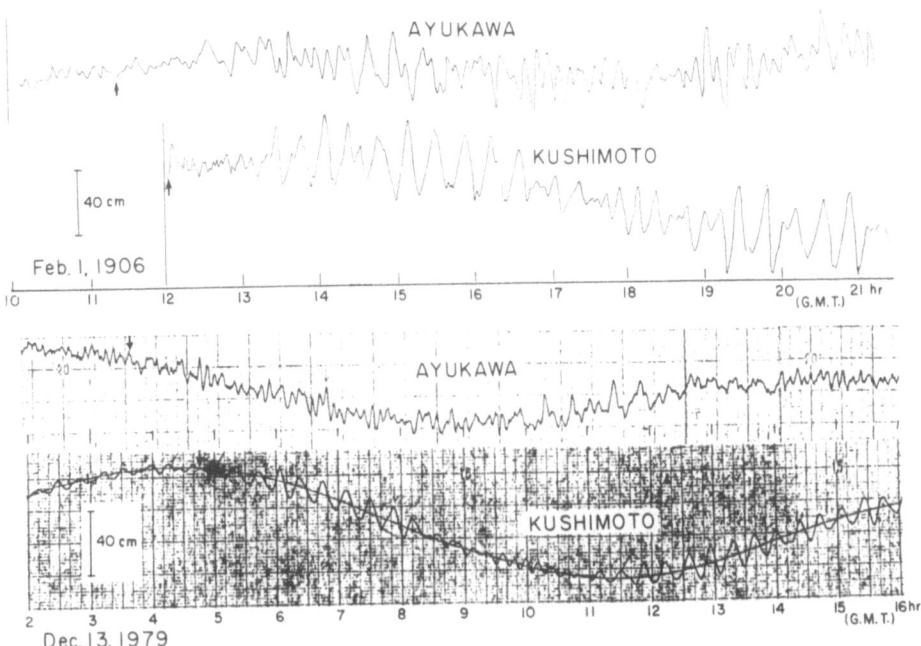

Fig. 6. Tide-gauge records of the Colombia-Ecuador tsunamis observed at Ayukawa and Kushimoto, Japan. Top: Ecuador tsunami of Jan. 31, 1906; Bottom: Colombia tsunami of Dec. 12, 1979.

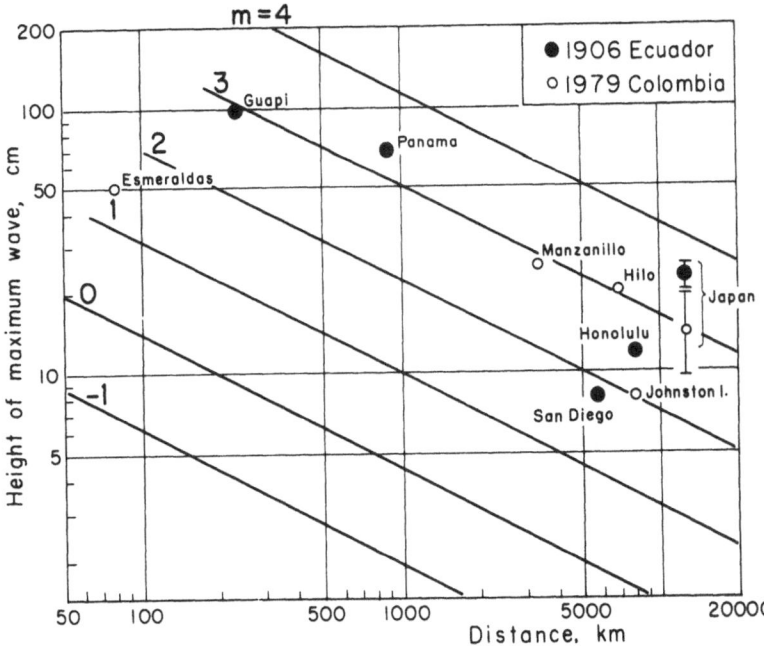

Fig. 7.   Comparison with semi-amplitudes of the 1906 Ecuador tsunami and those of the 1979 Colombia tsunami on amplitude-distance diagram.

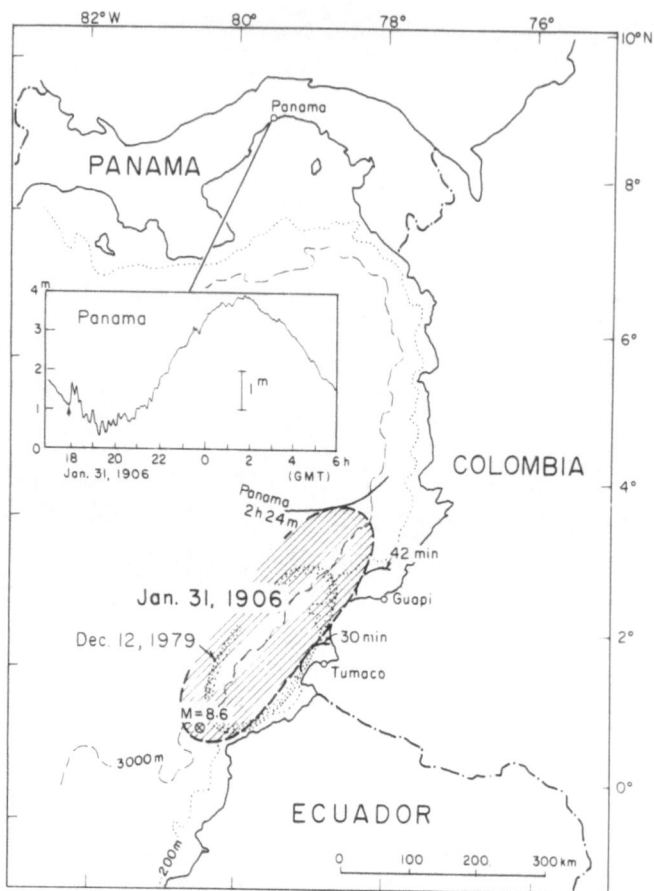

Fig. 8.   Estimated source area of the Ecuador tsunami on Jan. 31, 1906, and tide-gauge record. The last wave fronts correspond to travel times determined by a tide-gauge record and eyewitnesses.

5.   Source Areas of Historical Tsunamis

The Sanriku coast in northeastern Japan suffered damage from four historical tsunamis generated in the Peru-North Chile region since 1586. Figure 9 shows the distributions of inundation heights inferred from old documents. The heights at the Sanriku coast are larger than those at other regions, and the distribution patterns are similar to those of the 1960 Chile tsunami. Descriptions in newspapers of those days were very poor and did not report the North Chile tsunami of May 9, 1877, but many persons were drowned at Kujukurihama, Chiba Prefecture (MIYOSHI, 1962). The tsunamis in 1868 and 1877 also caused much damage to the Hawaiian Islands (IIDA *et al.*, 1967; PARARAS-CARAYANNIS, 1969). By these tsunamis 10–20 m in height, many coastal towns in the Peru-North Chile region were destroyed (e.g., BERNINGHAUSEN, 1962). In particular, the Peru tsunami on Oct. 28, 1746 was the most destructive at Lima, but it was not recorded in Japan.

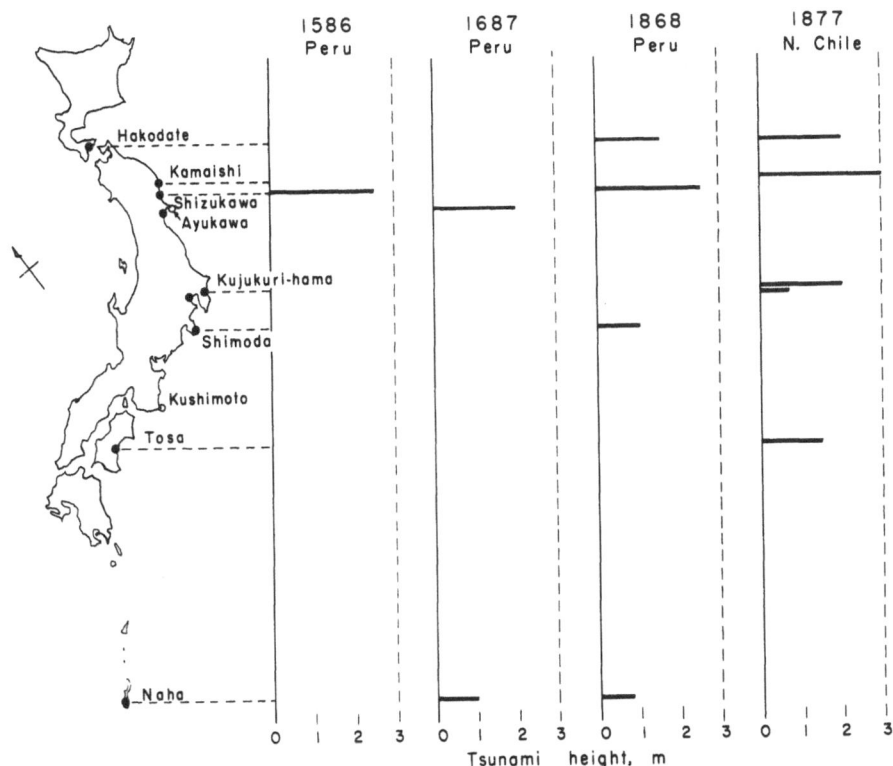

Fig. 9. Distributions of estimated inundation heights of historical Peru-North Chile tsunamis along the Japanese coast.

In Fig. 10, all the estimated source areas of tsunamis generated in the Colombia-Peru-North Chile regions since 1586 are shown on a bathymetric chart. Hatched areas represent the sources of tsunamis generated in this century. Four of the tsunamis during the period between 1913 and 1958 were not recorded in Japan. The source areas of recent tsunamis were estimated from the data of the aftershock areas together with some wave-front arrival times, while those of the historical tsunamis were inferred from the documents of inundation heights and seismic intensities bordering the epicentral regions. For example, tsunami inundation heights of 10–20 m and a seismic intensity I = 5 (JMA scale) for the 1868 and 1877 earthquakes were distributed along the coastal region of about 400 km (SOLOVIEV and GO, 1975). The lengths of large tsunamis in 1586, 1687, 1868, and 1877 may have been 400–500 km.

Source areas of these historical tsunamis line up along the Peru-Chile Trench. Source areas of recent tsunamis occupied a part of the historical tsunami sources. In the space distribution of tsunami sources, a remarkable gap can be seen in the region from South Peru to North Chile (14°–23°S), where no event has occurred for at least 105 years since the 1877 earthquake.

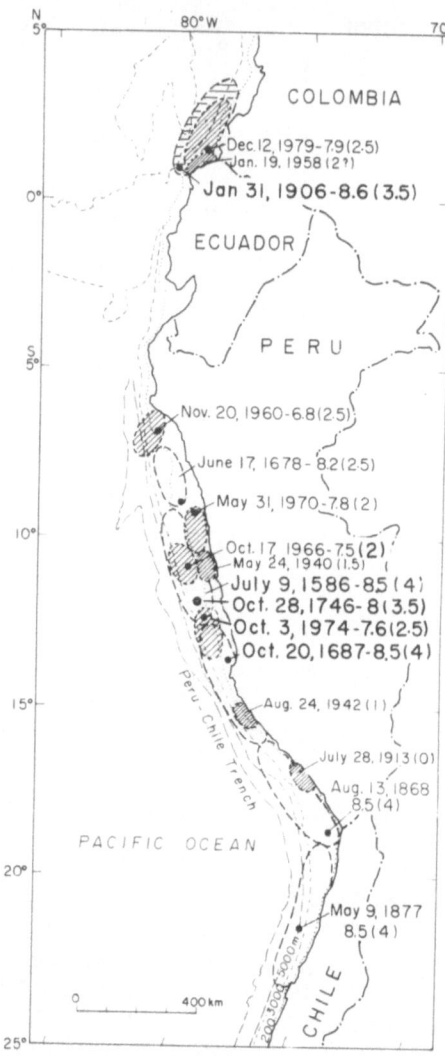

Fig. 10.   Estimated source areas of tsunamis in the Colombia-North Chile region. Hatched areas represent
source areas of tsunami generated in this century. Dates, earthquake magnitude, M, and tsunami magnitude,
(m), are indicated.

6.   Conclusion

Making use of tide-gauge records of the USCGS and Japan, the behavior of
Colombia-Peru tsunamis between 1960 and 1979 were investigated. Effects of the
tsunamis along the Japanese coast were evidently small. Tsunami magnitudes were m
= 2–2.5. Taking aftershock areas into account, the lengths of the tsunami sources were
estimated to be 120–270 km, and they were mostly located on the continental slope.

KELLEHER (1972) pointed out the existence of four seismic gaps in the Colombia-Peru-North Chile regions. Among them, the earthquake energy at two regions was later released by the 1974 Peru and the 1979 Colombia earthquakes. Among the remaining two gaps, the 600 km segment along the coast from Ecuador to North Peru is believed to be a region of low seismic activity since historic times. However, in the seismicity gap from Central Peru to North Chile, six historical tsunamis ($m = 3-4$) were generated and four of these hit the Japanese coast. Estimated source areas of these historical tsunamis were located on the steep continental slope, extending 400–500 km parallel to the Peru-Chile Trench. The source areas of recent tsunamis occupied only a part of the historical tsunami source areas and there are seismic gaps located off Chimbote ($8°-10°S$) and off Lima ($11°-13°S$). In particular, it is speculated that a region of relatively high tsunami risk lies off the coast from South Peru to North Chile ($14°-23°S$). The region from Central Peru to North Chile should be watched for future large earthquakes (M 8–8.5).

REFERENCES

ABE, K., Mechanisms and tectonic implication of the 1966 and 1970 Peru earthquakes, *Earth Planet Inter.*, **5**, 367–379, 1972.

BERNINGHAUSEN, WM. H., Tsunami reported from the west coast of South America, *Bull. Seismol. Soc. Am.*, **52**, 915–921, 1962.

HATORI, T., Study on distant tsunamis along the coast of Japan, Part 2: Tsunamis of South American origin, *Bull. Earthq. Res. Inst., Univ. Tokyo*, **46**, 345–359, 1968.

HATORI, T., Tsunami magnitude and seismic moment, *Zisin*, **31**, 25–34, 1978 (in Japanese with English abstract).

HATORI, T., Relation between tsunami magnitude and wave energy, *Bull. Earthq. Res. Inst., Univ. Tokyo*, **54**, 531–541, 1979 (in Japanese with English abstract).

HATORI, T., Colombia-Peru tsunamis that observed along the coast of Japan, 1960–1979, *Bull. Earthq. Res. Inst. Univ. Tokyo*, **56**, 535–546, 1981 (in Japanese with English abstract).

HONDA, K., T. TERADA, Y. YOSHIDA, and D. ISITANI, Secondary undulations of oceanic tides, *J. Coll. Sci. Imp. Univ. Tokyo*, **24**, 1–113, 1908.

IIDA, K., D. C. COX, and G. PARARAS-CARAYANNIS, Preliminary catalog of tsunamis occurring in the Pacific Ocean, Hawaii Inst. Geophys., Hawaii Univ., Data Report No. 5, HIG-67-10, 1967.

KELLEHER, J. A., Rupture zones of large South American earthquakes and some predictions, *J. Geophys. Res.*, **77**, 2087–2103, 1972.

MIYOSHI, H., Tsunami, *J. Oceanogr. Soc. Japan*, 20th Anniversary Vol., 265–271, 1962 (in Japanese with English abstract).

PARARAS-CARAYANNIS, G., Catalog of tsunamis in the Hawaiian Islands, WDCA-T, U. S. Dept. of Commerce, 1969.

PARARAS-CARAYANNIS, G., The tsunami of October 3, 1974 in Peru, *Newsletter*, International Tsunami Information Center, Hawaii, **8**, 18–21, 1975.

PARARAS-CARAYANNIS, G., Earthquake and tsunami of 12 December 1979 in Colombia, *Newsletter*, International Tsunami Information Center, Hawaii, **13**, 1–11, 1980.

SOLOVIEV, S. L. and C. H. GO, Catalog of tsunamis at the east coast of the Pacific Ocean, *Acad. Sci. USSR, Moscow*, 1–204, 1975 (in Russian).

YUMURA, T., List of tsunamis which accompanied distant earthquakes, *Technical Rep. Japan Meteorolog. Agency*, **8**, 247–255, 1961 (in Japanese).

WATANABE, H., Discriptive table of tsunamis in and near Japan, *Zisin*, **21**, 293–313, 1968 (in Japanese with English abstract).

Tsunamis—Their Science and Engineering, edited by K. Iida and T. Iwasaki, 185–204.

Study on the Earthquake and the Tsunami of September 20, 1498

Yoshinobu Tsuji

National Research Center for Disaster Prevention, Japan

(Received September 24, 1981; Revised November 10, 1981)

A series of enormous earthquakes with an interval of 100 to 150 years is known to occur off the coast of the southern part of the Japanese Islands (Fig. 1). In accompaniment with these earthquakes strong tsunamis were generated. The tsunami of Sep. 20, 1498 was the largest one. Recently, many historical documents were discovered, and a large amount of knowledge about this tsunami was obtained. The tsunami attacked the southern coast of the Kanto, Tokai, and Kinki districts. An inundation height of more than 10 meters was recorded at Kamakura on the coast of Sagami Bay, and several places on the coast of Suruga Bay. It is estimated that more than 20,000 people were killed by the tsunami. A seismic intensity of the VI (in JMA scale) is estimated for the east coast of the Kii Peninsula, the coast of Shizuoka Prefecture, and in the Kofu Basin. Aftershocks were observed in Kyoto and Nara Cities, and Yamanashi Prefecture for three years after the main shock.

1. Introduction

A series of enormous earthquakes with an interval of about 100 to 150 years is known to occur off the coast of the southern part of the Japanese Islands. The epicenters of all these earthquakes were situated on the northern slope of the Nankai Trough. The earthquake of Nov. 29, 684 is the oldest one known, and the pair of earthquakes on Dec. 7, 1944 (Tonankai Earthquake) and on Dec. 21, 1946 (Nankai Earthquake) are the latest ones. Accompanying each earthquake is a strong tsunami, and the people who live on the coast have always suffered from them. Among these tsunamis, that of Sep. 20, 1498 was the largest one. In Japan it is called the "Meio Earthquake Tsunami," because the year 1498 is the seventh year of the Meio Era in the Japanese calendar. This tsunami caused great damage to the coast of Kanto and Tokai districts, and of the Kii Peninsula (Fig. 1).

Until recent, we have had limited information about the 1498 Tsunami, because most of documents relating to it had been lost when, with the downfall of the feudal lord of the Tokai District, Yoshimoto Imagawa, in 1560, almost all documents in his territory were burnt. Only documents written by priests and some traditional information had been handed down about it.

In the last few years many historical documents pertaining to this tsunami were discovered, from which considerable information could be obtained.

The first attempt to collect documents on historical earthquakes and tsunamis was

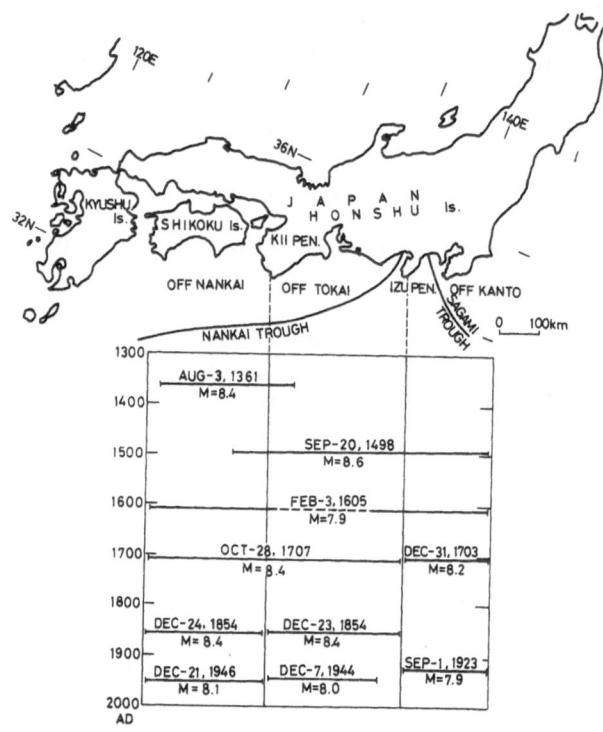

Fig. 1. Chronological diagram of huge earthquakes which occurred off the coast of the southern part of the Japanese Islands. The values of the magnitude of earthquakes are given by Usami (1979).

made by Tayama (1904). He published two volumes of data books in 1904, mounting to 1,201 pages in total. He introduced nineteen documents, concerning the Meio Earthquake, which occupied seven pages in his book.

MUSHA (1941–1949) revised and enlarged Tayama's book, and published four volumes of data books, amounting to 4,000 pages, in 1941 to 1949. The data for the 1498 Meio Earthquake are contained in the first volume, and thirteen new documents were introduced on the basis of those data. HATORI (1975) investigated the distribution of the inundation height of the tsunami along the coasts of the Kanto, Tokai, and Kinki Districts, and estimated the location of the tsunami source region.

In the last few years, a reinvestigation of historical documents has been made for the Tokai, Kinki, and Shikoku Districts, and a large amount of information was obtained (Tsuji, 1979, 1981a, b). Several documents were also introduced by IIDA et al. (1980) for the coast of Ise Bay, and by USAMI (1981) for the Kanto District and Niigata Prefecture.

Bibliographic discussions including investigation of the author of each document, background of the era of the writing, accurate meaning of the terms in the medieval ages in Japan, and the reliability of the document, are made by Tsuji (1980). Figure 2 shows the distribution of the places where the original documents were found with the indication of their introducers. Small circles, squares, rhomboids, triangles, and

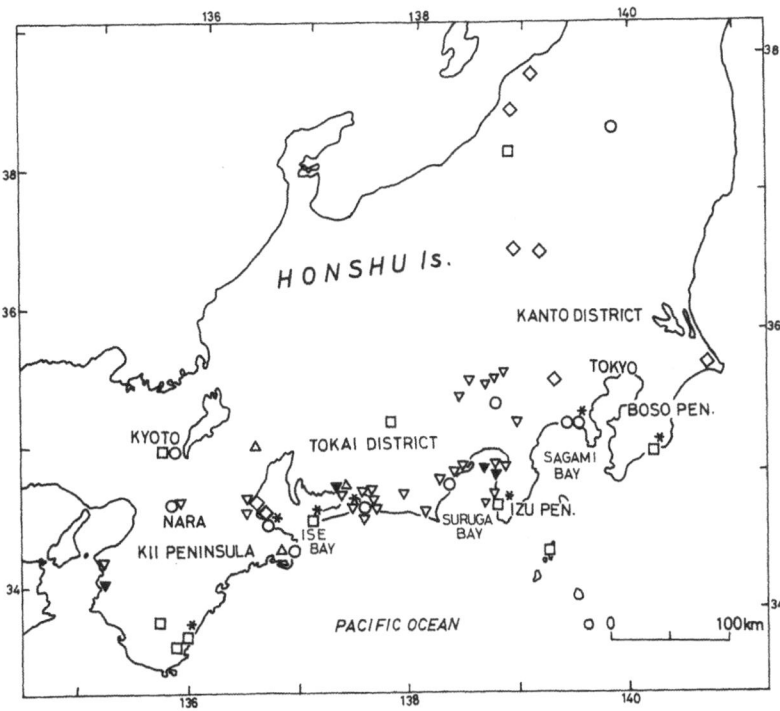

Fig. 2.   Distribution of places where the original documents on the 1498 Meio Earthquake Tsunami were found. Small circles, squares, rhomboids, triangles, and reversed triangles indicate the original data presented by TAYAMA (1904), MUSHA (1942), USAMI (1981), IIDA et al. (1980), and TSUJI (1979, 1981a, b), respectively. Stars show that places where the inundation height was estimated by HATORI (1975). Reverse black triangles show original documents recently found by the author.

reverse triangles denote whether the original documents were introduced by Tayama, Musha, Usami, Iida, or the author, respectively. Stars show that the inundation heights of the tsunami estimated by Hatori. Reverse black triangles indicate documents which were introduced in the present study for the first time.

Damage due to the direct effects of the 1498 earthquake was found not so heavy as that due to the tsunami. It is estimated that more than 20,000 people were killed by the tsunami. In the following section, we present events caused by the earthquake, and in Section 4 those by the tsunami.

2.   The 1498 Meio Earthquake

From descriptions in the original documents we can estimate the seismic intensity at each location. For example, descriptions of entirely destroyed houses or temples show that the seismic intensity VI to VII on the JMA scale. Descriptions of partial destruction of houses, toppling of gravestone, landslides, fissure in the ground, subsidence of land, abnormalities of gushing of hot springs, and changes in the course of rivers indicate an intensity of V or more. If there was no description of damage, and only descriptions of the disturbance of people, or the word "earthquake" with adjectives "big", "strong", "serious", or "heavy" we can estimate the intensity as IV or more. Thus, we can draw a map of the distribution of the seismic intensity for the 1498 Meio Earthquake (Fig. 3).

In Yamanashi Prefecture, the chief priest of the Fugenji Temple in Yamanashi City (A in Fig. 3) wrote the document "Oodaiki (The Chronicle of the Kingdom)", in which he states that Hozenji Temple in Kagami Village, which is in the town of Wakakusa at present, was destroyed, and the Kurokawa gold mine in the mountainous region of Enzan City suffered a serious collapse. Nakayama-Kogon'in Temple in the town of Ichinomiya was also destroyed.

The Buddhist priest Entsu said in his analects that the earthquake caused landslides, fissures on the ground, spouting of water from the earth in Hamaoka (B), and moreover, after a while, a huge tsunami came, which carried off everything, leaving nothing on the land. He wrote that innumerable persons were drowned.

Tokurenji Temple in the town of Tado (C) on the east coast of the Kii Peninsula was totally destroyed (IIDA *et al.*, 1980). It is estimated from several old documents that three temples in Tsu City (D) were destroyed. These documents do not indicate whether the destruction was caused by the earthquake or by the tsunami attack. In Shingu City (E)

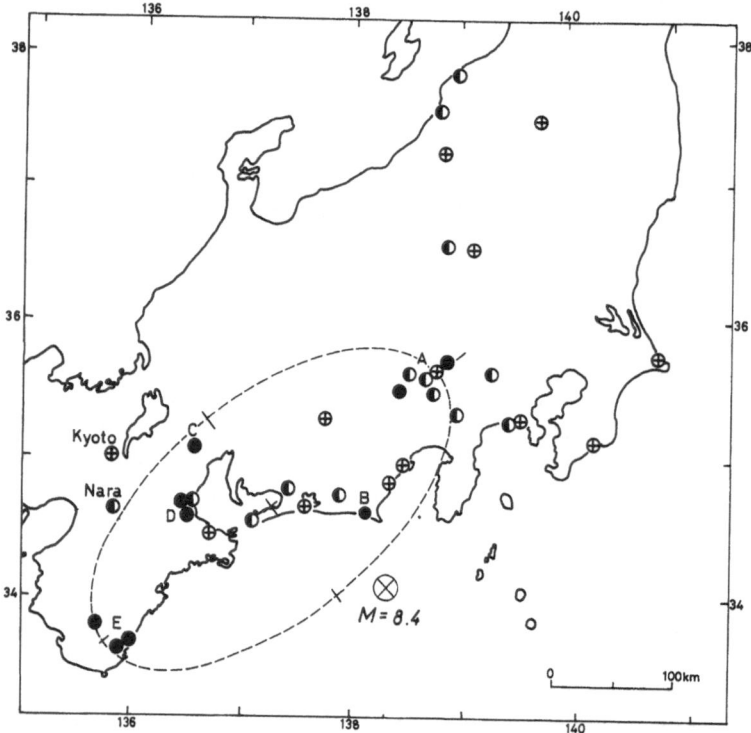

Fig. 3.   Distribution of seismic intensity of the Meio Earthquake. Circles with crosses, semi-blackened circles, and full black circles show the seismic intensity of the IV-th, V-th, and VI-th degree or more on the JMA Scale. As for letters A, B, . . E, see the text.

and its vicinity, a mansion of the lord of the district, the buildings of Seigandoji Temple, and of Hongu Shrine were destroyed due to the earthquake. In the mountainous region, the gushing of a hot spring called Mine-no-yu was interrupted for 42 days.

In Kyoto, which was the capital of Japan in those days, a court lady and six noblemen kept diaries. They state in their diaries that a large earthquake took place about 10 o'clock in the morning of Aug. 25 (Sep. 20 on the Gregorian calendar, Sep. 11 on the Julian calendar). Sir Sanetaka Sanjo wrote that he had never experienced such a severe earthquake in the 43 years of his life. But in no diary is there a description of destruction of houses or injured persons. The seismic intensity in Kyoto is estimated at the IV-th degree.

Nara was the center of the Buddhism at that time and was governed by Kofukuji Temple. The chief priest of the temple also kept a diary, and wrote that a building of the temple was slightly damaged. The intensity at Nara City is therefore estimated to be a weaker V-th degree.

Figure 4 shows the distributions of subsidence of the shore, landslides, fissures on the ground, changing of the course of the rivers, and abnormalities of hot springs.

In the town of Kawaguchiko (A in Fig. 4) on the northern foot of Mt. Fuji, there is a very old temple called Myohoji, which was founded more than 1,500 years ago. The chronicle of this temple says that three days after the earthquake a typhoon came through the district, and a great landslide occurred which burried three towns and more than fifty percent of their inhabitants were killed.

On the south coast of Boso Peninsula, Tanjoji Temple in the town of Amatsu-

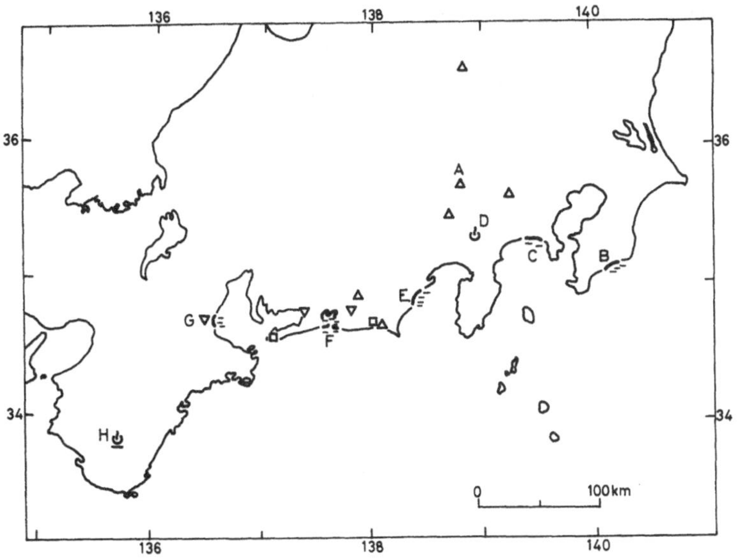

Fig. 4. Distributions of shoreline subsidence (indicated with a thick line with a minus), landslide (triangles), fissure on the ground (squares), changing of the course of river (reverse triangles), beginning of a gushing of hot spring (♨), and interruption of gushing of a hot spring (♨). As for letters A, B, . . H see the text.

Kominato (B) was swept away by the tsunami, the site of the temple became sea
bottom, and the buildings of the temple had to be reconstructed on a higher place.
HATORI (1975) estimated the inundation height as 4 to 5 meters there. It is not clear
whether the cause of regression of the shoreline was due to erosion by the tsunami, or
that of subsidence of the land due to the earthquake.

Off the coast of the Fujisawa (C), there is a small island called Enoshima. Twelve
years before the occurrence of the Meio Earthquake, the seabed around the island
suddenly heaved up and the island became connected to the mainland. Accompanying
the Meio Earthquake, the coast of this district subsided and since then the island has
again been separated from the mainland. The amount of subsidence was estimated at
about two meters.

Recently a local historian, Kyoji Hanno, who lives on the east foot of Mt. Fuji,
found a large number of old documents in the village of this district. A description in
one of the documents mentions that the hot spring called Yufune in the town of Oyama
began to gush after the earthquake (D).

On the coast of Yaizu City (E), the site of Rinsoin Temple subsided into the sea,
and the people were obliged to move the temple to a higher place. Subsidence also
occurred at the northern part of Lake Hamana (F in Fig. 4, and Fig. 5). The shoreline
regressed about one kilometer, and a town called Takase with 500 houses was
inundated. In former days, the lake had been separated from the sea and the water had
been fresh, but the separation bank was knocked down by the earthquake, and the lake
opened to the ocean. Four towns, Maisaka, Ubumi, Arai, and Hashimoto (a, b, h, and
g in Fig. 5) were washed away. At least eight shrines and two temples have recorded of
the Meio Tsunami, and all of them were washed away and reconstructed. According to
other documents it is said that the mouth of the lake opened because of the earthquake,
but the tsunami did not rush into the lake. The town on the shore of the lake was
submerged due to the high surge caused by a typhoon which attacked the Tokai
District the next year. Some inhabitants of the four towns on the bank moved to the
opposite shore and constructed the new village of Muragushi (c in Fig. 5). The tradition

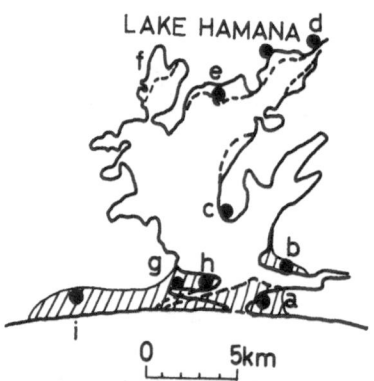

Fig. 5.   Detailed map of the Lake of Hamana. Hatched zone shows region submerged by the tsunami.
Broken and full lines show the shoreline before and after the tsunami. The crust of this district subsided with
the earthquake, and the shoreline regressed in several places. As for letters a, b, . . i, see the text.

of Hosoe Shrine in the town of Hosoe (d) says that it was founded in remembrance that the sacred jewels of the gods of a shrine in the town of Arai (g) were driven ashore on the coast. We also note that the rice field map of the ancient age for the innermost coast of the lake shows that the shoreline at the town of Mitsukabi (f) had been situated more southward in comparison with that of today, and that the earth subsided a few meters during the one thousand years in total.

At Tsu City (G in Fig. 4) the coastline regressed by two kilometers. Tsu City had been a prosperous port, but was seriously damaged and a part of the area inundated, and so since then the port declined. The town was obliged to move to a higher place for the peaceful living of the people. The pine grove on the coast became the sea bottom and after the earthquake only the top of the trees could be seen on the sea surface. The course of Ano river, which runs through the residential area of the city, changed at the same time. According to the description in old documents, the regression was not caused by the tsunami but by subsidence of the earth.

3. Aftershocks

In Kyoto and Nara, aftershock of the Meio Earthquake were recorded for three years. Figure 6 shows the number of aftershocks observed monthly in Kyoto and Nara, obtained on the basis of descriptions of all available diaries written in those cities. A rather strong shock occurred in the evening of Sep. 2 (Sep. 26 on the Gregorian calendar), and a strong shock occurred midnight of intercalary Oct. 28 (Dec. 9) of the year. Aftershocks were also felt in Yamanashi City (A in Fig. 4) for three years.

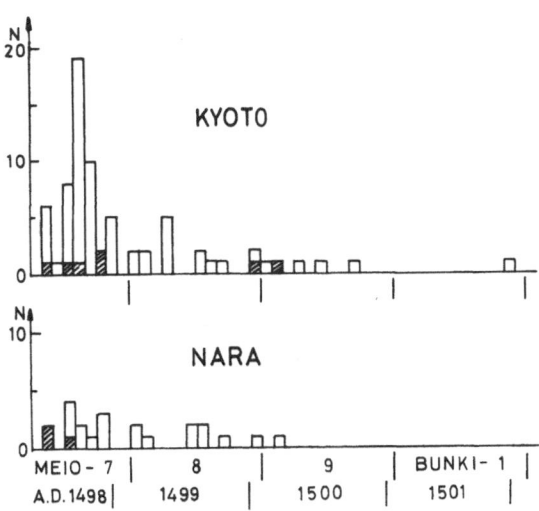

Fig. 6. Monthly number of aftershocks of the Meio Earthquake recorded in Kyoto and in Nara cities. Hatched areas indicate rather strong shocks.

4. Tsunami

Damage caused by the tsunami was considerably heavier than that by the earthquake. It is estimated that more than 20,000 people were killed by the tsunami in total. Figure 7 shows the calculated astronomical tide at the ports of Yokohama, Shimizu, Owase, and Wakayama, the locations of which are shown in Fig. 8 with squares and the letters in italics. An arrow on Fig. 7 shows the occurrence time of the earthquake. The tsunami was generated at low tide.

In the case that there are the explicit descriptions of the point to which the tsunami climbed, we can easily estimate the inundation height of the tsunami with considerable accuracy. On the other hand, some documents only contain descriptions of damage, for example, the washing away of towns and buildings, or of persons drowned. In the latter cases, we can not accurately estimate the inundation height. In this study we gave the presumed heights which were obtained by comparing the events associated with other tsunamis which have occurred in later times for those cases. The inundation height was estimated as 6 m or 5 m for the cases of on entirely washed out town, or a partially washed out town, respectively. Figure 8 shows the distribution of the estimated inundation height of the tsunami.

A short description on the tsunami appears in the chronicles written by the ancestors of a famous soy sauce maker in Choshi City (A). If no damage is recorded in it, it is assumed that the tsunami caused no harm to the city, and that the inundation height is less than two meters. The affairs of the town of Amatsu-Kominato (B) were already mentioned above.

At Kamakura (C), a temple located 14.5 meters above present sea level (measured by the construction section of Kamakura City Office) collapsed. A large statue of Buddha had been in the temple. Ever since the tsunami, this statue has been in the open air. The earth heaved up by about one meter due to the Kanto Earthquake of 1923, and another great earthquake occurred in 1703 in the same pattern as in 1923. It is probable that the level of the ground was two or three meters lower in 1498 than today. On the other hand, the run-up level of the sea water would have been two meters or higher than that of the

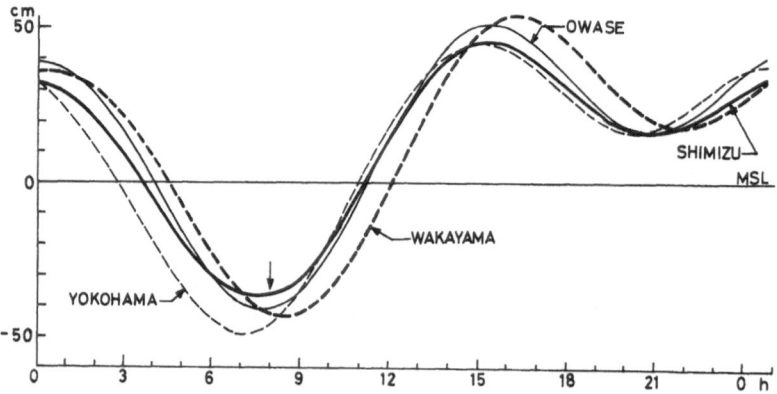

Fig. 7.   Astronomical tide at Yokohama, Shimizu, Owase, and Wakayama Ports. The locations of those ports are shown in Fig. 8 with squares.

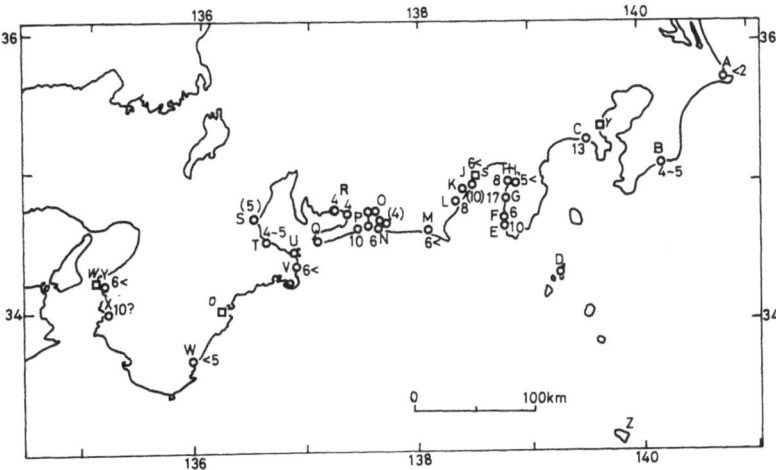

Fig. 8.   Distribution of the estimated inundation height of the Meio Tsunami. Square show the ports the astronomical tides of which are given in Fig. 7. As for the letters, see the text.

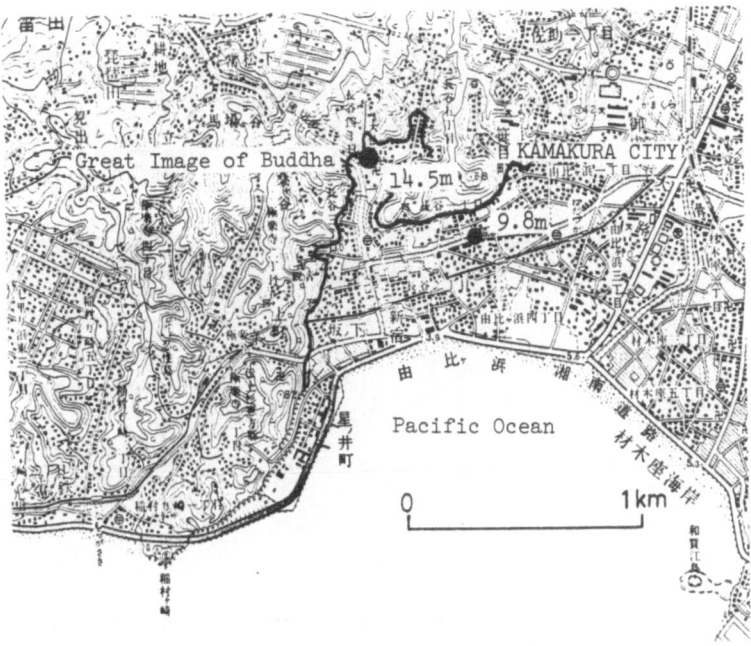

Fig. 9.   Inundation area in Kamakura. The great image of Buddha had been in a temple, which collapsed due to the Meio Tsunami. Since then it has been in the open air. Point C in Fig. 8.

194 Y. Tsuji

ground of the site of the temple, so that we can estimate the inundation height to be 13 meters or more. Hatori gave the height as 8 meters, but this was apparently underestimated.

On the shore of Hachijo Island (Z), a fisherman who preparing for sailing was drowned. Other damage was not recorded there, and the inundation height is estimated at two or three meters.

At the town of Nishina (E), sea water ran along the valley to Terakawa hamlet, which is situated two kilometers from the mouth of the river. A map on a scale of 1 to 25,000 shows that the height of the hamlet is about 10 meters above sea level. Tago Village (F), which is in the north of the town of Nishina, was also swept away, and after the tsunami the Tago Shrine was reconstructed on a small hill.

The tradition associated with Eigenji Temple in Odoi Village (G) was only recently discovered and says that more than thirty inhabitants were killed, all properties lost, and that after the tsunami people were dreadfully affected by a plague and a famine. The tradition also says that octopuses were found on the branches of the pine trees at the site of Hachiman Shrine in the village. The ground at the shrine is 14 m above sea level, and therefore the inundation height of the tsunami is estimated to be 17 meters or more.

Thirteen volumes of historical books "Zuzhu-Shiko (The History of Izu Peninsula)" were written by Funan Akiyama in 1798. A short description of the Meio Tsunami at Enashi Village (in Numazu City at present; H in Fig. 8) is mentioned in this

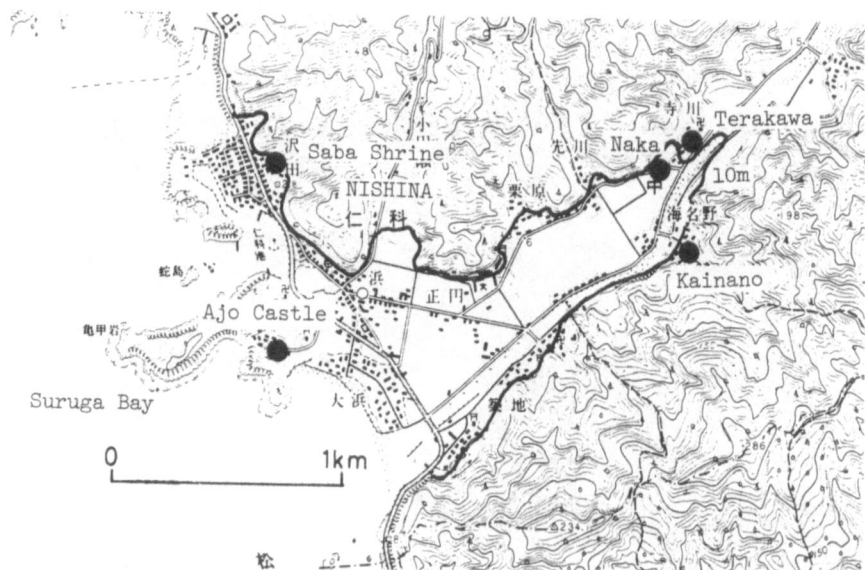

Fig. 10. Inundation area in the town of Nishina (Nishi-Izu at present). The Meio Tsunami climbed up to the hamlet of Terakawa, the level of which is more than 10 meters above the sea. The rice fields of Naka hamlet were entirely submerged. Saba Shrine was also damaged and was reconstructed after the tsunami. The lord of this district had lived in Ajo Castle, and after the attack of the tsunami he moved to Kainano hamlet for safe living (E in Fig. 8). .

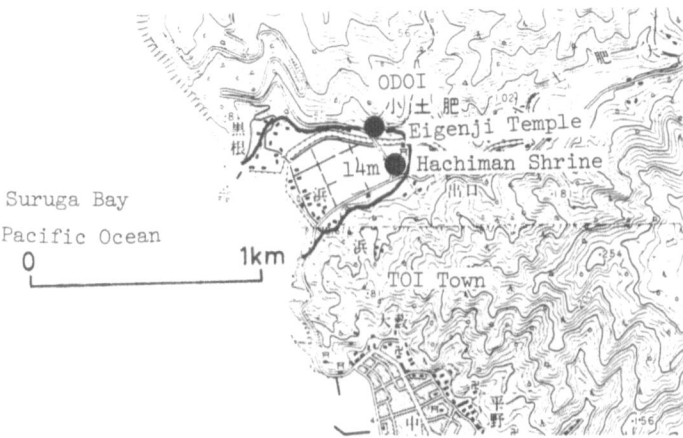

Fig. 11.   The tsunami climbed up to the branches of the pine trees in the site of Hachiman Shrine the level of which is 14 meters above the sea. The tradition of Eigenji temple in Odoi village (the town of Toi at present) says that more than 30 people were killed by the tsunami. G in Fig. 8.

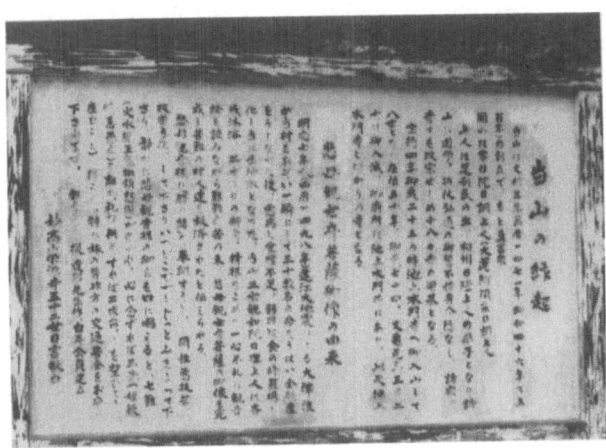

Photo 1.   Information plate of Eigenji temple in Odoi town. The founding year of the temple and damage due to the Meio Tsunami are written.

book, where says that the statue of the god in Kohoin temple 'performed a miracle' on the occasion of the Meio Tsunami. It was not clear what the concrete meaning of the word 'miracle' is. But only recently, another document was found (Photo 2) in this village. This document explains that most treasures and old documents of the village were washed away due to the Meio Tsunami in 1498. The priest of this temple, Mr. Taiken Kato, has stated that the word 'miracle' probably means that all houses in the village were washed away, but that only the main building of the temple, in which there was the image of the god, escaped from the disaster, and that only the inhabitants who took refuge in the building were saved from drowing. The level of the main building of

Photo 2.   A document which was just recently found in Enashi village, and which says that almost all the treasures and old documents of the village were washed away due to the tsunami of Aug. 25, in the seventh year of the Meio Era (Sep. 20, 1498, line with arrow).

the temple is 9 m above the sea, and therefore the inundation height is estimated 8 meters or more.

At Ashiho Village (I), a building of Tenjin Shrine was deposited onto the marine terrace directly after the tsunami. It is assumed that this village was seriously damaged by the tsunami.

The tradition of Kaichoji Temple in the Muramatsu Block of Shimizu City (J),

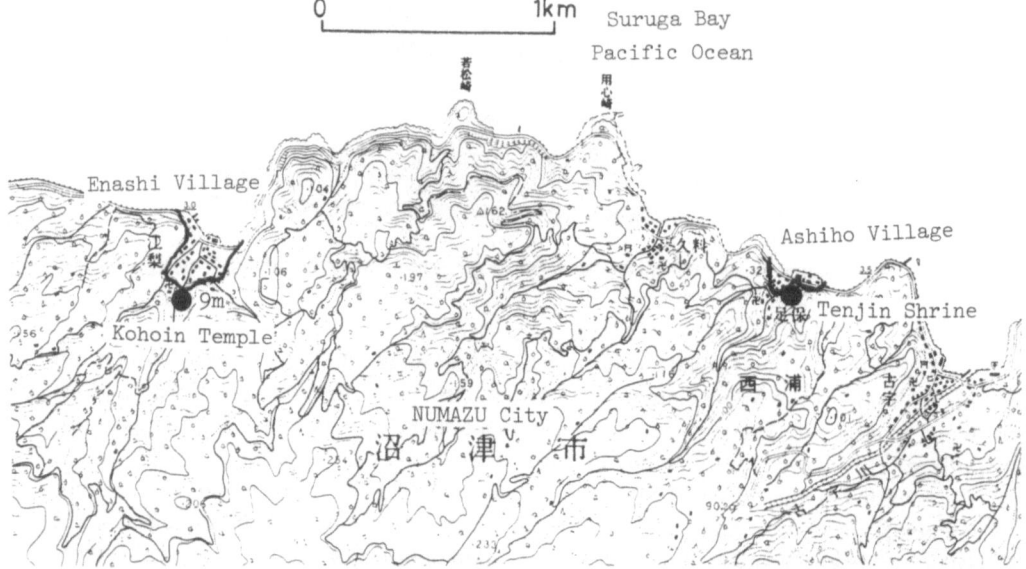

Fig. 12.   The Meio Tsunami washed out almost all houses in Enashi and Ashiho villages (H and I in Fig. 8).

says that houses, buildings of the temples, trees, cows, and horses were swept away, and that branch temples in Kogawa Village (in Yaizu City at present) were also washed away without leaving a trace (L). On the west of Suruga Bay, the tsunami climbed up to the following villages; Terada Village (10 m above sea level, K in Fig. 8) in Shizuoka City, Sangamyo Village (7 m), Egenoshima (3 to 4 m, swept away), Shibahara (7 m), and Kogawa Village (5 to 6 m, swept away) in Yaizu City (L). Sangamyo Village is located two kilometers from the shore. The inundation height was estimated at 7 to 8 meters in this district. A reliable document of Kyonenji Temple in Yaizu City describes that more than 500 houses were swept away, and several thousand people were killed including those in the neighboring villages.

Events in the town of Hamaoka (M) and on Lake Hamana have already been discussed in a previous section. The post town of Shirasuka (P) was also damaged and

Fig. 13.   Inundation area in Yaizu City. Legend says that ship were washed up to Sangamyo Village, and that all houses and buildings of temples in Kogawa and Egenoshima Villages were entirely swept away. L in Fig. 8.

after the tsunami seven families moved to the hill and constructed a new town. Two hundred and nine years after the occurrence of the Meio Tsunami, the huge tsunami of the Hoei Earthquake (1707) attacked the town (see Fig. 1), and since then almost all people of the town have moved to the hill. The height of the former town is 6 to 8 meters above sea level and therefore the inundation height is estimated at 8 meters or more.

Chronicles of the Jokoji Temple in Horikiri village near the tip of the Atsumi Peninsula (Q) describe that the earthquake caused fissure in the ground, and at the same time the tsunami struck the coast. No damage is described.

On the coast of Mikawa Bay (R), Susanoo Shrine in Muro-Yoshida Village (Toyohashi City at present), and Hakusan Shrine in Shiozu Village (Gamagori City at present) were washed away, and after the tsunami these shrines were moved to higher places. The course of the Toyokawa River altered due to the earthquake (IIDA et al., 1980).

About one thousand houses of Oominato Port (T) in Ise City on the east coast of the Kii Peninsula were entirely swept away, and no one escaped from drowing. Five thousand people were killed in the port, and at least 10,000 people were killed in the neighboring districts. A document mentions that the tsunami climbed up to Nagaya Hamlet which is located 3.5 kilometers from the shoreline (HATORI, 1975), and as its level is 4 meters above sea level, the inundation height is estimated at 4 to 5 meters. Hatori gave the height as 6 to 8 meters by consideration of the damage to houses and injuries, but this estimate is apparently somewhat large. As there is no hill behind the port and the people could therefore not find shelter near the town, this would be the

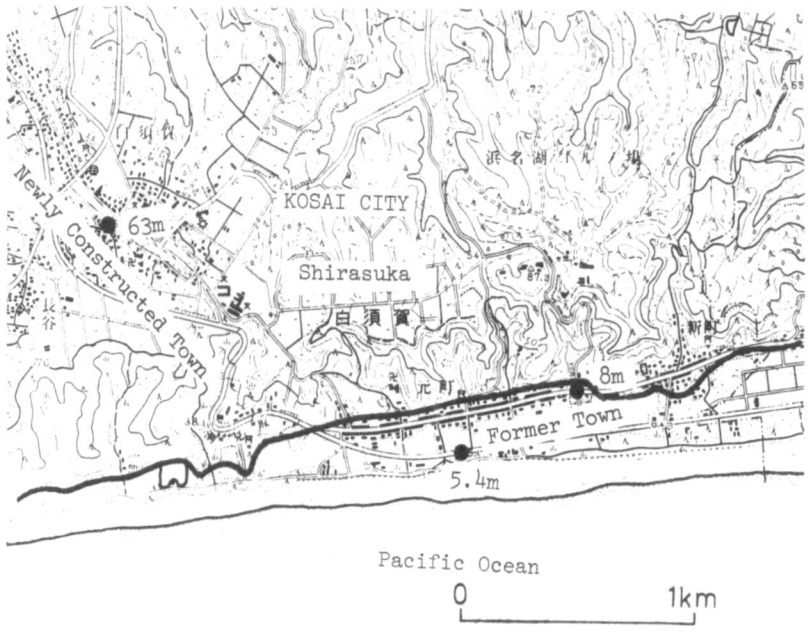

Fig. 14.   The post town of Shirasuka was also swept away (the former town). Some persons moved onto a hill and constructed a new town for safe living after the Meio Tsunami.

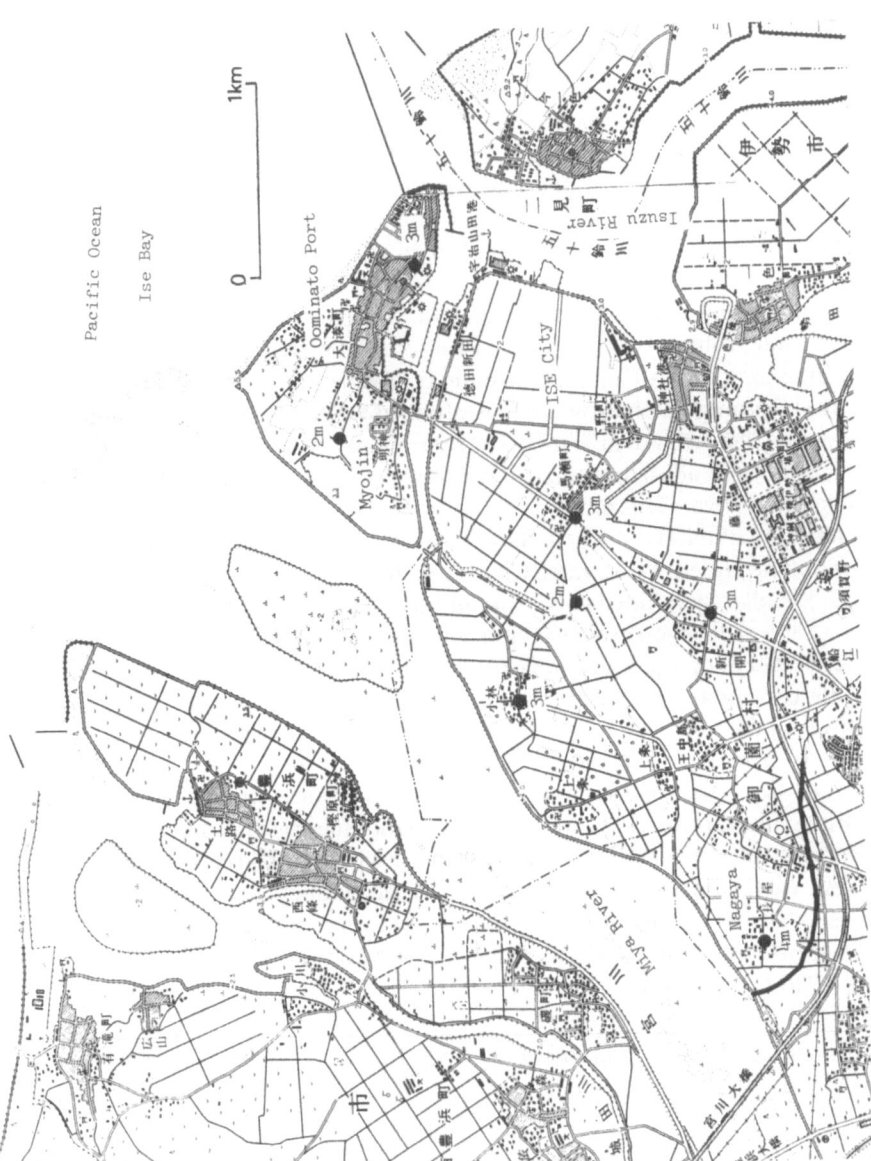

Fig. 15. One thousand houses with five thousand persons were drowned in Oominato Port in Ise City, on the west coast of Ise Bay. The tsunami climbed up to Nagaya Hamlet (HATORI, 1975) which is 4 meters above sea level.

reason why the greater part of the inhabitants were killed in spite of the fact that the tsunami height was not so high.

One historian has assumed that a village on Toshi Island (U) was ruined due to the Meio Tsunami, but it is only a presumption and not reliable.

Ootsu Village (V) in Toba City near the top of Shima Peninsula was also entirely swept away and the residents and one temple and two shrines relocated to the neighboring village of Kuzaki.

At Shingu City (W), the mansions of the lord of this district were destroyed. The destruction was caused not by the tsunami but by the earthquake. The mansions had been situated on a seaside plain called Tazuru-Hara, the height of which is 4 to 5 meters, and so the inundation height of the tsunami is estimated as less than 5 meters there.

A legend says that the tsunami climbed up the gate of Hachiman Shrine in Hirokawa Town (X) on the southwest coast of Kii Peninsula. The gate is located on ground 10 meters above sea level. The legend is not so reliable for our study.

Ninety-seven volumes of the local history books of Kii Province (Wakayama Prefecture at present), "Kii-Koku-Zoku-Fudoki (The Second series of the Description of the Natural Features of Kii Province)" were edited by Koko Niida, a great scholar of the Wakayama Clan, who died in 1839. In those books, a description is given of the prosperous port of Wada-Unoshima, which was located on the north shore of the mouth of the Kino River. This port was entirely swept away and the inhabitants moved to the south shore of the river, where they constructed new towns which comprise the central zone of the present Wakayama City. The relocation of the town in the Meio Era (1492–1501) is also mentioned in the traditions of five shrines and one temple in Wakayama City. A question arose that none of those documents mention the accurate date of the disaster and, moreover, none of them state explicitly that the high wave was caused by the earthquake, and therefore it is premature to make the conclusion that the washing away of the town was truly caused by the 1498 Meio Earthquake Tsunami. It is necessary to determine whether the surge was caused by the earthquake or by a typhoon. On the basis of bibliographical, meteorological, and oceanographical considerations, Tsuji (1980, 1981c) demonstrated that there is only a small possibility that the surge in these descriptions was caused by a typhoon. The cardinal grounds of his argument are (1) the sea off Wakayama City is deep (about 60 meters in mean), and therefore a high surge would hardly generated by a strong wind, and (2) no description of a strong storm appears in any diary written in Kyoto and Nara Cities during the Meio Era (1492–1501).

5. Estimations of the Magnitude of the Meio Earthquake and of the Location of the Epicenter

It is impossible to estimate the magnitude and the location of the epicenter of historical earthquakes with the accuracy of modern science. Errors within plus or minus 0.4 in magnitude and 50 km in the location of the epicenter, or so, seem to be unavoidable.

Muramatsu (1969) gave formula for the estimation of the magnitude of an earthquake as

$$\log_{10} S_{VI} = 1.36\,M - 6.66$$

where S_{VI} denote the area (km^2) of the region where a seismic intensity of VI-th degree is observed, and M is the magnitude. If we regard the elliptic shaped zone in Fig. 3 as a region of the VI-th degree, then we have $S_{VI} = 6 \times 10^4$ km^2 and $M = 8.4$, which would be an estimation on the lower side because the seaward area of the VI-th degree region is estimated conservatively.

A calculation of the mean value of the longitude of the tsunami at the observation points weighted with the inundation height at each point gives $\lambda = 138.3°$E. By consideration of the region of VI-th degree, the epicenter is estimated as $\lambda = 138.3°$E, and $\varphi = 34.1°$N. USAMI (1979) gave $M = 8.6$, and $\lambda = 138.2°$E, $\varphi = 34.1°$N. We may say that there is no significant difference between them.

6. Comparison with Other Earthquakes of the Same Series

As mentioned in the introduction, the 1498 Meio Earthquake is one in a series of huge earthquakes which has occurred off the south coast of the Tokai District with an interval of 100 to 150 years. Other earthquakes of the same series are the 1707 Hoei, the 1854 Ansei-Tokai, and the 1944 Tonankai Earthquakes. These earthquakes have several similarities and dissimilarities. Hereafter, for brevity we denote these earthquakes as H, A, and T respectively, and the 1498 Meio Earthquake as M.

One of the eminent common characteristics of these earthquakes is that after their occurrence, another huge earthquake was produced almost in the same time, or within a few years at the most, off the south sea region of the Kii Straits. The Kii Straits are located between Shikoku Island and Kii Peninsula, and the sea region is sometimes called the Nankai Sea. As examples, 32 hours after the occurrence of A, the Ansei-Nankai Earthquake was felt, and two years after T, the 1946 Nankai Earthquake took place. In the case of H, it is considered that two huge earthquakes occurred simultaneously, one in the south sea of the Tokai District, the other in the Nankai Sea region. In contrast to those earthquakes, for M it has been considered that only one earthquake in the sea south of the Tokai District occurred, and that it was not accompanied by an earthquake in the Nankai Sea region. However, it is found out in the present study for the first time that, Tsunami M also hit Wakayama City. Damage in Wakayama City indicates the possible existence of another earthquake in the Nankai Sea, independently. At least we can conclude that part of the tsunami source region extended to the southwest, off the Kii Peninsula. If documentation of the tsunami could be found for the coast of Shikoku Island, the hypothesis could be tested, but it may be hopeless to search for records on Shikoku Island because a serious battle took place there in 1585, and almost all old documents were lost.

It is well-known that in the area of Lake Hamana, the land subsided with the occurrence of earthquakes H and A. It was determined in this study that subsidence also took place with the occurrence of M. In addition to damage on the Tokai coast and on the west coast of the Kii Peninsula, serious damage was also produced in the Kofu Basin by H and A. We found a similar situation for M.

The distributions of the tsunami inundation heights of M, H, and A are very similar for the coast of Suruga Bay, the mouth of Lake Hamana, and the east coast of the Kii Peninsula. This shows that the earthquakes of M, H, and A were of the same nature. However, M has several particular characteristics which the others do not.

The inundation height of 13 meter in Kamakura, and subsidence on Enoshima Island show that the tsunami source areas extended to the south of the Kanto District. Only weak tsunamis with heights not higher than two meters were observed on the south coast of the Kanto District in H, A, and T.

It is worth noticing that a hot spring began to gush due to M at the town of Oyama, east of Gotemba City, where no abnormalities were observed, and at which little damage occurred for the case of a series of earthquakes of south of the Kanto District, for example, the 1703 Genroku and the 1923 Kanto Earthquakes (see Fig. 1), which broke out along the Sagami Trough. These facts show that the 1498 Meio Earthquake was caused not only by rupture at the Nankai Trough, but also at the Sagami Trough. It should be noted that the south coast of the Kanto District subsided in M, while it heaved up for the Genroku and Kanto Earthquakes. It is a particular characteristics of M that Niigata Prefecture, on the coast of the Japan Sea, also suffered, while no damage occurred for H, A, and T.

7.   Summaries

Newly found information concerning this earthquake and tsunami as follows;
    i)   the distribution of the inundation height of the tsunami, extends from the south coast of the Kanto District, where a tsunami with the height of 13 meter was recorded, to the west coast of the Kii Peninsula, where about a 6 meter wave height was recorded.
    ii)   Seismic intensity of V-th and VI-th degree occurred in Niigata Prefecture and in the Kofu Basin.
    iii)   The tsunami attacked the Atsumi Peninsula (Q in Fig. 7) immediately after the occurrence of the earthquake.
    iv)   Land Subsidence took place on the south coast of the Kanto District, in Yaizu City, on the coast of Lake Hamana, and on the shore of Tsu City.
    v)   A hot spring began gushing in Oyama Town, and the gushing of a hot spring in Minenoyu in the mountainous region on Kii Peninsula was interrupted.

Recently, numerical calculations of tsunamis have been made for many actual cases, and in some of them successful results were obtained. If a numerical calculation for the 1498 Meio tsunami is to be made in the future, the conditions above mentioned should be remembered.

8.   Discussion on the Method of Interpretation of the Original Documents

Some of the history books compiled by municipalities in recent years contain descriptions of the 1498 Meio Earthquake Tsunami. A number of these are written on the basis of local original documents, and are very valuable for our study. On the other hand, descriptions in others are written not on the basis of local documents, but on the chronological tables given by established textbooks of Japanese seismology. The latter scarcely give us original information and are not worth mentioning.

Another point that we should be careful of when we study historical earthquakes and tsunamis, is that we should not describe the matter on the basis of imagination. As for the present study, in all of records which were found on the west coast of Izu Peninsula, only damage due to the tsunami are described and there are no descriptions of the earthquake. In this case, we should frankly state that the seismic intensity on the west coast of Izu Peninsula is undetermined on bibliographical considerations. Thus the author did not give an assumed intensity on the distribution map (Fig. 3). Of course, we might easily assume the intensity on the coast of the V-th to VI-th degree from reasonable seismological considerations, but if we plot the assumed value, we would be guilty of the fabrication of data. We should also notice that the judgement of the present study is formed on the basis of old documents, and that the correctness of the judgement is subject to reliability of each document.

The author wishes to express his thanks to Mr. Shingo Sekino of Kuzura Port in Numazu City; to Mr. Taiken Kato, the priest of Kohoin temple in Enashi village in Numazu City; and to Mr. Sumio Shibata, the Director of the Historical Museum of Barrier Station of the town of Arai in Shizuoka Prefecture, for their assistance in collecting valuable information.

The author also thanks Profs. T. Usami, K. Kajiura, and Y. Nagata of University of Tokyo, for their valuable suggestions.

REFERENCES

HATORI, T., Sources of large tunamis generated in the Boso, Tokai and Nankai Regions in 1498 and 1605, *Bull. Earthq. Res. Inst.*, **50**, 171–185, 1975 (in Japanese).

IIDA, K., and Research Group on the Earthquake Damages in Tokai District, *Research and Comparison on Four Big Earthquakes of the Meio, the Hoei, the Ansei-Tokai, and the Tonankai*, pp. 210, Aichi Institute of Technology, 1980 (in Japanese).

MURAMATSU, I., Relationship between the distribution of the seismic intensity and the magnitude of earthquake, *Shizenkagaku*, Gifu Univ., **4**, 168–176, 1969 (in Japanese).

MUSHA, K., *Historical Materials on Japanese Earthquakes*, revised ed., Vols. 1–3, Shinsai Yobo Hyogikai, 1941–1943 (in Japanese).

MUSHA, K., *Historical Data on Japanese Earthquakes*, pp. 757, Mainichi Press, Tokyo, 1949. (in Japanese).

TAYAMA, M., *Historical Data on Japanese Earthquakes*, Rep. Imp. Earthq. Invest. Comm. 46, A—pp. 606, B— pp. 595, 1904 (in Japanese).

TSUJI, Y., *Data of Earthquakes and Tsunamis in Tokai District*, Vol. 1, pp. 436, Nat. Res. Center for disaster Prevention, 1979 (in Japanese).

TSUJI, Y., Bibliographical discussions on the 1498 Meio earthquake tsunami, *Kaiyokagaku*, **12**, 7, 504–526, 1980 (in Japanese).

TSUJI, Y., *Supplementary Data on Historical Earthquakes and Tsunamis in Japan*, pp. 41, Nat. Res. Center for Disaster and Prevention, 1981a (in Japanese).

TSUJI, Y., *Data of Earthquakes and Tsunamis in Kii Peninsula*, Nat. Res. Center for Disaster Prevention, 1981b (in Japanese).

TSUJI, Y., The Meio tsunami attacked Wakayama City, *Kagaku*, **51**, 5, 329–333, 1981c (in Japanese).

USAMI, T., Study of historical earthquakes in Japan, *Bull. Earthq. Res. Inst.*, **54**, 399–439, 1979.

USAMI, T., *New Collection of Materials for the History of Japanese Earthquakes*, Vol. 1, pp. 193, Earthq. Res. Inst., 1981 (in Japanese).

204   Y. Tsuji

Note added in proof

The present author (1980) had once mentioned that the Meio Tsunami hit the Niijima Island (D in Fig. 8). Prof. Usami indicated that this was an argument founded on a false basis in interpretation of original materials. Moreover, he suggested that there are many cases in which we can not describe historical affairs in a conclusive tone. The present author wishes to withdraw the argument on the harm on Niijima Island due to the Meio Tsunami. He also wishes to take note the suggestion with profound respect and humility.

Tsunamis—Their Science and Engineering, edited by K. Iida and T. Iwasaki, 205–211.

The Largest Tsunami in the Sanriku District

H. Miyoshi,* K. Iida,** H. Suzuki*** and Y. Osawa****

*3–3–33 Minami-Oizumi, Nerima-ku, Tokyo, 177, Japan
**Aichi Inst. of Tech., Yakusa-cho, Toyota, Aichi Pref., 470–03, Japan
***Sakari-cho, Ofunato City, Iwate Pref., 022, Japan
****Raga, Tanohata Village, Iwate Pref., 032–04, Japan

(Received September 20, 1981; Revised March 8, 1982)

Two typical tsunami earthquakes (1771 and 1896) with low frequency, the true magnitudes of which were almost 8, occurred off the coast of Japan. They produced the tremendous tsunamis. The recent of these is that of 1896. The resulting tsunami attacked the Sanriku District. The numbers of eyewitnesses are now decreasing rapidly. Traditions concerning this unusual tsunami are, however, even now vivid and accurate. We should therefore, hurry with our work of recording them. As a part of this work, we visited Raga, Tanohata Village, in the Sanriku District, and Y. Osawa, an inhabitant of this village, carried out the measurement of the distance between the locations of a huge rock before and after this tsunami. As a result of our survey, it was confirmed that the tsunami of 1896 was underestimated in the northern part of the Sanriku District, where the scientists seldom visited. The capability of conveyance by tsunami, seldom confirmed in the mainland of Japan, was confirmed in this study.

1. Introduction

It is said that in the Sanriku District, approximately 27,000 and 3,000 people were killed by the tsunamis of 1896 and 1933, respectively. The former figure is, however, very inaccurate, and a student of the local history, who had studied this problem in detail, suggested the figure of some 22,000 instead of 27,000 quite recently. Of course, the surveys concerning this tsunami of 1896 were, as a whole, very rough. Especially the northern part of the Sanriku District had been inaccessible till quite recently. We concluded, therefore, that even a brief report might be significant.

We tried some additional important considerations in this paper.

2. Survey in Tanohata Village (I)

While the southern part of Sanriku coast is a typically submerged coast, the northern part is an emerged one, where the coastal terraces have been deeply carved by rivers. It was therefore very difficult to visit the northern part till quite recently, and few surveys concerning the 1896 tsunami in the north could be accomplished.

Under these circumstances, in spite of the fact that the origin of the 1896 tsunami was a little farther north than that of the 1933 tsunami, and much nearer to the northern coast, and the northern part of the Sanriku District was therefore more severely hit in 1896, only one wave height of 23 m reached by the tsunami of 1896 was recorded (IGI, 1897).

We visited Raga, Tanohata in 1980 and found that the tsunami reached as high as 29.0 m in 1896, at the former house of Mr. Shimosaka. This was determined using a map on a scale of 1 to 25,000, and a hand level. As the slope between the Shimosaka house and the shoreline is considerably steep, correct survey method could be used. The previously reported height of 23 m is not reliable because it was obtained without a level.

The present Shimosaka house is shown in Fig. 1. The tsunami of 1896 left a piece of kelp on the eaves of the original house at this site. In the neighborhood of this site, there is another house, the residents of which had moored their boat in the front garden. This boat was washed away by the tsunami. We determined that this front garden is 28 m above mean sea level. This fact gives a strong supporting evidence of the wave height of 29 m at the Shimosaka house.

The estimated height of 29 m is, of course, contains small instrument errors. We will, however, assume the height of 29 m from now on. We can easily guess that at many locations the tsunami of 1896 had heights larger than 29 m. Along the indented coastline of the southern half of the Sanriku District, we can find only three V-shaped bays, those being Ryori Bay, Yoshihama Bay and Ryoishi Bay. IGI (1897) reported that the tsunami of 1896 ran up as high as 22 m, 24.4 m, and 17.7 m, respectively, in these three bays. We expect that higher values may be found in these three bays. Actually, MATSUO (1933) reported that the tsunami of 1896 ran up as high as 38.2 m right behind the head of Ryori Bay (shown in Fig. 2).

It was confirmed that a large block of rock was carried up as high as 20 m in Raga by the tsunami. Its surface is perforated by many holes made by organisms in the sea water. It seems rather paradoxical, when we consider that the suction of the receding water wave is stronger than the pressure of the surging wave. This fact suggests that the motion of sea water is not reversible even on the land of the V-shaped bay, that is, when

Fig. 1. The present Shimosaka's house, Raga, Tanohata Village. The second highest run-up in the mainland of Japan, reached by the 1896 tsunami caused by an earthquake, was observed here.

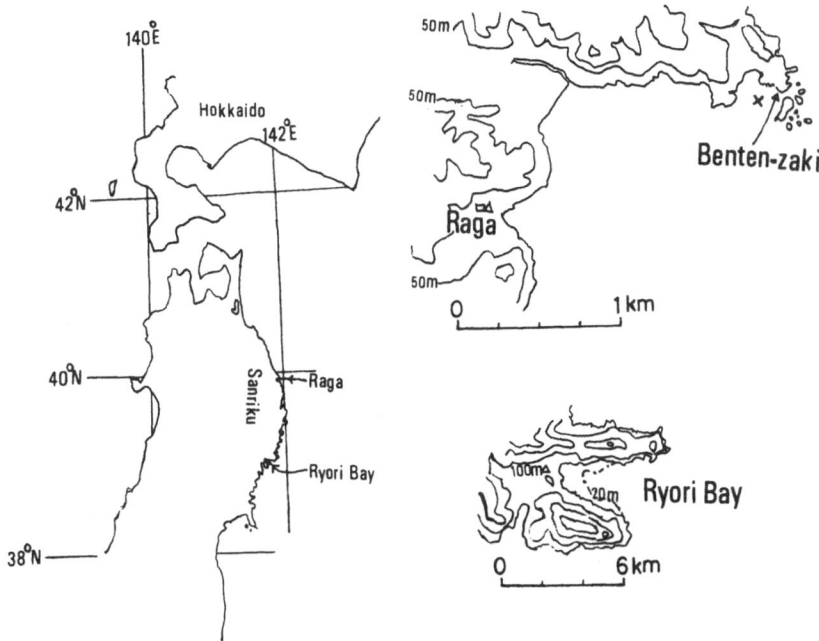

Fig. 2. Maps of two parts of the Sanriku District, showing locations and regional geographic settings of Raga and Ryori Bay. We want to call the reader's attention to the great difference between the scales of two maps. In the map of Raga, △ and □ show the locations of the Shimosaka house and that of another house, respectively. The arrow-head and × show the original and present locations of Hanaguri-iwa, respectively. The contours on the land are drawn every 50 m. In the map of Ryori Bay, △ shows the location where the tsunami of 1896 ran up as high as 38.2 m above sea level. △ and □ in the former map and △ in the latter are all situated right behind the heads of V-shaped bays. In the latter map, the contours on the land are drawn every 100 m. Only the submarine contour of 20 m is available. Along the line of the mouth of this bay connecting two capes, the sea depth is some 50 m.

the tsunami rushes toward the land, wave energy concentrates from the both sides. And when it recedes toward the sea, wave energy passes principally through the central part of the bay. Large blocks of rock are therefore apt to be carried up. The existence of such the blocks of rock shows us that the motion of sea water might be irreversible. And the difference between 29 m and 20 m, namely "9 m" in Raga is very instructive. MILLER (1960) did a similar description in his report on the "giant wave in Lituya Bay, Alaska," as follows: (At the highest point reached by sea water,) "the water rose about 20 feet higher than the highest overturned trees."

On Ishigaki Island, one of the southernmost islands of the Ryukyu Chain, we can find large blocks of coral, the specific gravity of which is 2.5 or so, carried up by the tsunami of 1771, as high as "40 m."

The maximum run-up of this tsunami was, according to tradition, approximately 85 m high. The most authorized table (Rika Nempyo) in Japan lists this "85 m" figure. There is another thought that when we consider, however, the above-mentioned "9 m" over "40 m," we can get some "50 m" as the maximum run-up of this tsunami.

MIYOSHI (1975, 1977) has published an account of what is believed to be the most violent tsunami (1771) recorded in world history as one caused by an earthquake, only to be followed by the 1896 tsunami in the Sanriku District. Both tsunamis were caused by tsunami earthquakes with low frequency. The warning problem concerning these tsunamis is important.

3.   Survey in Tanohata Village (II)

There is a place called Benten-zaki nearly 2 km to the northeast of Shimosaka house. There, we found that a rock of granite, the specific gravity and weight of which are 2.635 and 733 tons respectively, was translated horizontally by the tsunami of 1896 as far as 115 m from its original location (as shown in Fig. 2). In Fig. 3, we can see two rocks in the sea. The one on the left with a white spot on it is the rock of interest, and is called "Hanaguri-iwa" in the local area. We confirmed the translated distance of 115 m using two boats, a long piece of nylon cord which lacks elasticity, and so on (Fig. 4). Several village people assisted us.

There is a fundamental cognition among the specialists that the strength of the storm surge is more remarkable than that of the tsunami in the sea, and the structure in the sea which can weather the pressure of the storm surge can therefore easily stand against the torrent of tsunami. And someone supposes that the tsunami of 1896 translated this rock for only a part of this "115 m," and the storm surges after this tsunami, the other parts.

The name of "Hanaguri" is here considerably important. According to the inhabitants, this name suggests that this rock had, before the tsunami, played here a special role to fasten the small fishing boats to the land.

In spite of the fact that many storm surges had broken, as we can easily realize in Fig. 2, at the point shown by the arrow-head, this rock had never been moved, not to speak at the present location shown by ×. The tsunami of 1896 translated, therefore, this rock all the distance of this 115 m, which coincides with eyewitness evidence.

Fig. 3.   We can see two rocks in the sea. These rocks appear in the map on a scale of 1 to 25.000. The left-hand one, "Hanaguri-iwa" was translated horizontally by the tsunami of 1896 as far as 115 m from its original location.

Fig. 4.   The distance between the original location of the huge rock and the present location off Benten-zaki was measured on September 15, 1980. We used two boats and a long piece of nylon string (the dotted line) which lacked elasticity.

Here is a socially important collapse of the above-mentioned fundamental cognition among the specialists. In the extreme situations, as in Ryori Bay or around the inlet of Raga, the phenomena which common sense has not expected may present themselves.

Generally speaking, rocks around the shoreline of the Japanese mainland are considerably buried in sand and hardly could be carried by tsunamis. We were therefore surprised when we found that a coral block on Ishigaki Island, the specific gravity and weight of which are 2.5 and 750 tons respectively, must have been carried as far as 2,500 m from the nearest beach and as high as 30 m from sea level by the tsunami of 1771.* We spent a great effort to introduce new knowledge concerning the capability of conveyance by the tsunami, into the mainland of Japan, only because the coast of the mainland is abundant in sand.

There is little sand around an isolated island where coral grows well, and such a conveyance was done as discussed immediately above. There is also exceptionally little sand around Benten-zaki, and a huge rock could have been, in principle, carried by the 1896 tsunami.

Let us consider, here, a cube, the specific gravity and the weight of which are ρ and A tons, respectively. The area of one surface is given by

$$S = \left(\frac{A \times 10^6}{\rho}\right)^{2/3} \text{cm}^2 = A^{2/3}\rho^{-2/3} \times 10^4 \text{ cm}^2. \tag{1}$$

Its weight under the buoyancy of sea water is given by

$$G = A\left(\frac{\rho - 1.02}{\rho}\right) \times 10^6 \text{ g}. \tag{2}$$

Therefore,

$$S/G = A^{-1/3}\frac{\rho^{1/3}}{\rho - 1.02} \times 10^{-2} \text{ cm}^2/\text{g}. \tag{3}$$

When A and ρ are 733 and 2.635, respectively, S/G becomes 0.948×10^{-3} cm^2/g. Of course, rolling friction, not static friction which corresponds to the weight given by Eq. (2) is under consideration. This equation is not therefore used for check of sliding, but of rolling. This figure of 0.948×10^{-3} cm^2/g in the case of Hanaguri-iwa, should

*An objection to this guess was proposed from geology. We must, however, pay attention to the fact that geology itself cannot also trace the motion of the blocks.

be born in mind at the construction or reconstruction of a harbor.

It is a little difficult to adjust this thought to that of a technoligist. We are discussing on the situation that the torrent of the tsunami utterly covers the rock and is rolling it, as in the case of Hanaguri-iwa, which is, as we can see in Fig. 3, almost submerged.

TANIMOTO (1981), for example, described that the Kawaragi (Hachinohe, in the Sanriku District) breakwater collapsed over a total length of 338 m under the tsunami attack of 1968. It was concluded after field investigation that the main cause of collapse was the considerable difference of water levels between outside and inside of this breakwater. The dynamics in this case is a little different one.

Though we couldn't find any clue to the run-up height of the tsunami at Benten-zaki, because there was few inhabitants around it at that time, a tsunami which run up as high as 29 m at the Shimosaka house can easily simulated, to estimate the velocity of the flow around Benten-zaki. Between the velocity of the simulated flow of sea water, the kinetic energy, and the difference of water levels, the potential energy, might be there a close relation.

4. Conclusions

We must pay attention to the tsunamis caused by the earthquakes with low frequency, the magnitudes of which are almost 8. The human body feels as if their magnitudes were only 7.5 or so. Then people are apt to be off guard, when a tremendous tsunami does attack. Two typical examples of such earthquakes occurred in 1771 and 1896, off the coast of Japan. They sent tremendous tsunamis to Ishigaki Island and the Sanriku District, respectively. (On page 411 of the report by F. P. SHEPARD *et al.* (1950) concerning the tsunami of April 1, 1946, we find that this tsunami was not so remarkable.)

As the former occurred long ago, we can't trace its behavior in detail except by the many blocks of coral carried by it. The number of eyewitnesses of the latter event is decreasing rapidly. Traditions concerning this tsunami are, however, even now vivid and accurate. We should hurry with our work of gathering this information and correcting previous data which had been very roughly obtained. We have illustrated this procedure, by one example, the survey in Raga, Tanohata Village, through the help of many local inhabitants. Besides the significance of these studies for society, this work may be necessary to locate this unusual tsunami among the usual ones.

Finally, we'd like to add that we don't deny that the same dynamics as that of the Kawaragi breakwater might occur in the case of above-mentioned Hanaguri-iwa. In this case, the rock might start to move in a moment, corresponding to the collapse of the breakwater. But the subsequent translation might be controlled by Eq. (3). The friction during the translation is much smaller than that at the start.

On September 15, 1980, two boats were used to measure the distance between the original and the present locations of the rock, Hanaguri-iwa. M. Horikawa, S. Nakasaki, S. Shimosaka and H. Kumagai conducted this operation. K. Kakudate of the Tanohata Village Office, estimated the weight of Hanaguri-iwa. Special thanks are due for their cooperation.

REFERENCES

IGI, T., A report of the field investigation of the tsunami of 1896 in the Sanriku District, *Shinsai Yobo Chosa-kai Hokoku*, **11**, 5–33, 1897 (in Japanese).

MATSUO, H., A report of the field investigation of Sanriku Tsunami, *Doboku Shikenjyo Hokoku*, **24**, 83–112, 1933 (in Japanese).

MILLER, D. J., Giant waves in Lituya Bay, Alaska, *Geol. Survey Prof. Pap.*, **354-C**, 51–86, 1960.

MIYOSHI, H., Upper limits of the tsunamis, *Rep. Tokyo Univ. Fish.*, **10**, 19–29, 1975 (in Japanese).

MIYOSHI, H., The most amazing tsunami in history, *Newsletter ITIC*, **10**, 1–3, 1977.

SHEPARD, F. P., G. A. MACDONALD and D. C. COX, The tsunami of April 1, 1946, *Bull. Scripps Inst. Oceanogr.*, 391–527, 1950.

TANIMOTO, K., On the hydraulic aspects of tsunami breakwater in Japan, *Abst. Intern. Symp. IUGG Pap.*, 121–124, 1981.

Tsunamis—Their Science and Engineering, edited by K. Iida and T. Iwasaki, 213–224.

Digitization of Tsunamigrams

Sydney O. WIGEN

Institute of Ocean Sciences, Sidney, B. C., Canada
(Received September 3, 1981)

Analogue records from tide stations throughout the world contain a wealth of tsunami data, only a fraction of which has been utilized. Collecting these tsunamigrams, converting them to digital format and making them accessible to researchers will provide the basis for a more complete undertanding of the propagation and coastal interaction of tsunamis.

Equipment for machine digitization, procedures, and programs in use at the Institute of Ocean Sciences for such digitization are outlined. Some of the limitations of tide gauge stations as a source of tsunamigrams are discussed. Comments and response are invited from others with experience in tsunamigram digitization and from users of digital data.

1. Introduction

Within the limits of the Pacific, and throughout the world, a wealth of tsunami data exists within the analogue records from tide gauge stations. Only a fraction of this information has been extracted, and much less has been fully used.

Collecting these that past records or tsunamigrams, referencing them to time and height datums, converting them from analogue to digital format and making their information accessible to researchers, will provide the basis for a more complete understanding of the propagation and coastal interaction of tsunamis.

Machine digitization of analogue diagrams has been used for many years in Canada in the systematic processing of tide records. Similar techniques are now being applied at the Institute of Ocean Sciences, Patricia Bay for the digitization of tsunamigrams. The object of this paper is not to present a final definitive solution to tsunamigram digitization. Rather it is to share information on the present state of development at the Institute of Ocean Sciences, and what is now available for use. It is also to cite some of the problems, and to invite response from those who have experience with tsunami digitizing procedures, and any who have a need in their tsunami research for digital information.

For both tides and tsunamis the approach is similar, to digitize in $X-Y$ coordinates a sufficient number of points from the analogue trace that the significant details of the original record can be reproduced between these points by linear interpolation. For tidal processing, digitization is often carried out on a smoothed, hand-drawn curve, that discards short period non-tidal fluctuations (Fig. 1). For tsunamis the objective is to digitize as completely and accurately as possible, all fluctuations that can be discretely identified, particularly the high and low points, and

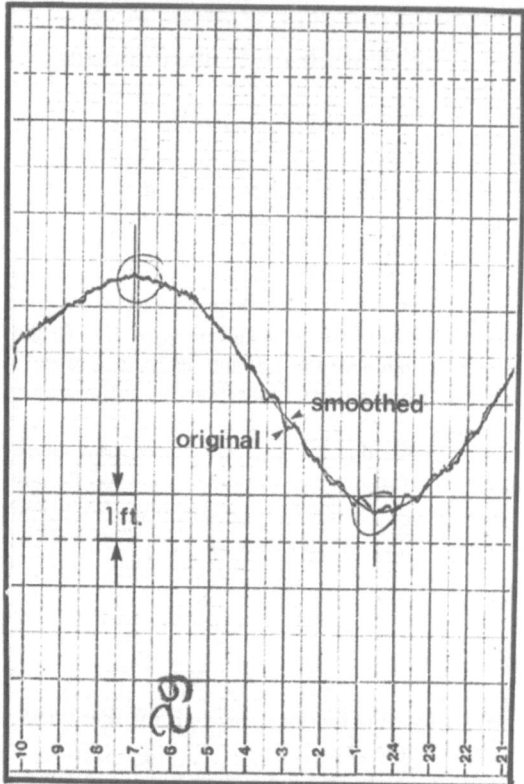

Fig. 1.   Analogue tide record, smoothed for tidal processing.

points of inflection. The principal limitations on accuracy then become the time and height scales of the source record, and the clarity of its analogue trace.

Before describing the processing of tsunamigrams we will consider some characteristics of tide stations relevant to the recording of tsunamis. The next two sections may be omitted by anyone concerned only with the digitizing.

2.   Tide Recording Stations, the Source of Tsunamigrams

From the inception of automatic tide recording, more than one hundred years ago, the most commonly used gauging system has consisted of a cylindrical stilling well erected in the tidal zone, having a small orifice below low tide allowing limited passage of water, and a float within the well rising and falling with the tide. A recording gauge registers the float elevation by a pen or pencil trace on paper, as a change of water elevation with time (Fig. 2).

Although other sensing systems are coming into increasing use for automated tide measurement, we are primarily concerned at this time with the float actuated gauge, as the source of most historical tsunami records.

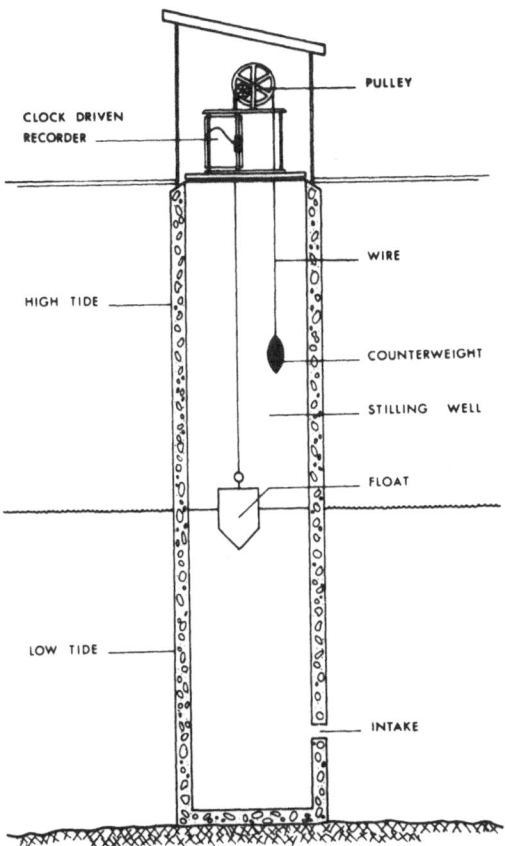

CLOCK DRIVEN
RECORDER

PULLEY

HIGH TIDE

WIRE

COUNTERWEIGHT

STILLING   WELL

FLOAT

LOW TIDE

INTAKE

Fig. 2.   Tide gauge station, with orifice intake.

Time control on analogue tide records has normally been provided by an observer noting local time on the diagram once daily. Prior to radio time signals, observers were dependent on mechanical clocks or chronometers. Precision electronic timers that have recently become available have not yet been generally applied to analogue tide recorders. Time errors inherent in the gauge operation may be of little consequence in the spectral analysis of a tsunamigram, but may significantly affect the recorded arrival time of a tsunami, and the determination of traval times from the area of generation.

Height control on analogue tide records is obtained by independent water level readings being taken on a tide staff or sight gauge tape, and identified with corresponding points on the diagram record. Zero elevations of the tide staff or sight gauge are in turn referenced by instrumental levelling to a reference bench mark or marks ashore. Water levels are normally reduced to a tidal datum defined relative to the primary bench mark. It is possible that the datum will have been changed one or more time during the history of operation of the gauge.

Analogue traces are commonly registered now on strip charts, with time measured on the X-axis, and heights on a perpendicular Y-axis. However some gauging systems have paper mounted on a cylindrical clock, to provide a 24-hour graph, and successive days of tide records may overlap on a single diagram. Time scales on diagrams are typically between 1 and 2.5 cm per hour, and height ratios, between vertical water changes and excursions on the *Y*-axis, between 1:10 and 1:100 (Fig. 3). Some gauge mechanisms have a worm gear drive for the pen movement, permitting reversal when it reaches either upper or lower limit on the *Y*-axis of the diagram.

Most tide stations were originally established to provide data for navigation, charting, and fishing. Eventually the tide records were applied in geodesy for mean sea level determination, and to studies such as ocean dynamics, coastal engineering, and

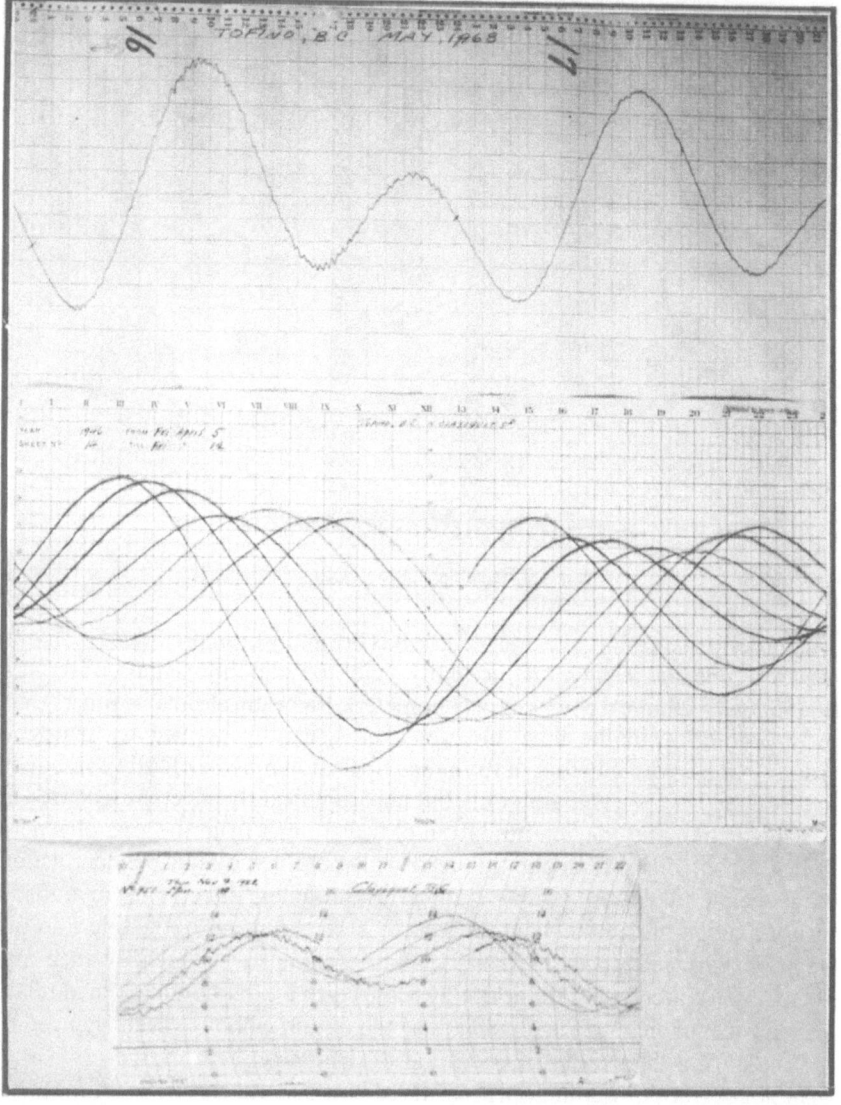

Fig. 3. Three types of analogue tide records programmed for digitization: C1, Strip chart; C2, 24-hour rectilinear; C3, 24-hour curvilinear.

crustal movements. The scientific and engineering applications of the tidal records led to higher standards of accuracy in gauging. However, tsunami recording was until recently an incidental byproduct of the gauging, and few instruments and installations have been designed specifically to provide an accurate record of water level changes during a tsunami.

3.   Tide and Tsunami Characteristics, and Significant Gauging Errors

Tides are a predictable phenomenon. The analysis of a time series of sea levels obtained from a gauge, relative to lunar and solar periods, yields harmonic constants from which past and future tide levels at any instant can be predicted. Storm waves, surf, and seiches are not in this sense predictable. Therefore, tide recording systems have been designed to damp out as much as possible of these high frequency fluctuations without significantly distorting the measurement of the tidal portion of the water level changes. In the float gauge this is accomplished hydraulically by having an orifice of small diameter or a long intake pipe as the entry to the stilling well. Ratio of orifice diameter to well diameter is usually between 1:10 and 1:30, with 1:20 being common. Ratio of surface area of orifice and stilling well corresponding to these diameters will be respectively, 1:100, 1:900, and 1:400. Taking the median value, flow velocity through the orifice will be 400 times as fast at the change of water outside the well if actual rate of rise or fall is to be recorded accurately. Inevitably there is some difference in absolute levels, since the potential energy between the two water levels provides the driving force to move the water through the orifice. For a tidal change of 1 metre per hour, a rate quite common on many coasts, the corresponding orifice velocity at 1:400 ratio would be 11 cm per second. If during a tsunami the rise of water were ten times as rapid, orifice flow would need to have a speed of more than 1 metre per second. During such high velocity flow turbulence and friction at the orifice further reduce its effective diameter.

Sediments and marine organisms may at times restrict the intake to the point that even the normal tide is visibly distorted (Fig. 4). These and other problems inherent in the conventional tide gauge, are examined by LENNON (1971).

In recent years the time constant of some stilling wells has been measured during annual tide gauge inspections. An instantaneous change of water level, Δh, is applied in the well. The time constant τ describes an exponential return to the original level given by $\Delta he(-t/\tau)$. At time $t = \tau$ the step input will have decayed to $1/e$ of the initial difference. Knowledge of the time constant is applicable in evaluating how the well may respond to a tsunami, and may provide the basis for describing the real water level changes outside the well that produced the gauge record.

NOYE (1976) and BRADDOCK (1980) have provide analyses of the response of tide gauges to tsunamis, and the possibility of making corrections to the tsunamigram. For historical records where the time constant is not known, such recreation is not possible. However a subjective assessment can be made. If normal tide records show the well to be responding actively to short period oscillations, as in Fig. 5, the major features of the tsunami are probably being reproduced. If the normal record shows little activity or an inhibited response, it is likely that a tsunamigram for any but a small tsunami will be

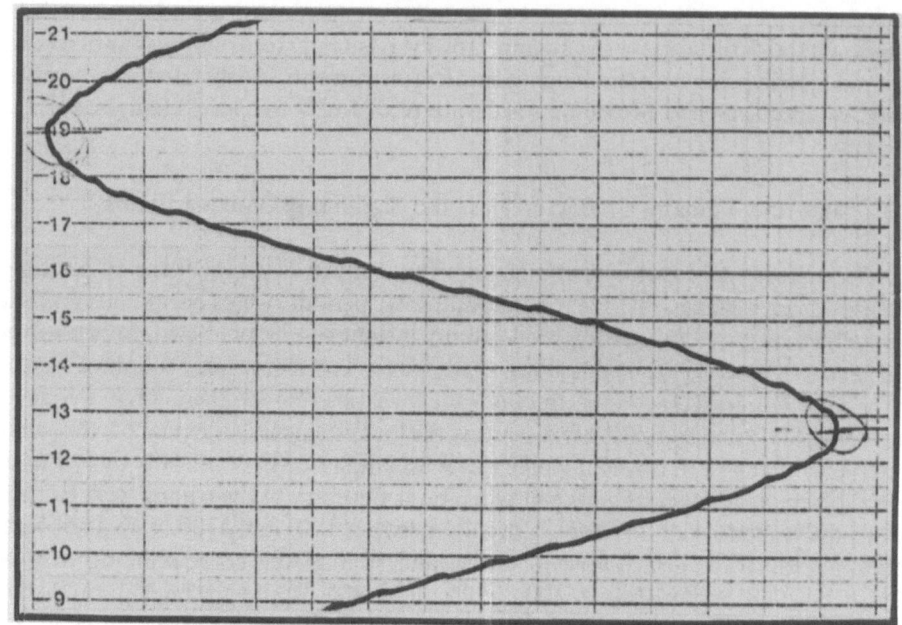

Fig. 4.   Tide record distorted by sediment clogging an intake pipe.

severely distorted.

4. Digitizing Equipment

Of the various digitizing systems available at the Institute of Ocean Sciences, the Gradicon Digitizer at the associated Pacific Geoscience Centre was selected for its high resolution as most suitable for processing tsunamigrams. This system contains four components, the Gradicon digitizing table, keyboard assembly, display/control console, and IBM 29 card punch (Fig. 6).

The table has a working surface of 90 cm × 120 cm, on which to secure the tsunamigram. Illumination comes through the table surface as well as from overhead lighting. A hand-held transparent cursor with cross hairs, Fig. 7, is used by an operator to follow the tsunamigram trace. A magnetic follower on the underside of the table tracks the cursor movements, and registers positions through the console in X and Y co-ordinates, with a resolution of .01 mm. Discrete points are digitized by the operator pushing a button on the cursor, and through the display/control console are automatically keypunched. Header cards for each data set are punched manually on the card punch, and control or flag symbols for the program are inserted through the keyboard assembly on the table.

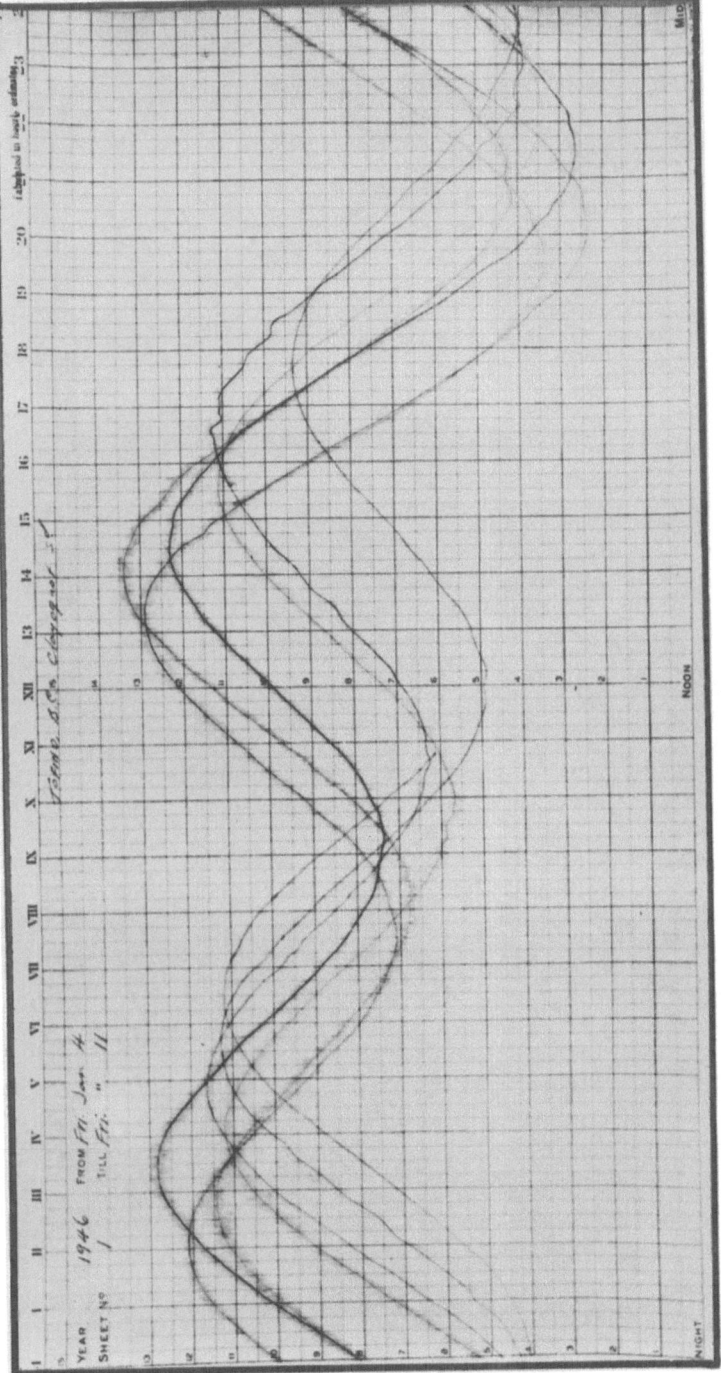

Fig. 5. Tide record with active response to short period waves.

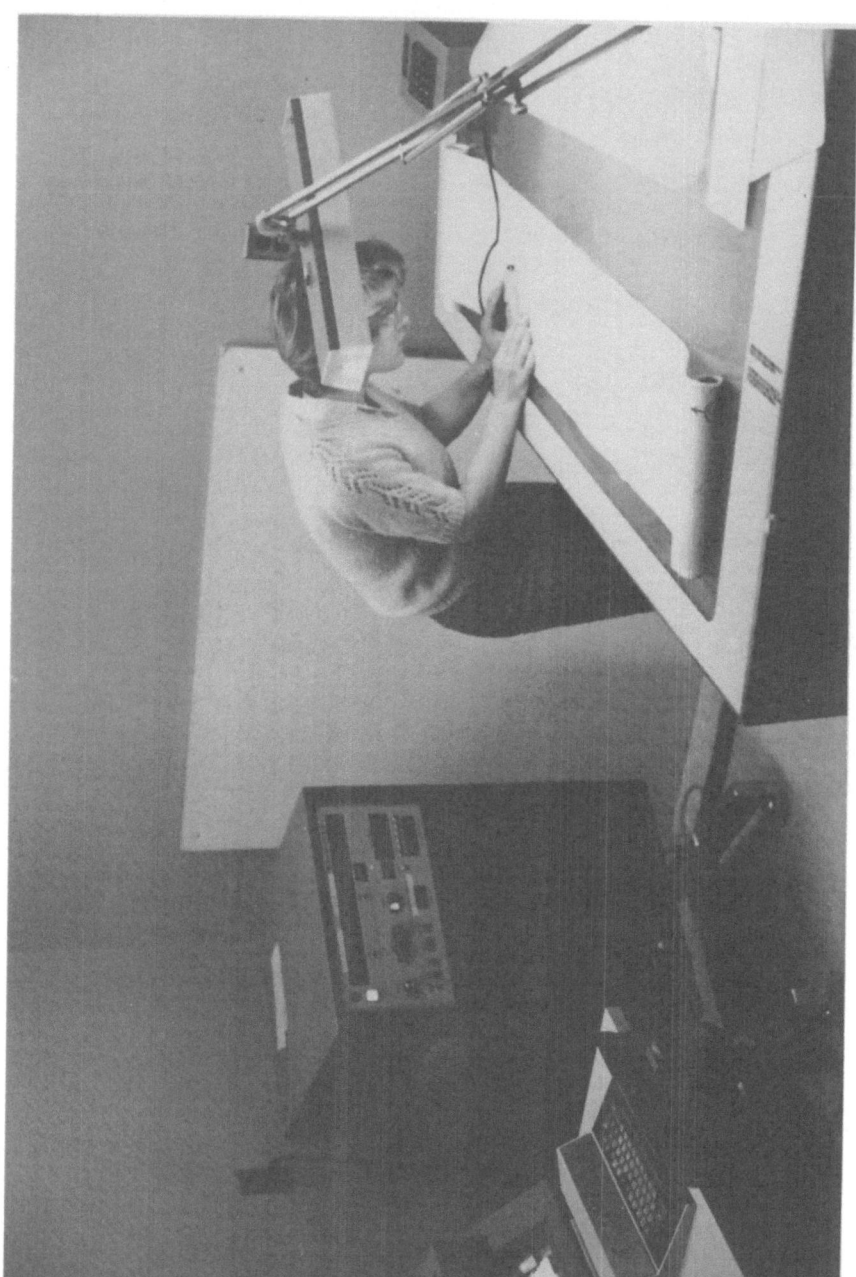

Fig. 6.   Gradicon digitizing system components, card punch, display/control console, digitizing table with cursor and keyboard assembly.

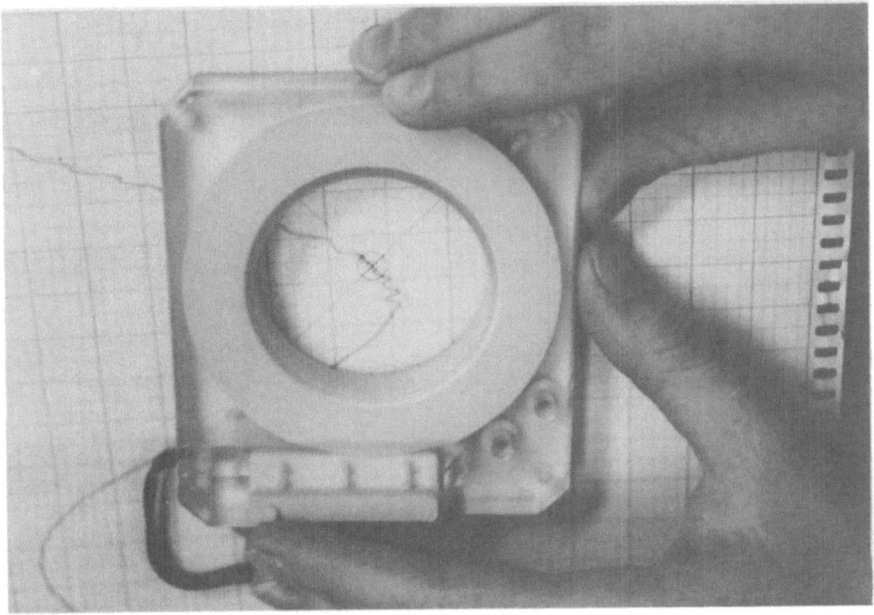

Fig. 7.   Digitizing cursor.

5.   Control Data for Digitization

Prior to digitization, control information is compiled that will be necessary in the processing of the digital data, and in its future access. Figure 8 shows the control sheet and the columns in summary:

Col. No.   Header Card 1

1	Card number 1
2–11	Tsunami origin, using the 10-digit INST reference (INTERNATIONAL TSUNAMI INFORMATION CENTER, 1975). Global quadrant in INST is defined by

Digit 1   Northern and eastern hemisphere

2   Northern and western hemisphere

3   Southern and western hemisphere

4   Southern and eastern hemisphere

Latitude and longitude are given in degrees and minutes, with fractional units deleted.

12–23	Time of tsunami origin, Universal Time.
24–49	Digitizing agency, person, date, and machine.
50–52	Resolution of digitizing machine, to .01 millimetres.
53–67	Computer processing program (available from Institute of Ocean Sciences).
68–69	Type of chart being digitized (as shown in Fig. 3).
71–80	Location of digitized data, after processing and cataloguing.

HEADER CARDS FOR UPDATE

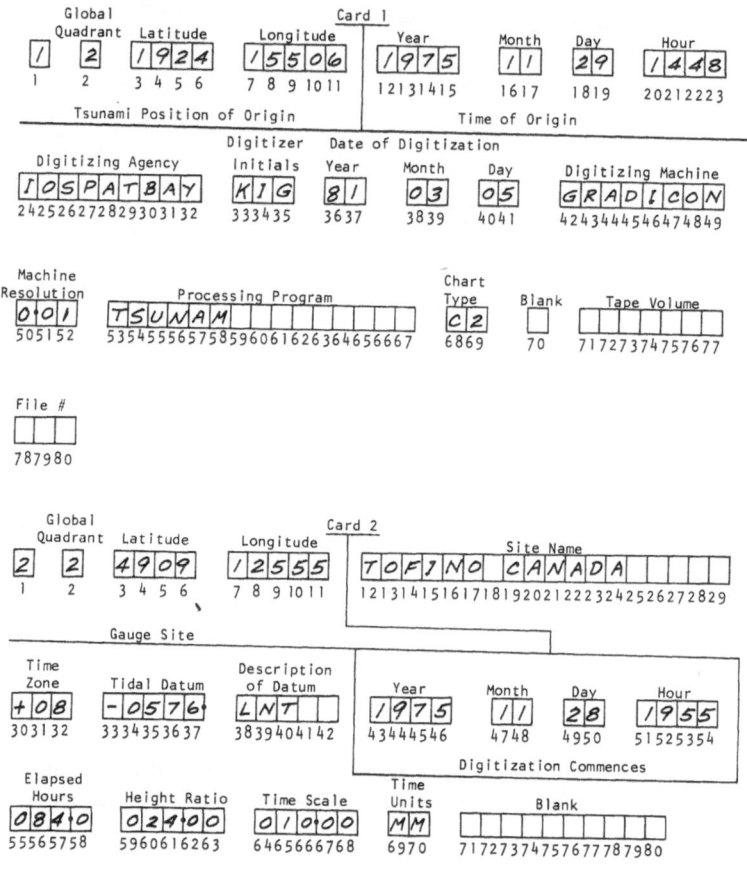

Fig. 8.   Control sheet for header cards.

Header Card 2

1   Card number 2
2–11   INST location of tide station.
12–29   Tide station, geographical name and country
30–32   Time zone of tsunamigram : + for west, and − for east of Greenwich Meridian.
33–37   Correction for tidal datum relative to zero of diagram; given in mm : + when datum is below diagram zero, and − when above diagram zero.
38–42   Description of datum; e.g., LNT is Lowest Normal Tide, the tidal and chart level adopted by the International Hydrographic Bureau.
43–54   Commencement of digitization, Universal Time, normally about 25 hours before apparent arrival of tsunami, is begun on an hourly time line on the tsunamigram. Time error of the graph, if known is applied. Thus for a gauge record that is 5 minutes fast, the hour 2000 is actually 1955.
55–58   Length of record to be digitized, to 0.1 hours. Digitization is carried through

to a point on the tsunamigram 25 hours after the tsunami ceases to be distinguishable, or 50 hours after the start of the tsunami, whichever is greater.

59–63   Height ratio, the number of cm of water level changes represented by 1 cm of diagram.

64–68   Time scale, length of one hour on the diagram, to .01 mm.

69–70   Time unit, normally mm/hr on all graphs now being processed, but could be degrees/hr on a circular chart.

6.   Digitizing and Processing

The operator secures the tsunamigram to the digitizing table with time increasing from left to right, and the diagram approximately parallel to the X-axis of the table. Depending on the type of recording system water levels on the diagram may increase directly or inversely with increasing digital value of the Y-axis. The processing program accepts either, as well as diagram reversals identified by the operator.

A prescribed routine is followed in putting the Gradicon digitizing system in service, and the two header cards are then punched. The operator uses the cursor to digitize control points on the corners of the diagram and the defined starting point of the tsunamigram. Then the cursor is moved carefully along the track of the tusunamigram, following all water level changes that can be discretely identified. All highs, lows and points of inflection are digitized, and sufficient points along any arc, so that the slope of the tsunamigram can be recreated as a series of straight lines between these points taken at random intervals. Each point is recorded as a 10-digit number, e.g. 0734216598 signifies co-ordinate $X = 73.42$ mm; $Y = 165.98$ mm. Typically for a tsunamigram of 75 hours, on a scale of 10 mm per hour the operator digitizes about 1,000 points.

One tsunamigram requires about 2 hours of concentrated effort by the operator. Accuracies are in the order of 0.2 mm if the tsunamigram trace is clear and well defined, or about 1 minute in time, for a scale of 10 mm/hour.

The processing program translates the data from card to tape storage, and carries out several routines. Using the diagram control points, digitized values are rotated to correspond to the X and Y axes of the table. Time of the initial digitized point is given a zero X value, and the X value of all subsequent points are corrected. If any time inversions occur, as may happen if the operator was digitizing a near vertical change in water level, the program interpolates so that successive X values are always increasing.

Successive days of data from a 24-hour diagram are sequenced, as are data from sequential diagrams. Time gaps can be inserted by the operator for missing sections of record. Time inversions are corrected. Heights are retained as Y values in units of 0.01 mm above diagram zero of the tsunamigram. Conversion to another datum may be made by using the datum reference and height scale from the header card. For the C 3 diagrams, X and Y values are converted from curvilinear to rectilinear co-ordinates.

A plot of the tsunamigram to original scale of the graph is made on a drum plotter. The trace is carefully compared visually to the original tsunamigram, and significant differences are marked. For minor errors, corrections are made to individual numbers in cards of the original data deck. For more extensive errors, sections of graph are redigitized and inserted into the card deck, with appropriate header cards.

For the C 3 diagrams, two plots are prepared; one following the curvilinear trace for the accuracy check, and the second, after correction, a rectilinear plot.

Some programming errors are still being disclosed by the plotting routine, and debugging is in progress. Systematic storage and cataloguing of data is being programmed, in order that tapes can be duplicated to provide data sets for researchers.

REFERENCES

Braddock, R. D., Response of a conventional tide gauge to a tsunami, *Marine Geodesy*, **4**. 223–236, 1980.
International Tsunami Information Center, International Numbering System for tides (INST), *ITIC Newsletter*, **VIII**, 2, 1975.
Lennon, G. W., A critical examination of the conventional tide gauge, Proceedings of Symposium on Tides, International Hydrographic Bureau, Monaco, UNESCO, Paris, 1971.
Noye, B. J., Recording of tsunamis by tide wells, Royal Society of New Zealand bulletin 15, Tsunami Research Symposium 1974, pp. 87–94, 1976.

TSUNAMI GENERATION AND PROPAGATION

Tsunamis—Their Science and Engineering, edited by K. Iida and T. Iwasaki, 227–240.

A Numerical Model for Tsunami Generation and Propagation

Philip L.-F. LIU and Jeff EARICKSON*

*School of Civil and Environmental Engineering,
Cornell University, Ithaca, NY, U.S.A.*

(Received September 13, 1981)

The generatoin and propagation of tsunamis as non-linear shallow-water waves, by general motions of an impermeable seabed, is investigated. A perturbation technique is used to derive systematically a set of governing equations of fluid motion; the basic shallow water approximation (i.e., the water depth is small compared with a characteristic horizontal length scale of the motion) is employed. The effects of dispersion and non-linearity are included in these equations. Different regimes of approximation in terms of the order of magnitude of ground motion are presented.

The resulting differential equations are solved by a finite-difference method. Numerical examples are given for two-dimensional cases. The validity and the accuracy of computed solutions in both the tsunami generation region of seabed motions and in the far-field propagation regions are shown by comparison to either analytical solutions or experimental data.

1. Introduction

The rigorous derivation of governing equations of long waves over a fixed uneven bottom has been developed (see, e.g. MEI and LEMEHAUTÉ, 1966; PEREGRINE, 1967; CHWANG and WU, 1976; etc.). The effect of an uneven bottom on water waves is of obvious engineering importance; for instance, equations of this kind are useful for the study of tsunami run-up. Extensive experimental work has also been carried out in corroboration with these theories (IPPEN and KULIN, 1955; KISHI and SAKEI, 1966; CAMFIELD and STREET, 1969; GORING, 1978).

It is, however, equally important to derive governing equations of long waves generated by ground movement. The immediate application is for the study of the generation of tsunamis (or seismic sea waves). Most previous analytical work on this subject has been restricted to constant water depth situations and has been based on a linearized wave theory (WEBB, 1962; KAJIURA, 1963; BRADDOCK *et al.*, 1973; etc.). The applicability of these results is limited to small sea floor deformations and the range of wave propagation for which the non-linear effects remain negligible.

Recently, some investigators (e.g. HWANG and DIVOKY, 1970; TUCK and HWANG, 1972) have employed a non-linear wave theory and have studied tsunami generation and

---

*Current address: U.S. Army, Waterways Experiment Station, Vicksburg, MS, U.S.A.

propagation in a variable water depth. Airy equations were employed in their formulation and dispersion effects are ignored, which could however be significant (Wu,
1981).

In the present paper, governing equations for long waves generated by ground
motion are derived systematically. The effects of dispersion and non-linearity are
included in the equations. Several limiting situations are also examined according to
the magnitude of the ground movement (see, Table 1).

The resulting equations are solved numerically by a finite difference method. Two
2-D examples are studied. One of them is the same as that investigated by Hammack
(1972). Present numerical results agree with Hammack's experimental data very well.

2. Analysis

2.1 Formulation

We first summarize the governing equations and boundary conditions for flow
motion in a liquid bounded above by the free surface F and bounded below by a solid
boundary S. Cartesian coordinates (x^*, y^*, z^*) are employed and fixed on the mean free
surface, $z^* = 0$, where z^* is positive upwards. The following dimensionless variables
are adopted here using h^* and L^* as vertical and horizontal length scale, respectively.

$$(\eta, x, y, z) = [L^*]^{-1}(\eta^*, x^*, y^*, z^*)$$
$$\phi = [L^* \sqrt{gh^*}]^{-1}\phi^*$$
$$t = [\sqrt{gh^*}/L^*]t^*$$
$$p = [\rho gh^*]p^*$$

(1)

where all variables with an asterisk represent physical quantities and $z^* = \eta^*$ is the free
surface elevation. Initially the fluid is at rest with the free surface and solid boundary
defined by $z = 0$ and $z = -H(x, y)$, respectively. For $t > 0$ the solid boundary moves

Table 1. Possible classes of water waves generated by various magnitudes of ground motion.

$\partial h/\partial t$	$O(\varepsilon)$		$O(\varepsilon^2)$		$O(\varepsilon^3)$		$O(\varepsilon^m)$**	
η	$\phi^{(2n)*}$	$\phi^{(2n+1)}$	$\phi^{(2n)}$	$\phi^{(2n+1)}$	$\phi^{(2n)}$	$\phi^{(2n+1)}$	$\phi^{(2n)}$	$\phi^{(2n+1)}$
$O(\varepsilon)$	$O(1)$	$O(\varepsilon)$	—	—	—	—	—	—
$O(\varepsilon^2)$	$O(\varepsilon)$	$O(\varepsilon)$	$O(\varepsilon)$	$O(\varepsilon^2)$	—	—	—	—
$O(\varepsilon^3)$	—	—	$O(\varepsilon^2)$	$O(\varepsilon^2)$	$O(\varepsilon^2)$	$O(\varepsilon^3)$	—	—
$O(\varepsilon^m)$	—	—	—	—	$O(\varepsilon^3)$***	$O(\varepsilon^3)$	$O(\varepsilon^{m-1})$	$O(\varepsilon^m)$

*$n = 0, 1, 2, \ldots$
**$m = 4, 5, 6, \ldots$
***for the case $m = 4$ only.

in a prescribed manner given by $z = -h(x, y, t)$. The governing equation for the resulting velocity potential is simply the Laplace equation

$$\nabla_3^2 \phi = 0 \qquad -h < z < \eta \tag{2}$$

where

$$\nabla_3 = \left(\frac{\partial}{\partial x}, \frac{\partial}{\partial y}, \frac{\partial}{\partial z} \right)$$

if it is assumed that the fluid is incompressible and the flow irrotational. The boundary conditions required on the moving solid boundary, $z = -h$ and the free surface, $z = \eta$, are summarized as follows (e.g. see WEHAUSEN and LAITONE, 1960):

$$\frac{\partial h}{\partial t} + \nabla h \cdot \nabla \phi + \frac{\partial \phi}{\partial z} = 0, \qquad z = -h(x, y, t) \tag{3}$$

$$\frac{\partial \eta}{\partial t} + \nabla \eta \cdot \nabla \phi = \frac{\partial \phi}{\partial z}, \qquad z = \eta(x, y, t) \tag{4}$$

$$\frac{\partial \phi}{\partial t} + \frac{1}{2}[|\nabla_3 \phi|^2] + \frac{\eta}{\varepsilon} = 0, \qquad z = \eta(x, y, t) \tag{5}$$

where the small parameter

$$\varepsilon = h^*/L^* \ll 1 \tag{6}$$

and $\nabla = (\partial/\partial x, \partial/\partial y, 0)$ is the horizontal gradient vector.

2.2 Perturbation analysis

For shallow water and slow variations of the bottom profile, i.e.

$$O(h, |\nabla h|, \left| \frac{\partial^2 h}{\partial x^2} \right|, \left| \frac{\partial^2 h}{\partial y^2} \right|, \text{etc.}) = 0(\varepsilon) \ll 1 \tag{7}$$

the following perturbation solutions are sought:

$$\phi(x, y, z, t) = \sum_{n=0}^{\infty} (z + h)^n \phi^{(n)}(x, y, t) \tag{8}$$

The perturbation scheme described hereafter was first introduced by LIN and CLARK (1959) in studying the shallow water theory. In the present analysis, emphasis will be focused on the additional effects of ground movements.

Substituting Eq. (8) into the Laplace Eq. (2) and the boundary condition (3), respectively, one obtains the following recursive relations among $\phi^{(n)}$,

$$\phi^{(n+2)} = -\frac{\nabla^2 \phi^{(n)} + (n+1)\phi^{(n+1)}\nabla^2 h + 2(n+1)\nabla h \cdot \nabla \phi^{(n+1)}}{(n+1)(n+2)[1 + |\nabla h|^2]} \tag{9}$$

and

$$\phi^{(1)} = -\frac{\dfrac{\partial h}{\partial t} + \nabla h \cdot \nabla \phi^{(0)}}{[1 + |\nabla h|^2]} \qquad (10)$$

It is noted that the two-dimensional version of the recursive relation (9) is the same as that obtained by Mei and LeMéhaute (1) in the course of studying long waves over a fixed uneven bottom. However, a new term $\partial h/\partial t$, which is due to the ground movement appears in Eq. (10). Using these recursive relations, one can express any $\phi^{(n)}$ in terms of $\phi^{(0)}$ and shall obtain, from Eq. (4) and (5), two equations for two unknowns $\phi^{(0)}$ and η.

Following LIN and CLARK's argument (1959), in order to obtain nontrivial results, one must make at least the linear terms in Eq. (4) and (5) comparable in order of magnitude. Hence, from Eq. (5) this suggest that

$$O(\phi) = O(\phi^{(0)}) = O(\eta/\varepsilon) \qquad (11)$$

since the leading term $\phi^{(0)}$ in Eq. (8) should be of the same order of magnitude as that of ϕ. Using Eq. (7), one can show, from Eq. (10) and (11) that

$$O(\phi^{(1)}) = O\left(\eta, \frac{\partial h}{\partial t}\right). \qquad (12)$$

Since we are only interested in the free surface flow generated by ground movement, $\partial h/\partial t$, it is obvious that, from Eq. (12), $O(\partial h/\partial t)$ should be at least in the same order of magnitude of η; i.e. $O(\partial h/\partial t) \geq O(\eta)$. Moreover, the perturbation solution (i.e. Eq. (8) also requires that $O(\phi^{(0)}) \geq O(\phi^{(1)})$. Therefore, using Eqs. (11) and (12), one can obtain a relation between η and $\partial h/\partial t$ as follows

$$O(\eta/\varepsilon) \geq O(\partial h/\partial t) \geq O(\eta). \qquad (13)$$

In Table 1 the ranges of possible solutions in terms of $\partial h/\partial t$ and resulting η are shown. Due to long wave approximation, the free surface displacement is bounded by $O(\varepsilon)$; i.e. $\eta \leq O(\varepsilon)$. When and if $\partial h/\partial t = O(\varepsilon^m), m > 3$, the flow problem is effectively steady state with little practical significance. In the remainder of this section, equations for $\phi^{(0)}$ and η are derived for the case

$$\eta = O(\varepsilon^3), \qquad \frac{\partial h}{\partial t} = O(\varepsilon^3) \qquad (14)$$

and

$$\begin{aligned}\phi^{(0)} &= \phi^{(2n)} = O(\varepsilon^2)\\ \phi^{(1)} &= \phi^{(2n+1)} = O(\varepsilon^3)\end{aligned} \qquad (15)$$

which corresponds to the waves of the cnoidal class (URSELL, 1953). Two other limiting cases; i.e. (1) $\eta = O(\varepsilon)$ representing very shallow waves and (2) $\eta = O(\varepsilon^m), m > 3$, representing moderately shallow waves of infinitesimal amplitude, can be deduced easily and are presented in Section 3.

Substituting Eq. (8) into Eqs. (4) and (5), one obtains

$$\frac{\partial \eta}{\partial t} + \nabla \phi^{(0)} \cdot \nabla \eta = \phi^{(1)} + 2(\eta + h)\phi^{(2)} + 3h^2\phi^{(3)} + 4h^3\phi^{(4)} + O(\varepsilon^6) \qquad (16)$$

$$\frac{\partial \phi^{(0)}}{\partial t} + \frac{1}{2}|\nabla \phi^{(0)}|^2 + \frac{\eta}{\varepsilon} + h\frac{\partial \phi^{(1)}}{\partial t} + h^2\frac{\partial \phi^{(2)}}{\partial t} = O(\varepsilon^5) \qquad (17)$$

up to the order of magnitude indicated. In order to maintain the degree of accuracy shown in Eqs. (16) and (17), one should express $\phi^{(1)}$, $\phi^{(2)}$, $\phi^{(3)}$, and $\phi^{(4)}$ in terms of $\phi^{(0)}$ and $\partial h/\partial t$ up to $O(\varepsilon^5)$, $O(\varepsilon^4)$, $O(\varepsilon^3)$, and $O(\varepsilon^2)$, respectively. These can be derived from Eqs. (9) and (10) and are given in Appendix A; Eqs. (A.1–A.5).

Upon substitution of Eqs. (A.1–5) into Eqs. (4) and (5), differentiated with respect to x and y, respectively, and denoting that

$$\boldsymbol{u} = \nabla \phi^{(0)} = (u, v) \qquad (18)$$

the resulting equations are

$$\frac{\partial \eta}{\partial t} + \nabla \cdot [\boldsymbol{u}(\eta + h)] - \frac{h^3}{6}\nabla \cdot (\nabla^2 \boldsymbol{u}) = \boldsymbol{A} \cdot \boldsymbol{u} + B\nabla \cdot \boldsymbol{u} + \frac{3}{2}h^2\nabla h \cdot \nabla^2 \boldsymbol{u} + C + O(\varepsilon^6) \qquad (19)$$

and

$$\frac{\partial \boldsymbol{u}}{\partial t} + \boldsymbol{u} \cdot \nabla \boldsymbol{u} + \frac{1}{\varepsilon}\nabla \eta - \frac{h^2}{2}\nabla^2\left(\frac{\partial \boldsymbol{u}}{\partial t}\right) = \nabla \cdot (h\nabla h)\frac{\partial \boldsymbol{u}}{\partial t} + 2h\nabla h \cdot \nabla\left(\frac{\partial \boldsymbol{u}}{\partial t}\right) + \boldsymbol{D} + O(\varepsilon^5) \qquad (20)$$

$$\boldsymbol{A} = |\nabla h|^2\nabla h + \frac{1}{2}h^2\nabla(\nabla^2 h) + 3h\nabla^2 h\nabla h \qquad (21)$$

$$B = \frac{3}{2}\nabla \cdot (h^2\nabla h) \qquad (22)$$

$$C = \{|\nabla h|^2 + h\nabla^2 h - 2\nabla h \cdot \nabla(\nabla^2 h) - 1\}\frac{\partial h}{\partial t} + 2h\nabla h \cdot \nabla\left(\frac{\partial h}{\partial t}\right) + \frac{h^2}{2}\nabla^2\left(\frac{\partial h}{\partial t}\right) \qquad (23)$$

$$\boldsymbol{D} = \frac{\partial^2 h}{\partial t^2}\nabla h + h\nabla\left(\frac{\partial^2 h}{\partial t^2}\right). \qquad (24)$$

It is obvious that C and $\boldsymbol{D}$ would be zero if the solid boundary were a fixed one and in particular, the resulting two dimentional equations (KdV type) reduce to those derived by MEI and LEMÉHAUTÉ (1966). Futhermore, if the bottom were horizontal, h = constant, the right hand sides of Eqs. (19) and (20) would have been zero, which cases were studied by LIN and CLARK (1959), and LONG (1964).

3. Limiting Cases

In this section the approximate equations are presented for different orders of magnitude of η (or $\partial h/\partial t$) as listed in Table 1. These equations could be derived in a manner similar to that used in Section 2.2. However, they may be more readily obtained by inspection of Eqs. (19) and (20).

3.1 $\eta = O(\varepsilon)$, $\dfrac{\partial h}{\partial t} = O(\varepsilon)$

It follows from Table 1 that

$$O(\phi^{(0)}) = O(\phi^{(2n)}) = O(1)$$
$$O(\phi^{(1)}) = O(\phi^{(2n+1)}) = O(\varepsilon). \tag{25}$$

Retaining only the leading terms in Eqs. (19) and (20), one obtains

$$\frac{\partial \eta}{\partial t} + \nabla \cdot [u(\eta + h)] = -\frac{\partial h}{\partial t} + O(\varepsilon^3) \tag{26}$$

$$\frac{\partial u}{\partial t} + u \cdot \nabla u + \frac{1}{\varepsilon}\nabla \eta = O(\varepsilon^2). \tag{27}$$

Note that these equations are still valid for cases where $\eta = O(\varepsilon^2)$ (see Table 1) except that the terms being neglected in Eqs. (26) and (27) are $O(\varepsilon^4)$ and $O(\varepsilon^3)$, respectively. The two-dimensional version of these equations was employed by Tuck and Hwang (1972) for studying long wave generations on a uniformly sloping beach due to ground movements. The effects of dispersion are excluded in these equations, which become significant in the region where the characteristic horizontal length scale is comparable with the local water depth. Equations (19) and (20) must then be used to describe the wave motion. Without considering the ground motion, i.e. $\partial h/\partial t = 0$, Eqs. (26) and (27) reduce to the ordinary Airy equations for shallow water waves over an uneven bottom.

3.2 $\eta = O(\varepsilon^m)$, $\dfrac{\partial h}{\partial t} = O(\varepsilon^m)$; $m = 4, 5, 6, \ldots$

From Table 1, it is shown that in these cases

$$\phi^{(0)} = \phi^{(2n)} = O(\varepsilon^{m-1}), \qquad m = 4, 5, 6, \ldots$$
$$\phi^{(1)} = \phi^{(2n+1)} = O(\varepsilon^m), \qquad m = 4, 5, 6, \ldots \tag{28}$$

The non-linear terms in Eqs. (19) and (20) are always smaller than all the remaining terms in the equations. Hence the linearized equations become

$$\frac{\partial \eta}{\partial t} + \nabla \cdot (uh) - \frac{h^3}{6}\nabla \cdot (\nabla^2 u) = A \cdot u + B\nabla \cdot u + \frac{3}{2}h^2\nabla h \cdot \nabla^2 u + C + O(\varepsilon^{m+3}) \tag{29}$$

$$\frac{\partial u}{\partial t} + \frac{1}{\varepsilon}\nabla \eta - \frac{h^2}{2}\nabla^2\left(\frac{\partial u}{\partial t}\right) = \nabla \cdot (h\nabla h)\frac{\partial u}{\partial t} + 2h\nabla h \cdot \nabla\left(\frac{\partial u}{\partial t}\right) + D + O(\varepsilon^{m+2}) \tag{30}$$

for $m = 4, 5, 6, \ldots$. $A, B, C,$ and D are given in Eqs. (21), (22), (23), and (24), respectively. It should be noted that the above equations are also valid for the case in which $\eta = O(\varepsilon^4)$ and $\partial h/\partial t = O(\varepsilon^3)$ (in this case, $\phi^{(0)} = O(\varepsilon^3)$; see Table 1).

It should be noted that although the equations pertinent to each of the three groups of long waves have different mathematical solutions, the features characterizing each group are all contained in the equations representing the waves of the cnoidal class. Therefore for numerical computations Eqs. (19) and (20) can be considered to give a uniformly valid picture of long waves generated by different types of ground motion.

4. Numerical Results for Two Dimensional Problems

For two-dimensional problems, Eqs. (19) and (20) become

$$\frac{\partial \eta}{\partial t} + \frac{\partial}{\partial x}[u(\eta + h)] - \frac{h^3}{6}\frac{\partial^3 u}{\partial x^2} = Au + B\frac{\partial u}{\partial x} + \frac{3}{2}h^2\frac{\partial h}{\partial x}\frac{\partial^2 u}{\partial x^2} + C \qquad (31)$$

and

$$\frac{\partial u}{\partial t} + u\frac{\partial u}{\partial x} + \frac{1}{\varepsilon}\frac{\partial \eta}{\partial x} - \frac{h^2}{2}\frac{\partial^3 u}{\partial t \partial x^2} = \frac{\partial}{\partial x}\left(h\frac{\partial h}{\partial x}\right)\frac{\partial u}{\partial t} + 2h\frac{\partial h}{\partial x}\frac{\partial^2 u}{\partial x \partial t} + D \qquad (32)$$

with

$$A = \left(\frac{\partial h}{\partial x}\right)^3 + \frac{1}{2}h^2\frac{\partial^3 h}{\partial x^3} + 3h\frac{\partial^2 h}{\partial x^2}\frac{\partial h}{\partial x}$$

$$B = \frac{3}{2}\frac{\partial}{\partial x}\left(h^2\frac{\partial h}{\partial x}\right)$$

$$C = \frac{\partial h}{\partial t}\left\{\left(\frac{\partial h}{\partial x}\right)^2 + h\frac{\partial^2 h}{\partial x^2} - 2\frac{\partial h}{\partial x}\frac{\partial^3 h}{\partial x \partial x^3} - 1\right\} + 2h\frac{\partial h}{\partial x}\frac{\partial^2 h}{\partial x \partial t} + \frac{h^2}{2}\frac{\partial^3 h}{\partial x^2 \partial t}$$

$$D = \frac{\partial^2 h}{\partial t^2}\frac{\partial h}{\partial x} + h\frac{\partial^3 h}{\partial x \partial t^2}. \qquad (33)$$

For a prescribed bottom topography and the detailed time history of ground motion, the resulting free surface fluctuations can be obtained by solving the above equations numerically. A finite difference scheme, which is similar to that developed by PEREGRINE (1967), is used in this paper. The corresponding finite-difference equations are given in Appendix B. The convergence and the stability of the numerical technique have been confined by examining solitary waves propagating over a constant depth and a mild soope (EARICKSON, 1980).

To demonstrate the validity of the governing equations derived in Section II. and to verify the accuracy of the finite difference method, a specific ground motion, which has been studied experimentally by HAMMACK (1972), is chosen

$$h(x, t) = h_0 - \zeta_0(1 - e^{-\alpha t})H(x^2 - 1), \qquad t \geq 0 \qquad (34)$$

where h_0 is the constant water depth prior to the ground motion. ζ_0 denotes the maximum bed displacement. $H(x^2 - 1)$ is the Heavyside step function; thus the ground motion is confined in the interval $-1 < x < 1$. It should also be pointed out that Eq. (34) is dimensionless and b^*, the half width of bed displacement, has been used as the horizontal length scale. Numerical computations are carried out by using the data presented in Table 2. In Fig. 1, free surface elevations at $x = 0$ and $x = 1$ are plotted for case (a) (impulsive motion). The comparison between present numerical results and experimental data are quite good. The corresponding free surface profiles

Table 2. Classification of ground movement prescribed by Eq. (34) and input data for computations of the corresponding free surface profiles.

	$h = \varepsilon = h^*/b^*$	ζ_0^*/h^*	$\zeta_0 = \zeta_0^*/b^*$	$t_c\sqrt{gh^*}/b^*$	$\alpha = 1.11b^*/t_c\sqrt{gh^*}$
(a) Impulsive motion		0.2	0.0164	0.069	16.0869
(b) Transitional motion	0.0820	0.1	0.0082	0.39	2.8462
(c) Creeping motion		0.3	0.0246	8.70	0.1276
(d) Large amplitude motion		0.5	0.0410	0.70	1.5857

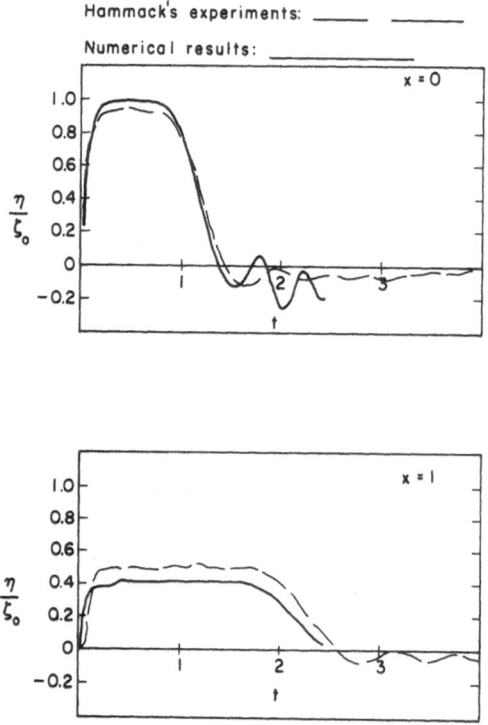

Fig. 1. Free-Surface fluctuations at $x = 0$ and $x = 1$ vs Time: numerical calculations vs Hammack's data, for an impulsive exponential block upthrust. $\Delta t = 0.002$, $\zeta_0 = 0.0164$, $\Delta x = 0.02$, $\alpha = 16.0869$, $\varepsilon = 0.082$, $t_0 = 0.069$.

are shown in Fig. 2. The leading solitary wave starts to form at $t = 1.0$. In Figs. 3 and 4, free surface elevations are plotted for cases (b) (transitional motion) and case (c) (creeping motion). Excellent agreement between laboratory data and numerical results is observed. It should be pointed out here that in these three cases the magnitude of $\partial h/\partial t$ varies from $O(\varepsilon)$ to $O(\varepsilon^2)$, Eqs. (31)–(33) are used for all computations without further approximation. As shown in Table 2, the case (d) represents large amplitude ground motion, $\zeta_0^*/h_0^* = 0.5$. Numerical results at $x = 15.76$ and $x = 33.8$ are

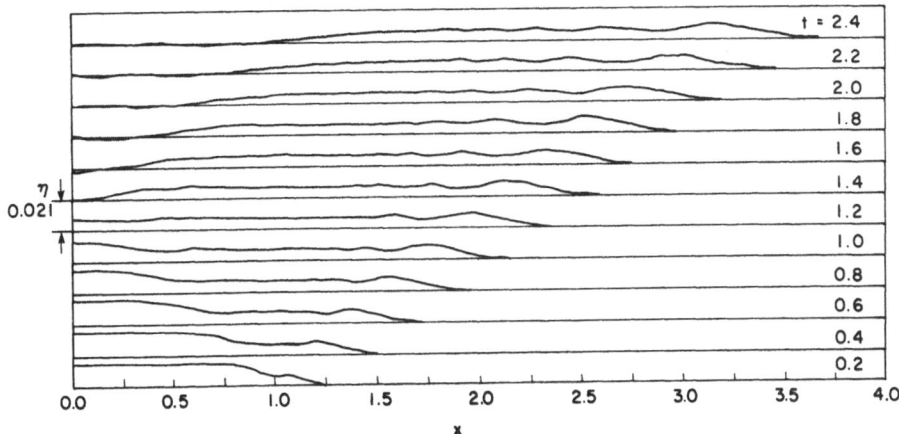

Fig. 2. Calculated free-surface profiles over time: impulsive exponential block upthrust. $\Delta t = 0.002$, $\zeta_0 = 0.0164$, $\Delta x = 0.02$, $\alpha = 16.0869$, $\varepsilon = 0.082$, $t_0 = 0.069$.

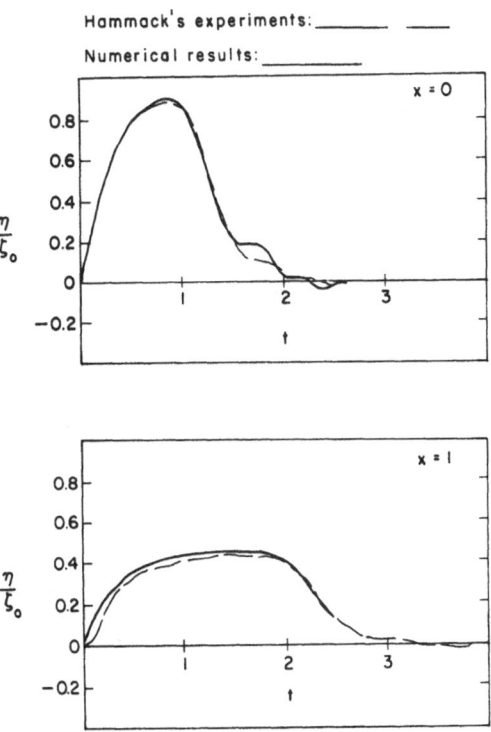

Fig. 3. Free-surface fluctuations at $x = 0$ and $x = 1$ vs Time: numerical calculations vs Hammack's data, for a transitional exponential block upthrust. $\Delta t = 0.002$, $\zeta_0 = 0.0082$, $\Delta x = 0.02$, $\alpha = 2.8461$, $\varepsilon = 0.0820$, $t_0 = 0.39$.

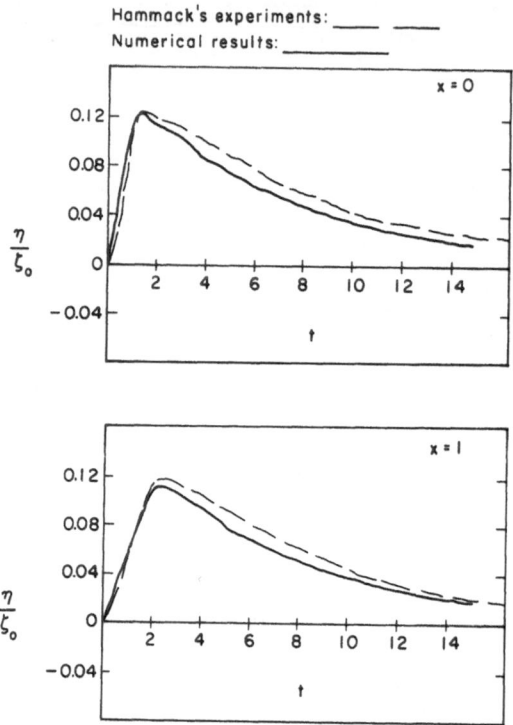

Fig. 4.   Free-surface fluctuations at $x = 0$ and $x = 1$ vs time: numerical calculations vs Hammack's data, for a creeping exponential block upthrust. $\Delta t = 0.01$, $\zeta_0 = 0.0246$, $\Delta x = 0.1$, $\alpha = 0.1275$, $\varepsilon = 0.0820$, $t_0 = 8.70$.

compared with experimental data in Fig. 5. It is seen that numerical results predict larger amplitude and faster wave speed. This could be attributed to the fact that the present model does not consider viscous effects. In Fig. 6, free surface profiles at different times are shown; the formation of solitons is clearly indicated.

5.   Concluding Remarks

The governing equations for long waves generated by ground movement are obtained herein. Equations (19)–(24) are shown to be uniformly valid in three different regimes: $O(\eta/\varepsilon^3)\, 3\, O(1)$; the corresponding magnitudes of the ground movement are shown in Table 1. Both nonlinear and dispersive effects are included in those equations. A finite-difference scheme is used for solving two-dimensional problems. Numerical results show very good agreement with available experimental data.

The current model has been applied to there-dimensional waves generated by ground upthrust of circular shape (Earickson, 1980). It is found that large negative disturbances are formed in the upthrust region, after the initial positive disturbance leaves the generation area.

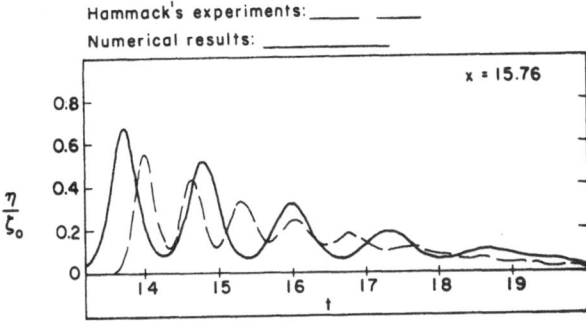

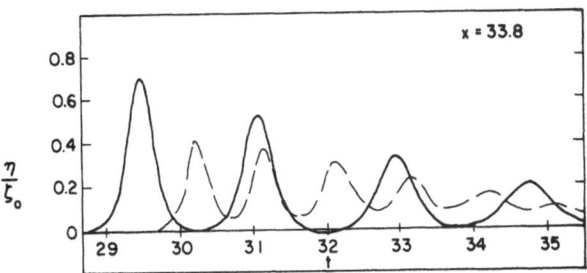

Fig. 5.   Free-surface fluctuations at $x = 15.76$ and $x = 33.8$ vs Time: numerical calculations vs Hammack's data, for wave propagation in the far-field. $\Delta t = 0.01$, $\zeta_0 = 0.041$, $\Delta x = 0.1$, $\alpha = 1.5857$, $\varepsilon = 0.0820$, $t_0 = 0.70$.

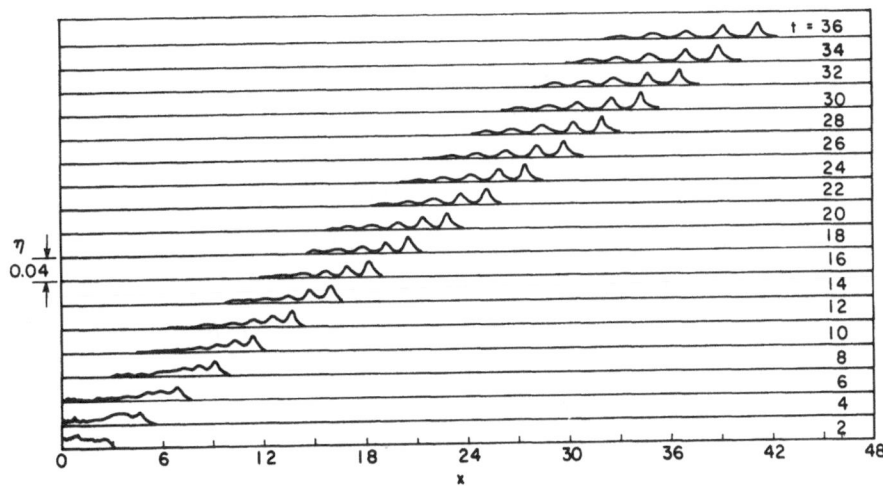

Fig. 6.   Calculated free-surface profiles over time: solitary wave formation in the propagation region. $\Delta t = 0.01$, $\zeta_0 = 0.041$, $\Delta x = 0.1$, $\alpha = 1.5857$, $\varepsilon = 0.0820$, $t_0 = 0.70$.

The research is in part supported by National Science Foundation through a Grant (NO. PRF-7815358) at Cornell.

Appendix A

In order to write Eqs. (19) and (20) in terms of $\phi^{(0)}$, the following formulae are needed

$$\phi^{(1)} = -[1 - |\nabla h|^2]\left[\frac{\partial h}{\partial t} + \nabla h \cdot \nabla \phi^{(0)}\right] + O(\varepsilon^6) \tag{A.1}$$

$$\phi^{(2)} = -\frac{1}{2}[1 - 3|\nabla h|^2]\nabla^2\phi^{(0)} + \frac{3}{2}\nabla^2 h\,(\nabla h \cdot \nabla \phi^{(0)}) + \frac{1}{2}\nabla^2 h\frac{\partial h}{\partial t} + \nabla h \cdot \nabla\left(\frac{\partial h}{\partial t}\right) + O(\varepsilon^5) \tag{A.2}$$

$$\phi^{(3)} = \frac{1}{2}\nabla h \cdot \nabla(\nabla^2\phi^{(0)}) + \frac{1}{2}\nabla^2 h\nabla^2\phi^{(0)} + \frac{1}{6}\nabla\phi^{(0)} \cdot \nabla(\nabla^2 h) + O(\varepsilon^4) \tag{A.3}$$

$$\phi^{(4)} = \frac{1}{24}\nabla^4\phi^{(0)} + O(\varepsilon^3). \tag{A.4}$$

These equations are derived from the recursive relation (i.e. Eqs. (9) and (10)) in cooperation with Eqs. (7), (14), and (15).

Appendix B

The finite-difference equations employed in the calculation described in Section 4 are as follows.

First the continuity equation, i.e. Eq. (31) is approximated by using forward differences in time and central difference in space,

$$\begin{aligned}
\eta_{i,j+1} = \eta_{i,j} + \nabla t\Bigg\{&\frac{1}{12}\left(\frac{h_{i,j}}{\Delta x}\right)^3 u_{i+2,j} + \left[\frac{B_{i,j}}{2\Delta x} + \frac{3}{2}\left(\frac{h_{i,j}}{\Delta x}\right)^2\right.\\
&\times \left.\left(\frac{\partial h}{\partial x}\right)_{i,j} - \frac{1}{6}\left(\frac{h_{i,j}}{\Delta x}\right)^3 - \frac{1}{2}(\eta_{i,j} + h_{i,j})\frac{1}{\Delta x}\right]u_{i+1,j}\\
&+ \left[A_{i,j} - 3\left(\frac{h_{i,j}}{\Delta x}\right)^2\left(\frac{\partial h}{\partial x}\right)_{i,j} - \left(\frac{\partial h}{\partial x}\right)_{i,j} - (\eta_{i+1,j}\right.\\
&\left. - \eta_{i-1,j})\frac{1}{2\Delta x}\right]u_{i,j} + \left[\frac{1}{2\Delta x}(\eta_{i,j} + h_{i,j}) + \frac{1}{6}\left(\frac{h_{i,j}}{\Delta x}\right)^3\right.\\
&\left. - \frac{B_{i,j}}{2\Delta x} + \frac{3}{2}\left(\frac{h_{i,j}}{\Delta x}\right)^2\left(\frac{\partial h}{\partial x}\right)_{i,j}\right]u_{i-1,j}\\
&- \frac{1}{12}\left(\frac{h_{i,j}}{\Delta x}\right)^3 u_{i-2,j} + C_{i,j}\Bigg\}
\end{aligned} \tag{B.1}$$

where $X_{i,j} = X(i\,\Delta x, j\,\Delta t)$, $i = 1, 2, \ldots$ and $j = 1, 2, 3, \ldots$. Δx is the nondimensional grid spacing and Δt the non-dimensional time step. Hence, the right hand side of Eqs. (B.1) is written in terms of all the variables at time $j\,\Delta t$ which are used for computing the free surface displacement at time $(j + 1)\Delta t$.

The momentum equation, i.e. Eq. (32), is also discretized in the similar way and is used to solve $u_{i,j+1}$,

$$\left[\frac{u_{i,j}}{4\Delta x} + M_{i,j}\right]u_{i+1,j+1} + N_{i,j}u_{i,j+1} + \left[P_{i,j} - \frac{u_{i,j}}{4\Delta x}\right]u_{i-1,j+1}$$

$$= \left[M_{i,j} - \frac{u_{i,j}}{4\Delta x}\right]u_{i+1,j} + N_{i,j}u_{i,j} + \left[P_{i,j} + \frac{u_{i,j}}{4\Delta x}\right]u_{i-1,j}$$

$$+ D_{i,j} + \frac{\eta_{i-1,j+1} - \eta_{i+1,j+1} + \eta_{i-1,j} - \eta_{i+1,j}}{4\varepsilon\Delta x} \tag{B.2}$$

with

$$M_{i,j} = -\left[\frac{1}{2}\left(\frac{h_{i,j}}{\Delta x}\right)^2 + 2\left(\frac{h_{i,j}}{\Delta x}\right)\left(\frac{\partial h}{\partial x}\right)_{i,j}\right]\frac{1}{\Delta t}$$

$$N_{i,j} = \left[1 + \left(\frac{h_{i,j}}{\Delta x}\right)^2 - \left(\frac{\partial h}{\partial x}\right)^2_{i,j} - h_{i,j}\left(\frac{\partial^2 h}{\partial x^2}\right)_{i,j}\right]\frac{1}{\Delta t}$$

$$P_{i,j} = \left[2\left(\frac{h_{i,j}}{\Delta x}\right)\left(\frac{\partial h}{\partial x}\right)_{i,j} - \frac{1}{2}\left(\frac{h_{i,j}}{\Delta x}\right)^2\right]\frac{1}{\Delta t}. \tag{B.3}$$

Note that Eq. (B.2) is written for time level $j + 1$, therefore the terms on the right hand side of the equation are known. The matrix equations for $u_{i,j+1}$, which are tridiagonal, can be solved by standard method. After obtaining up-dated $u_{i,j+1}$, the approximate continuity Eq. (B.1) is modified by replacing the velocity at previous time step $u_{i,j}$ by an average velocity, i.e. $(u_{i,j} + u_{i,j+1})/2$. The modified equation is then used to update the free surface profile at $j + 1$.

REFERENCES

BRADDOCK, R. D., P. VAN DEN DRIESSCHE, and G. W. PEADY, J. Fluid Mech. **59**, 817–828, 1973.

CAMFIELD, F. E. and R. L. STREET, Proc. ASCE, J. W. W. and Harbor Div., **95**, 1–22, 1969.

CHWANG, A. T. and T. Y. WU, Lecture Notes in Physics, Vol. 64, Springer Verlag, 80–90, 1976.

EARICKSON, J., MS thesis, Department of Environmental Engineering, Cornell University, 1980.

GORING, D. G., Report No. KH-R-38, W. M. Keck Laboratory, California Institute of Technology, 1978.

HAMMACK, J. L., Report No. KH-R-28, W. M. Keck Lab. of Hydraulics and Water Resources, California Institute of Technology, 1972.

HWANG, L.-S. and D. DIVOKY, J. Geophy. Res., **75**, 6802–6817, 1970.

IPPEN, A. T. and G. KULIN, M.I.T. Hydro. Lab. Tech Report No. 15, 1955.

KAJIURA, K., Bull. Earth. Res. Inst., **41** 535–571, 1963.

KISHI, T. and H. SAEKI, Proc. 10th Conf. Coastal Engng., **1**, 322–348, 1966.

LIN, C. C. and A. CLARK, Tsing Hua, J. Chinese Studies, Spec. No. 1, Nat. Sci., 1959.

LONG, R. R., J. Fluid Mech., **20**, 161–170, 1964.

MEI, C. C. and B. LÉMÉHAUTE, J. Geophy. Res., **71**, 393–400, 1966.

Peregrine, D. H. *J. Fluid Mech.*, **27**, 815–827, 1967.

Tuck, E. O. and L. -S. Hwang, *J. Fluid Mech.*, **51**, 449–461, 1972.

Ursell, F. *Proc. Camb. Phil. Soc.*, **59**, 685–694, 1953.

Webb, L. M., *Nat. Eng. Sci. Co. Tech Rep.* SN 57–2, 1962.

Wehausen, J. and E. Laitone, *Handbuck der Physik*, Vol. IX-3, Springer, 1960.

Wu, T. Y., *J. Engrg. Mech.*, *ASCE*, Vol. 107, No. EM3, 501–521, 1981.

Tsunamis—Their Science and Engineering, edited by K. Iida and T. Iwasaki, 241–250.
Copyright © 1983 by Terra Scientific Publishing Company (TERRAPUB), Tokyo.

One-Dimensional Dispersive Deformation of Tsunami with Typical Initial Profiles on Continental Topographies

Masakazu SHIBATA

*INA Civil Engineering Consultants Ltd.,
22-1 Suido-cho, Shinjuku-ku, Tokyo 162, Japan*

(Received August 30, 1981; Revised January 11, 1982)

A numerical simulation of one-dimensional tsunami propagation is performed by adopting the Peregrine equation as the governing equation. Three typical sea bottom topographies and two initial wave profiles are investigated. The dispersion terms included in the Peregrine equation are expected by theoretical argument to depress the wave height. Results of numerical calculations coincide with this expectation when the initial wave profile does not contain eminent short wave components. A detailed examination reveals that the finite element method, adopting only the water level and flow velocity as the nodal variables, applied to such a perturbational equation tends to become unstable when the element size is smaller than the water depth. Within the limits of applicability of this scheme it is confirmed that the dispersion effects of the perturbational terms are not small, that is, the maximum water level is considerably depressed by such a description.

1. Introduction

It is important in view of disaster preventation to predict the deformation of tsunami waves as they propagate in water. There exist several types of equations describing water wave propagation according to the characteristics of each wave. It is not always possible to adopt the most appropriate formula to describe a particular problem, however. When we examine all the possible governing equations leading to a numerical simulation of the problem, inevitably a compromise must be made between the cost of carrying out the numerical calculation and the accuracy of the solution. In the design of tsunami countermeasure devices, a two-dimensional numerical calculation is often performed to simulate the tsunami propagation based on an actual topography and a particular wave origin model. It is known that such a two-dimensional numerical simulation requires a considerable calculation time. Then we usually restrict that the pressure distribution in water is static,

$$\Delta p = \rho g \Delta z \qquad (1)$$

and that the resulting two-dimensional long-wave equation takes into account the contribution of the surface level deviation from the mean water level η only through the term

$$g \frac{\partial \eta}{\partial x}, \, g \frac{\partial \eta}{\partial y} \tag{2}$$

The tsunami is one of the most typical waves that show dispersion, so it is interesting to describe its deformation in more detail.

This paper adopts, as the governing equation, the one-dimensional Peregrine equation that takes into account a deviation from the static water pressure distribution. Boussinesq introduced a deviation from the static pressure distribution for the case of water of constant depth (Le Méhauté, 1976). The Peregrine equation is an extension to the case of varying depth (Peregrine, 1967). There are two important parameters associated with long waves. One is the ratio of amplitude to depth, ε, and the other is the ratio of depth to wavelength, σ. The basic assumption in the derivation is that the parameters ε and σ^2 are of the same order and are much less than unity. Then a perturbation expansion of the irrotational long-wave equation to order ε^2 leads to the Peregrine equation.

2. Numerical Formulation

The Peregrine equation in non-dimensional form is,

$$\frac{\partial u}{\partial t} + u \frac{\partial u}{\partial x} + \frac{\partial \eta}{\partial x} - \frac{1}{2} H \frac{\partial^3}{\partial x^2 \partial t}(Hu) + \frac{1}{6} H^2 \frac{\partial^3}{\partial x^2 \partial t} u = 0 \tag{3}$$

with

$$\frac{\partial \eta}{\partial t} + \frac{\partial}{\partial x}[(H + \eta)u] = 0. \tag{4}$$

Here H denotes the mean water depth. Non-dimensional variables are produced as follows:

$$(x, y, z) = H_0^{-1}(\tilde{x}, \tilde{y}, \tilde{z})$$

$$t = \left(\frac{g}{H_0}\right)^{1/2} \tilde{t}, \quad p = \tilde{p}/\rho g H_0$$

$$(u, v, w) = (g H_0)^{-1/2}(\tilde{u}, \tilde{v}, \tilde{w}). \tag{5}$$

Here the tilde $\sim$ denotes a dimensional variable, and H_0 is a constant offshore water depth.

The perturbation terms which generate dispersion effects are first examined analytically. In a flat bottom region, the dispersion relation is easily derived from the Peregrine equation when the non-linear terms are neglected.

$$\frac{c^2}{g H_0} = \frac{1}{1 + \frac{1}{3}\left[\frac{2\pi H_0}{\lambda}\right]^2}. \tag{6}$$

Here, c denotes the phase velocity and λ, a characteristic wavelength of the component of interest. This shows that the dispersion effects are eminent when λ is on the order of

H_0. Although the spatial scale of upheaval L_0 ($= 50$ km) is much larger than the water depth H_0 ($= 3$ km), abrupt variations included in the initial wave profile have wavelength components on the order of, or less than, the depth H_0 (see Fig. 1). These dispersion effects are expected to bring down the multicomponent wave height.

In a sloping region, we can Fourier-expand the governing equations in order to investigate their dispersion effects.

$$-i\omega u_n + i\frac{n\pi}{l}\eta_n + \sum_m i\frac{(n-m)\pi}{l}u_m u_{n-m} - \alpha^2 \cdot \left[\omega n\pi + \frac{4}{9}i\omega n^2\pi^2\right]u_n$$

$$-\frac{2}{3}\cdot\alpha^2 i\omega \sum_m \left[\left\{\frac{1}{m^2} + \frac{1}{im}\right\}(-1)^m(n-m)^2 + \frac{\omega}{im}(-1)^m(n-m)\right]u_{n-m}$$

$$-\frac{2}{3}\cdot\alpha^2 i\omega \sum_m \left[\left\{\frac{1}{m^2} + \frac{1}{im}\right\}(-1)^m(n+m)^2 + \frac{\omega}{im}(-1)^m(n+m)\right]u_{n+m} = 0 \quad (7)$$

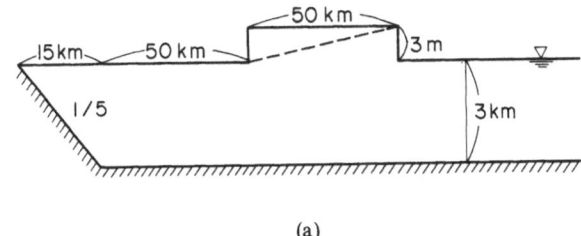

(a)

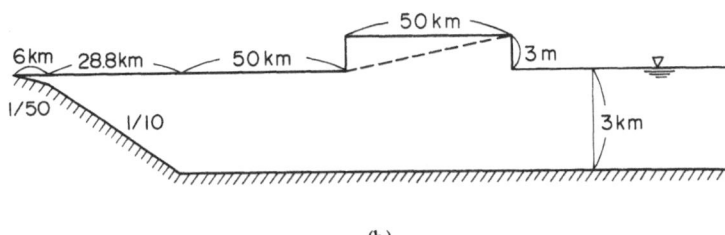

(b)

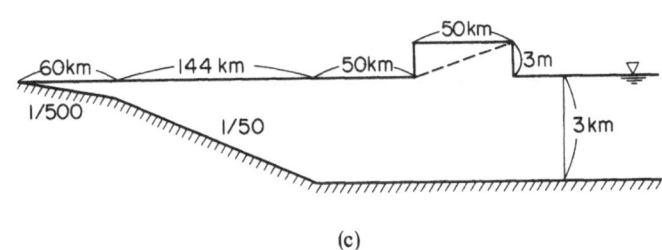

(c)

Fig. 1. Representative sea bottom topographies and initial wave profiles corresponding to similar upheavals of the sea bottom, (a) steep slope, (b) intermediate slope, (c) continental shelf.

$$-i\omega\eta_n + \frac{i\pi}{l}\sum_m (n-m)u_m\eta_{n-m} + \alpha[1 + in\pi]u_n + \alpha\frac{i\pi}{l}\sum_m (n-m)\eta_m u_{n-m}$$

$$+\alpha\sum_m \frac{(-1)^m}{m}[(n-m)u_{n-m} + (n+m)u_{n-m}] = 0. \tag{8}$$

Here the parameters α and $2\,l$ denote the gradient and the horizontal length of the slope, respectively. The integer n represents the component of a wavelength of interest and the m's are the components coupling with the component n. Here we find linear couplings due to slope, that is, the spatial dependence of the water depth H. Furthermore the coupling modes m extend essentially over the entire range of integers. That is, the coupling constants never diminish with increasing m. So we can not expect to analyze the dispersion effects further theoreticaly in the sloping region. This circumstance leads us to perform a numerical simulation of these governing equations.

Kawai and Watanabe formulated FEM scheme of KdV equation by adopting, as the nodal variables, the original variable u and its first and second derivatives, u' and u'', and obtained fairly coincident results with analytical solution (KAWAI, 1976). However, in a practical simulation, the numerical formulation should be examined from the view point of calculation time. Thus we develop a FEM scheme of the governing equation by adopting, as the nodal variables, only the original variables u and η, and then examine its applicability.

The governing equations are discretized with respect to the spatial variable x by the finite element method with second-order interpolation functions, $\phi_i(x)$

$$u(x,\,t) = \sum_i u_i(t)\phi_i(x) \tag{9}$$

$$\eta(x,\,t) = \sum_i \eta_i(t)\phi_i(x). \tag{10}$$

The temporal integration is performed by the fifth-order Runge-Kutta method which automatically controls the step size Δt, so that the fifth and sixth-order prototype Runge-Kutta schemes may predict the same results, in the meaning that the difference of the results of the two schemes divided by the initial value is less than a critical value which is taken here to be 0.01.

When a wave approaches shore, the parameter ε gradually increases. The equation adopted here no longer remains applicable to the problem when the wave breaks. A rough criterion of applicability of the equation is that the ratio of the amplitude to the initial depth of water at the position of the crest be less than 0.6 (PEREGRINE, 1967). The calculation is therefore terminated when this point is reached.

3.   Results of Calculation

Representative sea bottom topographies and initial wave profiles corresponding to similar upheavals of the sea bottom investigated here are shown in Fig. 1. These three types of sea bottom topographies are typical in the Pacific Ocean side of the Japan island arc. A steep slope shown in Fig. 1(a) is found in Suruga Bay of Shizuoka Prefecture, an intermediate slope shown in Fig. 1(b) is found off the Sotobo Coast, and a continental shelf shown in Fig. 1(c) is found off the Sanriku Coast of the Tohoku district. In the three figures, a solid line represents an initial wave profile corresponding to a homogeneous sea bottom upheaval, and a broken line corresponds to a linear upheaval.

Figure 2 shows maximum water levels calculated for six cases which correspond to the three continental topographies and the two initial profiles. Here, (a), (b) and (c) denote the continental topographies: steep slope, intermediate slope and continental shelf, and L and H denote the linear and homogeneous initial wave profiles, respectively. A solid line represents the full Peregrine equation and a broken line represents the equation without the dispersion terms. Triangles on the abscissa represent positions where the bottom undergoes a change in slope, for example, a shelfbreak.

These results show that 1) the more moderate sea bottom slope produces the stronger wave height amplification, 2) the initial condition of the homogeneous upheaval brings stronger wave height amplification than that of the linear upheaval, and that 3) the theoretical arguments on the dispersion effects are well confirmed; the dispersion terms are seen to depress the wave height in linear upheaval, but for conditions of homogeneous upheaval this proves not to be the case.

4. Theoretical Consideration of the FEM Scheme

We are then led to examine the FEM scheme adopted here in more detail. The Peregrine equation includes a perturbation term which is on the order of σ^2 multiplied by $\partial u/\partial t$. Formulated in the FEM scheme, matrix elements corresponding to the terms above and $\partial u/\partial t$ turn out to be, in dimensional form,

$$-\frac{1}{3}H^2\frac{\partial^3 u}{\partial x^2 \partial t}\left(\simeq \frac{1}{3}\left(\frac{H}{\lambda}\right)^2 \dot{u} \ll \dot{u}\right)$$

$$\rightarrow \frac{1}{3}\sum_i \int_{\Delta x_e} dx\, H^2(x)\frac{d\phi_i(x)}{dx}\frac{d\phi_j(x)}{dx}\dot{u}_i(t) + \cdots$$

$$\rightarrow \frac{1}{3}\frac{2H_0}{x_e}\sum_i \int_{-1}^{1} d\xi\, \frac{H^2(\xi)}{H_0^2}\frac{d\phi_i(\xi)}{d\xi}\frac{d\phi_j(\xi)}{d\xi}\dot{u}_i(t) + \cdots \qquad (11)$$

$$\frac{\partial u}{\partial t} \rightarrow \sum_i \int_{\Delta x_e} dx\,\phi_i(x)\phi_j(x)\cdot \dot{u}_i(t)$$

$$\rightarrow \frac{\Delta x_e}{2H_0}\sum_i \int_{-1}^{+1} d\xi\,\phi_i(\xi)\phi_j(\xi)\cdot \dot{u}_i(t). \qquad (12)$$

Here ξ is the local coordinate defined for each matrix element. The matrix element integrals with respect to ξ appearing there are both on the order of unity. It is seen that the two terms, when converted into matrix elements, do not conserve their relative

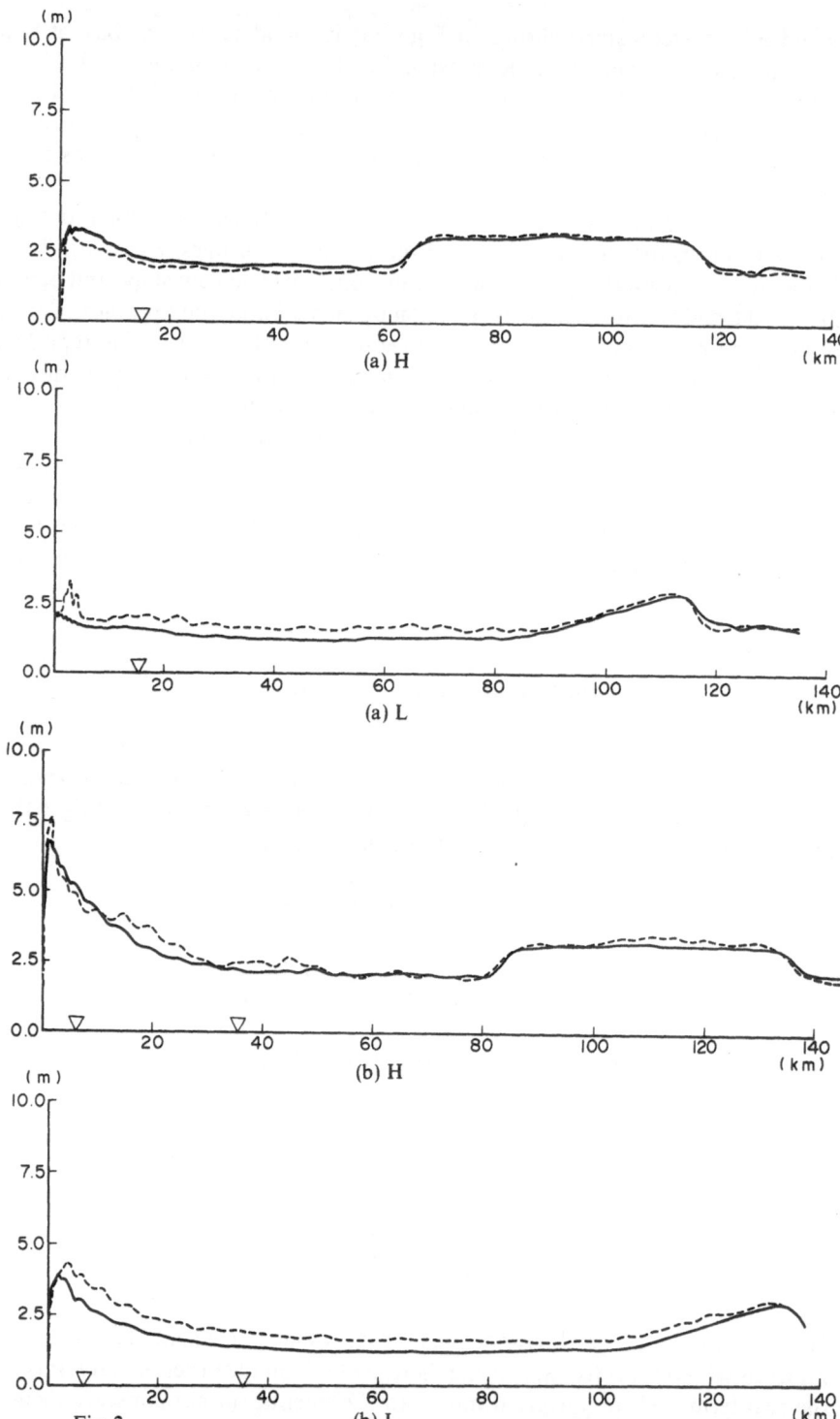

Fig. 2

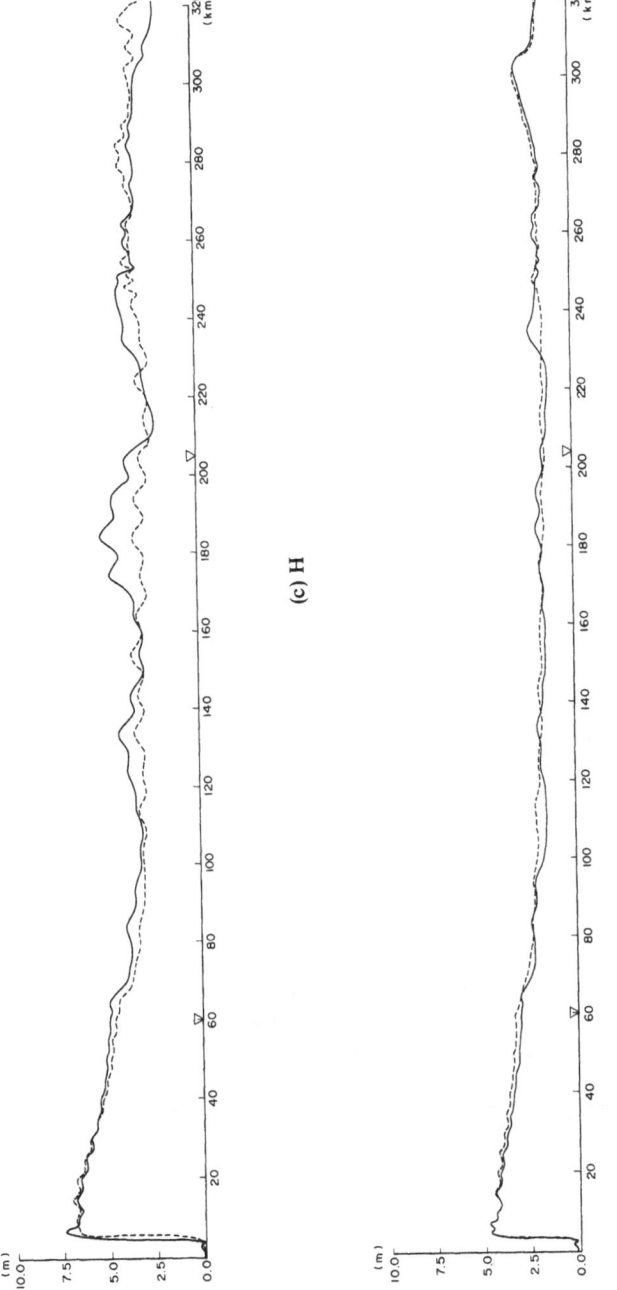

(c) H

(c) L

Fig. 2. Maximum water levels calculated for six cases which correspond to the three continental topographies and the two initial profiles, (a) steep slope, (b) intermediate slope, (c) continental shelf, H homogeneous upheaval, L linear upheaval.

magnitudes. That is, when the water depth H_0 is much larger than the element size Δx_e, the matrix element of the perturbation term (Eq. (11)) becomes much larger than that of the umperturbed term $\partial u/\partial t$ (Eq. (12)). This circumstance causes the FEM scheme to become unstable. Goto and Shuto pointed out that a certain finite difference scheme adopted for the Peregrine equation became unstable in deep water (GOTO and SHUTO, 1980). The calculation presented here is the FEM analogy of this.

It is interesting to examine why this happens in the FEM scheme. Number the local nodal points such as

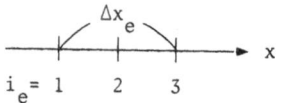

Fig. 3.   Numbering of local nodal points.

Then the interpolation functions take expressions as, by the local coordinate ξ,

$$\phi_1^e(\xi) = \frac{1}{2}\xi(\xi - 1)$$

$$\phi_2^e(\xi) = -(\xi + 1)(\xi - 1)$$

$$\phi_3^e(\xi) = \frac{1}{2}\xi(\xi + 1). \tag{13}$$

Hence the element matrix $A^e_{ij} = \int_{-1}^{1} d\xi\, d\phi_i^e/d\xi \cdot d\phi_j^e/d\xi$ is calculated as

$$[A^e] = \begin{bmatrix} \dfrac{7}{6}, & -\dfrac{4}{3}, & \dfrac{1}{6} \\[2mm] -\dfrac{4}{3}, & \dfrac{8}{3}, & -\dfrac{4}{3} \\[2mm] \dfrac{1}{6}, & -\dfrac{4}{3}, & \dfrac{7}{6} \end{bmatrix} \tag{14}$$

The element matrices are superposed on the entire matrix as shown in Fig. 4. Odd-numbered rows correspond to the nodal points at the ends of each element, while even-numbered rows correspond to at the center. If we regard $\dot{u}_i^e(t)$ also as a function of x, and Taylor-expand it with respect to x around the central nodal point $i^e = 2$, we have

$$\dot{u}_1^e = \dot{u}_2^e - \frac{\Delta x_e}{2}\frac{\partial \dot{u}_2^e}{\partial x} + \frac{1}{2}\left(-\frac{\Delta x_e}{2}\right)^2\frac{\partial^2 \dot{u}_2^e}{\partial x^2} + \cdots$$

$$\dot{u}_2^e = \dot{u}_2^e$$

$$\dot{u}_3^e = \dot{u}_2^e + \frac{\Delta x_e}{2}\frac{\partial \dot{u}_i^e}{\partial x} + \frac{1}{2}\left(\frac{\Delta x_e}{2}\right)^2\frac{\partial^2 \dot{u}_2^e}{\partial x^2} + \cdots. \tag{15}$$

Assume that the nodal point numbers 1, 2, 3 in the above element correspond to the row numbers $2k - 1$, $2k$, $2k + 1$ in the entire matrix. Then the products of A and $\dot{u}$,

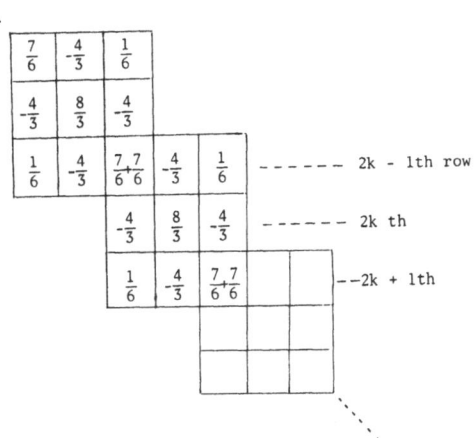

Fig. 4. Entire matrix.

$\sum_j A_{ij}\dot{u}_j$ behave as

$$\sum_j A_{2k,j}\dot{u}_j = -\frac{4}{3}\left(\frac{\Delta x_e}{2}\right)^2\frac{\partial^2 \dot{u}_2^e}{\partial x^2} + O\left((\Delta x_e)^3\frac{\partial^3 \dot{u}_2^e}{\partial x^3}\right) \qquad i = 2k \qquad (16)$$

$$\sum_j A_{2k-1,j}\dot{u}_j = \frac{\Delta x_e}{2}\frac{\partial \dot{u}_2^{e-1}}{\partial x} - \frac{\Delta x_e}{2}\frac{\partial \dot{u}_2^e}{\partial x} + O\left((\Delta x_e)^2\frac{\partial^2 \dot{u}_2^e}{\partial x^2}\right) \qquad i = 2k-1. \qquad (17)$$

Here, the suffix e-1 denotes the left-hand-side neighbor of the e-th element of interest.

Further Taylor-expanding the two leading terms at the right-hand-side of Eq. (17), both Eqs. (16) and (17) turn out to be on the order of $(\Delta x_e)^2\partial^2 \dot{u}_2^e/\partial x^2$, that is, on the order of $(\Delta x_e/\lambda)^2 \dot{u}_2^e$. Due to the symmetry of the matrix described above, the expansion of $\dot{u}$ begins to contribute to the equation at the second derivative term with respect to x. Then the product of the matrix and the second-derivative of $\dot{u}$ is thus seen to be on the order of $(H/\lambda)^2$ times $\dot{u}$, that is, σ^2 times $\dot{u}$.

We can now understand that the perturbational term $-1/3H^2\partial^3 u/\partial x^2\partial t$ generates numerical instability when the element size Δx_e is smaller than the water depth H, and that this instability comes from the FEM discretization scheme itself.

5. Concluding Remarks

Arguments in Section 4 showed that the element size Δx_e in the formulation is restricted as $\Delta x_e \gtrsim H$. On the other hand, it is known that 10 to 20 nodal points per wavelength are necessary to describe wave propagation accurately, that is, $\lambda \geq (5$ or $10)\cdot \Delta x_e$. So this scheme can not operate accurately unless the condition $\lambda \geq (5$ or $10)\cdot H$ is satisfied For conditions of homogeneous upheaval the initial wave profile includes more short wavelength components than that for conditions of linear upheaval, and then the results of calculations did not coincide with the theoretical expectation

(Section 3, item 3)). The same arguments lead to the conclusion that the qualitative statement 2) may also be a result of the limits of applicability of this scheme. This is also the case when the sea bottom is rather steep, that is, when the discretized nodal points can not follow the variation of depth because the element size is restricted as $\Delta x_e \gtrsim H$. Hence in the case of steep slope, the results of the calculation may be less accurate than those in the other cases.

Although the applicability of the FEM formulation adopted here is restricted in the sense described, in the several cases investigated here where this scheme is applicable, it appears that the dispersion effects of the perturbational terms are not small. That is, the maximum water level is considerably depressed by these terms. Such effects play an important role in calculations of tsunami propagation.

REFERENCES

Goto, T. and N. Shuto, Comparison of tsunami runup calculation methods and the wave front conditions, Proc. of 27th Japanese Conference on Coastal Engineering, pp.80–84, 1980 (in Japanese).

Kawai, T. and M. Watanabe, Analysis of a solitary water wave problem by the method of weighted residuals, Second International Conference on Finite Element Method in Flow Problems, 731–741, 1976.

Léméhauté, B., *An Introduction to Hydrodynamics and Water Waves*, Springer-Verlag, New York, 1976.

Peregrine, D. H., Long waves on a beach, *J. Fluid. Mech.*, **27**, 815–827, 1967.

Tsunamis—Their Science and Engineering, edited by K. Iida and T. Iwasaki, 251–263.

Focusing and Reflection of a Cylindrical Solitary Wave

Allen T. Chwang and Henry Power*

Institute of Hydraulic Research,
The University of Iowa, Iowa City, Iowa, U.S.A.

(Received August 13, 1981)

Due to the finite size of the source region, the earthquake-generated tsunamis may be represented by cylindrical waves in the ocean. As the wave reaches the coastal region, the focusing and reflection processes will significantly affect the maximum amplitude of a cylindrical wave which is dispersive as well as nonlinear.

The focusing and reflection of a cylindrical solitary wave in waters of constant depth are analyzed by applying the inner-outer expansions technique to the cylindrical Boussinesq equations. It is found that in the outer region the wave amplitude is inversely proportional to the square root of the radial distance from the center. The maximum wave amplitude at the center or at a vertical cylindrical wall has been obtained. The time at which this maximum amplitude is reached is also determined. It is also found that the incident wave does not reflect immediately at the center or at the wall as predicted by the linear wave theory. Rather, the wave suffers a time delay, called the phase lag, during the reflection process. This phase lag increases with a decrease of the initial wave amplitude. The analytic solutions obtained in this paper are found to be in qualitative agreement with the numerical results of Chwang and Wu (1976).

1. Introduction

Due to the finite size of the source region, the earthquake-generated tsunamis may be represented by cylindrical waves in the ocean. Although the leading wave of a tsunami may be analyzed by the linear wave theory in the deep ocean, the nonlinearity becomes equally important as the dispersion effect after the leading wave of a tsunami climbs up the continental shelf and reaches the coastal region where the focusing and reflection processes will drastically increase the maximum amplitude of a cylindrical wave.

Hershkowitz and Romesser (1974) observed experimentally that a cylindrical ion-acoustic solitary wave propagating toward the origin travels somewhat faster than the corresponding one-dimensional solitary wave, and that the square root of the maximum soliton amplitude multiplied by the width of the wave is a constant. Maxon and Viecelli (1974) have derived a modified Korteweg-de Vries (KdV) equation for axisymmetric cylindrical waves propagating in a collisionless plasma. They solved this modified KdV equation numerically and found that the wave amplitude grows initially

---

*Present address: Instituto de Mecanica de Fluidos, Facultad de Ingenieria, Universidad Central de Venezuela, Caracas, Venezuela.

like $r^{-1/2}$ as r decreases. Ogino and Takeda (1976) pointed out that Maxon and Viecelli's numerical solution is not valid when the wave approaches the origin due to the focusing and reflection processes. They studied the propagation of a cylindrical ion-acoustic solitary wave based on a numerical integration of the cylindrical Boussinesq equations, and they obtained an approximate solution for the early stage of propagation, which behaves like Maxon and Viecelli's solution.

Chwang and Wu (1976) investigated the propagation and focusing of a cylindrical solitary wave in water of constant and variable depth. They solved these problems by integrating the cylindrical Boussinesq equations, adopting a finite-difference scheme. For the case of a wave propagating in water of constant depth, their solution consists of a main positive wave whose maximum amplitude initially grows like $r^{-1/2}$ as r decreases and a small but persistent negative wave which follows the main wave. When the crest of the incident wave reaches the center ($r = 0$), it remains there for a period of time, during which the wave reaches its maximum amplitude, afterwards the crest of the wave begins to move away from the center as the wave changes its direction.

Cumberbatch (1978) found an approximate similarity solution for the cylindrical KdV equation which resembles the main part of the wave. The maximum amplitude of his solution grows like $r^{-2/3}$ as r decreases, while the width decreases like $r^{1/3}$. Miles (1978) also found a similarity solution with the wave amplitude grows like $r^{-2/3}$ as r decreases and the main wave is followed by a train of always positive or negative waves whose shape is approximately given by the square of an Airy function.

Ko and Kuehl (1979) found a small-time perturbation solution of the cylindrical KdV equation. The major conclusion of their work is the existence of a very small continual transfer of energy from the main wave to the trailing structure. Johnson (1979) proposed a change of variables in order to transform the cylindrical KdV equation into the Kadomtsev-Petviashvili equation so that the integral equation solution of the latter, based on the inverse scattering transform, may be applied.

The objective of the present paper is to develop an analytical solution for the propagation and focusing of a cylindrical solitary wave in water of constant depth based on the inner and outer expansions of the cylindrical Boussinesq equations.

2. Formulation of the Problem

We shall consider in this paper the focusing and reflection of a cylindrical solitary wave in water of constant depth. The fluid is assumed to be incompressible and inviscid, and the wave motion irrotational. Hence a velocity potential ϕ^* exists. The waves are assumed to be axisymmetric at any time t^*. In a cylindrical coordinate system (r^*, θ^*, z^*), with the z^*-axis pointing vertically upwards, the flow region is bounded below by a solid bottom at $z^* = -h$, and above by a free surface $z^* = \eta^*$ (r^*, t^*), which measures the displacement from the undisturbed free surface at $z^* = 0$. The fluid is supposed to be unbounded in the radial direction ($0 \leq r^* < \infty$).

We shall non-dimensionalize the wave amplitude η^*, the velocity potential ϕ^*, and the independent variables r^*, z^*, t^* by the following:

$$r = k_0 r^*, \quad z = z^*/h, \quad t = k_0 (gh)^{1/2} t^*,$$

$$\eta = \eta^*/a_0, \quad \phi = k_0 h \phi^*/(a_0(gh)^{1/2}), \tag{1}$$

where $k_0 (= 2\pi/\lambda_0, \lambda_0$ being the characteristic wavelength) is the characteristic wave number, a_0 the characteristic wave amplitude and g the constant acceleration of gravity. The velocity potential ϕ satisfies the Laplace equation in its dimensionless form

$$\frac{\mu^2}{r}(r\phi_r)_r + \phi_{zz} = 0 \quad (-1 < z < \varepsilon\eta), \tag{2}$$

where

$$\varepsilon = a_0/h \quad \text{and} \quad \mu = k_0 h. \tag{3}$$

The kinematic boundary condition at the free surface is

$$\phi_z = \mu^2(\eta_t + \varepsilon\phi_r\eta_r) \text{ at } z = \varepsilon\eta, \tag{4}$$

and the dynamic boundary condition there is

$$\mu^2(\phi_t + \eta) + \frac{\varepsilon}{2}(\mu^2\phi_r^2 + \phi_z^2) = 0 \quad \text{at} \quad z = \varepsilon\eta. \tag{5}$$

At the solid bottom we require

$$\phi_z = 0 \quad \text{at} \quad z = -1. \tag{6}$$

The two dimensionless parameters ε and μ defined in (3) are assumed to be small. However, their ratio $(\varepsilon/\mu^2 = Ur)$, now known as the Ursell number, is assumed to be of the order of one, $Ur = 0(1)$, for weakly nonlinear, long waves with frequency dispersion. Under these assumptions and with terms of the order of $0(\varepsilon\mu^2, \mu^4)$ being neglected, Eq. (2) to (6) will lead to the cylindrical Boussinesq equations (see CHWANG and WU, 1976)

$$\eta_t + \frac{1}{r}((1 + \varepsilon\eta)ru)_r = 0, \tag{7a}$$

$$u_t + \eta_r + \varepsilon u u_r - \frac{\mu^2}{3}\left(\frac{1}{r}(ru)_r\right)_{rt} = 0, \tag{7b}$$

where u is the depth-averaged radial velocity,

$$u(r, t) = \frac{1}{1 + \varepsilon\eta}\int_{-1}^{\varepsilon\eta} \phi_r dz. \tag{8}$$

The cylindrical Boussinesq equations, (7a) and (7b), contain nonlinear as well as dispersive terms, and they are valid for weakly nonlinear, long waves in water of constant depth. Since Eq. (7a) and (7b) can be used to account for both the radially outgoing wave (in the increasing r direction) and the radially incoming wave (in the decreasing r direction), they are suitable for studying the focusing and reflection of a cylindrical solitary wave.

We shall seek a solution of the cylindrical Boussinesq equations, (7a) and (7b), when the incident wave, which is located far away from the origin initially, has a cylindrical solitary wave profile given by

$$\eta = \operatorname{sech}^2 \left\{ \frac{\sqrt{3\varepsilon}}{2\mu} \left(r - r_0 + \left(1 + \frac{\varepsilon}{2} \right)(t - t_0) \right) \right\} \text{at } t = t_0, \tag{9}$$

where the initial time is taken to be t_0 instead of zero, the initial wave crest is located at $r = r_0$ ($r_0 \gg 1$), and the incident wave is propagating in the decreasing r direction.

Before we proceed to determine the analytic solution, let us first examine the numerical results obtained by CHWANG and WU (1976). Figure 1 shows a typical

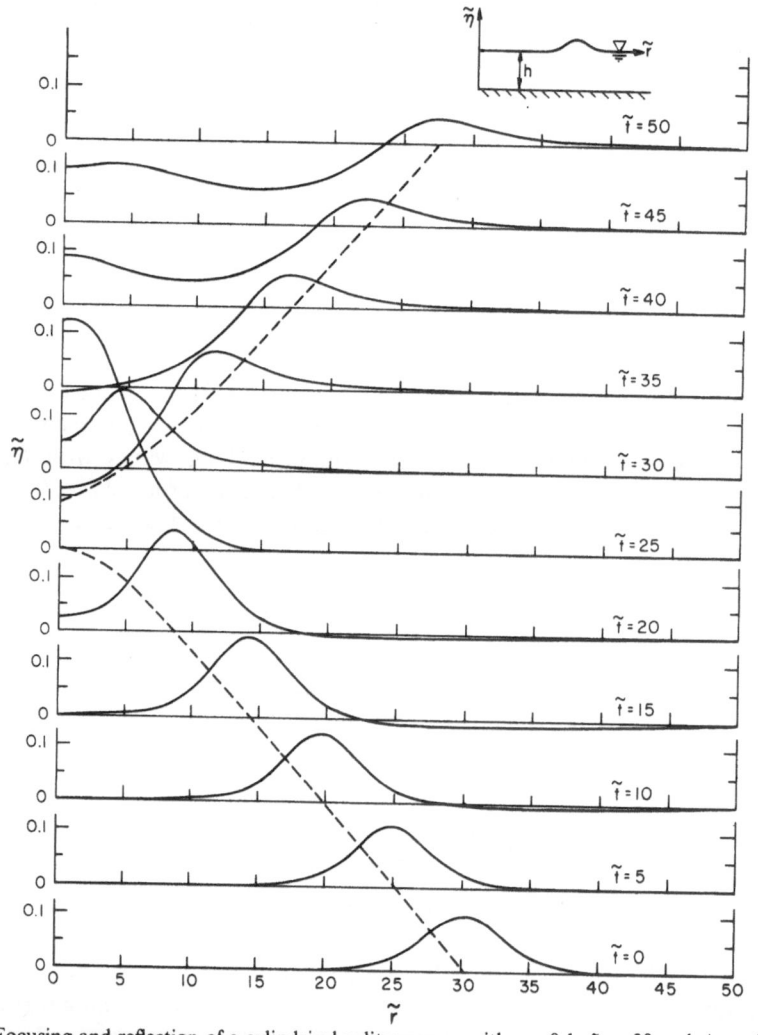

Fig. 1.   Focusing and reflection of a cylindrical solitary wave with $\varepsilon = 0.1$, $\tilde{r}_0 = 30$ and $\Delta t = \Delta r = 0.2 \mu$. (CHWANG and WU (1976)).
----location of wave crest.

numerical result presented by Chwang and Wu where the incident wave crest is located at $\tilde{r}_0 = r_0^*/h = 30$ and the initial wave amplitude is $\varepsilon = 0.1$. In Fig. 1 the dimensionless time $\tilde{t}$ is defined by

$$\tilde{t} = (t^* - t_0^*)(g/h)^{1/2}. \tag{10}$$

We note from Fig. 1 that before the crest of the incoming wave reaches the center ($\tilde{r} = 0$), the wave profile remains essentially similar to the initial one given by (9), with the wave amplitude ($\tilde{\eta} = \eta^*/h$) increasing like $\tilde{r}^{-1/2}$ as $\tilde{r}$ decreases, and the wave travels at a constant phase velocity $(1 + \varepsilon/2)$ as indicated by the dotted line connecting the wave crest locations at different times. We also note that the surface elevation behind the incident wave is slightly below the undisturbed water surface because of the "leakage flow" (CHWANG and WU, 1976), which is a characteristic of axisymmetric (two-dimensional) waves. As the wave enters the region close to the center where the focusing and reflection processes take place, its amplitude as well as its phase velocity increase drastically. When the wave crest reaches the center ($\tilde{r} = 0$), it does not reflect back immediately as it would based on the classical linear wave theory. Rather, the wave crest remains at the center for a certain time duration, called the "phase lag" (POWER and CHWANG, 1980), to complete the reflection process. During this time delay, the wave amplitude continues to increase until it reaches a maximum value, then it starts to decrease until the end of this time delay at which the wave crest begins to move away from the center and propagates in the increasing $\tilde{r}$ direction. Eventually, as the wave is sufficiently far away from the center, its phase velocity decreases to a constant $(1 + \varepsilon/2)$. The reflected wave then maintains at this constant phase velocity as it moves further away from the center while the wave amplitude decreases like $\tilde{r}^{-1/2}$ as $\tilde{r}$ increases. There is a negative wave immediately following the leading, radially-outgoing positive wave. This trailing negative wave then evolves into a train of oscillatory waves.

Based on the above observation, we may divide the flow region into two sub-regions, an inner region close to the center and an outer region far away from the center. In the inner region the reflection process takes place, and in the outer region the wave propagates only in one direction, in the decreasing r direction for the incident wave or in the increasing r direction for the reflected wave.

3. Outer Solution

In the outer region where the cylindrical wave propagates only in one direction, the Boussinesq Eq. (7a) and (7b) may be simplified. We assume that if we follow the incident wave, its wave profile will change very slowly in the outer region. We introduce a new pair of variables σ and τ by

$$\sigma = r + t, \quad \tau = \varepsilon t. \tag{11}$$

In terms of σ and τ, Eq. (7a) and (7b) may be written as

$$\eta_\sigma + \varepsilon\eta_\tau + \frac{\varepsilon}{\varepsilon\sigma - \tau}\left(u + \frac{\varepsilon\sigma - \tau}{\varepsilon}u_\sigma + \varepsilon\eta u + (\varepsilon\sigma - \tau)(\eta u)_\sigma\right) = 0, \tag{12}$$

$$u_\sigma + \varepsilon u_\tau + \varepsilon u u_\sigma + \eta_\sigma - \frac{\mu^2}{3}\left(\frac{\varepsilon}{\varepsilon\sigma - \tau}\left(\frac{\varepsilon\sigma - \tau}{\varepsilon}u\right)\right)_{\sigma\sigma} = 0, \tag{13}$$

where terms of the order of $0(\varepsilon\mu^2, \mu^4)$ are neglected. If both η and u vanish as σ approaches to plus or minus infinity for all values of τ, the lowest order solution of (12) and (13) is

$$\eta = -u. \tag{14}$$

Subtracting (13) from (12) and making use of (14), we have

$$-\varepsilon\eta_\tau + \frac{3}{2}\varepsilon\eta\eta_\sigma + \frac{1}{2}\frac{\varepsilon}{\varepsilon\sigma - \tau}\eta + \frac{\mu^2}{6}\eta_{\sigma\sigma\sigma} = 0(\varepsilon\mu^2, \mu^4). \tag{15}$$

In terms of the variables r and t, Eq. (15) becomes

$$-\eta_t + \frac{1}{2}\frac{\eta}{r} + \left(1 + \frac{3}{2}\varepsilon\eta\right)\eta_r + \frac{\mu^2}{6}\eta_{rrr} = 0(\varepsilon\mu^2, \mu^4), \tag{16}$$

which is the cylindrical Korteweg-de Vries (KdV) equation (Miles, 1978) for a wave propagating in the decreasing r direction. For a wave propagating in the increasing r direction, the minus sign in front of the first term in (16) has to be replaced by a plus sign, and Eq. (14) must be replaced by $\eta = u$.

The solution of Eq. (16) was given by Power (1981) as $u = \pm\eta$,

$$\eta = \left(\frac{t_0}{t}\right)^{1/2} \operatorname{sech}^2\left\{\frac{\sqrt{3\varepsilon}}{2\mu}\left(\frac{t_0}{t}\right)^{1/4}\right.$$

$$\left. ((r - r_i) \mp (t - t_0) \mp \varepsilon t_0^{1/2}(t^{1/2} - t_0^{1/2}))\right\} \qquad (t \geq t_0), \tag{17}$$

where the upper signs denote a wave propagating in the increasing r direction, whereas the lower signs denote a wave propagating in the decreasing r direction; r_i is the location of initial wave crest at $t = t_0$.

4. Inner Expansions

In order to find the inner expansions of the cylindrical Boussinesq equations, we first expand the region near $r = 0$ by introducing an inner variable ξ,

$$r = \varepsilon\xi. \tag{18}$$

With this change of variable Eq. (7a) and (7b) become

$$\varepsilon\left(N_t + \frac{1}{\xi}(\xi U N)_\xi\right) + \frac{1}{\xi}(\xi U)_\xi = O(\varepsilon^2\mu^2, \varepsilon\mu^4), \tag{19a}$$

$$\varepsilon(U_t + U U_\xi) + N_\xi - \frac{\mu^2}{3\varepsilon}\left(\frac{1}{\xi}(\xi U)_\xi\right)_{\xi t} = O(\varepsilon^2\mu^2, \varepsilon\mu^4), \tag{19b}$$

where N and U denote η and u respectively in the inner region. We now assume a perturbation solution of the type

$$N = N^{(0)} + \varepsilon N^{(1)} + \varepsilon^2 N^{(2)} + \cdots \tag{20a}$$

and

$$U = U^{(0)} + \varepsilon U^{(1)} + \varepsilon^2 U^{(2)} + \cdots. \tag{20b}$$

Substituting Eqs. (20a) and (20b) into Eqs. (19a) and (19b), we find the zeroth-order problem to be

$$N^{(0)}_\xi = 0 \quad \text{and} \quad \xi U^{(0)}_\xi + U^{(0)} = 0. \tag{21}$$

The solution of (21) is

$$N^{(0)} = f(t) \quad \text{and} \quad U^{(0)} = g(t)/\xi, \tag{22}$$

where $f(t)$ and $g(t)$ are arbitrary functions of t only. Since the velocity is finite at the center $r = 0$, $g(t)$ must vanish. Therefore

$$U^{(0)} \equiv 0. \tag{23}$$

Corresponding to (23), the velocity of the leading-order outer solution must approach to zero as r qoes to zero. Hence by (9) and (17), the leading-order outer solution is

$$
\begin{aligned}
\eta^{(0)} \\
= \left(\frac{t_0}{t}\right)^{1/2} \operatorname{sech}^2 &\left\{ \frac{\sqrt{3\varepsilon}}{2\mu}\left(\frac{t_0}{t}\right)^{1/4}\left((r - r_0) + (t - t_0) + \varepsilon t_0^{1/2}\,(t^{1/2} - t_0^{1/2})\right)\right\} \\
+ \left(\frac{t_0}{t}\right)^{1/2} \operatorname{sech}^2 &\left\{ \frac{\sqrt{3\varepsilon}}{2\mu}\left(\frac{t_0}{t}\right)^{1/4}\left((r + r_0) - (t - t_0) - \varepsilon t_0^{1/2}\,(t^{1/2} - t_0^{1/2})\right)\right\}.
\end{aligned}
\tag{24}
$$

which reduces to the initial wave profile (9) as t approaches to t_0. Consequently, the leading-order outer solution is a linear combination of the incident wave whose initial crest position at $t = t_0$ is at $r = r_0$ and a reflected wave whose initial crest position at $t = t_0$ is at an imaginary point located at $-r_0$.

From the matching condition

$$\lim_{\xi \to \infty} N^{(0)} = \lim_{r \to 0} \eta^{(0)},$$

we obtain $N^{(0)}$ as

$$N^{(0)} = f(t) = 2\left(\frac{t_0}{t}\right)^{1/2} \operatorname{sech}^2\left\{\frac{\sqrt{3\varepsilon}}{2\mu}\left(\frac{t_0}{t}\right)^{1/4}(t + \varepsilon t_0^{1/2}\,(t^{1/2} - t_0^{1/2}))\right\}, \tag{25}$$

where we have assumed that $t_0 = -r_0$ without loss of generality. By Eqs. (19), (20), and (23), the first-order inner problem becomes

$$N^{(0)}_t + \frac{1}{\xi}(\xi U^{(1)})_\xi = 0, \tag{26a}$$

$$N_\xi^{(1)} - \frac{\mu^2}{3\varepsilon}\left(\frac{1}{\xi}(\xi U^{(1)})_\xi\right)_{\xi t} = 0. \tag{26b}$$

Integrating (26b) and using (26a), we have

$$N^{(1)} = H(t) - \frac{\mu^2}{3\varepsilon} N_{tt}^{(0)}, \tag{27}$$

where $H(t)$ is an arbitrary function of t. Integrating (26a), we obtain

$$U^{(1)} = -\frac{\xi}{2} N_t^{(0)}. \tag{28}$$

The second-order inner problem is obtained by substituting Eqs. (20a) and (20b) into (19a) and (19b) and by using Eqs. (23) and (25). Hence

$$(\xi U^{(2)})_\xi = -\xi N_t^{(1)} - N^{(0)}(\xi U^{(1)})_\xi, \tag{29a}$$

$$N_\xi^{(2)} - \frac{\mu^2}{3\varepsilon}\left(\frac{1}{\xi}(\xi U^{(2)})_\xi\right)_{\xi t} = -U_t^{(1)}. \tag{29b}$$

By (26a) and (29a),

$$\frac{1}{\xi}(\xi U^{(2)})_\xi = -N_t^{(1)} + N^{(0)} N_t^{(0)}, \tag{30}$$

which is a function of t only. Therefore, by (28), (29b), and (30), we have

$$N_\xi^{(2)} = \frac{\xi}{2} N_{tt}^{(0)}. \tag{31}$$

The mathematical condition for N to be a maximum at $r = 0$ is $N_\xi = 0$ and $N_{\xi\xi} < 0$ at $\xi = 0$.

In the present case, we have

$$N_\xi = N_\xi^{(0)} + \varepsilon N_\xi^{(1)} + \varepsilon^2 N_\xi^{(2)} + O(\varepsilon^3) = \frac{\varepsilon^2}{2}\, \xi N_{tt}^{(0)} + O(\varepsilon^3), \tag{32}$$

$$N_{\xi\xi} = \frac{\varepsilon^2}{2} N_{tt}^{(0)} + O(\varepsilon^3). \tag{33}$$

We note from (32) that N_ξ is always zero at the center $\xi = 0$, as it should be if N is continuous there. Based on (25) and (33), we note that the crest of the incident wave reaches the center when $N_{\xi\xi}$ changes from positive to negative at $t = t_1$ and it leaves the center when $N_{\xi\xi}$ changes from negative to positive at $t = t_2$. Hence

$$N_{tt}^{(0)} = 0 \text{ at } t = t_1 \text{ and } t = t_2, t_2 > t_1. \tag{34}$$

The phase lag, which is the duration the crest of the wave remains at the center to complete the reflection process, is

$$\Delta t = t_2 - t_1. \tag{35}$$

Between t_1 and t_2 the crest of the wave remains at the center and

$$N^{(0)}_{tt} < 0 \quad (t_1 < t < t_2). \tag{36}$$

The maximum value of N is obtained from (25) as

$$N_m = 2\left(\frac{t_0}{t_m}\right)^{1/2} \operatorname{sech}^2\left\{\frac{\sqrt{3\varepsilon}}{2\mu}\left(\frac{t_0}{t_m}\right)^{1/4}\left(t_m + \varepsilon t_0^{1/2}\left(t_m^{1/2} - t_0^{1/2}\right)\right)\right\} + O(\varepsilon), \tag{37}$$

where t_m is the time at which the maximum amplitude N_m is reached,

$$N^{(0)}_t = 0 \text{ at } t = t_m \ (t_1 < t_m < t_2). \tag{38}$$

If there is a vertical, cylindrical wall at $r = r_1$, the problem can be solved in a similar manner as discussed in this section; however, the inner variable ξ would be defined by

$$r - r_1 = \varepsilon\xi \tag{39}$$

rather than by Eq. (18). Equation (25) becomes

$$N^{(0)} = 2\left(\frac{t_0}{t}\right)^{1/2} \operatorname{sech}^2\left\{\frac{\sqrt{3\varepsilon}}{2\mu}\left(\frac{t_0}{t}\right)^{1/4}\left(r_1 + t + \varepsilon t_0^{1/2}\left(t^{1/2} - t_0^{1/2}\right)\right)\right\}. \tag{40}$$

The phase lag is still determined by Eqs. (34) and (35). However, $N^{(0)}$ is replaced by (40). The maximum value of N now becomes

$$N_m = 2\left(\frac{t_0}{t_m}\right)^{1/2} \operatorname{sech}^2\left\{\frac{\sqrt{3\varepsilon}}{2\mu}\left(\frac{t_0}{t_m}\right)^{1/4}\left(r_1 + t_m + \varepsilon t_0^{1/2}\left(t_m^{1/2} - t_0^{1/2}\right)\right)\right\} + O(\varepsilon), \tag{41}$$

where t_m is determined by (38) and (40).

Figure 2 shows the wave profiles between times $\tilde{t} = 0$ and $\tilde{t} = 45$ when the initial wave amplitude is $\varepsilon = 0.1$, the initial wave crest is located at $\tilde{r}_0 = 30$, and the cylindrical wall is at $\tilde{r}_1 = 10$. Figure 2 is obtained based on the numerical scheme of CHWANG and WU (1976). We note from Fig. 2 that the essential features we observed in Fig. 1 ($\tilde{r}_1 = 0$) appear also in Fig. 2, and these two figures are quite similar.

5. Discussion of Results

The maximum wave amplitude ($\tilde{N}_m = N^*_m/h$) at the vertical wall $\tilde{r} = \tilde{r}_1$ and the maximum wave amplitude at the center, which corresponds to the case of no walls $\tilde{r}_1 = 0$, are shown in Fig. 3 versus the initial wave amplitude. In Fig. 3, $\tilde{r}_0$ is chosen to be 30 and $\tilde{r}_1$ is 10. The solid lines represent the analytic results given by (41) and the dotted lines are obtained from the numerical results based on CHWANG and WU's (1976) scheme. We note from Fig. 3 that the maximum wave amplitude increases linearly as the initial wave amplitude ε for $\tilde{r}_1 = 10$, and in this case the analytic solution agrees very well with the numerical result. However, in the case of no walls ($\tilde{r}_1 = 0$), there is only qualitative agreement between the analytic solution and the numerical result; probably we have to calculate higher-order terms in order to have an accurate analytic

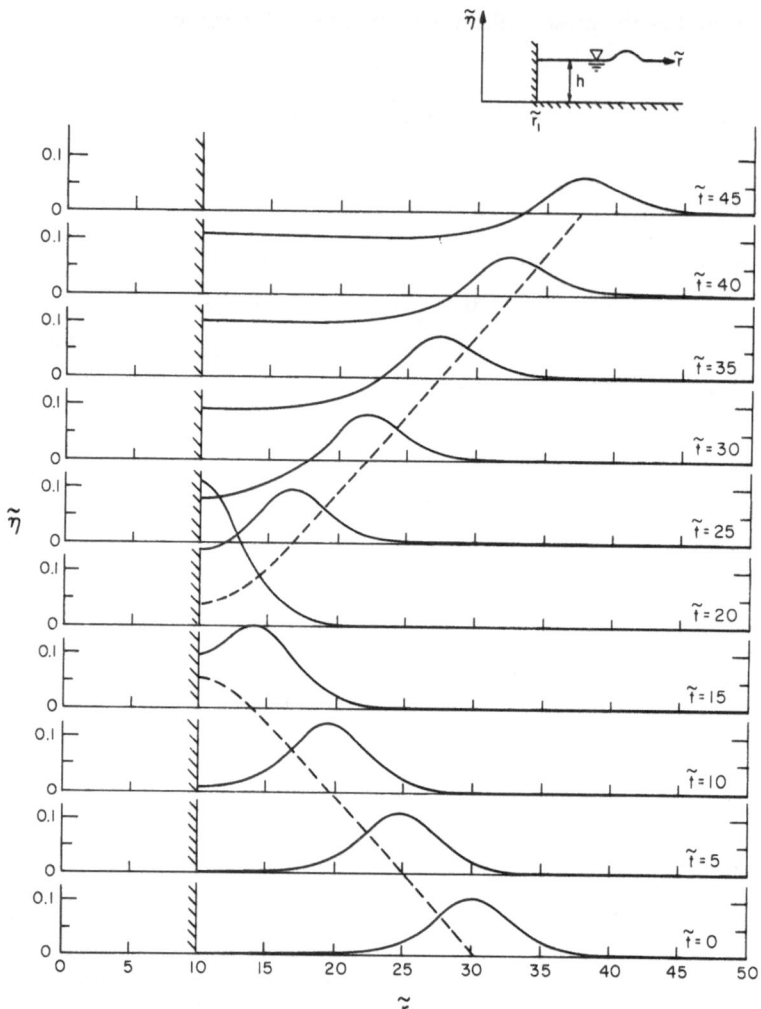

Fig. 2. Reflection of a cylindrical solitary wave due to a vertical cylindrical wall at $\tilde{r}_1 = 10$ with $\varepsilon = 0.1$, $\tilde{r}_0 = 30$ and $\Delta t = \Delta r = 0.2\mu$ (CHWANG and WU (1976)).
--- location of wave crest.

solution since the maximum wave amplitude is quite high and the focusing and reflection effects are very strong at the center.

The time at which the wave amplitude reaches its maximum value at $\tilde{r} = \tilde{r}_1$, calculated from Eqs. (38) and (40), is plotted versus the initial wave amplitude as solid lines in Fig. 4. The numerical results, represented by dotted lines, are also shown in Fig. 4 for comparison. The dimensionless time $\tilde{t}_m$, $\tilde{t}_m = (t_m^* - t_0^*)(g/h)^{1/2}$, decreases as ε increases because a wave of large amplitude propagates faster than a wave of small amplitude. The agreement between the analytic solution and the numerical results for both $\tilde{r}_1 = 0$ and $\tilde{r}_1 = 10$ cases is fairly well.

Finally, the phase lag during reflection, as given by Eq. (35) with $\Delta \tilde{t} = \Delta t^* (g/h)^{1/2}$,

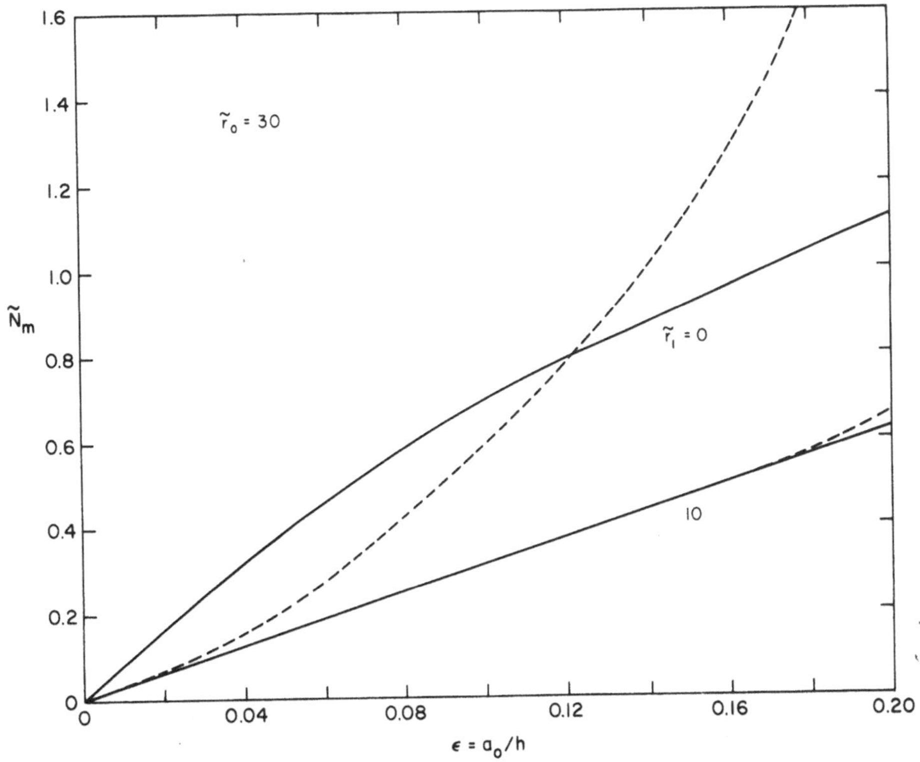

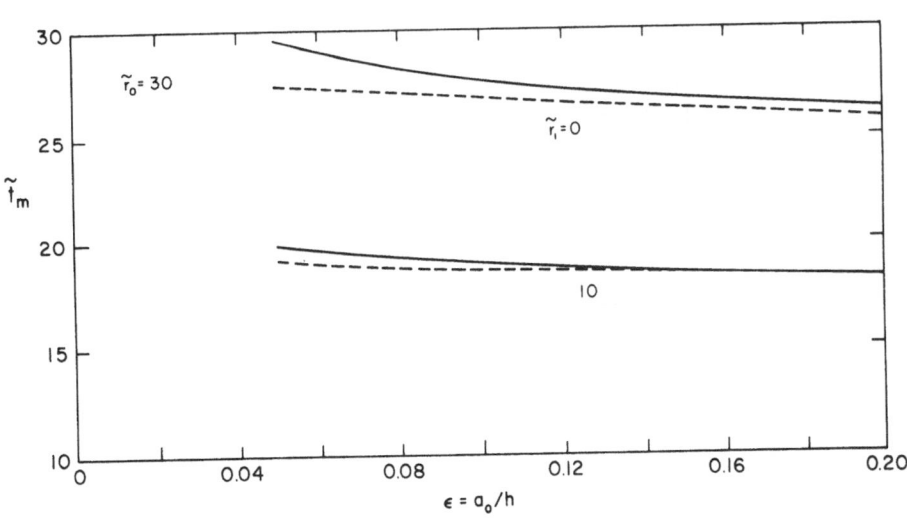

Fig. 3. Maximum wave amplitude $\tilde{N}_m$ vs the initial wave amplitude ε. ——, analytic solution; -----, numerical result.

Fig. 4. The time at which the wave reaches its maximum amplitude, $\tilde{t}_m$, vs the initial wave amplitude ε. ——, analytic solution, -----, numerical result.

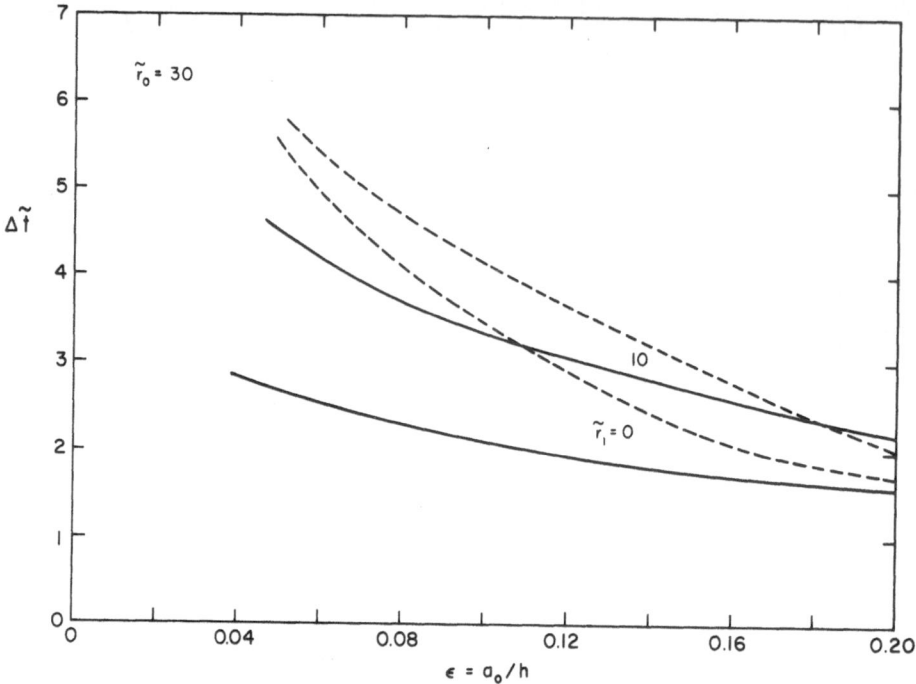

Fig. 5.   Magnitude of phase lag $\Delta \tilde{t}$ vs the initial wave amplitude ε. ———, analytic solution ; -----, numerical result.

is shown in Fig. 5 together with that obtained by the numerical method for both cases $\tilde{r}_1 = 0$ and $\tilde{r}_1 = 10$. We note from Fig. 5 that the phase lag is inversely proportional to the initial wave amplitude, which is in the same trend as that for planar solitary waves (POWER and CHWANG, 1980). There is only qualitative agreement between the analytic solution and the numerical results. The agreement gets better for large values of ε, $\tilde{r}_1$ being fixed. It seems that higher-order terms in both the inner and outer expansions are needed in order to have accurate analytic solutions and better quantitative agreement between the analytic and numerical results.

This work was sponsored by the National Science Foundation under Grant ENG 79–10049 and CME 80–20564.

REFERENCES

CHWANG, A. T. and T. Y. WU, Cylindrical solitary waves, Proc. IUTAM Symp. Waves on water of variable depth, Canberra, pp. 80–90, 1976.
CUMBERBATCH, E., Spike solution for radially symmetric solitary waves, Phys. Fluids, 21, 374–376, 1978.
HERSHKOWITZ, N. and T. ROMESSER, Observations of ion-acoustic cylindrical solitons, Phys. Rev. Lett., 32, 581–583, 1974.
JOHNSON, R. S., On the inverse scattering transform, the cylindrical Korteweg-de Vries equation and similarity solutions, Phys. Lett., A72, 197–199, 1979.

Ko, K. and H. H. Kuehl, Cylindrical and spherical Korteweg-de Vries solitary waves, *Phys. Fluids*, **22**, 1343–1348, 1979.

Maxon, S. and J. Viecelli, Cylindrical solitons, *Phys. Fluids*, **17**, 1614–1616, 1974.

Miles, J. W., An axisymmetric Boussinesq wave, J. Fluid Mech., **84**, 181–191, 1978.

Ogino, T. and S. Takeda, Computer simulation and analysis for the spherical and cylindrical ion-acoustic solitons, J. Phys. Soc. Japan, **41**, 257–264, 1976.

Power, H., On cylindrical solitary waves, Ph. D. thesis, The University of Iowa, 1981.

Power, H. and A. T. Chwang, Reflection of a planar solitary wave, *Bull. Am. Phys. Soc.*, **25**, 1079, 1980.

Tsunamis—Their Science and Engineering, edited by K. Iida and T. Iwasaki, 265–274.

Tsunami Generation as Finite Depth Cauchy-Poisson Problem or Long Wave Problem

Takashi ICHIYE

Department of Oceanography, Texas A & M University
College Station, TX, U.S.A.

(Received September 9, 1981)

Generation of tsunami by a bottom disturbance has been treated either as the shallow water case of the Cauchy-Poisson initial wave problem by taking the vertical velocity of the bottom movement as a boundary value of the velocity potential or as the long wave problem with the vertical acceleration at the bottom as the forcing function. The latter method may be easily applied to practical prognostic and diagnostic problems of tsunami propagation. When the horizontal dimension of the bottom disturbance is of the same order of magnitudes as the water depth, some deviations from the long wave model become serious far from the source. A solution of the linearized Euler equations of the long waves is not always the same as that of the Cauchy-Poisson problem with the long wave approximation. In order that the linear long wave Eulerian equations may be valid, two conditions should be satisfied: linearity and hydrostatic conditions. If further condition for simultaneous bottom and surface movement (SBSM) is satisfied, the sea level displacement equals the bottom displacement. The three conditions can be expressed in terms of four non-dimensional parameters $\alpha = t_c (gH)^{\frac{1}{2}} b^{-1}$, $\beta = \eta_0/H$, $\gamma = t_c (g/H)^{\frac{1}{2}}$, $\delta = H/b$ where t_c, b, and η_0 are characteristic time scale, horizontal distance and vertical height of the bottom deformation, respectively. The linearity condition can be satisfied for $O(\alpha) \lesssim 1$ and $\alpha\beta \ll 1$ (case I) or $\alpha \gg 1$ (case II) where the latter case is called creep bottom motion. The SBSM condition is satisfied for $\alpha \ll 1$ (case I) or $\alpha \gg 1$ (case II). The hydrostatic condition is satisfied for $\gamma^2 \beta^{-1} \gg 1$ (case I) or $\gamma\delta\beta^{-1} \gg 1$ (case II).

1. Introduction

Among many hydrodynamical problems related to tsunami, its generation may be considered as already solved or even trivial. This is because the problem was treated by many authors since SANO and HASEGAWA in 1915 and also because tsunami is apparently a long (shallow water) wave with least dispersion and without influence by Coriolis' or other planetary effects such as an atmospheric long wave. However, when the problem is scrutinized both mathematically and physically, it remains still paradoxical.

Mathematically speaking, a Cauchy-Poisson (hereafter C-P) problem for tsunami generation is not solved completely for a finite depth and a finite horizontal dimension of bottom deformation because the Fourier integral involved ·is not expressed in a closed form. On the other hand, it is not warranted to approximate the Eulerian system of hydrodynamic equations by those of shallow water waves within a generation area while tsunami is generated by the bottom deformation, though such approximation makes subsequent analysis much simpler both analytically and numerically. From an experimental point of view, the generation problem still imposes a tremendous difficulty, since it is not feasible to determine the actual sea level change or bottom deformation within and near a generation area. Therefore it seems to be rather important to investigate dynamics of tsunami generation more critically in order to understand behaviors of tsunami waves both in the open sea and coastal areas.

2.   Cauchy-Poisson Problem

Results of theoretical investigation on tsunami generation were reviewed in WILSON (1964). These are based on C-P models of different geometrical conditions. Although WILSON (1964) did not mention, SANO and HASEGAWA in 1915 obtained a solution of a C-P problem for sudden bottom deformation which is concentrated at a point and they calculated the integral numerically. SYONO (1935) generalized the problem for a finite area of bottom deformation but for the final calculation of sea level change he assumed the ratio of bottom deformation dimension to water depth (hereafter called aspect ratio) either extremely large or small, thus leading to the shallow water wave or deep water wave, respectively. TAKAHASHI (1943, 1948) further generalized the problem for a two-dimensional source area and for bottom defor-mation represented by an arbitrary function of time and space. Again in the final calculation of sea level change he assumed an extremely large aspect ratio. ICHIYE (1950) tried to determine the sea level analytically with an intermediate aspect ratio for the bottom motion which begins rapidly but decays gradually. He did not succeed except for distances far from the source region. Later ICHIYE (1958) calculated the sea level change near the source region numerically for a one-dimensional bottom deformation.

In order to compare the C-P approach and long wave treatment, a linear case of uniform depth is discussed.

In the C-P problem velocity potential Φ satisfies the Laplace equation

$$\partial^2 \Phi / \partial x^2 + \partial^2 \Phi / \partial z^2 = 0 \tag{1}$$

where x and z are the horizontal and vertical (upwards) coordinates. The boundary conditions at the sea surface and at the bottom are

$$[\partial \Phi / \partial t + 2\mu \Phi]_{z=H} = \zeta \tag{2}$$

$$(\partial \Phi / \partial t)_{z=0} = \eta = f(x)F(t) \tag{3}$$

respectively, where t is time, ζ is the surface elevation from the equilibrium sea level, μ is fictitious viscosity, $f(x)$ and $F(t)$ are arbitrary functions for bottom deformation η, and the bottom and the surface are designated as $z = 0$ and $z = H$, respectively.

An analytical solution of (1) under conditions (2) and (3) is given in terms of ζ by

$$\zeta = \frac{1}{2\pi} \int_0^\infty (\cosh kH)^{-1} dk \int_{-\infty}^t F(t) \cos \gamma (t - s) ds \int_{-\infty}^\infty f(\lambda) e^{ik(x - \lambda)} d\lambda \quad (4)$$

where

$$\gamma = (gk \tanh kH - \mu^2)^{\frac{1}{2}}$$

(TAKAHASHI, 1948; ICHIYE, 1958). TAKAHASHI (1948) further derived a simple form of (4) by assuming that

$$F(t) = 0 \quad \text{for } t < 0, t > T \quad (6)$$

$$f(t) = 1/T \quad \text{for } 0 < t < T \quad (7)$$

leading to

$$\zeta = [\xi(t) - \xi(t - T)]T^{-1} \quad \text{for } t > T \quad (8)$$

$$\zeta = [\xi(t) - \xi(0)]T^{-1} \quad \text{for } 0 < t < T \quad (9)$$

where

$$\xi(t) = \frac{1}{2\pi} \int_{-\infty}^\infty \frac{\sin \gamma t \, dk}{\gamma \cosh kH} \int_{-\infty}^\infty f(\lambda) e^{ik(x - \lambda)} d\lambda \quad (10)$$

and

$$\gamma = (gk \tanh kH)^{\frac{1}{2}}.$$

The integrals about k in (10) causes a difficulty since it includes γ. To obtain approximation for (10), a Gaussian form for $f(x)$ is used

$$f(x) = \exp(-x^2/a^2). \quad (11)$$

Then we have

$$\xi(t) = \frac{a}{4\sqrt{\pi}} \int_{-\infty}^\infty \frac{\sin \gamma t}{\gamma \cosh kH} \exp\left\{-\frac{1}{4}(ka)^2 + ikx\right\} dk. \quad (12)$$

Wave number k and time are scaled by H^{-1} and $(g/H)^{-1/2}$, respectively. Then a non-dimensional form of (12) becomes

$$\xi(t) = \frac{a}{4\sqrt{\pi}} \int_{-\infty}^\infty \frac{\sin h\tau}{h \cosh m} \exp\left\{-\frac{1}{4}(Em)^2 + im\rho\right\} dm \quad (13)$$

where $E = a/H$ is the aspect ratio and $\rho = x/H$, $h = (m \tanh m)^{\frac{1}{2}}$, $\tau = t(g/H)^{-\frac{1}{2}}$.

If the factor $(m \tanh m)^{1/2}$ is replaced by m, the integral of (13) represents a long wave propagation with a speed $(gH)^{1/2}$. This replacement is possible if the amplitude of the integrand decreases rapidly with increasing values of m, since then the magnitude of the integrand contributes substantially only in the range $|m| \ll 1$. In that case (13) becomes

$$\xi_0(t) = \frac{a}{4\sqrt{\pi}} \int_{-\infty}^\infty m^{-1} \sin m \exp\left\{-\frac{1}{4}(Em)^2 + im\rho\right\} dm. \quad (14)$$

In order that (14) may approximate (13), the ratio I

$$I = m \exp\left\{-\frac{1}{4}(Em)^2\right\} \{(m \tanh m)^{\frac{1}{2}}\cosh m\}^{-1} \tag{15}$$

should rapidly decrease to zero as $|m|$ increases from zero. The curves of I against m are plotted in Fig. 1 for $E = 1, 10$, and 50. This figure indicates that the value of I decreases to 0.1 at about $m = 0.04, 0.3$, and 2 for $E = 50, 10$, and 1, respectively. Therefore the approximation(14) may be valid for $E = 50$ but definitely not for $E = 1$.

The integral of (13) can be evaluated for large values of $|\rho|$ and τ by use of Kelvin's stationary phase method (LAMB, 1932; ICHIYE, 1950; SIROVICH, 1971). In this method, the stationary phase is at m which satisfies

$$dh(m)/dm = \pm \rho/\tau \quad \text{for} \quad |\rho| \gg 1 \text{ and } \tau \gg 1. \tag{16}$$

where

$$h(m) = (m \tanh m)^{\frac{1}{2}}. \tag{17}$$

For a Cauchy-Poisson problem of infinite depth this becomes

$$\frac{1}{2}m^{-\frac{1}{2}} = \pm \rho/\tau \text{ for } |\rho| \gg 1 \quad \text{and} \quad \tau \gg 1 \tag{18}$$

with H being replaced by some characteristic length such as a predominant wave length. The right hand side of (16) and (18) is plotted in Fig. 2. For $|\rho| > \tau$ there is no stationary phase point for Eq. (16). This indicates that the major surface disturbance does not exist beyond the distance of $(gH)^{1/2}t$ from the source region or that the major part of the wave propagates with a speed less than $(gH)^{1/2}$. Also the amplitude of the integrand (13) decreases rapidly with m for a large aspecto ratio E. Therefore, the major part of the wave exists near the stationary phase point closest to $m = 0$. In other words, the major part propagates with a speed of a long wave.

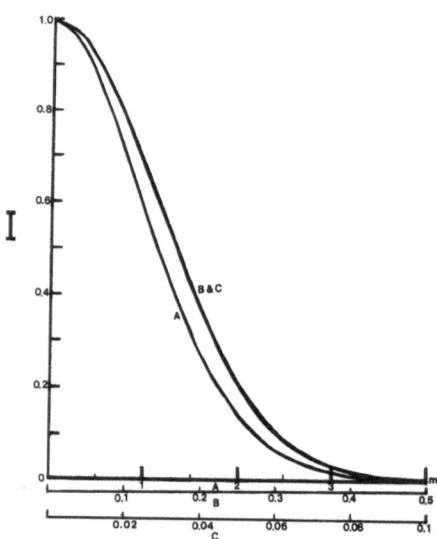

Fig. 1.   Ratio I vs m for $E = 1, 10$, and 50. Curve A, B, or C is for $E = 1, 10$, or 50, respectively. The scale of abscissa is different for different curves.

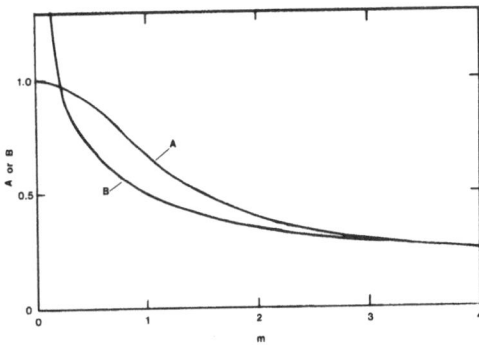

Fig. 2.   Curves for $dh(m)/dm$ and $1/2\, m^{-1/2}$ vs m. Curve A or B is for the former or the latter, respectively.

If E is small, the amplitude of the integrand of (13) remains almost the same for larger values of m, thus the resultant wave is proportional to

$$J(m_0)\, 2\pi\, \{\tau h''(m_0)\}^{-1} \exp i\left[h(m_0)\tau \pm m_0\rho + \frac{1}{4}\pi\, \text{sign}\, \{h''(m_0)\} \right] \qquad (19)$$

where m_0 is the stationary phase point, $J(m)$ is the amplitude of the integrand of (13). Some features of expression (19) were described in ICHIYE (1950) for different forms of $J(m_0)$ from Eq. (13). It is seen from Eq. (18) that the wave form is not of solitary wave type but sinusoidal, representing dispersive effects.

3.   Tsunami as a Long Wave

When tsunami is treated as a long wave, equations of motion and continuity become very simple. If the non-linear and dissipation terms are neglected and only one dimensional source is considered, these become

$$\partial u/\partial t = -g\partial\zeta/\partial x \qquad (20)$$

$$\partial\zeta/\partial t = \partial(Hu)/\partial x + \partial\eta/\partial t \qquad (21)$$

where u is the horizontal velocity and η is the bottom displacement. The system of Eqs. (20) and (21) are amenable to numerical calculation even for a two dimensional source and variable depth (AIDA, 1980).

Equation (20) is based on the hydrostatic pressure and Eq. (21) indicates that the surface and bottom displacement contributes to the water transport flux. Validity of these assumptions and of neglect of the non-linear effects will be discussed in the next section. Here it is shown that solutions of Eq. (20) and (21) are not the same as those of the Cauchy-Poisson problem of long wave approximation (noted as C-P LA) with the same boundary condition.

In order to compare with the Cauchy-Poisson problem, H is assumed constant. Elimination of u among (20) and (21) leads to

$$\partial^2\zeta/\partial t^2 = c^2\, \partial^2\zeta/\partial x^2 + \partial^2\eta/\partial t^2 \qquad (22)$$

where $c^2 = gH$. A special solution of this equation can be given by

$$\zeta = (2c)^{-1} \int_0^t \int_{x-a(t-\tau)}^{x+a(t-\tau)} \partial^2 \eta / \partial \tau^2 \, d\tau d\lambda \tag{23}$$

where λ and τ are dummy variables for x and t, respectively (Arsenin, 1981). If the bottom displacement η is given by an expression (3), (23) is expressed by

$$\zeta = -(2c)^{-1} \left(\frac{dF}{dt}\right)_{t=0} \int_{x-ct}^{x+ct} f(\lambda) d\lambda$$

$$+ \frac{1}{2} \int_0^t (dF/d\tau)[f\{x + c(t-\tau)\} + f\{x - c(t-\tau)\}] d\tau. \tag{24}$$

The derivative $dF/d\tau$ within the integral can be replaced with a delta function. This can be proved by a solution with Laplace transform of the function $F(t)$. It is seen that the solution (24) with a delta function is not the same as the solution expressed by (8), (9), and (10) of C-P LA as obtained by Takahashi (1948). The difference is attributed to the condition that Eqs. (20) and (21) are based on the assumptions of the hydrostatic condition but the C-P LA is based on the condition for simultaneous bottom and surface movement (SBSM). The long wave approximation is based on the linearity condition of both equations and boundary conditions but the C-P approach is based on linearity of the latter only.

4. Scaling

In order to determine conditions valid for shallow water wave assumptions, it is necessary to scale the complete Euler equations of motion and continuity. The dissipation terms are neglected for the sake of simplicity, though this may not be justified, particularly for a source in a shallow water. However, the dissipation mechanism can be understood only after the behavior of the bottom movement in an earthquake is known. The scaling of the Euler equations is adapted from Hammack (1973). Mass and Vastano (1978) derived some criterion for dispersive and non-dispersive propagation of tsunami by scaling the Euler equations differently.

Two dimensional equations of motion and continuity are

$$u_t + uu_x + wu_z + \rho^{-1} P_x = 0 \tag{25}$$

$$w_t + uw_x + ww_z + \rho^{-1} P_z + g = 0 \tag{26}$$

$$u_x + w_z = 0 \tag{27}$$

where w is the vertical velocity, ρ is the density, P is the pressure and subscripts mean partial derivative. Other notations are the same as defined before. The equilibrium sea level is taken at $z = 0$ and the bottom at $z = -H$. Then the sea surface and bottom boundary conditions are

$$w = \zeta_t + u\zeta_x \quad \text{on} \quad z = \zeta \tag{28}$$

$$w = \eta_t + u\eta_x \quad \text{on } z = -H + \eta. \tag{29}$$

Three characteristic scales besides the depth H are magnitude of vertical displacement of the bottom η_0, horizontal scale of the bottom motion b and its time scale t_c. The ratio α is defined by

$$\alpha = t_c (gH)^{\frac{1}{2}} b^{-1} \tag{30}$$

which is called time-scale ratio (TSR), since it represents the time scale of the bottom motion in terms of the time interval for a disturbance to travel the distance b with a speed of a long wave. The other parameters are defined by

$$\beta = \eta_0/H \tag{31}$$

$$\chi = t_c (g/H)^{\frac{1}{2}} \tag{32}$$

$$\delta = H/b \tag{33}$$

where δ is an inverse aspect ratio.

Two cases of scaling Eqs. (25) to (29) can be possible, according to time scale of bottom motion. The case I is for moderate α with $\beta \ll 1$. In this case the following scaling is used

$$x = bx^*, \; z = Hz^*, \; t = t_c t^*, \; u = [\eta_0 (gH)^{\frac{1}{2}}/H]u^*,$$
$$w = (\eta_0/t_c)w^*, \; \zeta = \eta_0\zeta^*, \; P = \rho gHP^*. \tag{34}$$

In order to show ranges of parameters in which the linearity, hydrostatic and SBSM conditions are valid only Eq. (26) and (27) may be scaled. The linearity condition determined from (26) and (27) can also be valid to (25), (28), and (29). Scaled form of (26) and (27) is given by

$$w_t + \alpha\beta u w_x + \beta w w_z + \kappa^2\beta^{-1} (P_z + 1) = 0 \tag{35}$$

$$u_x + \alpha^{-1}w_z = 0 \tag{36}$$

respectively, where the asterics are dropped. Equation (35) indicates that the linearity condition can be satisfied if $\beta \ll 1$ and $\alpha\beta \ll 1$. The hydrostatic condition can be satisfied if $\kappa^2\beta^{-1} \gg 1$ in (35). Equation (36) indicates that the SBSM condition ($w_z \approx 0$) can be satisfied if $\alpha \ll 1$.

If the time scale of the bottom motion is large, different scaling should be used instead of (34). This is called case II (or creep by HAMMACK, 1973). In this case the scaling

$$x = t_c (gH)^{\frac{1}{2}}x^*, \; z = Hz^*, \; t = [b(gH)^{-\frac{1}{2}}]t^*,$$
$$u = (b\eta_0/t_c H)u^*, \; w = (\eta_0/t_c)w^*,$$
$$\zeta = [b\eta_0/t_c (gH)]\zeta^*, \; \eta = [b\eta_0/t_c (gH)^{\frac{1}{2}}]\eta^*, \; P = \rho gHP^* \tag{37}$$

are used. Then (26) and (27) become

$$w_t + \alpha^{-2}\beta u w_x + \alpha^{-1}\beta w w_z + \kappa\beta^{-1}\delta^{-1} (P_z + 1) = 0 \tag{38}$$

$$u_x + \alpha w_z = 0. \tag{39}$$

Therefore with $\beta \ll 1$, the linearity condition is satisfied if $\alpha \gg 1$. The hydrostatic condition can be satisfied if $\kappa \gg 1$ as in case I. The SBSM condition is now satisfied for $\alpha \gg 1$ from (39).

Synthesizing the above results, it is noted that the long wave Eq. (20) and (21) can be used if the linearity and hydrostatic conditions are satisfied, whereas if the SBSM condition is satisfied, the sea level change is the same as the bottom displacement. Two examples are shown to illustrate these conditions: deep water and shallow water examples.

For the shallow water example, H is taken as 0.1 km and $\beta = 10^{-2}$, corresponding to 1m bottom displacement. Table 1 indicates the values of α for time $t_c = 1$ sec to 10^5 sec and $b = 1$ km to 10^2 km. It is noted that the linearity condition is always satisfied for these ranges of t_c and b if the criterion of negligible smallness is taken at 1% level in Eq. (35), (36) or (38), (39). The hydrostatic condition is also valid for all these ranges except for $t_c = 1$ sec, which is marginal. On the other hand the SBSM condition is not satisfied except for rather extreme cases such as sudden and wide motion or slow and spacially concentrated motion.

For the deep water, H is taken as 4 km and $\beta = 10^{-3}$, corresponding to 4 m bottom displacement. The critical parameters α, $\kappa^2 \beta^{-1}$ and $\kappa (\beta \delta)^{-1}$ are listed in Table 2 for the same ranges of t_c and b as in Table 1. Again this table indicates that the

Table 1.   Shallow water ($H = 0.1$ km, $\beta = 10^{-2}$).

	t_c					
sec	1	10	10^2	10^3	10^4	10^5
min			1.7	16.7	167	1667
hr					2.8	27.8
			$t_c (gH)^{\frac{1}{2}}$			
km	3.1×10^{-2}	$\times 10^{-1}$	$\times 1$	$\times 10$	$\times 10^2$	$\times 10^3$

a)   Linearity and SBSM conditions $\alpha = t_c (gH)^{1/2}/b$

b (km)						
1	3.1×10^{-2}	$\times 10^{-1}$	$\times 1$	$\times 10$	$\times 10^2$	$\times 10^3$
10	3.1×10^{-3}	$\times 10^{-2}$	$\times 10^{-1}$	> 1	$\times 10$	$\times 10^2$
100	3.1×10^{-4}	$\times 10^{-3}$	$\times 10^{-2}$	$\times 10^{-1}$	$\times 1$	$\times 10$

———— Case I ————   ← Case II ————

b)   Hydrostatic condition

Case I $\kappa^2 \beta^{-1}$

	10	10^3	10^5	10^7

Case II $\kappa^2 \beta^{-1} \delta^{-1}$

δ				
1		3.1×10^2	$\times 10^3$	$\times 10^4$
10^{-1}			$\times 10^4$	$\times 10^5$
10^{-2}				$\times 10^6$

*Underlined are cases of invalid conditions.

Table 2.   Deep water ($H = 4$ km, $\beta = 10^{-3}$).

	t_c					
sec	1	10	10^2	10^3	10^4	10^5
	$t_c (gH)^{\frac{1}{2}}$					
km	2×10^{-1}	$\times 1$	$\times 10$	$\times 10^2$	$\times 10^3$	$\times 10^4$

a) Linearity and SBSM conditions $\beta = t_c (gH)^{1/2} D$

b (km)						
1	2×10^{-1}	$\times 1$	$\times 10$	$\times 10^2$	$\times 10^3$	$\times 10^4$
10	2×10^{-2}	$\times 10^{-1}$	$\times 1$	$\times 10$	$\times 10^2$	$\times 10^3$.
100	2×10^{-3}	$\times 10^{-2}$	$\times 10^{-1}$	$\times 1$	$\times 10$	$\times 10^2$

——————————Case I————————→|←— Case II ————

b) Hydrostatic condition

	Case I	$\kappa^2 \beta^{-1}$		
	2.5×1	$\times 10^2$	$\times 10^4$	$\times 10^6$

	Case II	$\kappa^2 \beta^{-1} \delta^{-1}$		
δ				
1	5×10^2	$\times 10^3$	$\times 10^4$	$\times 10^5$
10		5×10^4	$\times 10^5$	$\times 10^6$
10^2			5×10^6	$\times 10^7$

*Underlined are cases of invalid conditions.

linearity and hydrostatic conditions are valid for these ranges of t_c and b except $t_c = 1$ sec where the hydrostatic condition is not valid. The SBSM condition is not valid except rather extreme cases as in the shallow water example.

5.   Conclusion

The tsunami generation can be determined by the linearized long wave equations which are much simpler to treat either analytically or numerically. These equations are valid for reasonable ranges of time and space scales of the bottom motion. The C-P LA is not always valid for tsunami generation, though it does not require linearity conditions except for boundary conditions. However, the long wave approximation cannot describe the dispersion of tsunamis when these waves propagate a long distance from the source region or when the bottom motion is of the order of seconds. Such cases can be treated by numerically calculating the C-P integral without the long wave approximation

The work is supported by ONR Contract N00014-80-(-0113) and NASA Wallops Flight Center Contract NAS6-2884.

REFERENCES

Aida, I., Tsunami simulation based on earthquake fracture models, *Marine Sciences/Monthly. Tokyo*, **12**(7), 485–494, 1980 in Japanese.

Arsenin, V. Ya., *Method of Mathematical Physics and Special Function* (Translated in Japanese from Russian by T. Kubo), pp. 51–53, Morikitashoten, Tokyo, 1981.

Hammack, J. L., A note on tsunamis: Their generation and propagation in an ocean of uniform depth, *J. Fluid Mech.*, **60**(4), 769–799, 1973.

Ichiye, T., On the theory of tsunami, *Oceanogr. Mag. Tokyo*, **2**(3), 83–100, 1950.

Ichiye, T., A theory on the generation of tsunamis by an impulse at the sea bottom, *J. Oceanogr. Soc. Japan*, **14**(2), 41–44, 1958.

Lamb, H., *Hydrodynamics*, pp. 395–398, Cambridge University Press, Cambridge, 1932.

Mass, W. J. and A. D. Vastano, An investigation of dispersive and non-dispersive long wave equations applied to oceans of variable depth, Tech. Rept. Ref. 78-8-T, 83 pp., Dept of oceanog., Texas A & M Univ., 1978.

Sano K. and K. Hasegawa, On waves caused by an impulse at the sea bottom, *Bull. of Central Meteorological Obs. Japan*, **2**, 1–13, 1915.

Sirovich, L., *Techniques of Asymptotic Analysis*, pp. 86–106, Spring-Verlag, New York, 1971.

Syono, S., On the waves caused by a sudden deformation of finite portion of the bottom of a sea of uniform depth, *Geophys. Mag. Tokyo*, **10**(1), 21–41, 1935.

Takahashi, R., On seismic sea waves caused by deformation of the sea bottom, *Bull. Earthq. Res. Inst.*, **20**, 375–398, 1943 (in Japanese).

Takahashi, R., On seismic sea waves caused by deformations of the sea bottom, the third report, The one-dimensional source, Bull. Earthq. Res. Inst., **25**, 5–9, 1948.

Wilson, B. W., Generation and dispersion characteristics of tsunamis, in *Studies on Oceanography*, edited by K. Yoshida, pp. 375–398, University of Tokyo Press and University of Washington Press, Seattle, 1964.

TOPOGRAPHIC EFFECTS ON TSUNAMI WAVES

Tsunamis—Their Science and Engineering, edited by K. Iida and T. Iwasaki, 277–291.

Numerical Simulation of Historical Tsunamis Generated off the Tokai District in Central Japan

Isamu AIDA

Earthquake Research Institute, University of Tokyo, Tokyo 113, Japan

(Received August 20, 1981; Revised December 1, 1981)

The 1944 Tonankai and the 1854 Ansei-Tokai tsunamis, both of which were generated off the Tokai district in central Japan, are numerically simulated by making use of bottom displacement fields inferred from seismic fault models proposed by various investigators. The reliability of these simulations is examined by comparing simulated wave heights with observed values, and it is found that the simulations based on the seismic fault model are satisfactory.

The source models for the 1707, 1605, and 1498 tsunamis, which caused severe damage in this district, are next constructed by the trial and error method to be consistent with the distribution of tsunami inundation heights described in old records, although information is rather vague. Among the five tsunamis investigated here, the 1498 Meio tsunami seems ᴗo have been the worst event and should be noted from the viewpoint of hazard prevention.

Finally, taking the Meio tsunami as an example, a zoning map of destructive forces is prepared for the inundation at the Shimizu Harbor area. To do this, a tsunami and possible run-up is simulated for Shimizu Harbor with a local simulation model. It is found that breakwaters and reclaimed land which have been constructed recently are effective in reducing tsunami heights within the harbor for tsunamis of the Meio type.

1. Introduction

Several years ago, the possibility was suggested that a destructive earthquake could occur in the near future in the vicinity of the Pacific coast of the Tokai district, central Japan. According to this suggestion, the postulated seismic source area of this earthquake covers the whole of Suruga Bay. Since the seismic source is located very close to land, seismic intensities in the Tokai region would be among the strongest of past earthquakes.

A destructive tsunami might be generated accompanying this earthquake. The prediction of the expected tsunami behavior has become a subject of intense research effort. However, much larger tsunamis than that expected for this earthquake were experienced in this region according to old records. Therefore, it is very important to investigate not only the behavior of the tsunami expected in the near future, but also the strongest tsunami experienced in the past.

First, numerical simulations for tsunamis which have occurred in recent years (1944 and 1854) are carried out and compared with available data to verify the

277

reliability of the use of a seismic fault model as a tsunami source. Second, source models for older tsunamis are searched for by simulating tsunamis which reasonably well explain coastal inundation heights described in old records. This enables an estimation of inundation heights along the coast where the behavior of the tsunami was not described in records, and offers a useful guide for the planning of preventive measures against tsunami disaster. In addition, the estimated earthquake source model may give important clues of geophysical interest concerning sizes and geographic locations of earthquakes occurring in this region.

Furthermore, tsunami inundations on land are simulated by a local model, the input of which is obtained by the wider model for which the estimated tsunami source is used. The inundated area and the inundation heights on land can be examined by this simulation. The distribution of hydraulic pressure due to the inundation water current on land is known to be well correlated with the tsunami destructive power. Therefore, a distribution map of the hydraulic pressure may provide useful data for microzoning of tsunami hazards.

2.   Numerical Method

A tsunami is described by shallow water equations, since the wave length is very large comparing with the water depth.

$$\frac{\partial q_x}{\partial t} = -gD\frac{\partial \zeta}{\partial x} - f\frac{q_x Q}{D^2} - \left(\frac{q_x}{D}\frac{\partial q_x}{\partial x} + \frac{q_y}{D}\frac{\partial q_x}{\partial y}\right) \tag{1}$$

$$\frac{\partial q_y}{\partial t} = -gD\frac{\partial \zeta}{\partial y} - f\frac{q_y Q}{D^2} - \left(\frac{q_x}{D}\frac{\partial q_y}{\partial x} + \frac{q_y}{D}\frac{\partial q_y}{\partial y}\right) \tag{2}$$

$$\frac{\partial \zeta}{\partial t} = -\frac{\partial q_x}{\partial x} - \frac{\partial q_y}{\partial y} + \frac{\partial \xi}{\partial t}. \tag{3}$$

Here, q_x and q_y are the x- and y-components of volume transports integrated to the bottom, ζ is the water surface elevation, ξ is the vertical displacement of the bottom, h is the water depth, f is the quadratic friction coefficient of the bottom, $D = h + \zeta - \xi$ and $Q = (q_x^2 + q_y^2)^{1/2}$. The second and third terms on the right-hand side in Eqs. (1) and (2) were neglected in the wide models of simulations including the tsunami source area.

A vertical displacement of the sea bottom at the time of tsunami generation is assumed to be approximated by the permanent elastic displacement of an idealized seismic fault model (rectangular faulting with uniform slip), which can be computed by the formulation of MANSINHA and SMYLIE (1971). The duration time required to complete the bottom displacement is assumed to be much smaller than the propagating time of a shallow water wave across the tsunami source area, so the initial elevation of the water surface is identical to the bottom displacement ξ.

The generated tsunami is simulated by means of Eqs. (1) to (3) which are solved by a space staggered finite difference scheme with a regular grid size of 5 km over most of the open sea area. At selected coastal regions shallower than about 200 meters, the grid size is decreased in four steps down to 1/16 of the regular grid size. The adoption of a telescopic grid scheme improves the high frequency wave characteristics which critically depend on the grid size, and improves the accuracy of geometrical modelling of the coastal area. The computing time step is 0.05 minutes which satisfies the numerical stability condition over the whole area.

The boundary condition at the coast is q_x or $q_y = 0$. At the open boundary appearing artificially due to the restriction of the computational area, the following formula related to long progressive waves is assumed to hold:

$$q_x = \pm (c^2 \zeta - q_y^2)^{1/2}$$

or   (4)

$$q_y = \pm (c^2 \zeta - q_x^2)^{1/2}.$$

Here c is a long-wave velocity and signs at the right hand side are taken to satisfy the condition that q_x or q_y are in the outward direction from the computing area when ζ is plus.

3. Verification of the Numerical Simulations

3.1 The 1944 Tonankai tsunami

The 1944 Tonankai tsunami attacked the Tokai district shown by hatching in the inset of Fig. 1. Tide-gauge records of the tsunami were obtained at Mera, Ito, Uchiura, Matsuzaka, Shimotsu and Tosa-Shimizu. The computation area is 965×400 km^2, shown in Fig. 1, which includes small areas surrounded by dotted lines near the tide-gauge stations where telescopic variations are introduced.

For this earthquake, simulations are carried out making use of the vertical bottom

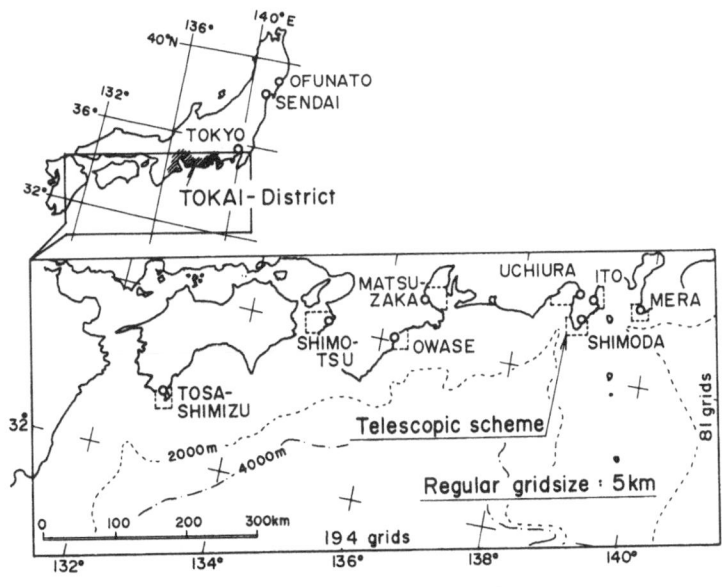

Fig. 1. Map showing the computation area.

displacment fields inferred from parameters of four different fault models published so far (AIDA, 1979). Figure 2 shows these fault models (KANAMORI, 1972; ANDO, 1975; INOUCHI and SATO, 1975; ISHIBASHI, 1976, 1981) in which rectangles indicate the horizontal projection of the fault plane and arrows indicate the dislocation of the hanging wall side of the fault. Dip angles of fault planes are 10 to 30 degrees in the north-west direction.

Two parameters K and κ are used to express the reliability of a model (AIDA, 1978). Let x_i and y_i be the respective amplitudes of the observed tsunami and the simulated wave at the i'th station, and let the ratios $K_i = x_i/y_i$ be calculated for n stations. The first parameter K is the geometrical mean of K_i, which is defined as $\log K = (1/n) \sum_{i=1}^{n} \log K_i$. The factor K represents a correction to be applied to the slip displacement of a fault model. The second parameter κ is a logarithmic standard deviation defined by the following formula, which is a measure of the variation of K_i, namely, the accuracy of the height simulation.

$$\log \kappa = \left[(1/n) \sum_{i=1}^{n} (\log K_i)^2 - (\log K)^2 \right]^{1/2}. \tag{5}$$

The reliability of the present simulations is examined for the amplitudes of the first and sceond half cycles of the wave-time histories, a_1 and a_2.

Taking the locations of tide-gauge stations on the abscissa, the ratios K_i/K are plotted in Fig. 3. Since Models I and II give large variation of K_i and Model IV gives a systematic difference between three eastern stations and three western stations, these

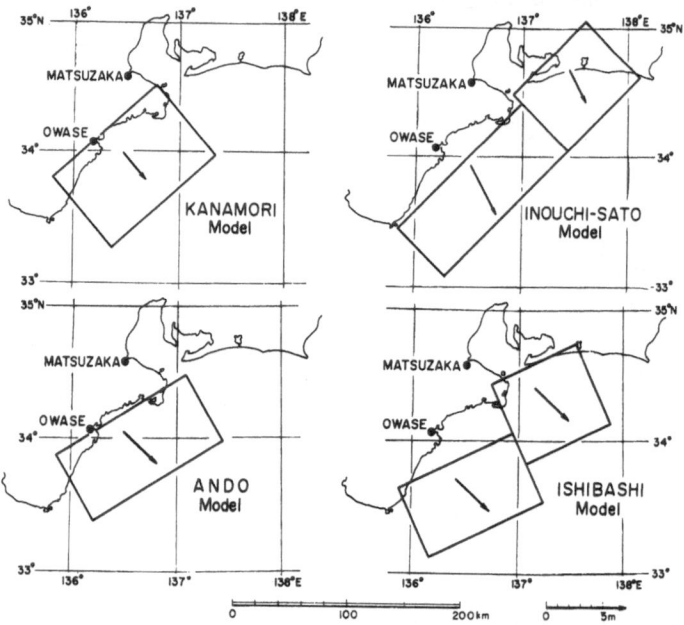

Fig. 2.   Seismic fault models determined by several investigators. Rectangles drawn by thick lines are horizontal projections of the fault planes. Arrows indicate the direction and the magnitude of the relative slip of the hanging wall side of the fault.

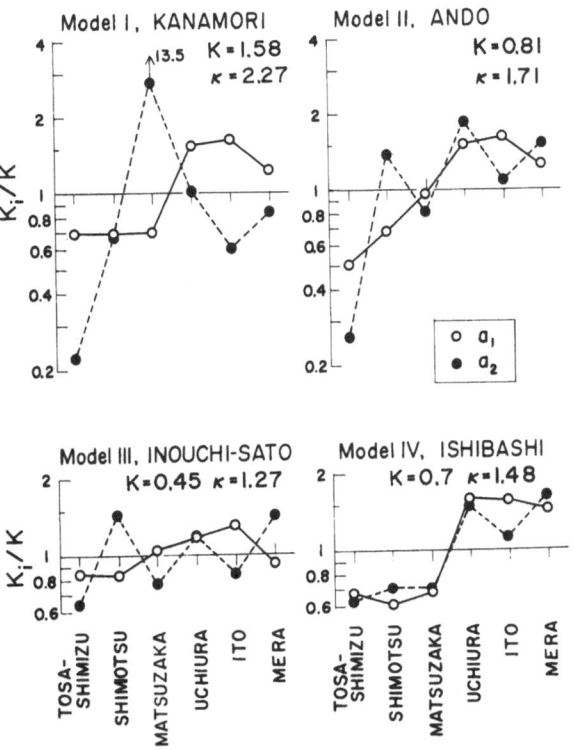

Fig. 3. Variation of K_i/K for 6 tide-gauge stations, where K_i is the ratio of the observed amplitude to the computed one with K the geometric mean of K_i. a_1 and a_2 indicate the values for the first and the second half cycles of the tsunami waves.

models are not adequate as tsunami sources. Model III seems to be the best model since κ is the smallest; $\kappa = 1.27$ in average for a_1 and a_2. But the K factor of this model is 0.45. This shows that the dislocation assumed in this fault is too large by a factor of about 2.

The simulation applying the factor K to the fault dislocation resulted in the very good agreement between the computed elevation-time histories and actual tsunami records as shown in Fig. 4.

3.2 The 1854 Ansei-Tokai tsunami

A fault system for the 1854 Ansei-Tokai earthquake has been published by ISHIBASHI (1976, 1981) as shown in Fig. 5. It is composed of two parts. The southwestern part experienced the Tonankai earthquake 90 years after the 1854 Ansei-Tokai earthquake and thus the strain energy was believed to have been released again. The northeastern part has experienced no event in the 127 years since the 1854 earthquake. Therefore, it is speculated that the next earthquake should occur in the northeastern region.

Vertical bottom displacement fields inferred from this fault dislocation are displayed by solid (uplift) and dashed (subsidence) contour lines in Fig. 5. This pattern will be used as the source model for the 1854 Ansei-Tokai tsunami.

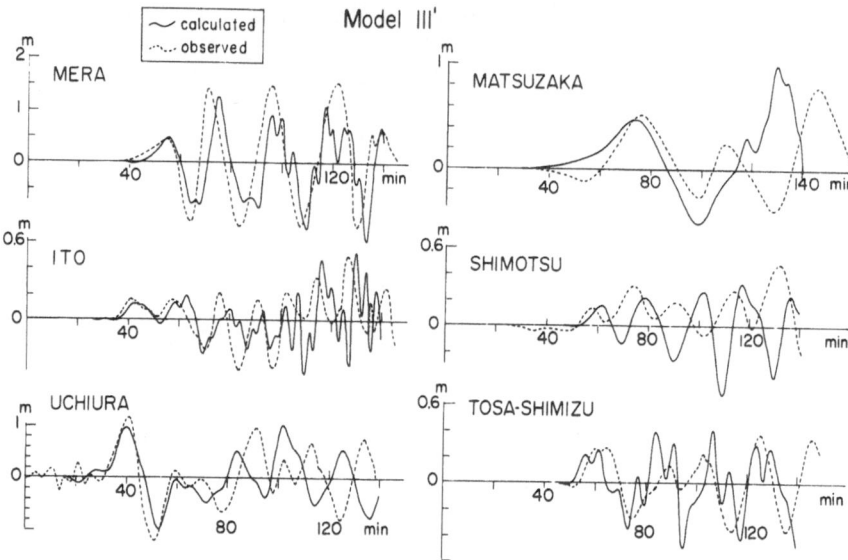

Fig. 4.   The computed time history of water elevation at 6 tide-gauge stations (solid line) and the observed tsunami records (dotted line).

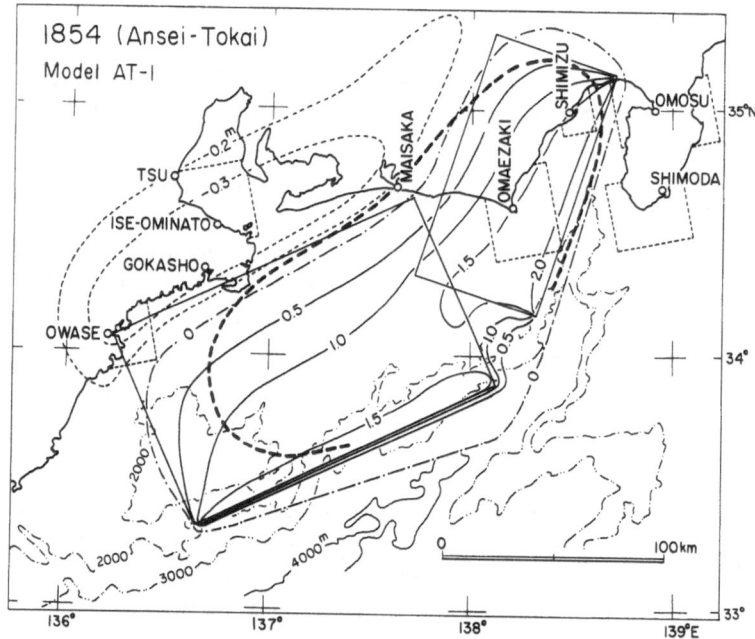

Fig. 5.   Seismic fault model for the 1854 Ansei-Tokai earthquake proposed by ISHIBASHI (1976, 1981). Vertical bottom displacement fields are shown by solid and broken contour lines. A thick broken line indicates the tsunami source area estimated by HATORI (1976). Telescopic grid schemes are used within areas surrounded by thin dotted lines.

Actual tsunami inundation heights along the coast have been estimated from tsunami behavior described in old documents (HATORI, 1976, 1977), which are called the observed values hereafter. For comparison with the observed tsunami heights, the simulated wave heights are determined by either of the following two methods depending on the location.

A)   In places covered by the telescoped grid system, the computed maximum elevation above the still water level was used without modification. Shimizu and Omaezaki were added to the telescoped system and Shimotsu and Tosa-Shimizu were removed.

B)   On coastal locations covered only by the coarse grid, the maximum elevation was estimated from the wave height (double amplitude) H_0 computed on the 200 meters isobath by applying an amplification factor. The amplification factor in shallow water is determined from a simulation for the 1944 tsunami such that the ratio of surveyed inundation heights on the coast and H_0 computed on the 200 meters isobath represent the amplification factor for that location.

Figure 6 shows the observed heights (short bars) and the heights calculated by two methods, A (double circle) and B (single open circle). Simulated values have been corrected for the tide level and vertical ground displacement at the particular time and location. The agreement is fairly good, with ratios of observed and simulated heights in the range of 0.8 to 1.2.

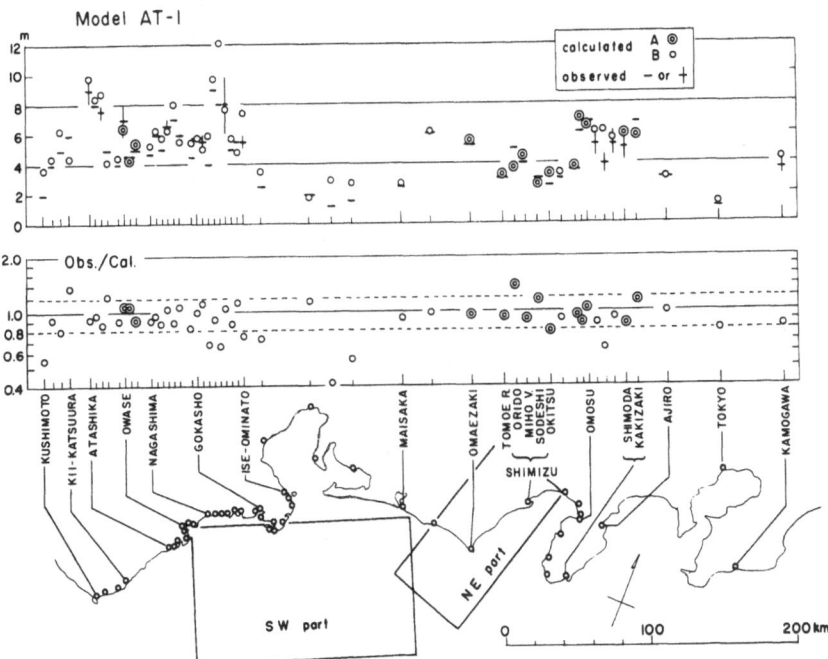

Fig. 6.   Results of the simulation for the 1854 Ansei-Tokai tsunami. Upper: Distribution along the coast of estimated heights of the real tsunami (Obs) and calculated tsunami heights (Cal) by the procedures A (area with telescopic scheme) and B (area with ordinary grid-size). Lower: Ratios of values of the real tsunami (Obs) to calculated values (Cal).

The reliability parameters become $K = 0.99$ and $\kappa = 1.16$ for locations where method A was applied, and $K = 0.92$ and $\kappa = 1.2$ for methods A and B.

3.3   Discussion of reliability

Many numerical simulations have been carried out for tsunamis off Hokkaido and Tohoku districts (AIDA, 1977a, 1978). Table 1a shows the reliability parameters of these tsunami simulations including those of the 1944 Tonankai tsunami. In this group, the verification of the simulation was made using the amplitudes of the actual tsunami records for the first and second half cycles of waves, a_1 and a_2.

Reliability parameters which were obtained from the measured or estimated run-up heights and the simulated heights by method B are shown in Table 1b which includes results from the 1854 Ansei-Tokai tsunami.

According to these tables, a measure of variation of simulated values κ is about 1.4

Table 1a.   Reliability parameters, K and κ, of tsunami simulations compared with actual tsunami records.

No.	Earthquake		n	a_1		a_2	
				K	κ	K	κ
1	1952 Mar 4	Tokachi-Oki	4	1.09	1.37	0.84	1.71
2	1968 May 16	Tokachi-Oki	6	1.33	1.61	1.47	1.62
3	1968 Jun 12	Iwateken-Oki	6	4.45	1.64	3.48	1.61
4	1969 Aug 12	Kurile Is.	6	1.25	1.20	1.06	1.21
5	1973 Jun 17	Nemuro-Oki	6	2.28	1.27	1.16	1.47
6	1944 Dec 7	Tonankai	6	1.05	1.17	0.98	1.39
7	1968 Apr 1	Hyuganada	5	1.14	1.23		
8*	1968 Jun 12	Iwateken-Oki	6	1.09	1.69	0.82	1.66
9	1978 Jun 12	Miyagiken-Oki	10	1.02	1.20	0.98	1.54
	Average				1.36		1.52

*; modified parameters of No. 3.

Table 1b.   Reliability parameter κ of tsunami simulations compared with measured or estimated run-up heights.

No.	Earthquake			n	κ
101	1611 Dec 2	Keicho 16	Sanriku-Oki	7	1.47
102	1793 Feb 7	Kansei 5	Miyagi-Oki	7	1.49
103	1854 Dec 23	Ansei 1	Tokaido-Oki	50	1.20
104	1856 Aug 23	Ansei 3	Hachinohe-Oki	9	1.37
105	1896 Jun 15	Meiji 29	Sanriku-Oki	10	1.37
106	1897 Aug 5	Meiji 30	Miyagi-Oki	13	1.6
107	1933 Mar 3	Syowa 8	Sanriku-Oki	26	1.59
	Average				1.44

on the average, without a significant difference between the two tables. In the present simulations, we obtained $\kappa = 1.17$ (a_1) and 1.39 (a_2) for the 1944 tsunami, and $\kappa = 1.2$ for the 1854 tsunami. These values seem to be quite satisfactory since they are considerably smaller than the average. That is, it has been verified that the tsunami source derived from a seismic fault model can be used in the simulation of tsunami originating in the Tokai district just as well as in the Tohoku district.

4. Source Models for Other Historical Tsunamis

Besides the two tsunamis already stated, very large tsunamis attacked the Tokai district in 1498, 1605, and 1707.

There are a considerable number of old records relating to the 1707 Hoei tsunami (HATORI, 1977). Coastal inundation heights of this tsunami inferred from these records are almost the same as that of the 1854 Ansei-Tokai tsunami. The question as to whether or not the seismic fault plane extended to the head of Suruga Bay has not been resolved yet (HATORI, 1977; MOGI, 1977; ISHIBASHI, 1977). Results of several simulations in which different locations were assumed for the fault in Suruga Bay showed clearly that a model used for the 1854 tsunami could explain quite well the behavior of the 1707 tsunami on the coast in the vicinity of Suruga Bay (AIDA, 1981). Therefore, the source model for the 1707 tsunami is presumed here to be the same as the model for the 1854 tsunami.

On the other hand, records describing the 1498 and the 1605 tsunamis are so scarce that an estimation of inundation heights based on these is somewhat uncertain. For these tsunamis, several probable models are assumed, taking into account the geographical arrangements of the Suruga and Nankai troughs and also the speculations (HATORI, 1975; ISHIBASHI, 1978) of the source area. The most reasonable source parameters are selected on the basis of a comparison between computed inundation heights and descriptions in old records.

In summary, seismic faults are shown in Fig. 7 which are adequate as source models for five tsunamis in this region. Rectangles denote the horizontal projections of fault planes of the fault locations. Other fault parameters are tabulated in Table 2. In Table 2, seismic moments are also calculated from fault areas and dislocations assuming a rigidity of 5×10^{11} dyne $\cdot$ cm^{-2}. These historical earthquakes are found to be of the largest class in and near Japan.

Using the fault parameters listed in Table 2, reliability parameters for these tsunamis are as follows, $K = 1.04$ and $\kappa = 1.36$ for the 1605 tsunami, and $K = 1.12$ and $\kappa = 1.42$ for the 1498 tsunami.

5. Regional Distribution of Tsunami Heights

To determine the general characteristics of the various tsunamis, the distribution of wave heights, H_0, along the 200 meters isobath is plotted in Fig. 8. On the western half, namely in the Kumanonada region, wave heights H_0 are almost the same for all tsunamis. However, on the eastern half, namely in the Enshunada region, the differences of H_0 from tsunami to tsunami are considerable. Among these tsunamis, the

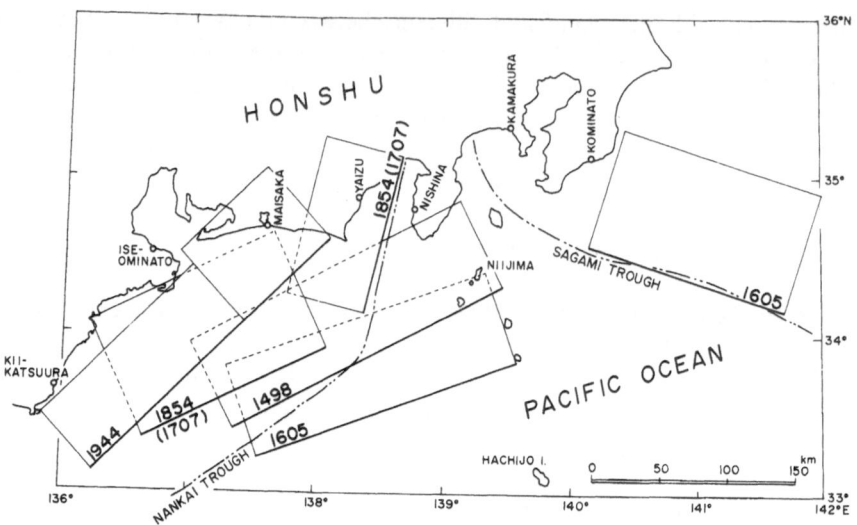

Fig. 7.   Seismic fault models generated tsunamis off the coast in the Tokaido district which were estimated in the present study. The parameters are tabulated in Table 2.

Table 2.   Estimated parameters of tsunami source models.

Earthquake	Fault	L (km)	w (km)	δ (°)	ϕ (°)	u_d (m)	u_s (m)	M_0 (dyne-cm)
1944 Tonankai	NE part	84	78	30	N45W	−1.4	−0.5	1.6×10^{28}
	SW part	154	67	30	N45W	−2.0	−0.7	
1854 Ansei-Tokai ⎫ 1707 Hoei ⎬	NE part	115	70	34	N72W	−3.8	−1.3	4.6×10^{28}
	SW part	150	100	24	N25W	−3.7	1.6	
1605 Keicho	Off Boso	150	100	30	N17E	−3.1	6.3	$11. \times 10^{28}$
	Off Tokai	200	80	30	N20W	−7.2	3.6	
1498 Meio		220	80	30	N28W	−7.6	2.6	$7. \times 10^{28}$

L, fault length; w, width; δ, dip angle; ϕ, dip direction; u_d, dip-slip component (normal +); u_s, strike-slip component (right lateral +); M_0, seismic moment (rigidity 5×10^{11} dyne · cm^{-2}).

1498 Meio tsunami shows the highest values which are about 1.5 times that of the 1854 tsunami in many places.

   Tsunami heights (maximum water elevation) at several places in four typical bays are shown in Fig. 9. At Owase Bay in the western part, the height differences among various tsunamis are very small. However, at Shimizu, Uchiura, and Shimoda Bays in the eastern part, coastal tsunami heights depend strongly on the particular model. At

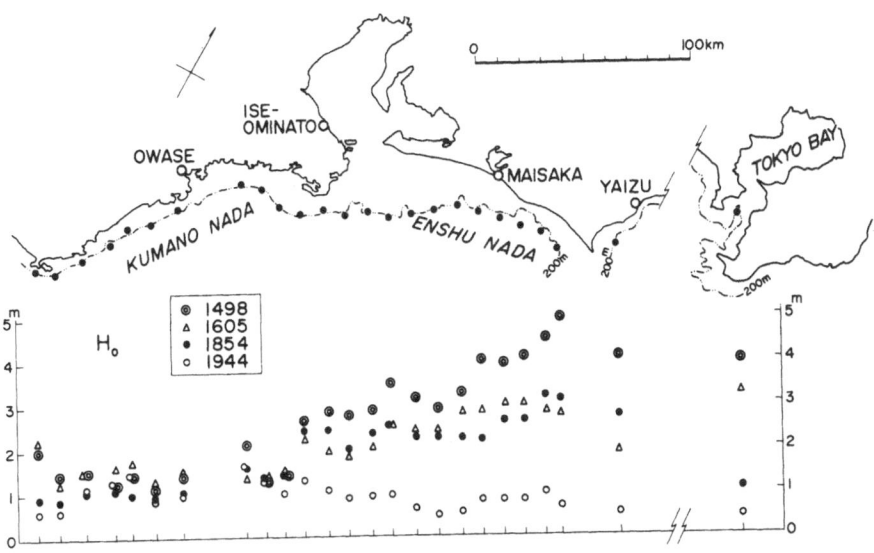

Fig. 8. Distributions of computed tsunami heights H_0 along the 200 meter isobath (double chain line) for the 1498, 1605, 1854 and 1944 tsunamis.

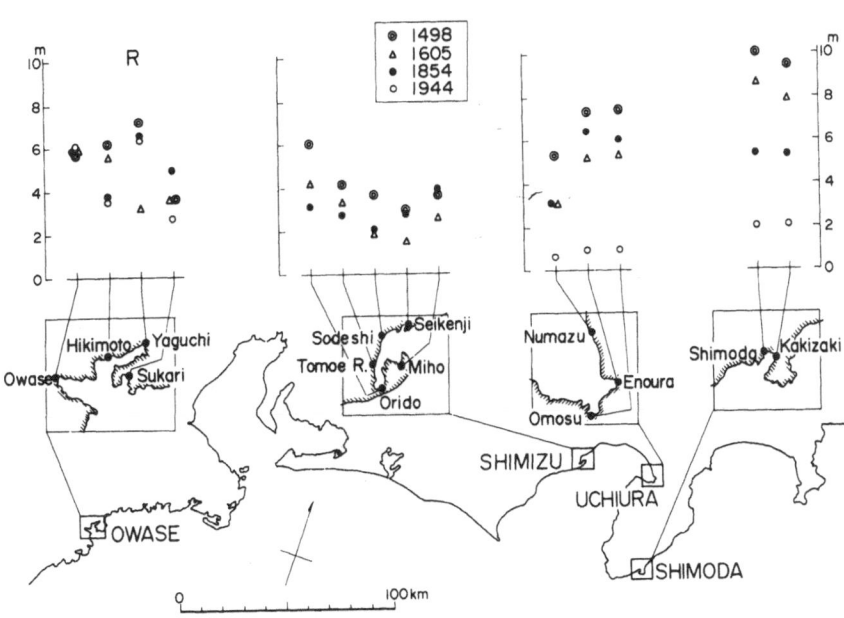

Fig. 9. Computed heights R of each tsunami in four harbors.

Shimizu and Shimoda Bays, the heights of the 1498 tsunami are about 1.5 times that of the 1854 tsunami. According to these simulations, it appears that the 1498 Meio tsunami was the highest tsunami ever experienced in this region.

6.   Simulation of Tsunami Inundation

Simulations of tsunamis inundating over land have been carried out by the author on the basis of the shallow water equations, Eqs. (1) to (3) (AIDA, 1977b).

The water front is determined by checking the boundary between a wet and dry meshes, where the water velocity is related to the square root of the height of water front with a coefficient of 0.5. Energy losses caused by a step change at the original shoreline are replaced by effective energy losses due to bottom friction in one grid interval, in which the effective friction coefficient is assumed to be 0.03. In an urban area consisting of houses, energy losses due to obstacles on land are represented by an effective friction coefficient of 0.02.

The friction coefficient in a bay proper is taken to be 0.005 except where the water depth is less than 5 meters and on land where the coefficient is taken as 0.01.

In the present paper, the inundation corresponding to the Meio tsunami is simulated for Shimizu Harbor. The input wave form is given at the boundary outside of the harbor, for which the results of the wide model discussed in the previous section are utilized.

Figure 10 shows the envelope of the maximum water surface elevation of the tsunami invading Shimizu Harbor with the old topography. The natural shoreline in

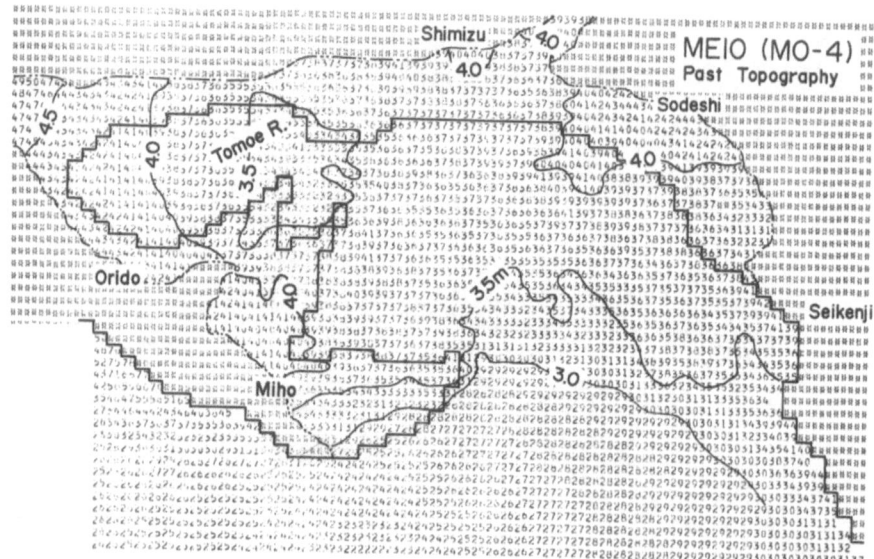

Fig. 10.   Distribution of the maximum water elevation in Shimizu Harbor computed by a local inundation model, in which the input waves were computed by the 1498 tsunami model, and the old topography was estimated from available records.

the harbor is shown by thick solid lines and the ultimate inundated water front is shown by chain lines. Water surface elevations are 3 meters high near the mouth of the harbor and about 5 meters on the ground at the bay head. The value at the bay head is somewhat smaller than the case of non-inundated model, but at other places, water surface levels are nearly equal to that of non-inundated model.

Recently, a breakwater has been constructed near the harbor entrance and the topography of the inner harbor has been greatly changed by reclamation works. The behavior of tsunamis of the same type for the present topography simulated under identical input condition as for the old topography is shown in Fig. 11. The result shows clearly that the water elevation is reduced by 50 cm at the bay head and 80 cm at most other places. In particular, the decrease of 150 cm is found at Shimizu, in the uppermost portion of the figure. Thus, the construction works of breakwaters and land reclamation in the harbor appear to be effective for reducing inundation heights in the present case.

It has been shown that the percentage of destroyed houses at local area of a town is well correlated to the hydraulic pressure due to the inundation water current calculated from simulations (AIDA, 1977b). The hydraulic pressure is represented by the product of the square of the water current velocity and the water elevation above the ground surface.

The distribution of the maximum representative pressure in m^3/sec^2 is indicated by contour lines in Fig. 12. Taking the experience of previous investigations into account, more than 50 % of the wooden houses in the area shown by hatching would be destroyed. It is clearly seen that in this region, the riverside, bay head and tip of the peninsula are locations of high risk.

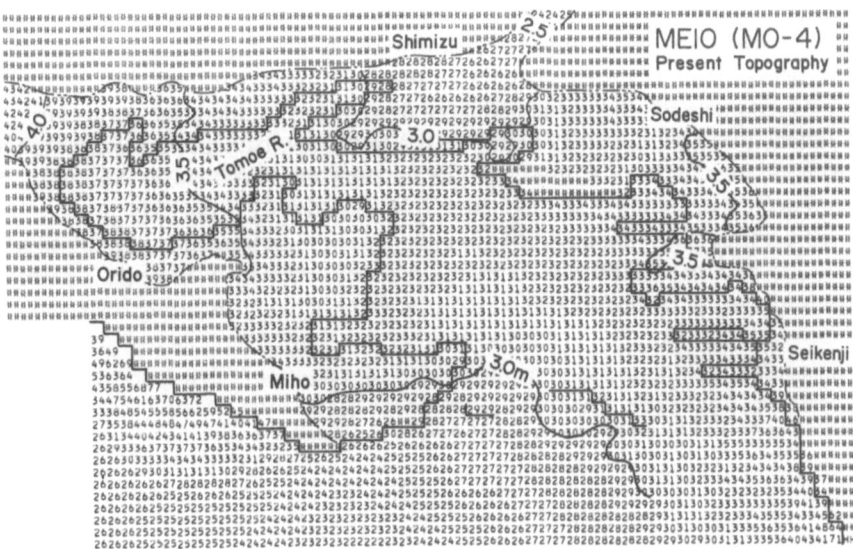

Fig. 11.   The same as for Fig. 10, but with the present topography used in the simulation.

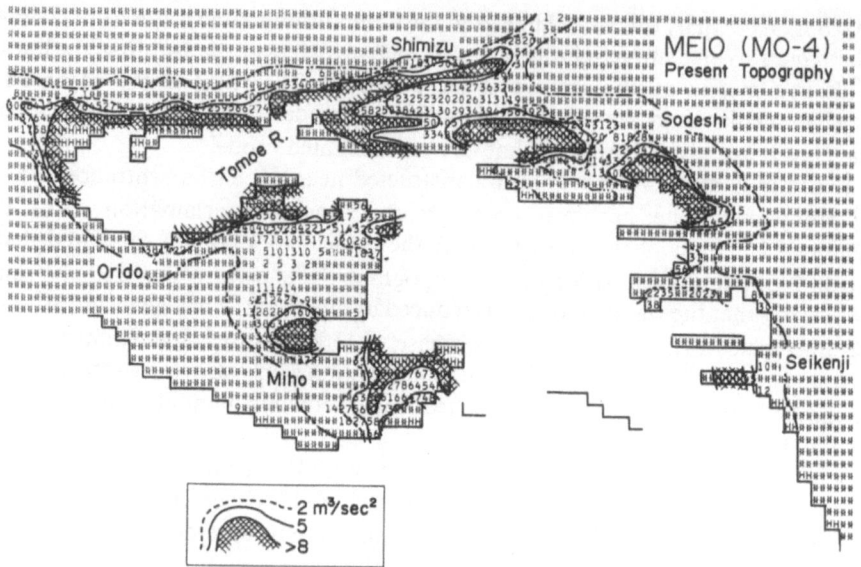

Fig. 12.   Distribution of the maximum value of [current velocity]^2 × [inundation height above the ground surface] as a representative quantity of hydraulic pressure of inundating water on dry land in Fig. 11.

7.   Conclusions

A simple earthquake fault model is determined as a first approximation in which for simplicity a rectangular fault and a uniform distribution of a fault slip are assumed. In a simulation of a tsunami based on a simple fault model with a uniform slip over a rectangular fault area, the uncertainty of the result amounts to a factor of 1.2 to 1.5. Some improvements of the approximation may be possible by selection of source models such as the 1854 Ansei-Tokai and the 1944 Tonankai tsunamis.

Source models of older tsunamis were determined by the same procedure. The most destructive tsunami among them was the 1498 Meio tsunami. It is estimated that this tsunami would have caused run-up as high as 10 meters at Shimoda, but unfortunately, a description of this tsunami in old records is missing.

Using the tsunami source model obtained here, a local inundation simulation was carried out and a zonation of tsunami hazards attempted through use of the distribution map of hydraulic pressure of the water current invading overland.

I would like to thank Prof. Kajiura, Earthquake Research Institute, University of Tokyo, for helpful suggestions. The computations were carried out by the Computer Center, University of Tokyo and this research was supported by the scientific research fund of the Ministry of Education.

REFERENCES

AIDA, I., Simulations of large tsunamis occurring in the past off the coast of the Sanriku district, *Bull. Earthq. Res. Inst.*, **52**, 71–101, 1977a (in Japanese).

AIDA, I., Numerical experiments for inundation of tsunamis—Susaki and Usa, in Kochi Prefecture—, *Bull. Earthq. Res. Inst.*, **52**, 441–460, 1977b (in Japanese).

AIDA, I., Reliability of a tsunami source model derived from fault parameters, *J. Phys. Earth*, **26**, 57–73, 1978.

AIDA, I., A source model of the tsunami accompanying the Tonankai Earthquake of 1944, *Bull. Earthq. Res. Inst.*, **54**, 329–341, 1979 (in Japanese).

AIDA, I., Numerical experiments of historical tsunamis generated off the coast of the Tokaido district, *Bull. Earthq. Res. Inst.*, **56**, 367–390, 1981 (in Japanese).

ANDO, M., Source mechanisms and tectonic significance of historical earthquakes along the Nankai trough, Japan, *Tectonophysics*, **27**, 119–140, 1975.

HATORI, T., Sources of large tsunamis generated in the Boso, Tokai and Nankai regions in 1498 and 1605, *Bull. Earthq. Res. Inst.*, **50**, 171–185, 1975 (in Japanese).

HATORI, T., Documents of tsunami and crustal defomation in Tokai district associaged with the Ansei Earthquake of Dec. 23, 1854, *Bull. Earthq. Res. Inst.*, **51**, 13–28, 1976 (in Japanese).

HATORI, T., Field investigation of the Tokai tsunamis in 1707 and 1854 along the Shizuoka coast, *Bull. Earthq. Res. Inst.*, **52**, 407–439, 1977 (in Japanese).

INOUCHI, N. and H. SATO, Vertical crustal deformation accompanied with the Tonankai earthquake of 1944, *Bull. Geogr. Surv. Inst.*, **21**, 10–18, 1975.

ISHIBASHI, K., Re-examination of a great earthquake expected in the Tokai district : Possibility of the 'Suruga Bay earthquake,' *Abstracts, Seismol. Soc. Japan*, 1976 No. 2, 30–34, 1976 (in Japanese).

ISHIBASHI, K., Did the rupture zone of the 1707 Hoei earthquake not extend to deep Suruga Bay?, *Rep. Subcomm. Tokai distr., Coord. Comm. Earthq. Predict., Geogr. Surv. Inst.*, 69–78, 1977 (in Japanese).

ISHIBASHI, K., On the source region of the 1605 Keicho earthquake : Doubt on the prevailing view, *Abstracts, Seismol. Soc. Japan*, 1978 No. 1, 164, 1978 (in Japanese).

ISHIBASHI, K., Specification of a soon-to-occur seismic faulting in the Tokai district, Central Japan, based upon seismotectonics, *Earthquake Prediction—An International Review, Maurice Ewing Ser. IV*, edited by D. Simpson and P. Richards, Am. Geophys. Un., 297–332, 1981.

KANAMORI, H., Tectonic implications of the 1944 Tonankai and 1946 Nankaido earthquakes, *Phys. Earth Planet. Inter.*, **5**, 129–139, 1972.

MANSINHA, L. and D. SMYLIE, The displacement fields of inclined faults, *Bull. Seismol. Soc. Am.*, **61**, 1433–1440, 1971.

MOGI, K., An interpretation of the recent tectonic activity in the Izu-Tokai district, *Bull. Earthq. Res. Inst.*, **52**, 315–331, 1977 (in Japanese).

Tsunamis—Their Science and Engineering, edited by K. Iida and T. Iwasaki, 293–301.
Copyright © 1983 by Terra Scientific Publishing Company (TERRAPUB), Tokyo.

Finite Element Method for Tsunami Wave Propagation in Tokai District, Japan

Kumizi IIDA,* Takao SUZUKI,** Kazuo INAGAKI,*** and Kenichi HASEGAWA***

Aichi Institute of Technology, Toyota, Japan
**The Chubu Electric Power Company Inc., Nagoya, Japan*
***Unic Corporation, Tokyo, Japan*

(Received August 25, 1981; Revised November 30, 1981)

The finite element method is used for the analysis of the tsunami wave propagation. Following the conventional Galerkin procedures, the finite element method is applied to discretize space variables of velocity and water elevation. To discretize the time evolution, an improved two step Lax-Wendroff method is employed. The numerical computation is made for the heights of historical large scale tsunamis such as the Ansei-Tokai Tsunami of 1854 and the Tonankai Tsunami of 1944 in Japan and the Ishibashi model is applied for the fault model of these tsunamigenic earthquakes. The investigation area is located on the Pacific coast of the Tokai district and covers an area of about 378,000 square kilometers. The analysis domain is divided into triangular finite elements, the number of which is 69,068. The sea-bottom deformation is estimated by the fault model of earthquakes.

The numerical results are compared with the tidal observation records and the inundation heights, and explain satisfactorily the distribution of the real observed tsunami heights along the coast.

1. Introduction

In the Tokai district, which faces the Pacific Ocean, a number of earthquakes have occurred and occassionally a large scale tsunami has attacked the coast. In this paper, a numerical analysis of tsunami applying the finite element method is presented based on the shallow water equations. The result is compared with the tidal observation records and the inundation heights. The numerical study is made for the Ansei-Tokai Tsunami of 1854 and the Tonankai Tsunami of 1944 and the model of ISHIBASHI (1977) is applied for the fault model of these tsunamigenic earthquakes.

2. Basic Equations

Tsunami wave propagation has already been analyzed numerically by AIDA (1969), HWANG and DIVOKY (1970) using the finite difference method. Recently, several finite element methods of shallow water wave equations have been published (KAWAHARA *et al.*, 1978).

Following the conventional Galerkin procedures, the finite element method is applied to discretize space variables of velocity and water elevation. To discretize the time evolution, an improved two step Lax-Wendroff scheme (INAGAKI et al., 1980) is employed.

Tsunami wave propagation in the ocean is governed by the linearized Euler equation of motion,

$$\frac{\partial u_i}{\partial t} + g\eta_{,i} = 0, \tag{1}$$

and equation of continuity,

$$\frac{\partial}{\partial t}(\eta - b) + \{(h + \eta - b)u_i\}_{,i} = 0, \tag{2}$$

where u_i and η denote velocity and tide elevation, and h and g are the mean sea depth and gravitational acceleration, respectively. The sea bottom upheaval in the seismic center region is represented by b. Here and henceforth, the usual summation convention with repeated indices is used.

As boundary conditions on the velocity, the following is assumed,

$$u = \eta \sqrt{\frac{g}{(h+\eta)}} \qquad \text{on the seaward boundary,} \tag{3}$$

$$u_n = 0 \qquad \text{on the shore boundary,} \tag{4}$$

where u_n denotes the velocity component normal to the shoreline.

The finite element governing equations are interpolated with linear polynomial functions Φ, and are written as follows,

$$M_{\alpha\beta}\ddot{u}_{i\beta} + gN_{i\alpha\beta}\eta_\beta = 0, \tag{5}$$

$$M_{\alpha\beta}\dot{\eta}_\beta + K_{\alpha\beta j\gamma}(H_\beta u_{\gamma j} + H_\gamma u_{\beta j}) = 0, \tag{6}$$

where

$$\left.\begin{array}{l} M_{\alpha\beta} = \int_V (\Phi_\alpha\Phi_\beta)\,dV, \\[4pt] N_{\alpha\beta i} = \int_V (\Phi_\alpha\phi_{\beta,i})\,dV, \\[4pt] K_{\alpha\beta j\gamma} = \int_V (\Phi_\alpha\Phi_{\beta,j}\Phi_\gamma)\,dV, \\[4pt] H = h + \eta - b. \end{array}\right\} \tag{7}$$

To solve Eqs. (5) and (6), numerical integration in time has to be performed. Applying the two step Lax-Wendroff scheme, the following equations can be introduced.

First step

$$\left.\begin{array}{l} \bar{M}_{\alpha\beta}u_\beta^{n+\frac{1}{2}} = M_{\alpha\beta}u_\beta^n - \dfrac{\Delta t}{2}gN_{\alpha\beta i}\eta_\beta^n, \\[10pt] \bar{M}_{\alpha\beta}\eta_\beta^{n+\frac{1}{2}} = M_{\alpha\beta}\eta_\beta^n - \dfrac{\Delta t}{2}(K_{\alpha\beta j\gamma}H_\beta^n u_{\gamma j}^n + K_{\alpha\beta j\gamma}H_\gamma^n u_{\beta j}^n). \end{array}\right\} \tag{8}$$

Second step

$$
\left.
\begin{aligned}
\bar{M}_{\alpha\beta} u_{\beta}^{n+1} &= M_{\alpha\beta} u_{\beta}^{n} - \Delta t\, g\, N_{\alpha\beta i}\eta_{\beta}^{n+\frac{1}{2}}, \\
\bar{M}_{\alpha\beta}\eta_{\beta}^{n+1} &= M_{\alpha\beta}\eta_{\beta}^{n} - \Delta t\,(K_{\alpha\beta j\gamma} H_{\beta}^{n+\frac{1}{2}} u_{\gamma j}^{n+\frac{1}{2}} + K_{\alpha\beta j\gamma} H_{\gamma}^{n+\frac{1}{2}} u_{\beta j}^{n+\frac{1}{2}}).
\end{aligned}
\right\}
\tag{9}
$$

Here, the superscript n denotes the time step and Δt is the time increment. A bar–indicated a lumped coefficient matrix.

When Eqs. (8) and (9) are applied to the tsunami wave propagation analysis, a dissipation of energy results. To prevent such dissipation, the mass matrix $M_{\alpha\beta}$ is transformed as follows:

$$
\tilde{M}_{\alpha\beta} = (1 - \gamma) M_{\alpha\beta} + \gamma \bar{M}_{\alpha\beta} \qquad (0 \le \gamma \le 1).
\tag{10}
$$

Introducing the Eq. (10), the mass matrix can be expressed approximately by a type of lumped mass matrix $\bar{M}_{\alpha\beta}$ and Eq. (10) effectively excludes energy dissipation. The value of the parameter γ was taken as 0.8 based on many test examples applied to a rectangular model region.

3. Numerical Results

The study area is situated on the coast of the Pacific and covers an area of approximately 378,000 square kilometers. The analysis domain is divided into triangular finite elements and each side of an element is from 625 to 10,000 meters long (Fig. 1). The number of nodes is 35,265 and the number of elements is 69,068. The time intergration interval is 6.0 seconds and the duration time of bottom upheaval is 30 seconds. Applying the theory of MANSINHA and SMYLIE (1971), the bottom upheaval according to ISHIBASHI's (1977) model can be estimated.

Here and henthforth, the following symbols are used: a, inundation height; η, calculated wave height over mean sea level; a°, tidal observation record.

3.1 Tonankai Tsunami

In this case, numerical results are compared with tidal observation records (Fig. 2). The calculated wave heights show good agreement with the observed tide levels and the average of (η/a°) is 0.85.

At Uchiura, the time history of the calculated tide level is compared with the tidal observation record (Fig. 3), and it seen that both show good agreement on the first half period.

The calculated wave height over mean sea level is compared with the inundation height by IIDA (1977) as shown in Fig. 4. In the portion between No. 8 and No. 16, inundation heights are 5 to 10 meters. These areas have a very complex configuration. Considering the distribution of the ratio η/a, this ratio is about 0.5 to 0.3 indicating the calculated wave height is rather small compared with the inundation height. The reasons are that, a) the finite element mesh applied to this case study does not approximate the complex configuration sufficiently along the coast, and b) the inundation height is phisically different from the wave height. For example, in the area between No. 3 and No. 20 there exists the inner part of a little bay, and this little bay was not included in the finite

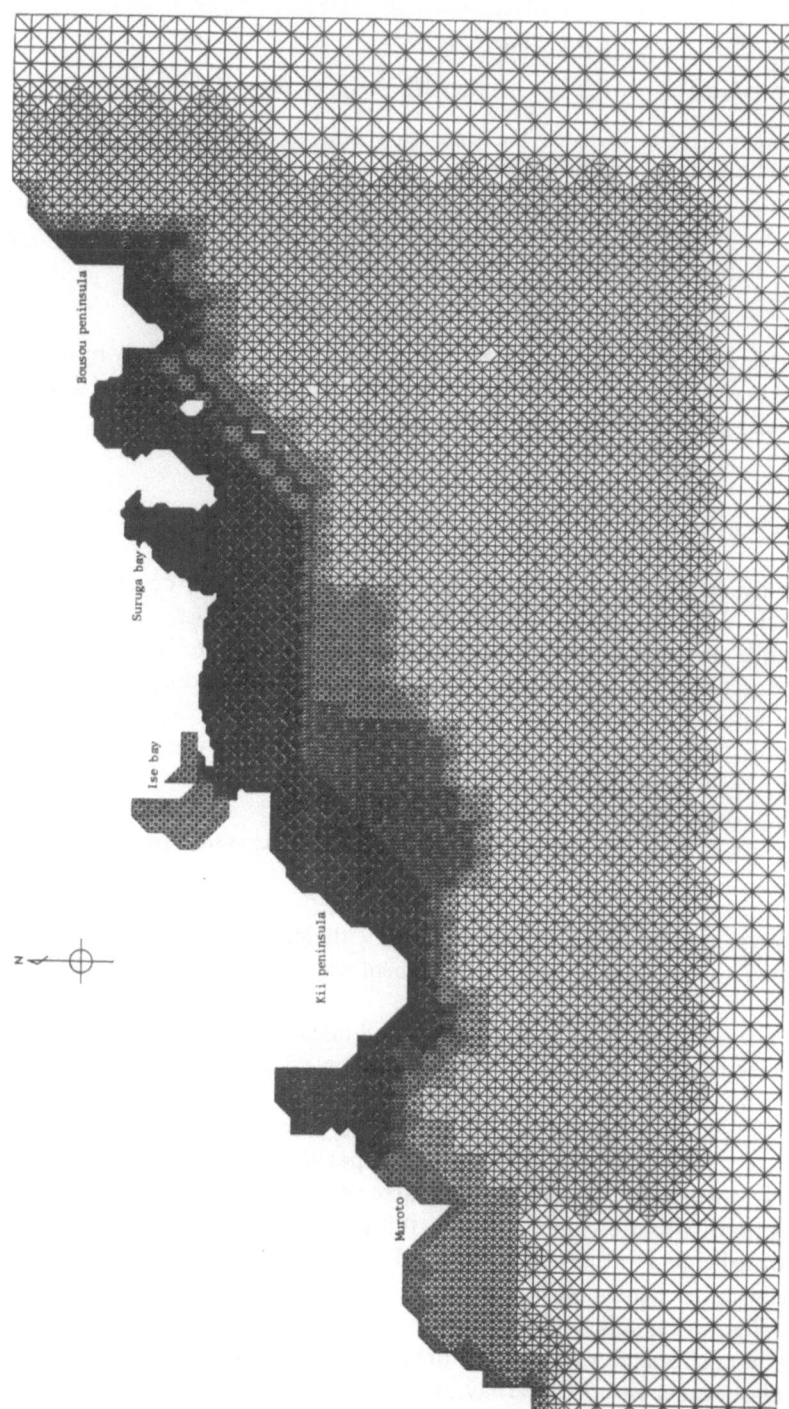

Fig. 1.   Finite element mesh for the Tokai district.   Number of node points: 35,265.   Number of elements: 69,068.

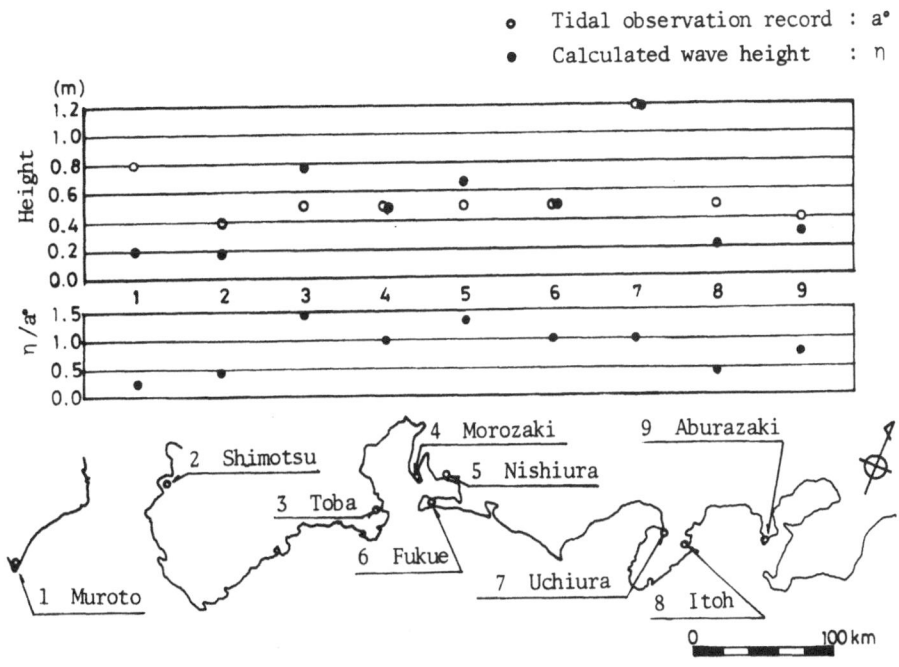

Station	Tidal observation record : a° (m)	Calculated wave height : η (m)	η / a°
Muroto	0.8	0.20	0.25
Shimotsu	0.4	0.18	0.45
Toba	0.53	0.78	1.47
Morozaki	0.5	0.48	0.96
Nishiura	0.5	0.66	1.32
Fukue	0.5	0.51	1.02
Uchiura	1.2	1.17	0.98
Itoh	0.5	0.22	0.44
Aburatsubo	0.4	0.29	0.73
average	—	—	0.85

Fig. 2. The comparison between tidal observation record and calculated wave height of the Tonankai tsunami of 1944.

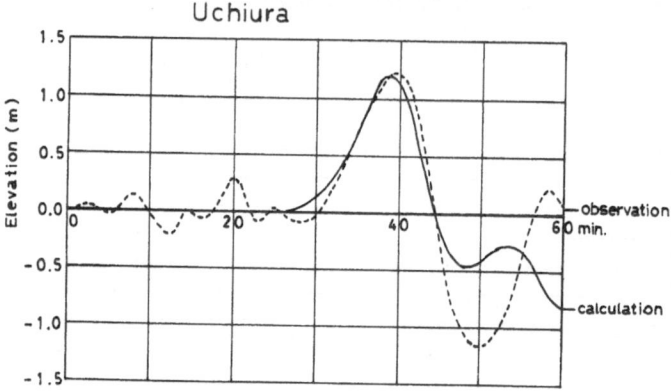

Fig. 3.   The time history of tide level at Uchiura.

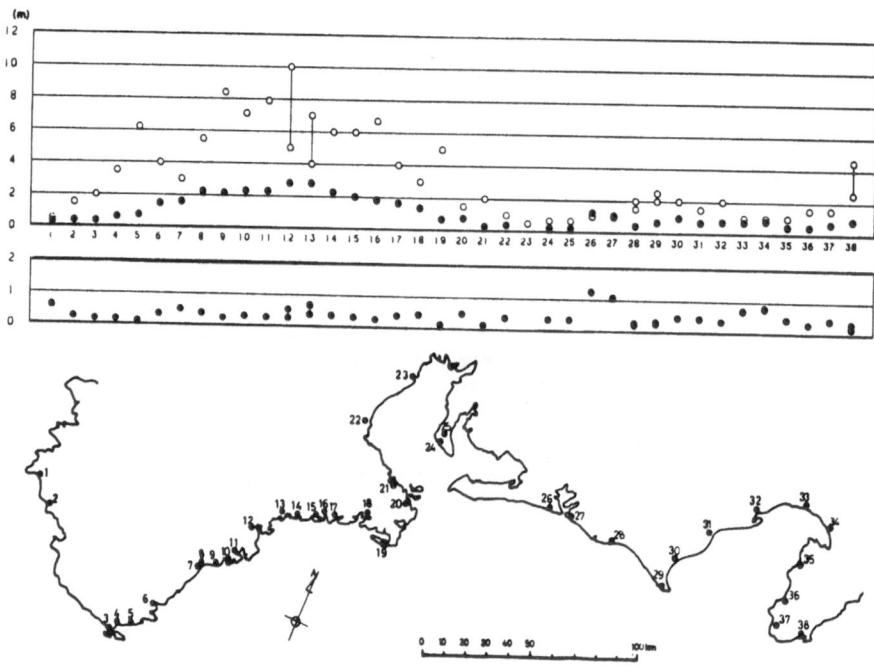

Fig. 4.   The comparison between inundation height by IIDA (1977) and calculated wave height of the Tonankai Tsunami of 1944.

element mesh.

In order to examine the amplification at the shore caused by the variation of water depth and configuration, the wave motion in Owase Bay is calculated using a fine spaced finite element mesh. The boundary conditions, water level and velocity component are obtained from the former numerical result.

Figure 5 shows the time history of the calculated tide level along the boundary to

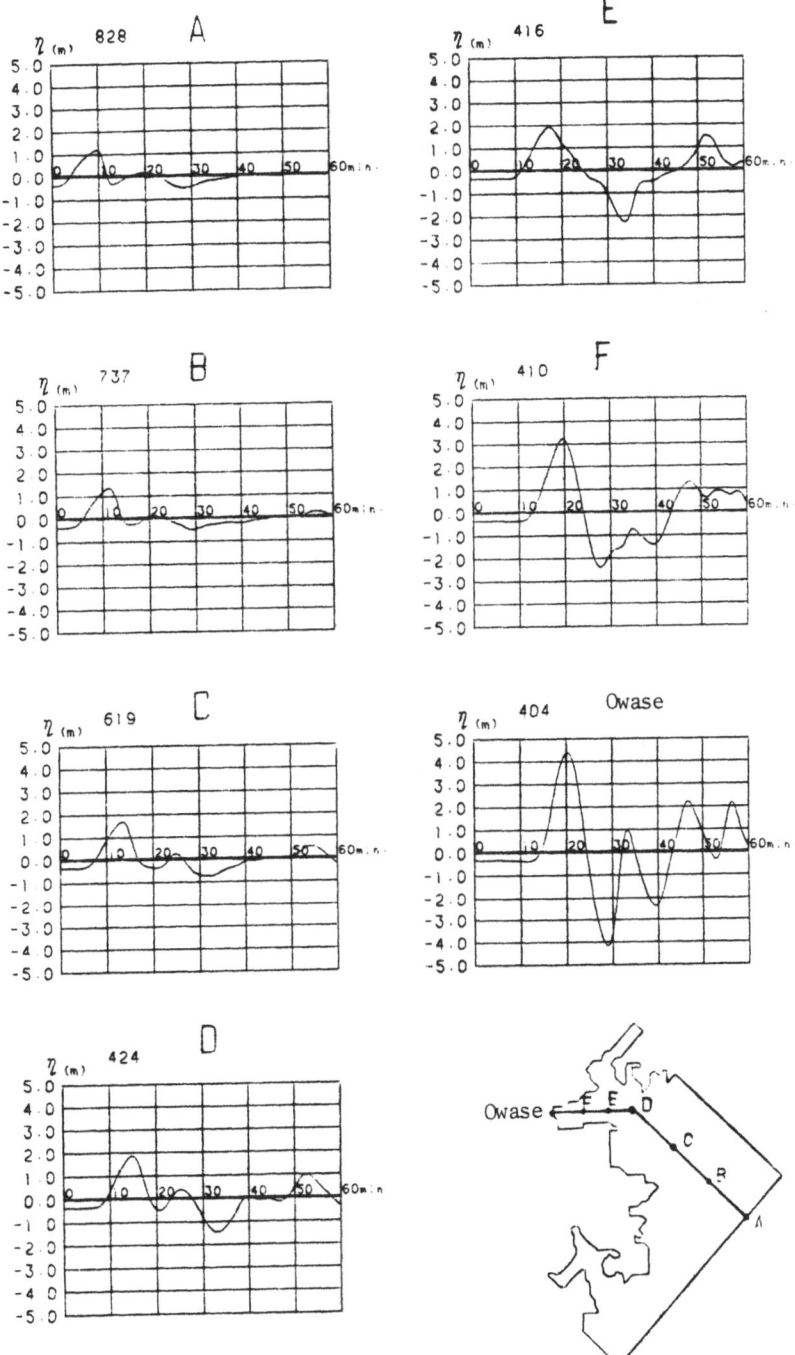

Fig. 5.   Time history of calculated tide level (Owase Bay) of the Tonankai Tsunami of 1944.

the inner part at Owase Bay. As the result of the numerical examination, calculated wave height at the shoreline amplifies to 3.7 times of the wave height at the bay mouth. Therefore, the wave height at Owase Bay would be 4.5 meters. The actual inundation height is 5.0 to 10.0 meters. It is concluded that, if the near shore configuration is divided into a fine spaced finite element mesh, it is possible to recover the actual inundation height.

3.2   Ansei-Tokai Tsunami

A comparison of the calculated wave height and the observed inundation height by HATORI (1980) is shown in Fig. 6. The wave height (η) is corrected by multiplying the ratio (a/η) calculated for the Tonankai Tsunami. Thus as can be seen from Fig. 6, the numerical results agree well with the observed inundation heights along the coast of the Tokai district in the case of the Ansei-Tokai Tsunami.

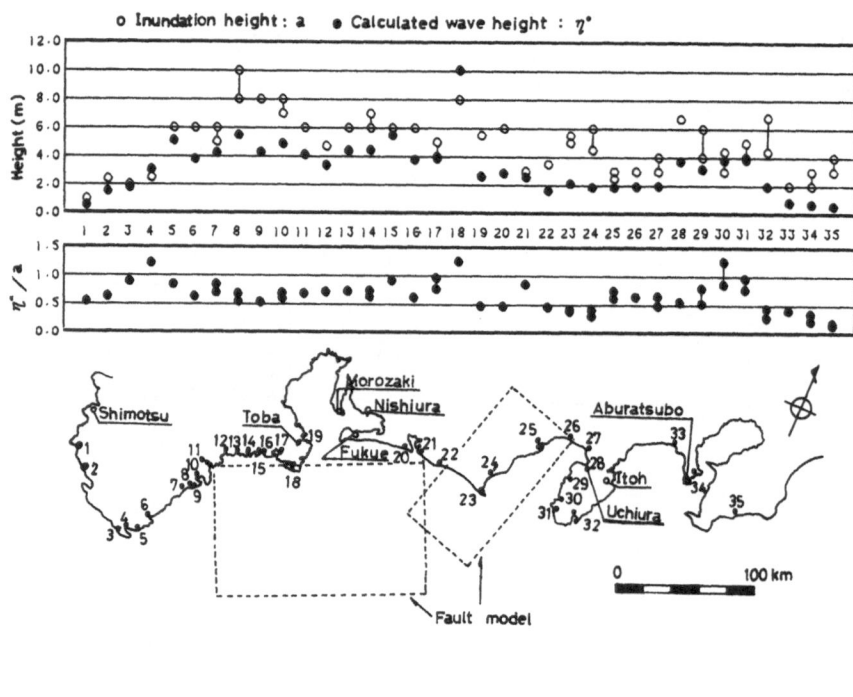

$$\eta^\circ = \eta \cdot \frac{a_T}{\eta_T}$$

η : calculated wave height due to Ansei-Tokai Tsunami

η_T : calculated wave height due to Tonankai Tsunami

a_T : inundation height due to Tonankai Tsunami

Fig. 6.   The comparison between inundation height by HATORI (1980) and calculated wave height of the Ansei-Tokai Tsunami of 1854.

4.   Concluding Remarks

Application of the improved two step Lax-Wendroff finite element method in tsunami wave propagation analysis, enabled a calculation of the wave height by successfully excluding energy dissipation.

REFERENCES

AIDA, I., Numerical experiments for the tsunami propagation—the 1964 Niigata Tsunami and the 1968 Tokachi Tsunami, *Bull. Earthq. Res. Inst., Univ. of Tokyo,* **47**, 673–700, 1969.

HATORI, T., Wave heights from the field investigation of the 1707 Hoei Tsunami and the 1854 Ansei Tsunami, *Marine Sci.,* **12**, 495–503, 1980 (in Japanese).

HWANG, Li-san and D. DIVOKY, Tsunami generation, *J. Geophys. Res.,* **75**, No. 33, 6802–6817, 1970.

IIDA, K., Earthquake damage and intensity distribution of the Tonankai Earthquake of December 7, 1944, Report of Earthquake Disaster Prevention Committee of Aichi Prefecture, Japan, pp. 1–120, 1977 (in Japanese).

INAGAKI, K., T. MAEHARA, M. KOBAYASHI, and M. KAWAHARA, An application of improved two step Lax-Wendroff scheme to the equation of long wave in shallow sea water, paper presented at the Second Symposium of Finite Method Analysis for Flow, 1980 (in Japanese).

ISHIBASHI, K., Re-examination of a great earthquake expected in the Tokai district, Central Japan—possibility of the Suruga Bay Earthquake, *Rep. Coord. Comm. Earthq. Predict.,* **17**, 126–132, 1977 (in Japanese).

KAWAHARA, M., N. TAKEUCHI, and T. YOSHIDA, Two step explicit finite element method for tsunami wave propagation analysis, *Inter. J. Numerical Methods in Eng.,* **12**, 331–351, 1978.

MANSINHA, L. and D. E. SMYLIE, The displacement fields of inclined faults, *Bull. Seismol. Soc. Am.,* **61**, 1433–1440, 1971.

Tsunamis—Their Science and Engineering, edited by K. Iida and T. Iwasaki, 303–314.

Effects of the Continental Shelf on Harbor Resonance

Philip L.-F. LIU

School of Civil and Environmental Engineering,
Cornell University, Ithaca, NY, U.S.A.

(Received September 10, 1981; Revised January 25, 1982)

Effects of the continental shelf on harbor oscillations are examined. Linear long-wave equations are used as the basis of the present study. Obliquely incident waves are assumed to be periodic in time with infinitesimal amplitudes.

For the convenience of analytical investigations, the harbor is assumed to be rectangular in shape, and the width of the harbor is small in comparison with the incident wave length. Constant water depth is assumed inside the harbor. The continental shelf is modeled by a step (water depth increases suddenly at some distance away from the coastline).

The problem is solved by the method of asymptotic matching. The harbor and the ocean are treated separately. Two solutions are brought together by requiring the continuity of normal mass flux and free surface elevation across the harbor mouth. Therefore, the Green's function for a semi-infinite ocean with a step-shaped water depth is required and developed in this paper.

The amplification factor is investigated in terms of the following parameters: (1) the depth ratio, h_1/h_2; h_1 = water depth on the continental shelf and h_2 = water depth in the ocean basin, and (2) the ratio of the width of the continental shelf, L, to the length of the harbor, l. It is found that portions of the radiated wave energy may be trapped on the continental shelf and hence increase the resonance response inside the harbor. The resonant modes of the continental shelf can be tuned in with that of the harbor and it results in huge oscillations inside the harbor.

1. Introduction

Harbor oscillations due to incident waves have been extensively studied in the framework of inviscid linear long-wave theory (see, e.g. MILES, 1974; RAICHLEN, 1979). Efficient numerical methods have also been developed for harbors of arbitrary shape and depth (LEE, 1971; HWANG and TUCK, 1970; OLSEN and HWANG, 1971; BERKHOFF, 1972; CHEN and MEI, 1974). These theories all assumed that the topographies vary only inside or/and near the entrance of a harbor. In other words, the effects of a continental shelf on the harbor oscillation have been ignored.

Recently, MOMOI (1976) studied the scattering of long waves at the mouth of estuaries bordering a continental shelf. He demonstrated clearly that radiated wave energy can be trapped on the continental shelf and it can affect the wave motions inside estuaries. In this paper we investigate analytically the interaction of harbor oscillation and wave propagation on a continental shelf. For simplicity, the harbor is assumed to

303

be a rectangular basin with a width which is smaller than the wave length of incident waves. The continental shelf is modeled as a step.

Solutions are obtained by the method of matched asymptotic (MILES, 1971). A Green's function for an oscillatory point source in a semi-infinite domain with a step-shaped topography is developed. Results are presented for different values of the width of continental shelf and depth ratio.

2. Formulation

2.1 The shallow water equations

We consider a harbor and ocean system as illustrated in Fig. 1. The water depth inside the harbor is the same as that on the continental shelf. There is a continental break at $x = L$ where water depth changes abruptly from h_1 to h_2 ($h_2 > h_1$). The water depth is uniform in the alongshore direction, i.e. $h = h(x)$. For simplicity, we also assume that the harbor is a long narrow rectangular basin with a dimension $l \times 2a$. The linear shallow water equations are employed herein to describe the velocity $\hat{u}(x, y, t)$, $\hat{v}(x, y, t)$ and the free surface displacement $\hat{\zeta}(x, y, t)$. We shall, however, study only incident waves and responses which are periodic in time, with radian frequency ω. Let us represent the solutions as

$$\hat{u}(x, y, t) = \mathrm{Re}\{u(x, y)e^{-i\omega t}\},$$
$$\hat{v}(x, y, t) = \mathrm{Re}\{v(x, y)e^{-i\omega t}\},$$
$$\hat{\zeta}(x, y, t) = \mathrm{Re}\{\zeta(x, y)e^{-i\omega t}\}. \tag{1}$$

The governing equations become

$$\frac{\partial \zeta}{\partial x} = -\frac{i\omega}{g}u \tag{2a}$$

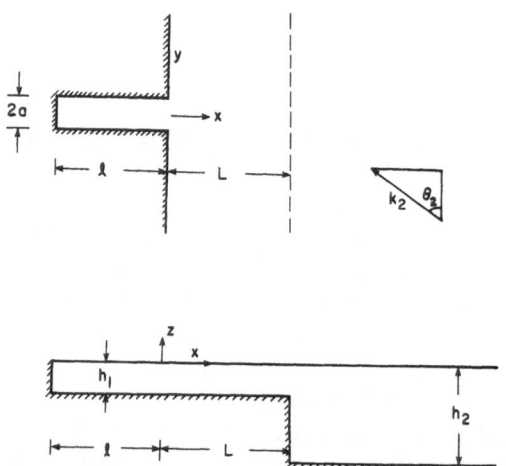

Fig. 1. Sketch of harbor geometry and continental shelf topography.

$$\frac{\partial \zeta}{\partial y} = -\frac{i\omega}{g} v \tag{2b}$$

$$-i\omega\zeta + \frac{\partial}{\partial x}(hu) + \frac{\partial}{\partial y}(hv) = 0, \tag{2c}$$

where (2a) and (2b) are momentum equations and (2c) denotes the conservation of mass. Equations (2) can be combined into a single equation for ζ

$$\frac{\partial^2 \zeta}{\partial x^2} + \frac{\partial^2 \zeta}{\partial y^2} + \frac{1}{h}\frac{dh}{dx}\frac{\partial \zeta}{\partial x} + \frac{\omega^2}{gh}\zeta = 0, \tag{3}$$

where $h = h(x)$ has been used.

The incident plane wave with a unit amplitude at infinity can be written as

$$\zeta_{inc} = \exp[-ik_2(x\sin\theta_2 - y\cos\theta_2)], \tag{4}$$

where

$$k_2 = \frac{\omega}{\sqrt{gh_2}} \tag{5}$$

is the wave number and θ_2 is the angle of incidence.

The boundary conditions are that the normal velocity vanishes along the sidewalls of the harbor and along the coastline. Since we shall treat the harbor and ocean separately, the normal flux and the surface elevation must be continuous across the mouth of the harbor. Some continuity conditions should also be applied at the continental break. Finally, the radiation condition requires that radiated waves must be outgoing waves at infinity.

2.2 Solution form inside the harbor

Assuming that the harbor is long and narrow, $O(k_1 a) \ll 1$ and $O(k_1 l) = 1$, the harbor response is essentially one-dimensional except within a distance $O(k_1 a)$ from the entrance (MILES and MUNK, 1961; UNLUATA and MEI, 1973),

$$\left.\begin{array}{l} \zeta(x) = T\cos k_1(x + l) \\ u(x) = -(igk_1 T/\omega)\sin k_1(x + l), \end{array}\right\} x < 0 \tag{6}$$

where $T = A[\cos k_1 l - iZ\sin k_1 l]^{-1}$, $k_1 = \omega/\sqrt{gh_1}$, $\tag{7}$

A = wave amplitude of incident and reflected waves at the entrance of the harbor, and Z = ocean impedance.

It may be remarked that the real part of the ocean impedance corresponds to the radiation damping while the imaginary part corresponds to the mass reactance. The objective of the rest of the paper is to find the incident-reflected wave amplitude, A, and the impedance Z, corresponding to the prescribed topography.

3. Solutions Outside the Harbor

The free surface displacement outside the harbor can be divided into two parts

$$\zeta = \zeta^I + \zeta^R, \quad x > 0 \tag{8}$$

where ζ^I denotes the surface elevation of the incident and reflected wave system in the absence of a harbor, and ζ^R is the surface displacement of radiated waves due to the presence of a harbor. Both ζ^I and ζ^R depend on the bottom topography outside of the harbor.

3.1 The incident-reflected waves

We first consider the wave field without the presence of a harbor. Using ζ_1^I and ζ_2^I as the surface displacements in the regions $0 \leq x < L$ and $L \leq x$, respectively, we may simplify the governing Eq. (3) to be

$$\frac{\partial^2 \zeta_j^I}{\partial x^2} + \frac{\partial^2 \zeta_j^I}{\partial y^2} + \frac{\omega^2}{gh^j} \zeta_j^I = 0, \quad j = 1, 2 \tag{9}$$

since in each region the water depth $h_j (j = 1,2)$ is a constant. The boundary condition along the coastline $(x = 0)$ and the jump conditions along the edge of the step $(x = L)$ can be expressed as follows:

$$\frac{\partial \zeta_1^I}{\partial x} = 0, \qquad\qquad x = 0 \tag{10a}$$

$$h_1 \frac{\partial \zeta_1^I}{\partial x} = h_2 \frac{\partial \zeta_2^I}{\partial x}, \qquad x = L \tag{10b}$$

$$\zeta_1^I = \zeta_2^I, \qquad\qquad x = L. \tag{10c}$$

The Sommerfeld radiation condition is also applied at infinity. After some straightforward manipulations, the solution to (9) and (10) can be readily obtained as

$$\zeta_1^I = E \cos(xk_1 \sin\theta_1) \exp(iyk_1 \cos\theta_1), \quad 0 \leq x < L \tag{11}$$

$$\zeta_2^I = [B \exp(ixk_2 \sin\theta_2) + \exp(-ixk_2 \sin\theta_2)] \exp(iyk_2 \cos\theta_2), \quad x \geq L \tag{12}$$

where

$$k_1 \cos\theta_1 = k_2 \cos\theta_2 \tag{13}$$

$$A = 2[\cos(k_1 L \sin\theta_1) - i\varepsilon \sin(k_1 L \sin\theta_1)]^{-1} \exp(-ik_2 L \sin\theta_2) \tag{14}$$

$$B = \frac{\cos(k_1 L \sin\theta_1) + i\varepsilon \sin(k_1 L \sin\theta_1)}{\cos(k_1 L \sin\theta_1) - i\varepsilon \sin(k_1 L \sin\theta_1)} e^{-2ik_2 L \sin\theta_2} \tag{15}$$

$$\varepsilon = \frac{k_1 h_1 \sin\theta_1}{k_2 h_2 \sin\theta_2}. \tag{16}$$

It is obvious that (11) indicates the existence of a standing wave in the on-offshore direction on the continental shelf, $0 \leq x < L$. The wave amplitude near the harbor entrance, $k_1 x$, $k_1 y \ll 1$, can be approximated as

$$|\zeta_1^I(0,0)| = |A| = 2[\cos^2(k_1 L \sin\theta_1) + \varepsilon^2 \sin^2(k_1 L \sin\theta_1)]^{-1/2}. \qquad (17)$$

The resonant frequency or shelf length are approximately at the zeros of

$$\cos(k_1 L \sin\theta_1) = 0, \quad k_1 L \sin\theta_1 = \sin\theta_1 = \frac{\pi}{2}, \frac{3\pi}{2}, \dots. \qquad (18)$$

The corresponding resonant peak is

$$\max |A| \cong 2/\varepsilon. \qquad (19)$$

The the parameter ε, which is the measure of relative water depth in two regions, corresponds to the radiation damping of the continental shelf. For normal incidence $\theta_1 = \theta_2 = \pi/2$, then $\varepsilon = \sqrt{h_1/h_2}$. The maximum amplitude becomes $\max |A| \approx 2\sqrt{h_2/h_1}$.

It may be noted that for normal incidence, the lowest mode requires the continental shelf length be one-quarter of the wave length of the waves on the shelf. However, for the obliquely incident waves, $\theta_2 < \theta_1 < \pi/2$, the required wave length for lowest resonant mode becomes shorter (see (18)).

3.2 The radiated waves

The radiated wavefield can be expressed as a superposition of sources distributed along the harbor entrance, $|y| < a$. Thus

$$\zeta^R(x,y) = \frac{i\omega}{g} \int_{-a}^{a} V(y')\zeta^G(x,y;y')dy', \quad x \geq 0 \qquad (20)$$

where $V(y)$ is the amplitude of the normal velocity at the harbor entrance. The Green's function $\zeta^G (x, y; y')$ satisfies the governing Eq. (9) with the boundary condition

$$\frac{\partial\zeta^G}{\partial x} = \delta(y - y'), \quad x = 0, \quad |y'| < a. \qquad (21)$$

The jump conditions along the edge of a continental break, described by (10b,c) also apply to the Green's function ζ^G, as does the radiation boundary condition at infinity.

To find the Green's function, we introduce the Fourier transform of ζ^G and its inverse transform as follows:

$$\bar{\zeta}(x,\beta;y') = \int_{-\infty}^{\infty} \zeta^G(x,y;y')e^{i\beta y}\,dy, \qquad (22a)$$

$$\zeta^G(x,y;y') = (2\pi)^{-1}\int_{-\infty}^{\infty} \bar{\zeta}e^{-i\beta y}\,d\beta. \qquad (22b)$$

The governing equations and boundary conditions for $\bar{\zeta}$ on the (x,β) plane become

$$\frac{\partial^2 \bar{\zeta}_j}{\partial x^2} = \left(\beta^2 - \frac{\omega^2}{gh_j}\right)\bar{\zeta}_j; \quad j = \begin{cases} 1, 0 \leq x < L \\ 2, x \geq L \end{cases} \qquad (23)$$

$$\frac{\partial \bar{\zeta}_1}{\partial x} = e^{i\beta y'}, \quad x = 0, \quad |y'| < a \tag{24}$$

$$\bar{\zeta}_1 = \bar{\zeta}_2, \quad x = L \tag{25a}$$

$$h_1 \frac{\partial \bar{\zeta}_1}{\partial x} = h_2 \frac{\partial \bar{\zeta}_2}{\partial x}, \quad x = L. \tag{25b}$$

At infinity, $x \to \infty$, the solution should behave as

$$\bar{\zeta}_2 \propto \exp(i\alpha_2 x), \tag{26a}$$

where

$$\alpha_2 = (k_2^2 - \beta^2)^{1/2}. \tag{26b}$$

Equation (26a) denotes an outgoing wave (in the positive x-direction) for the cases $|\beta| < k_2$. On the other hand, when $|\beta| > k_2$, α_2 becomes an imaginary constant. The expression (26a) becomes an evanescent mode, which indicates that a part of the radiated wave energy is trapped on the continental shelf. A sketch of the radiated wave ray pattern is shown in Fig. 2. The critical angle at which the trapping occurs is defined as

$$\tan\theta_{cr} = \frac{\alpha_1}{\beta} = \frac{\sqrt{k_1^2 - \beta^2}}{\beta} = \frac{\sqrt{k_1^2 - k_2^2}}{k_2} = \sqrt{\frac{h_2}{h_1} - 1}. \tag{27}$$

The solution to (23)–(26) can be readily obtained and be expressed as

$$\bar{\zeta}_1 = \frac{(\alpha_1 h_1 - \alpha_2 h_2)\cos\alpha_1 x}{\alpha_1(\alpha_1 h_1 \sin\alpha_1 L + i\alpha_2 h_2 \cos\alpha_1 L)} e^{i(\alpha_1 L + \beta y')} - \frac{i}{\alpha_1} e^{i(\beta y' + \alpha_1 x)} \tag{28a}$$

$$\bar{\zeta}_2 = \frac{\alpha_1 h_1}{\alpha_1(\alpha_1 h_1 \sin\alpha_1 L + i\alpha_2 h_2 \cos\alpha_1 L)} e^{i[\beta y' + \alpha_2(x - L)]}, \quad x \geq L \tag{28b}$$

where

$$\alpha_1 = (k_1^2 - \beta^2)^{1/2}. \tag{29}$$

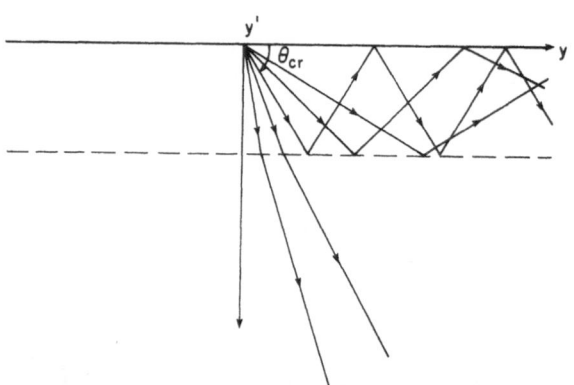

Fig. 2.   Sketch of radiated waves from harbor to continental shelf.

The Green's function $\zeta^G(x,y;y')$ can be found from the Fourier inverse transform. Thus

$$
\zeta_G(x,y;y') = \begin{cases}
-\dfrac{i}{\pi}\displaystyle\int_0^\infty \dfrac{(\alpha_1 h_1 - \alpha_2 h_2)e^{i\alpha_1 L}\cos\alpha_1 x \cos\beta(y-y')}{\alpha_1(\alpha_2 h_2 \cos\alpha_1 L - i\alpha_1 h_1 \sin\alpha_1 L)}\,d\beta \\[4mm]
-\dfrac{i}{2}H_0^{(1)}(k_1 r'), \qquad 0 \le x < L & (30a) \\[4mm]
-\dfrac{i}{\pi}\displaystyle\int_0^\infty \dfrac{\alpha_1 h_1 e^{i\alpha_2(x-L)}\cos\beta(y-y')}{\alpha_1(\alpha_2 h_2 \cos\alpha_1 L - i\alpha_1 h_1 \sin\alpha_1 L)}\,d\beta, \quad x \ge L & (30b)
\end{cases}
$$

where the following identity has been used (GRADSKTEYN and RYZHIK, 1980; p. 736)

$$
\int_0^\infty \frac{1}{\alpha_1}e^{i\alpha_1 x}\cos\beta(y-y')\,d\beta = \frac{\pi}{2}H_0^{(1)}(k_1 r'); \quad r' = \sqrt{x^2 + (y-y')^2} \tag{31}
$$

which corresponds to the Green's function for the case with constant depth h_1.

3.3 Ocean impedance

For a small harbor entrance, $k_1 a \ll 1$, if we assume that the currents, $V(y)$ in (20), are nearly uniform, the ocean impedance Z can be approximated as (MILES, 1971, 1974)

$$
Z = \frac{ik_1}{2a}\int_{-a}^a \int_{-a}^a \zeta^G(0,y;y')\,dy'\,dy \tag{32}
$$

Substitution of (30a) into (32) yields

$$
Z = Z_p + Z_0, \tag{33}
$$

where

$$
Z_0 = \frac{ik_1}{2a}\int_{-a}^a \int_{-a}^a \left[-\frac{i}{2}H_0^{(1)}(k_1|y-y'|)\right]dy'dy \simeq k_1 a\left\{1 + \frac{2i}{\pi}\left[\ln(\gamma k_1 a) - \frac{3}{2}\right]\right\}, \tag{34}
$$

$$
Z_p = \frac{2k_1}{\pi a}\int_0^\infty \frac{\sin^2\beta a(\alpha_1 h_1 - \alpha_2 h_2)(\alpha_2 h_2 \cos\alpha_1 L + i\alpha_1 h_1 \sin\alpha_1 L)e^{i\alpha_1 L}}{\beta^2 \alpha_1[(\alpha_2 h_2)^2\cos^2\alpha_1 L + (\alpha_1 h_1)^2\sin^2\alpha_1 L]}\,d\beta, \tag{35}
$$

and $\ln\gamma = 0.577216\ldots$ is the Euler's constant. It should be noticed that (34) is the well-known solution for the case of uniform water depth $h = h_1$. On the other hand, Z_p is the contribution due to the appearance of the continental shelf.

From (29) and (26b) it is clear that α_1 becomes an imaginary constant for $|\beta| > k_1$ and α_2 changes to an imaginary constant for $|\beta| > k_2$. We may introduce the following notations

$$
\begin{aligned}
\gamma_1 &= (\beta^2 - k_1^2)^{1/2} = -i\alpha, \quad k_2 < k_1 < |\beta| \\
\gamma_2 &= (\beta^2 - k_2^2)^{1/2} = -i\alpha_2, \quad\quad k_2 < |\beta|
\end{aligned} \tag{36}
$$

and rewrite (35) in terms of a new dummy variable $\zeta = \beta h_1$ in the following form

$$Z_p = \frac{2k_1 h_1^2}{\pi a} \sum_{m=1}^{3} I_m, \tag{37}$$

where

$$I_1 = \int_0^{k_2 h_1} \frac{\sin^2(\zeta a/h_1)(\alpha_1 h_1 - \alpha_2 h_2)(\alpha_2 h_2 \cos\alpha_1 L + i\alpha_1 h_1 \sin\alpha_1 L)}{\zeta^2 \alpha_1 h_1 [(\alpha_1 h_1)^2 \sin^2\alpha_1 L + (\alpha_2 h_2)^2 \cos^2\alpha_1 L]} d\zeta \tag{38a}$$

$$I_2 = P \int_{k_2 h_1}^{k_1 h_1} \frac{\sin^2(\zeta a/h_1)(\gamma_2 h_2 + i\alpha_1 h_1)}{\zeta^2 \alpha_1 h_1 [\alpha_1 h_1 \sin\alpha_1 L - \gamma_2 h_2 \cos\alpha_1 L]} d\zeta \tag{38b}$$

$$I_3 = -i \int_{k_1 h_1}^{\infty} \frac{\sin^2(\zeta a/h_1)(\gamma_1 h_1 - \gamma_2 h_2)}{\zeta^2 \gamma_1 h_1 [\gamma_1 h_1 \sinh\gamma_1 L + \gamma_2 h_2 \cosh\gamma_1 L]} d\zeta. \tag{38c}$$

It should be noticed that I_2 is a Cauchy principle integral since it becomes improper when ζ is the root of

$$\tan\alpha_1 L - \gamma_2 h_2/\alpha_1 h_1 = 0, \quad k_2 h_1 < \zeta < k_1 h_1. \tag{39}$$

Moreover, the real part of the integrand of I_2 and the integrand of I_3 have integrable singularities at $\zeta = k_1 h_1$. The integrands of (38) are plotted in Fig. 3.

4. Results and Discussions

Analytical solutions are obtained for fixed harbor parameters, $l/h_1 = 1.20952$ and $a/h_1 = 0.1175$. To investigate the effects of the continental shelf, the depth ratio, h_1/h_2, and the width of the continental shelf are varied. Only the normal incident waves are examined here.

In Figs. 4 and 5, the amplification factors of the wave amplitude at the end of the harbor ($x = -l$) are shown for $h_1/h_2 = 0.25$ and $L/h_1 = 1.0$ and 5.0, respectively. For comparison, solutions for a constant water depth case ($h_1/h_2 = 1.0$) are also presented. The appearance of a continental shelf seems to increase significantly the magnitudes of resonant peaks due to smaller radiation damping. The phase shift at the resonant frequency is small.

It should be pointed out that in Figs. 4 and 5 the wave amplitude is normalized by A, which is the amplitude at the harbor mouth. In the case of constant water depth, A equals 2 for all wave numbers. However, with the appearance of a continental shelf, A oscillates as a function of the wave number (Fig. 6). The magnitude of A increases as the depth ratio, h_1/h_2, decreases. The frequency of oscillations increases as L/h_1 increases. In Fig. 7 the amplification factor is replotted by including the effects of A. The resonant modes on the continental shelf clearly affect the harbor resonance.

For some combinations of depth ratio h_1/h_2, and width of the continental shelf, the resonant frequency of the shelf mode may become the same as that of the harbor oscillation. When this situation occurs, the wave amplification becomes very large. This is demonstrated in Fig. 8, where $|\zeta|_H$ represents the wave amplitude of the lowest resonant mode inside the harbor.

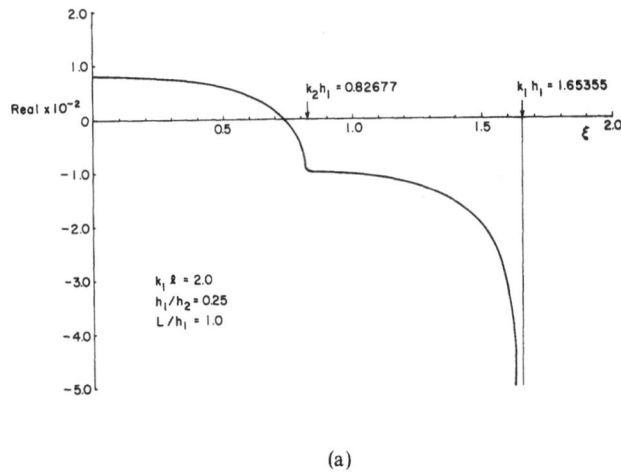

(a)

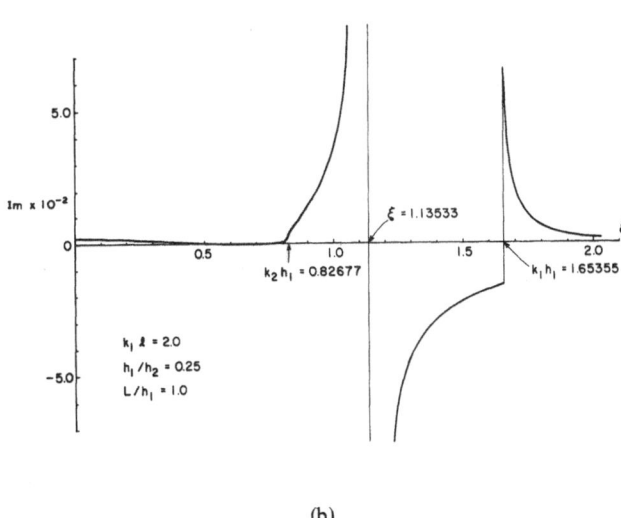

(b)

Fig. 3.   (a) Real part of and (b) imaginary part of the integrand of Eq. (38).

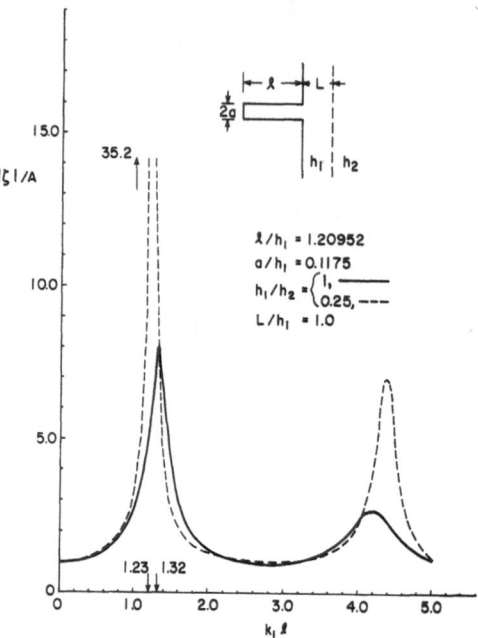

Fig. 4. Harbor responses for constant water depth $h_1/h_2 = 1.0$ and for narrow continental shelf width, $L/h_1 = 1.0$.

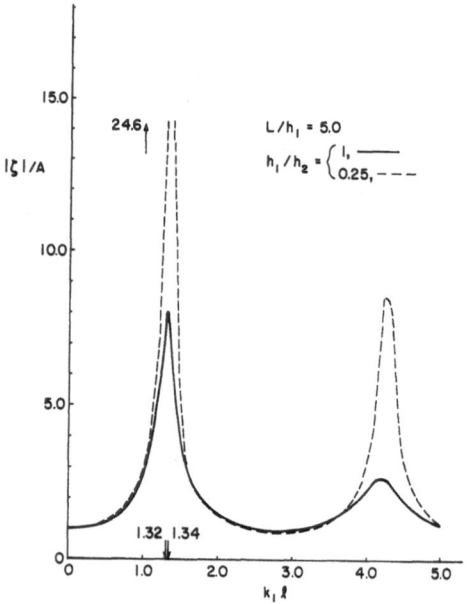

Fig. 5. Harbor responses for constant water depth and for wider width of a continental shelf, $L/h_1 = 5.0$.

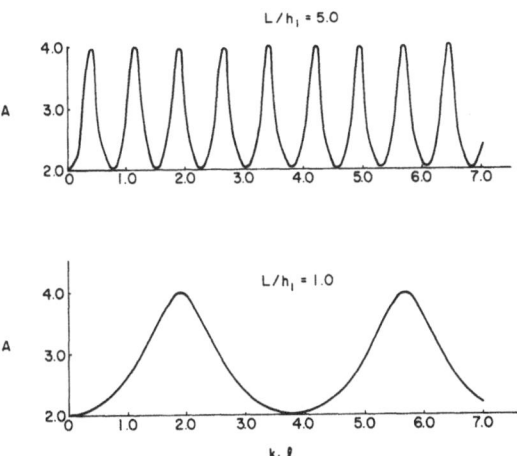

Fig. 6.   Wave amplitude at the harbor entrance without considering the existance of the harbor.

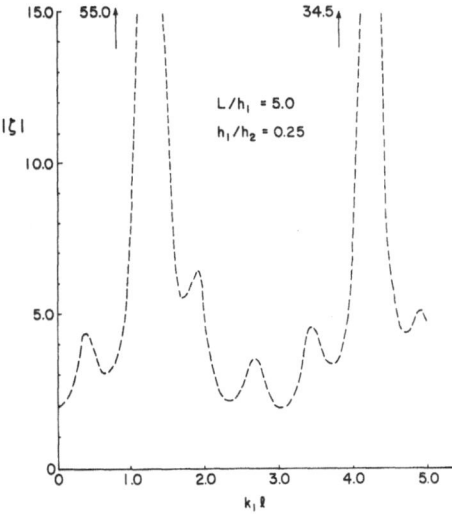

Fig. 7.   Harbor responses normalized by the incident wave amplitude at infinitiy.

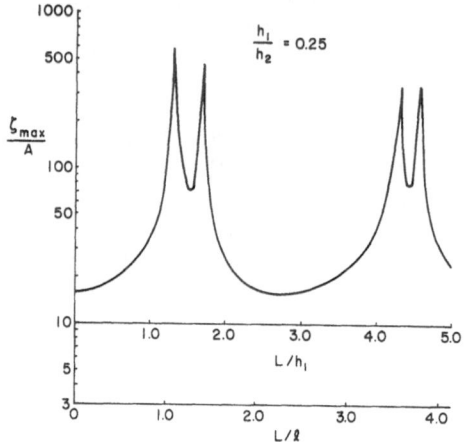

Fig. 8. Maximum harbor responses for different widths of continental shelf.

This work was carried out when the author was visiting W. M. Keck Laboratory at Caltech as a Guggenheim Fellow. Fruitful discussions with Professors F. R. Raichlen, J. J. Lee and T. Wu are acknowledged. The researh is, in part, supported by National Science Foundation through a Grant (No. PFR-7815358) at Cornell.

REFERENCES

BERKHOFF, J. C. W., Computation of combined refraction-diffraction, Proc. 13th Coastal Engrg. Conf., A.S.C.E., 471–490, 1972.

CHEN, H. S. and C. C. MEI, Oscillation and wave forces in a man-made harbor in the open sea, Proc. 11th Symp. Naval Hydrodyn., 573–594, 1974.

GRADSHTEYN, I. S. and I. M. RYZHIK, Table of Integrals, Series, and Products, Academic Press Inc., 1980.

HWANG, L. S. and E. O. TUCK, On the oscillation of harbours of arbitrary shape, J. Fluid Mech., 42, 447–464, 1970.

LEE, J. J., Wave-induced oscillations in harbours of arbitrary geometry, J. Fluid Mech., 45, 375–394, 1971.

MILES, J. W., Resonant response of harbors: an equivalent circuit analysis, J. Fluid Mech., 46, 241–265, 1971.

MILES, J. W., Harbor seiching, Ann. Rev. Fluid Mech., 6, 17–35, 1974.

MILES, J. W. and W. MUNK, Harbor paradox, J. Waterways, Harbors and Coastal Engineering Division, A.S.C.E., 87, WW3, 111–131, 1961.

MOMOI, T., Scattering of long waves at the mouth of estuaries bordering on a continental shelf, Part I and II, J. Phys. Earth, 24, 1–25 and 237–250, 1976.

OLSEN, K. and L-S. HWANG, Oscillations in a bay of arbitrary shape and variable depth, J. Geophys. Res., 76, 5048–5064, 1971.

RAICHLEN, F., Bay and harbor response to tsunamis, Tsunamis, Proc. the National Science Foundation Workshop, 188–221, 1979.

UNLUATA, U. and C. C. MEI, Excitation of long waves in harbors—an analytical study, Technical Report No. 171, Parson Laboratory, Department of Civil Engineering, M.I.T., 1973.

Tsunamis—Their Science and Engineering, edited by K. Iida and T. Iwasaki, 315–327.

Tsunami Response of the Tsugaru Straits

Susumu TAKAHASHI and Isao YAKUWA

*Department of Engineering Science, Faculty of Engineering,
Hokkaido University, Sapporo, Japan.*

(Received August 20, 1981; Revised March 23, 1982)

Two-dimensional numerical calculations were carried out by using two model tsunamis designed at the same place with the Tokachi-Oki tsunami in 1968 but with dimensions about twice, in order to obtain the tsunami response of the Tsugaru Straits and Mutsu Bay. To investigate the periods of secondary undulations of the calculated tsunami motions, the tsunami wave was analyzed by FFT method.

As a result of spectral analyses, many secondary undulations appeared and most of their periods were approximately in agreement with ones obtained from measurement of actual tsunamis along the strait.

In order to specify the mode of each undulation and the region of its growth, the model tsunami waves were separated into secondary undulations by using suitable bandpass filters. Distributions of wave height of individual waves drawn in the strait revealed that the ratio between the maximum wave heights caused by two model tsunamis at any point was not necessarily proportional to that of the initial level deviations at the source of the model tsunamis.

The existence of an undulation of very long period over 400 min was found in these calculations.

1. Introduction

The Tsugaru Straits, located between the main Japanese island of Honshu and the northern island of Hokkaido, connect the Pacific Ocean with the Japan Sea. The floor forms a trench having a depth of over 200 meters in the major part. In the western mouth of the strait lies a narrow ridge. The length of this strait is about 130 km and the widths os the eastern and the western mouth are about 50 km and 20 km, respectively. Mutsu Bay, about 60 km across, is joined to the Tsugaru Straits through a narrow neck. Along the coastline of the Tsugaru Straits and Mutsu Bay, there are six tide gauge stations at Yamasedomari, Hakodate, Oshima-Fukushima, Matsumae, Aomori, and Ohminato, as shown in Fig. 1.

In the 1960 Chilean tsunami, wave motion was observed at all stations except Yamasedomari, but records of other tsunamis such as the 1963 Kuril Island tsunami, the 1968 Tokachi-Oki tsunami, and the 1973 Off-Nemuro Peninsula tsunami, were obtained at only three stations or less. Therefore, details of tsunami response in the strait have not been fully analyzed.

According to the analysis of a limited number of observation records, only the

315

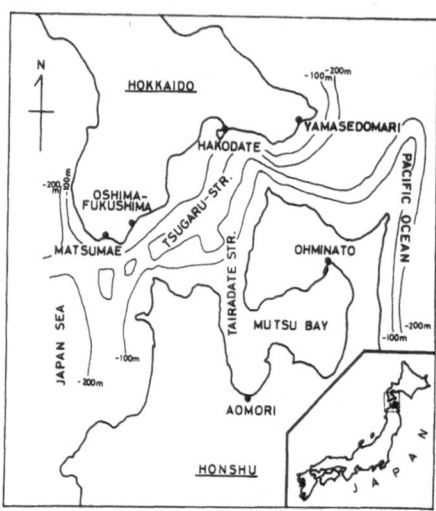

Fig. 1.   Location map of the Tsugaru Straits.

next results have been obtained. First, the maximum tsunami height in the strait appears near Hakodate for every tsunami. Secondly, the spectra of the wave motion are remarkably different at each tide gauge station, even for the same tsunami. On the other hand, at the same station, characteristics of the spectra of the various tsunamis are nearly the same as shown in Fig. 2 (in this figure, a tendency for the spectral intensity of the part of longer period to increase is caused by tidal components coexisted with the tsunami motion). These are also summarized in Table 1. Table 1 shows that there are long predominant periods of 302.4, 152.4, and 130 min. However, such a long predominant period except continental shelf resonance can not be found for any actual tsunami at the stations along the Pacific coast.

It is difficult to answer questions for the above results with the limited data as to in what region secondary undulations grow and what kind of modes the oscillations take. Therefore, computer simulations were carried out using two model tsunamis in order to determine more clearly the tsunami response of the Tsugaru Straits and Mutsu Bay.

2.   Basic Equations and Grid Scheme

By integrating the current velocity along the vertical z-axis, the long wave equations of motion and the continuity equation are transformed into two-dimensional common equations as

$$\frac{\partial M}{\partial t} = fN - \frac{1}{(H+\zeta)}\left(\frac{\partial}{\partial x}(M^2) + \frac{\partial}{\partial y}(MN)\right) - (H+\zeta)g\frac{\partial \zeta}{\partial x} + \frac{\gamma^2}{(H+\zeta)^2}M(M^2+N^2)^{\frac{1}{2}}$$

$$\frac{\partial N}{\partial t} = -fM - \frac{1}{(H+\zeta)}\left(\frac{\partial}{\partial x}(MN) + \frac{\partial}{\partial y}(N^2)\right)$$

$$\qquad - (H+\zeta)g\frac{\partial \zeta}{\partial y} + \frac{\gamma^2}{(H+\zeta)^2}\ N(M^2+N^2)^{\frac{1}{2}}$$

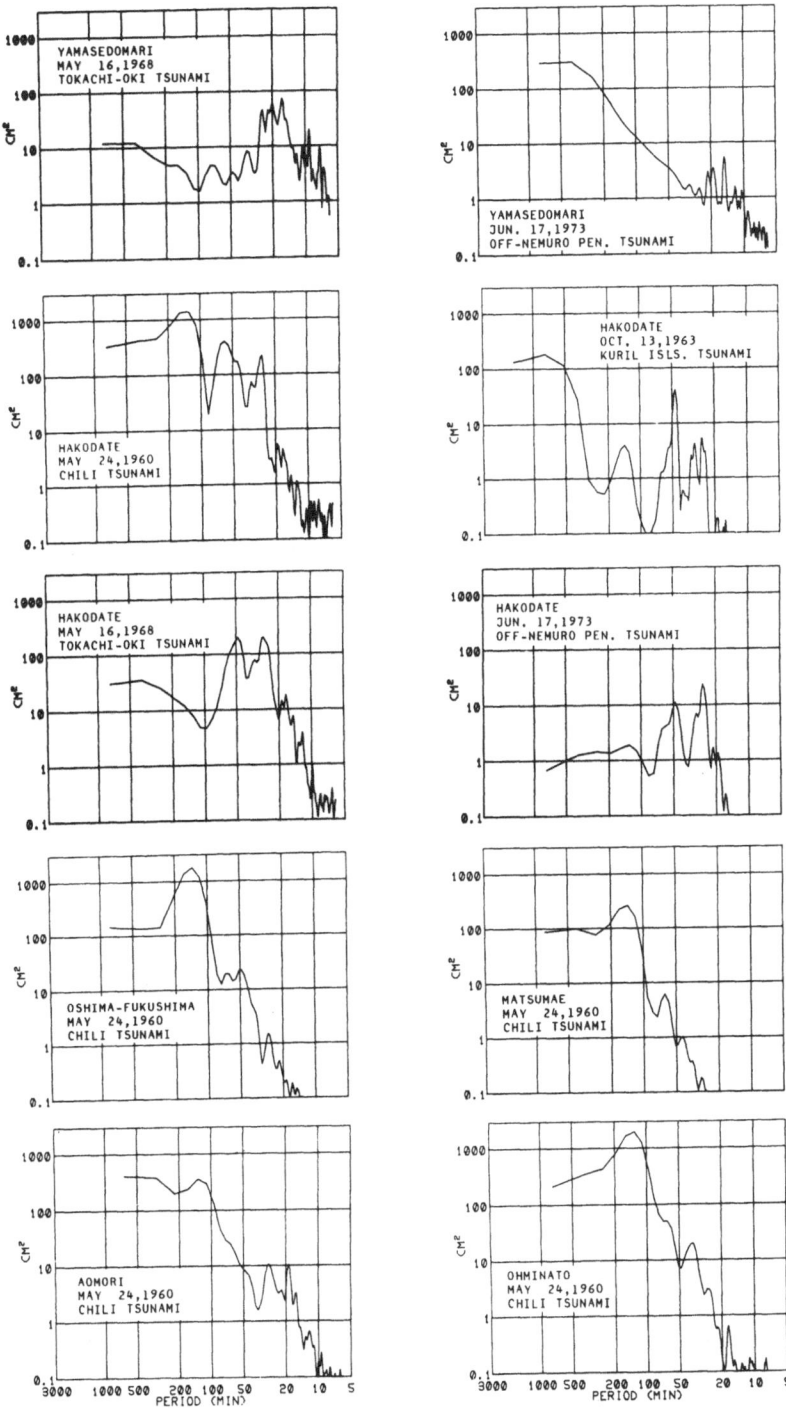

Fig. 2. Fourier transform of actual tsunamis at six tide gauge stations.

Table 1.　Principal spectra obtained from tsunami records.

	May 23, 1960 Chili	Oct. 13, 1963 Kuril Isls.	May 16, 1968 Tokachi Oki		Jun. 17, 1973 Off Nemuro Penin.	
YAMASEDOMARI			76.8 48.0 34.9 24.8 21.9	20.2 16.3	60.0 47.4 32.1 21.4 20.0	18.8
HAKODATE	152.4 135.0 54.9 46.3 26.5	139.6 56.9 46.5 26.5	132.0 55.0 45.5 28.7		137.4 58.2 54.0 48.8 26.0	
OSHIMA-FUKUSHIMA	136.6 68.3 48.2 26.4					
MATSUMAE	137.2 63.3 43.3					
AOMORI	131.6 28.6 18.8				302.4 137.5 100.8 52.1 30.9	19.4
OHMINATO	137.3 68.7 39.3 29.4					

(Min)

$$\frac{\partial \zeta}{\partial t} + \frac{\partial M}{\partial x} + \frac{\partial N}{\partial y} = 0 \quad M = \int_{-\zeta}^{H} u\,dz \quad N = \int_{-\zeta}^{H} v\,dz$$

where u and v denote velocities along the x-axis and y-axis, respectively, H, ζ, and γ^2 represent the water depth, elevation of the water surface and coefficient of friction at the sea bottom, and f and g denote the Coriolis parameter and gravitational acceleration.

Conditions at the ocean boundaries are $M = g^{\frac{1}{2}}(H + \zeta)^{\frac{1}{2}}\zeta$ along the y-axis, $N = g^{\frac{1}{2}}(H + \zeta)^{\frac{1}{2}}\zeta$ along the x-axis, and $M = 0$ along the y-axis, $N = 0$ along the x-axis at the coastline. A value of 0.0026 is used for the friction coefficient γ^2.

In the simulations, the basic equations were replaced by difference equations neglecting the frictional term at the sea bottom, non-linear term in the region of the Pacific Ocean and the Coriolis term in the Tsugaru Straits and Mutsu Bay. Figure 3 illustrates the computational grid scheme and the tsunami height given at the source area of the model tsunami. The size of the grid used in the region of the Pacific Ocean including the source area was 10×10 km, and 1/3 the length of this region in the strait.

Numbers in Fig. 3 show the elevation of the water surface in centimeters from the still water level in the model tsunami. A time step of 10 sec was used in these calculations and the simulations were continued for a time span of 1,650 min after tsunami generation. The model tsunami source area was shaped as an ellipse having its major axis directed to the mouth of the Tsugaru Straits. Length of the major and minor axes were 380 km and 170 km, respectively. Two model tsunamis of different surface elevations were designed. One was a weak model in which the tsunami energy could be estimated as 2.10×10^{20} erg, and the other was a strong model, the tsunami energy of 1.32×10^{21} erg. The maximum upward deviation of sea level at the center of the source area was taken as 0.6 meter in the former case (see Fig. 3) and five times that in the latter case. The location of the source area of the model tsunamis is almost the same as that of the 1968 Tokachi-Oki tsunami calculated by KAJIURA et al. (1968) and the JAPAN METEOROLOGICAL AGENCY (1969), but the size of the area is about two times as large as that of the Tokachi-Oki tsunami in spite of having the same order of energy. The effect of tidal motion on the tsunami response is not considered in this study. Figure 4 shows the amplitude spectrum of the tsunami elevation at a point near its source area in the weak model tsunami.

3. Results of Numerical Experiment for Weak Model Tsunami

Figure 5 shows the results of calculations obtained by the Fourier transform at six

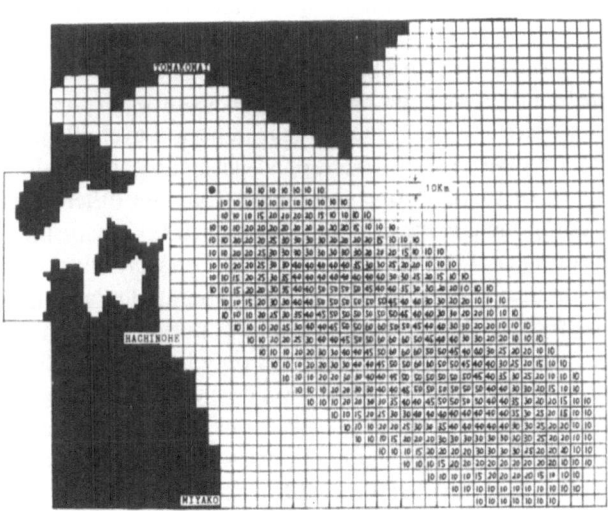

Fig. 3.   Grid scheme and source of a model tsunami.

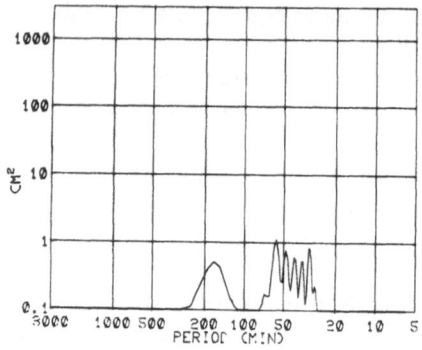

Fig. 4.   Fourier transform of weak model tsunami at a dotted grid near the source area as shown in Fig. 3.

grid points corresponding to the above mentioned tide gauge stations in the weak model tsunami. As shown in Fig. 5, the spectra have many peaks corresponding to periods of 307, 140, 102, 55, 51, 46 min, and so on. The peak at 140 min occurs at all grid points in the Tsugaru Straits and Mutsu Bay. Moreover, this peak is also found at outer points of the strait near the mouths. The existence of this peak has been already confirmed at five tide gauge stations, except Yamasedomari as shown in Fig. 2. In the actual case, this peak may not have been found at Yamasedomari because its amplitude is nearly equal to zero at the mouth of the Pacific Ocean side. The predominant periods of 307, 102, and 51 min are found out at all grids in Mutsu Bay. The peaks of period 102 min and 51 min correspond to the third and sixth harmonics of the oscillation of period 307 min. Therefore, occurrence of these peaks must be caused by resonance in Mutsu Bay.

The validity of these results is confirmed by analysis of the 1973 Off-Nemuro Peninsula tsunami at Aomori as shown in Fig. 6. Since the areas of the Tsugaru Straits and Mutsu Bay are narrow to resonate to the oscillation of period over 100 min, such oscillations can not grow there. On the other hand, the resonant oscillation of period under 100 min can be fully developed either in the strait or in the bay.

Many notable spectra are found in the range of period over 20 min, but almost all peaks of secondary undulations are overlapped with those of neighboring periods. Thus, it is difficult to distinguish one resonance from the others. To separate the tsunami wave into component waves, digital bandpass filters having unit gain and zero phase shift after the design of ORMSBY (1961) were used. As a result of calculation using this method, the resonance and its modes could be separated to a considerable extent.

Figure 7 shows a relief map of an undulation of period 307 min at 231 min after model tsunami generation. This wave progresses only into Mutsu Bay. In the Pacific Ocean and northern parts of the Japan Sea, level deviations by this wave are positive, but almost all other areas have negative levels at that time. In this case, the mode of undulation is not clear from this diagram by reason of its monotonic wave height distribution. Therefore, the level deviations at five time stages are shown in Fig. 8. The mode of oscillation can be determined by using this diagram. Contour lines in Fig. 8 between the area of positive and negative levels or boundary of the area of positive or

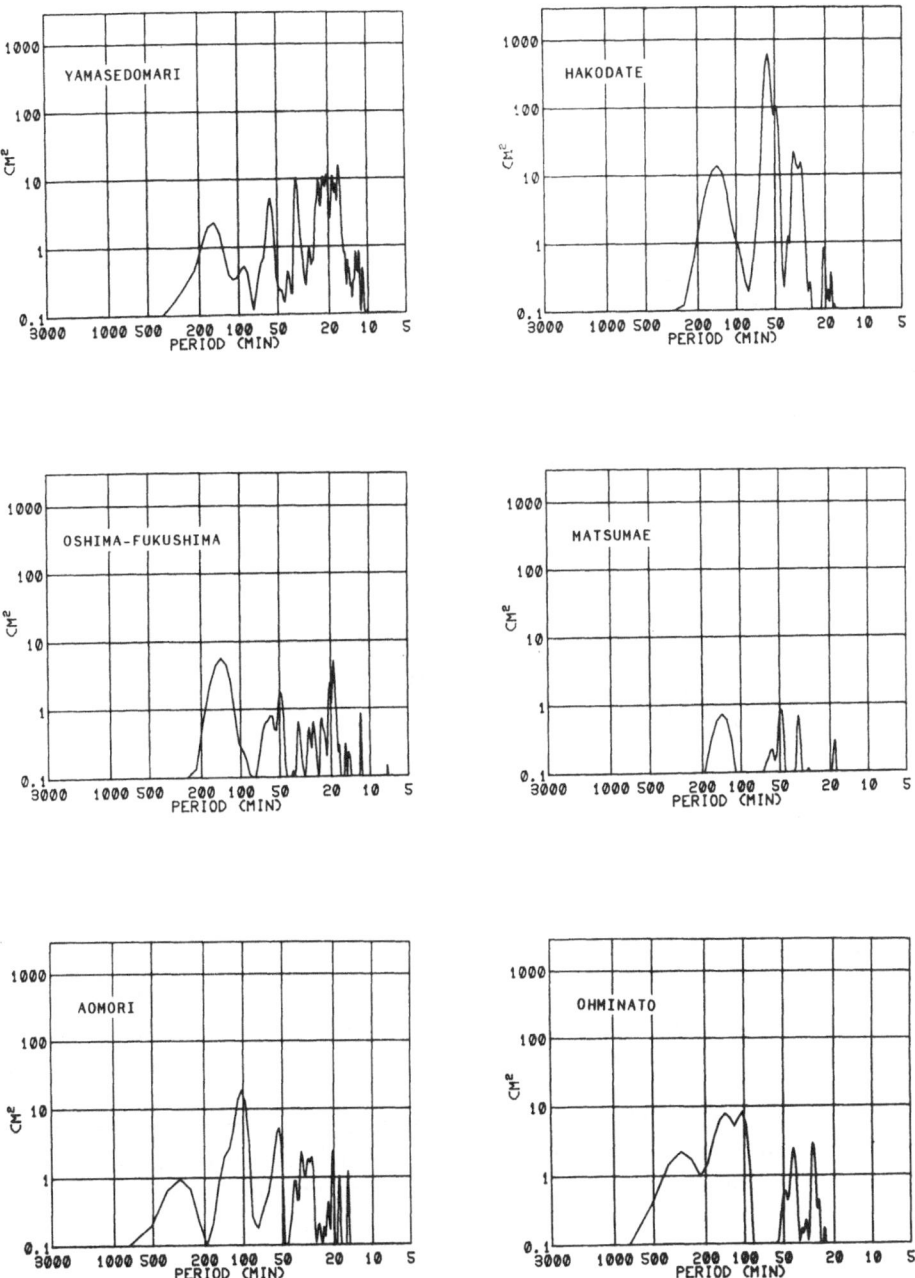

Fig. 5.   Fourier transform of weak model tsunami at the grids correspond to the six tide gauge stations.

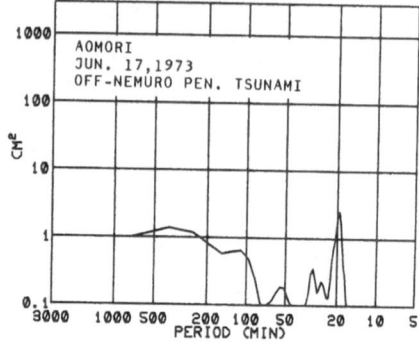

Fig. 6.   Fourier transform of the 1973 Off-Nemuro Pen. tsunami at Aomori.

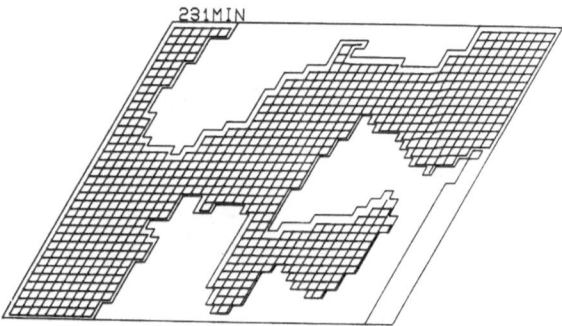

Fig. 7.   Relief of a period of 307 min (231 min after tsunami generation).

negative level mean that the level deviations are zero there. Arrows show the shifting direction of each contour line. The deviations from still water level are denoted by the symbols " + " or " − ". Symbols of " + + " and " − − " indicate large deviations in the sea level.

At a time 231 min after tsunami generation, there appear two lines. One of the lines comes into the strait from the eastern mouth and the other goes out from the strait to the Japan Sea. However, the latter case is not certain because major parts of the line do not appear on the computing domain. Then, the level rises from minus to zero in the strait and from " − − " to " − " in Mutsu Bay.

At the next stage, 285 min after tsunami generation, boundary lines advance westward from the eastern mouth through the strait. A new line approaches to the eastern mouth from the Ocean. All boundary lines that appeared at the eastern mouth and the neck of Mutsu Bay maintain their positions until the levels of both sides become equal at about 310 min after tsunami generation. Then another line intrudes into the strait from the western mouth. Accordingly, the level at the western part in the strait rises rapidly, and that at the eastern part slowly.

At about 385 min after tsunami generation, the levels in the strait and the Japan Sea again pass through zero, while in Mutsu Bay, the level is still positive. At a time

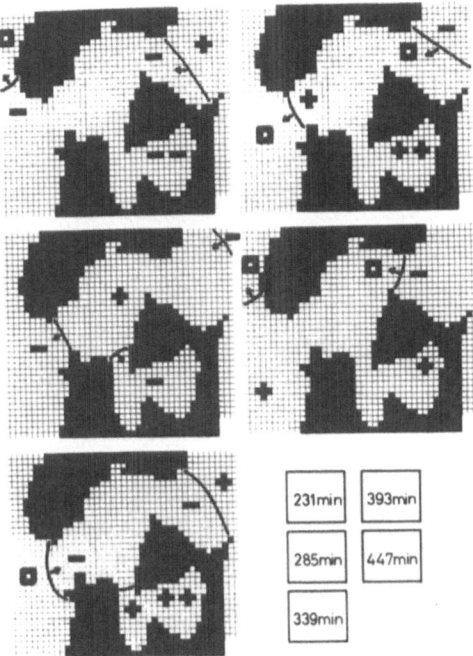

231min	393min
285min	447min
339min	

Fig. 8. Oscillation mode of a period of 307 min.

393 min after tsunami generation, a level deviation of gentle slope occupies the whole length of the strait and the two boundary lines advance westward. Distribution of the deviation is reversed to that at the time of 231 min. At this stage, a half cycle of period 307 min is over and the oscillation with the opposite phase occurs in the remaining half cycle. A maximum amplitude always appears at Ohminato.

Figure 9 shows the undulation of period 140 min. Nodal lines arise inside of the eastern mouth of the strait and the center of Mutsu Bay. The position of a line in Mutsu Bay does not change with time, while on the other hand, a line inside the eastern mouth moves periodically from east to west in the range of about 60 km.

The Pacific Ocean and the eastern part of Mutsu Bay are oscillating with inverse phase. The loops of this undulation always arise along the eastern coastline of Mutsu Bay including Ohminato, and also on a line connecting Aomori with Hakodate across the Tsugaru Straits. Other two nodal lines exist at the outer region of the eastern mouth and in the Japan Sea. The maximum amplitude always appears at Ohminato in the same way as for the case of 307 min.

An example of the undulation of period 102 min is shown in Fig. 10. This undulation corresponds to the third harmonic of the oscillation of the period 307 min and develops strongly only in Mutsu Bay. In the transitional stage of its growth, a nodal line arises inside the neck of Mutsu Bay and then gradually moves to the center of the bay with changes of direction from lateral to longitudinal. In the stationary stage, level deviations on the western side of the bay take the same phase as in the strait. The

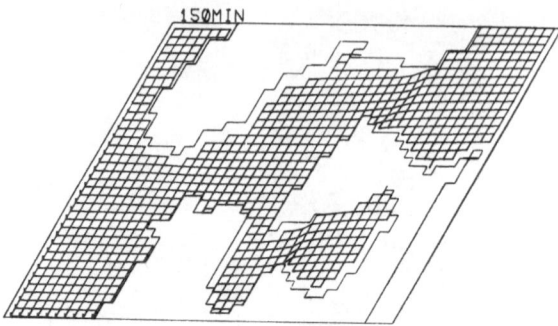

Fig. 9.   Relief of a period of 140 min (150 min after tsunami generation).

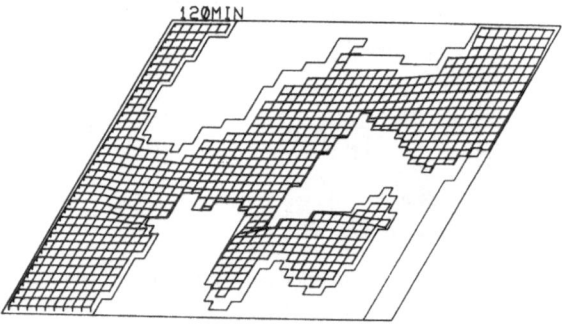

Fig. 10.   Relief of a period of 102 min (120 min after tsunami generation).

phase distribution of this oscillation in the strait occurs similarly as in the case of 140 min.

A group of the undulations in the range of periods from 46 min to 65 min is shown in Fig. 11. In Mutsu Bay, only one undulation of period 51 min occurs independently of others. This corresponds to the second mode of an undulation of period 102 min, because it has two nodal lines in the bay. In the strait, wave lengths in oscillations of period of about 60 min are nearly equal to the full length of the strait. Therefore,

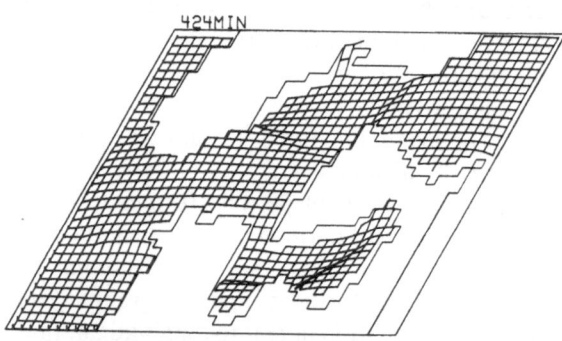

Fig. 11.   Relief of periods of 46–65 min (424 min after tsunami generation).

secondary undulations in this range may usually occur in the strait.

The results of filtering the model tsunami in the range of periods from 28 min to 45 min are shown in Fig. 12. In this range, the observed principal oscillation has a wave length equal to two-thirds of the full length of the strait and its period is about 40 min.

In the range of periods from 20 min to 28 min, the regions outside the narrow part near the eastern mouth and along the coastline of Hokkaido westward from Hakodate form a main resonant zone as shown in Fig. 13. On the other hand, in Mutsu Bay and in the western part of the strait, it appears that the undulations depend only on the location, namely the shape of the coastline and water depth.

4.   Influence of Tsunami Height on Secondary Undulations

In order to determine whether the tsunami response is influenced by change of the tsunami height with other conditions held constant, a simulation was performed using a water level elevation five times that of the weak model tsunami at the same source area. The results are shown in Table 2 in comparison with the weak model tsunami. Except for the predominant period of 102 min in Mutsu Bay, many spectral peaks of amplitudes in periods over 51 min grow to more than five times those in the weak model. In particular, one peak achieves nearly ten times that of the weak model at Oshima-Fukushima and Matsumae. At Hakodate, it grows only 3.6 times.

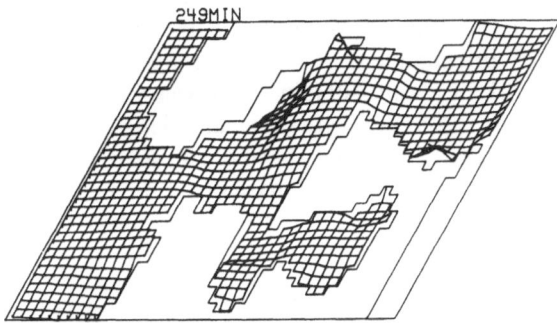

Fig. 12.   Relief of periods of 28–45 min (249 min after tsunami generation).

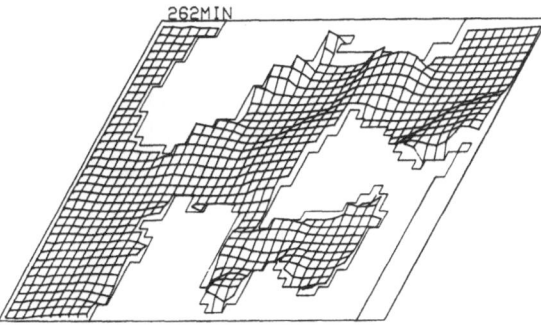

Fig. 13.   Relief of periods of 20–28 min (262 min after tsunami generation).

Table 2. The amplitude ratios of strong tsunami to weak tsunami model.

Period (MIN)	307	140	102	56	51	50	46	30
Yamasedomari		6.8						4.0
Hakodate		5.8		3.6			3.7	
Oshima-Fukushima		5.7		9.5		4.6		
Matsumae		5.6		9.5		4.5		4.3
Aomori	4.6	6.3	4.0		6.7			
Ohminato	5.8	6.0	4.1		5.0			

Oshima-Fukushima and Matsumae are located near the western mouth of the strait apart from the model tsunami source area as shown in Fig. 1. Therefore, tsunami motion at these points may be influenced by many kinds of non-linear wave interactions that occur along the strait. In the strait, the amplitudes of the fundamental secondary undulation of period 56 min can be estimated as 0.1 cm and 0.04 cm at the two points in the weak model tsunami as shown in Fig. 5. An amplitude in the strong tsunami model was estimated as nearly ten times as large as one in the weak model. This result shows that the undulation of period 56 min grew by energy transfer from the other undulations through the non-linear effect. Wave heights of peaks of periods shorter than 50 min in the strait do not reach five times those in weak model tsunami.

Figure 14 shows an example of the spectrum for the strong model tsunami at Yamasedomari. A peak of about 440 min which was not found in the weak model appears in this case. The resonant region of the undulation of such a long period may include not only the Tsugaru Straits but also more broad outer region than the case of period 307 min. The existence of this peak has not yet been confirmed by any tsunami observation record.

5. Conclusion

The following conclusions may be derived from the results developed in this paper.

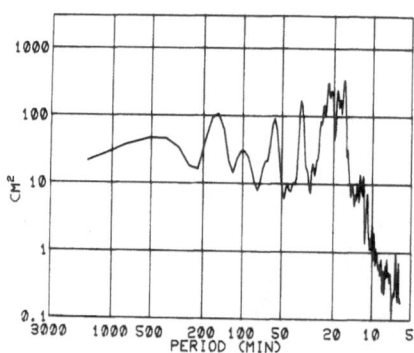

Fig. 14. Fourier transform of strong model tsunami at a grid point correspond to Yamasedomari.

1) As the results of Fourier transform of the tsunami waves in two model tsunamis in the Tsugaru Straits and Mutsu Bay, many secondary undulations having periods from 20 min to 440 min were estimated.

2) The region of growth and oscillating mode of each undulation were cleared by applying suitable digital bandpass filters to the calculated tsunami waves.

3) By comparison of the amplitude of each undulation in the weak model tsunami with one in the strong model tsunami which has a water level elevation five times as high as the former at the source area, it has become clear that the ratios of enlargement of wave height for each undulation in the strait are distributed from 3.6 to 9.5. These difference of amplitude ratios is caused by non-linear term in the equation of motion.

4) A predominant period of 440 min, that had not ever been obsreved at any tide gauge station, has been derived by the simulation for the strong model tsunami.

The authers express greatful thanks to the members of the Hokkaido Development Bureau, the Aomori Port Construction Office of the 2nd Port and Harbor Construction Bureau and the Japan Meteorological Agency for placing tsunami records at their disposal.

REFERENCES

JAPAN METEOROLOGICAL AGENCY, 1969. Report on the Tokachi-oki Earthquake, 1968, Technical report of the Japan Met. Agency, No. 68, p. 11, 1969 (in Japanese).

KAJIURA, K., T. HATORI, I. AIDA, and M. KOYAMA, 1968. A survey of the tsunami accompanying the Tokachi-Oki earthquake of May 1968, *Bull. Earthq. Res. Inst.*, **46**, 1373–1382, 1968 (in Japanese).

ORMSBY, J. F. A., Design of numerical filters with applications to missile data processing, *J. Assoc. Computing Machinery*, July, 440–466, 1961.

Tsunamis—Their Science and Engineering, edited by K. Iida and T. Iwasaki, 329–337.

Amplification of Linear Long Waves in Bays

Akira MANO

Department of Civil Angineering, Tohoku University, Sendai, 980, Japan

(Received January 13, 1982; Revised February 22, 1982)

Transient characteristics of response in bays where the depth and/or the width decrease parabolically toward the bayhead is investigated by Fourier analysis with the linear long wave approximation. The system function of the bays is derived, and it is demonstrated that the original five parameters expressing the bay topography, α, β, l, b_0, and d_0 are reduced to two, q and v. It is shown that q determines the trend, X, of the response curve and v determines the undulation on the trend. The former expresses an energy concentration effect and the latter a resonance effect.

The amplification of a transient process is estimated and calculated by the numerical integration of the Fourier transform in the linear system. It is found that the amplification of a single crested wave is approximated by $2\,XT$, where T is the transmission coefficient at the baymouth. Furthermore, the process of reaching the steady state at the fundamental resonant mode is obtained. For higher frequency regions the energy concentration effect is shown to become dominant.

1. Introduction

When tsunami waves intrude into bays such that the depth and/or the width decrease toward the bayhead, the wave height usually increases remarkably. It is well known that the mechanism of amplification is mainly due to energy concentration and resonance. The former has been discussed by Green's law. However, this law cannot be applied to a bayhead where the depth or the width diminishes to zero. Furthermore, application of this law should be modified to taking into account the frequency of the input waves. Concerning resonance, there are many solutions of the steady state oscillations for bays with various topography. However, in many cases, tsunami records indicate that the maximum height appears at first or the second wave. Since resonance appears when infinite sets of reflected wave are superimposed in phase, it is doubtful whether the solutions can be directly applied to tsunamis.

In this study, transient response characteristics of tsunamis are investigated for bays where the depth and the width decrease parabolically toward the bayhead.

2. Bay Model and System Function

Let us consider a one-dimensional bay connected with a semi-infinite open sea of uniform depth, d_0. The depth and the width of the bay are assumed to be $d = d_0 (x/l)^\alpha$ and $b = 2 b_0 (x/l)^\beta$ respectively, where x is taken from the bayhead $x = 0$, to the baymouth $x = l$ (see Fig. 1). Inside the bay the general solution for harmonic oscillation given by HOMMA (1933) is rewritten as follows:

$$\zeta(\omega^*; x^*) = x^{*pq}[BH_q^{(1)}(\omega^*x^{*p}) + CH_q^{(2)}(\omega^*x^{*p})] \tag{1}$$

where ζ is the amplitude, $\omega^* = \omega T_0$, ω is the angular frequency, $T_0 = l/p\sqrt{gd_0}$ which indicates the propagation time between the baymouth and the bayhead, $x^* = x/l$, $p = 1 - (\alpha/2)$, $q = (1 - \alpha - \beta)/(2 - \alpha)$ for $\alpha \neq 2$, g is the gravitational accaleration, B and C are integration constants to be determined by the boundary conditions, and $H_q^{(1)}$ and $H_q^{(2)}$ are Hankel functions of q th order.

Applying the matching condition at the baymouth by IPPEN and GODA (1963), which is based on the averaged amplitude just outside the baymouth, and the perfect reflection condition at the bayhead, the solution for normal incident wave with unit amplitude is derived:

$$\zeta(\omega^*; x^*) = \frac{2x^{*pq}J_{-q}(\omega^*x^{*p})}{J_{-q}(\omega^*) - J_{1-q}(\omega^*)\psi_2 + iJ_{1-q}(\omega^*)\psi_1} \tag{2}$$

where J_{-q} and J_{1-q} are Bessel functions of the $-q$th and $(1 - q)$th order respectively, and

$$\psi_1 = \psi_1(v\omega^*) = \frac{2}{\pi}v\omega^*\int_0^{v\omega^*}\frac{\sin^2 u}{u^2\sqrt{v^2\omega^2 - u^2}}du \tag{3}$$

$$\psi_2 = \psi_2(v\omega^*) = \frac{2}{\pi}v\omega^*\int_{v\omega^*}^{\infty}\frac{\sin^2 u}{u^2\sqrt{u^2 - v^2\omega^2}}du \tag{4}$$

with

$$v = pb_0/l. \tag{5}$$

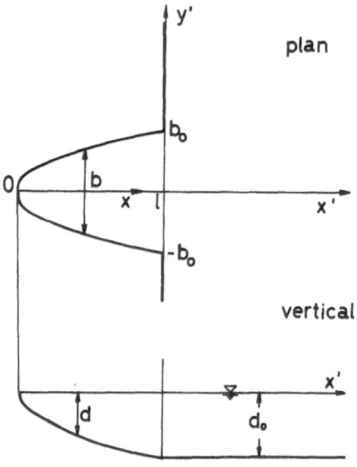

Fig.1.   Bay model.

The system function, $H(\omega^*; x^*)$, which expresses the complex amplification at x^*, has the same form as Eq. (2). At the bayhead,

$$H(\omega^*;0) = \lim_{x^* \to 0} H(\omega^*;x^*)$$

$$= \frac{2}{\left(\dfrac{\omega^*}{2}\right)^q \Gamma(1-q)[J_{-q}(\omega^*) - J_{1-q}(\omega^*)\psi_2 + i J_{1-q}(\omega^*)\psi_1]} \tag{6}$$

where Γ is the gamma function. This function describes bays with typical topographies, such as, rectangular ($\beta = 0$), U-shaped ($\beta = 0.5$), V-shaped ($\beta = 1$), and uniform depth ($\alpha = 0$), uniformly sloping depth ($\alpha = 1$), etc., which comes from the generality of Homma's solution (1).

Equation (6) has one independent variable, ω^*, and two parameters, q and v. The original parameters α, β, l, d_0, and b_0 which express the bay topography are now reduced to two.

The variable ω^* is a dimensionless angular frequency and agrees with the characteristic variables of bays or harbours which have been obtained before; kl for rectangular harbours, $4\pi l/L$ for harbours with uniform bottom slope, where k and L are the wave number and the wavelength, respectively, in an open sea. This expression of ω^* is more useful, because the frequency of the input waves is considered to be approximately independent of the depth.

The parameter q is a function of α and β which define the topography inside bays. It does not include the condition at baymouth which relates to resonance. As seen from Eq. (6), it determines the trend of the system function. Many bays with one value of q has same trend, for example $q = 0$ includes cases of $\alpha = 1$, $\beta = 0$; $\alpha = 0$, $\beta = 1$ and $\alpha = 0.5$, $\beta = 0.5$. The parameter v is a function of b_0/l and p, that is, α and determines the undulation on the trend.

Figure 2 shows the modulus of the system function for $q = 0$ and for $v = 0.1, 0.2$, 0.4, and $v \to \infty$. These curves have a linear trend on a log-log plot for large ω^*. The difference due to v appears most remarkably at the resonant peak value, E_r, and also at the resonant frequency, ω_r^*. With decrease of v, the peak grows and the frequency

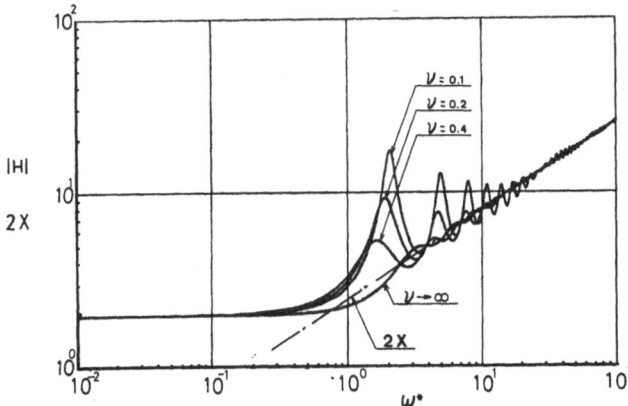

Fig.2.   Amplification curves in steady state oscillation for $q = 0$ and for $v = 0.1, 0.2, 0.4, \infty$ and the trend $2X$.

decreases.

Figure 3 showing the case for $q = 0.5$ represents similar characteristics. The slope of the trend is 0.

In the high frequency region $\omega^* \gg 1$, Eq. (6) has two asymptotic forms;

$$H(\omega^*;x^*) \simeq 2x^{*p(q-\frac{1}{2})}\cos\left[\omega^*x^{*p} + \frac{(2q-1)\pi}{4}\right]e^{-i\theta} \quad \text{for } \omega^*x^{*p} \gg 1 \qquad (7)$$

$$H(\omega^*;x^*) \simeq 2X(\omega^*)\left[1 - \frac{\omega^{*2}x^{*2p}\Gamma(1-q)}{4\Gamma(2-q)}\right]e^{-i\theta} \quad \text{for } \omega^*x^{*p} \ll 1 \qquad (8)$$

where

$$X(\omega^*) = \left(\frac{2}{\omega^*}\right)^{q-1/2}\pi^{1/2}/\Gamma(1-q) \qquad (9)$$

$$\theta(\omega^*) = \omega^* + (2q-i)\pi/4. \qquad (10)$$

Since the amplitude in Eq. (7), $x^{*p(q-\frac{1}{2})}$, can be rewritten as $x^{*(-\frac{\alpha}{4}-\frac{\beta}{2})}$, this equation represents the amplification by Green's law. That is, the law dominates for $x^* \gg 1/\omega^{*1/p}$, and the applicable region of the law enlarges to the bayhead with increase of ω^*. However, near the bayhead Eq. (8) dominates and the trend mentioned above is expressed by $2X(\omega^*)$. In these two equations, the phase of the system function, θ, is linear in ω^* with a phase shift of $(2q-1)\pi/4$. This means that a phase deformation does not occur. The factor two in the equations indicates amplification by perfect reflection at the bayhead. From the reasons mentioned above, $X(\omega^*)$ can be called energy concentration function at the bayhead, In particular, for $q = 0.5$, which includes a rectangular bay, $X(\omega^*)$ is unity and energy concentration does not occur.

The resonant frequency ω^*_r is given by the root of the real part of the denominator in Eq. (6),

$$J_{-q}(\omega^*_r) - J_{1-q}(\omega^*_r)\psi_2(\omega^*_r) = 0. \qquad (11)$$

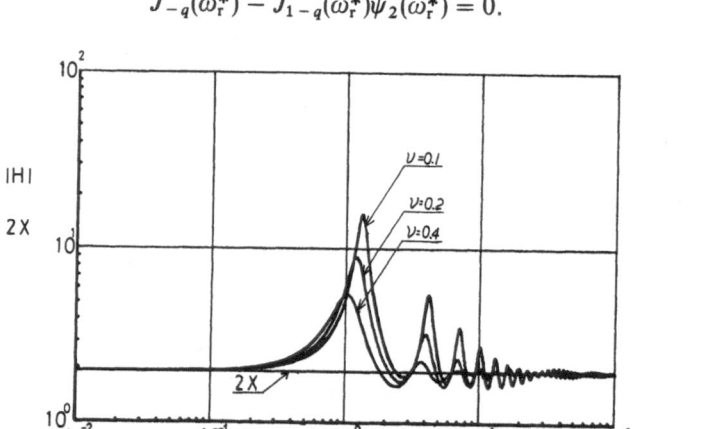

Fig. 3.　Amplification curves in steady state oscillation for $q=0.5$ and for $v=0.1, 0.2, 0.4$ and the trend $2X$.

Therefore, the resonant peak value, E_r, is given by,

$$E_r = \frac{2}{(\omega_r^*/2)^q \Gamma(1-q) J_{1-q}(\omega_r^*) \psi_1(v\omega_r^*)}. \tag{12}$$

For the region $v\omega_r^* \gg 1$, $\psi_1(v\omega_r^*)$ is approximated as from IPPEN and GODA (1963),

$$\psi_1(v\omega_r^*) \simeq v\omega_r^* + O(v^3\omega_r^{*3}). \tag{13}$$

Applying this approximation to Eq. (12),

$$E_r \simeq \frac{2}{v(\omega_r^*/2)^{q+1} \Gamma(1-q) J_{1-q}(\omega_r^*)}. \tag{14}$$

Since the variation of ω_r^* with v is small, the resonant peak value is inversely proportional to the parameter v. That is, v determines the resonant effect.

The function $X(\omega^*)$ with the condition $q < 0.5$ diverges as $\omega^* \to \infty$. Since this analysis is based on the assumption of linear long waves, this result can not be applied to such large ω^*.

Now consider a necessary condition of the long wave assumption, that the incident wave is a long wave $d_0/L < 1/25$. Using the relation $\omega^* = lk/p$,

$$\omega^* < 0.25 \, (l/d_0)/p. \tag{15}$$

For example, considering Ryori Bay on the Sanriku coast of Japan ($l \simeq 4{,}000$ m, $d_0 \simeq 50$ m, $\alpha \equiv 1$; $p = 0.5$), this condition is written as $\omega^* < 40$. This restriction is usually satisfied by tsunami waves.

As for the linear wave assumption, the ratio of the wave height to the depth might be discussed. However, one value of q does not determine a unique depth because q is a function of both α and β. Further discussion of limitation conditions should be made for individual bays with their respective values of α and β.

3. Amplification in the Transient Process

We discuss the amplification of the first crest and the amplifying process at the fundamental resonant mode. As a first step, we introduce an approximate solution from the results in the previous section, then compare them with the numerical solution of the Fourier integral. In order to obtain the frequency characteristics, the input wave to the baymouth, $f(t^*)$, is assumed to be of the form,

$$f(t^*) = \begin{bmatrix} \sin(\omega_j^* t^*) & \text{for } 0 \leqq t^* \leqq T_c^* \\ 0 & \text{otherwise} \end{bmatrix} \tag{16}$$

where the star superscript indicates a quantity non-dimensionalized by T_0, t^* is the time, $\omega_j^* = 2\pi/T_j^*$, ω_j^* and T_j^* are the frequency and the period, respectively, of the input wave, $T_c^* = N T_j^*/2$ is the duration time and N as integer. Furthermore, $g(t^*)$ denotes the output at bayhead, and g_{2m-1} and g_{2m} are the amplitude of the crest and the trough, respectively, in the mth cycle.

Wave amplification in the bays is caused mainly by energy consentration, reflection at the bayhead, and reflection and transmission at the baymouth. Suppose that the function $X(\omega^*)$ represents the amplification between the baymouth and the bayhead and ignore the partial reflection inside the bay. Applying the transmission coefficient, T, at the baymouth defined by NISHIMURA and HORIKAWA (1973), the amplification of the first crest at the bayhead, g_1, is estimated by

$$g_1 = 2XT \qquad \text{for } T_j^* \leq, \text{ that is } \omega_j^* \geq 2\pi/8 \simeq 0.8 \qquad (17)$$

where $T = 2/\sqrt{(1+\psi_1)^2 + \psi_2^2}$, and the restriction implies the condition that the first crest is not affect by the re-reflected wave at the baymouth. Since $T \to 1$ as $\omega^* \to \infty$, g_1 also approaches $|H_0(\omega^*;0)|$ as $\omega^* \to \infty$. The greatest difference between g_1 and $|H_0(\omega^*; 0)|$ appears at the fundamental resonant mode. At this mode the phase of the crest which propagates directly through the baymouth agrees with the phase of the crest which reflects at the baymouth. Introducing the attenuation coefficient of free oscillation, Q, then g_n can be written,

$$|g_n| = 2XT + |g_{n-1}|\exp(-\pi/2Q). \qquad (18)$$

For large n, as $|g_n| = |g_{n-1}| = E_r$, Q is given by

$$Q = \frac{\pi}{21 \, n \, (E_r/(E_r - 2XT))}. \qquad (19)$$

Applying the initial condition, Eq. (17), the recurrence relation, Eq. (18) is solved

$$|g_n/E_r| = 1 - \exp(-N\pi/2Q). \qquad (20)$$

Accurate output is obtained by the Fourier transform applying the well-known relationship in linear systems

$$G(\omega^*) = H(\omega^*)F(\omega^*) \qquad (21)$$

where $F(\omega^*)$ and $G(\omega^*)$ are the spectra of $f(t^*)$ and $g(t^*)$ respectively. Using Eqs. (16) and (21), the output $g(t^*)$ is derived

$$g(t^*) = \frac{2}{\pi}\int_0^\infty E(\omega^*)\left[\frac{\sin(\omega^* - \omega_j^*)T_c^*}{\omega^* - \omega_j^*}\sin(\phi(\omega^*) - (\omega^* - \omega_j^*)T_c^*)\right.$$
$$\left. - \frac{\sin(\omega^* + \omega_j^*)T_c^*}{\omega^* + \omega_j^*}\sin(\phi(\omega^*) - (\omega^* + \omega_j^*)T_c^*)\right]\cos(\omega^* t^*)d\omega^* \qquad (22)$$

where $E(\omega^*)$ and $\phi(\omega^*)$ are the modulus and phase, respectively, of the system function $H(\omega^*)$. Substituting the function given by Eq. (6), the integration is carried out numerically. Since the order of $H(\omega^*)$ here is $O(\omega^{*-q+1/2})$ and $F(\omega^*)$ is $O(\omega^{*-2})$ for large ω^*, the integration converges for $q > -1/2$.

Figure 4 shows the amplification of the first crest for the case of $q = 0.5$ and $v = 0.2$. Black circles g_1 are the numerical results given by Eq. (22) and the broken line is $2XT$ in Eq. (17). For the purpose of comparision, the modulus of the system function, $|H(\omega^*;0)|$, is also drawn as a solid curve. The numerical results agree with $2XT$ for $\omega^* > 0.8$ and approach to $|H(\omega^*; 0)|$ for $\omega^* < 0.8$.

Figure 5 shows the case for $q = 0$ and $v = 0.1$. The numerical results are approximated by $2XT$ and the difference decreases with increase of ω^*.

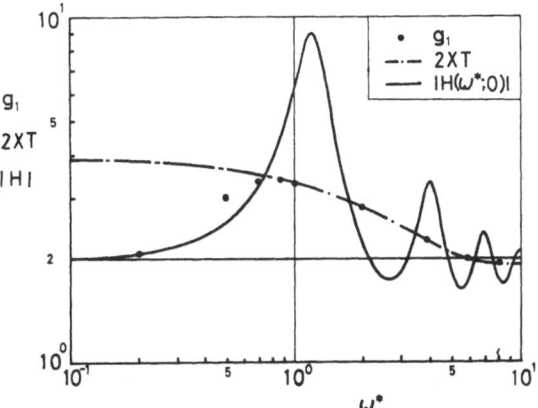

Fig. 4. Amplification of the first crest for $q = 0.5$ and $v = 0.2$.

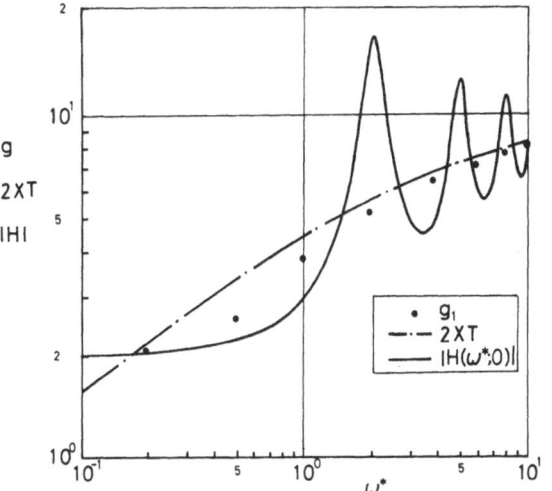

Fig. 5. Amplification of the first crest for $q = 0$ and $v = 0.1$.

Figure 6 shows the resonant process at the fundamental resonant mode for $q = 0.5$ and $v = 0.2$. Here black circles are the numerical results for the forced oscillation, $g_n (n \leqq N)$, and the symbol $+$ indicates the right side of Eq. (20). Good agreement is obtained. Free oscillations $g_n (n > N)$ are also drawn by white circles. For example, consider $N = 2$, that is, the case of an input with one crest and one trough; it is seen that significant amplification appears at the first crest of forced oscillation and at the second crest of free oscillations and they are comparable. Figure 7 shows the case for $q = 0$ and $v = 0.2$. The numerical results are approximated by Eq. (20). These comparisons comfirm that the function $X(\omega^*)$ represents the approximate amplification between the baymouth and the bayhead.

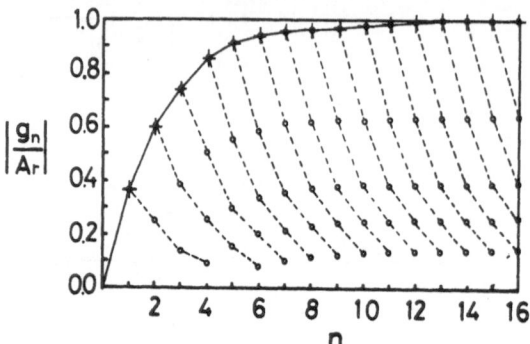

Fig. 6. Amplification of the resonant process at the fundamental resonant mode for $q = 0.5$ and $v = 0.2$.
●, numerical results g_n; +, Eq. (20); and ○, free oscillation.

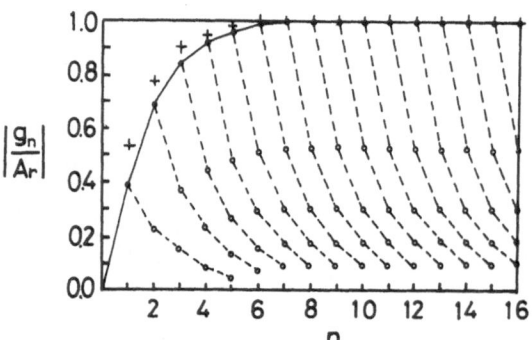

Fig. 7. Amplification of the resonant process at the fundamental resonant mode for $q = 0$ and $v = 0.2$.
●, numerical results g_n; +, Eq. (20); and ○, free oscillation.

4. Conclusions

The results obtained in this paper are as follows;

1) The system function of bays where the depth and/or the width decrease parabolically toward the bayhead is derived. It is found that the five parmeters expressing the bay topography, α, β, l, b_0, and d_0 are reduced to two, q and v. This means that many bays with the same q and v are equivalent in linear long wave theory.

2) The parameters q and v represent the energy concentration effect and the resonant effect respectively.

3) Inside a bay, Green's law dominates for $x^* \gg 1/\omega^{*1/p}$, and $2X(\omega^*)$ for $x^* \ll 1/\omega^{*1/p}$.

4) At the bayhead, amplifications for an incident wave with a single crest is approximated by $2 XT$ for $\omega^* > 0.8$, and approaches to $|H(\omega^*; 0)|$ for $\omega^* < 0.8$.

5) For higher frequency regions, the energy concentration effect dominates and the difference between the steady state oscillation and the transient oscillation is small.

6) The greatest difference appears at the fundamental resonant mode. The process of achieving steady state by continuous forcing oscillations is approximated by Eq. (20).

The author wishes to express his sincere gratitude to Professor Toshio Iwasaki for his valuable guidance.

REFERENCES

HOMMA, M., On the transformations of long wave, J. Japan Soc. Civil Eng., **9**, 1933 (in Japanese).
IPPEN, A. T. and Y. GODA, Wave induced oscillations in harbors, Rep. No. 59, Hydrodynamics Lan., M. I. T., 1963.
NISHIMURA, H. and K. HORIKAWA, On the boundary conditions at the bay entrance in the analyses of bay oscillation, *Coastal Ang. Japan*, **16**, 29–39, 1973.

Tsunamis—Their Science and Engineering, edited by K. Iida and T. Iwasaki, 339–358.

Seiches in Bays Forming a Coupled System

Masito NAKANO and Naohiro FUJIMOTO

*Faculty of Marine Science and Technology, Tokai University, Orido,
Shimizu-shi, Shizuoka-ken, Japan*

(Received August 25, 1981; Revised November 19, 1981)

The seiches occurring in two adjacent bays, Moroiso Bay and Koaziro Bay, situated in Miura Peninsula at the mouth of Tokyo Bay, are very regular, and the phenomenon of "beat" appears in their amplitudes of oscillation. Concerning this phenomenon, one of the authors presented a theoretical explanation by considering two adjacent rectangular bays of the same size and form as well as of uniform depth (NAKANO, 1932). It was also assumed that a kind of coupling takes place between the two bays through a portion of water flowing across the mouth-line of each bay. In the present paper, an extension of the previous theory has been made, and the validity of the theory has been proved in a series of hydraulic model experiments. According to the extended theory, the "coupling coefficient" $\sigma(0 < \sigma < 1)$, which is a non-dimensional parameter representing the degree of coupling between the two bays, is expressed by

$$\sigma^2 = \frac{4\pi^2}{g}\frac{Ld}{h}\frac{T_s}{T_bT_p^2}\frac{1}{\sin^2\dfrac{2\pi L}{T_pc}}\left\{1 - \left(\frac{T_pc}{4\pi L}\sin\frac{4\pi L}{T_pc}\right)^2\right\}^{1/2},$$

while the corresponding expression in the previous theory (which is applicable to a special case where the length of the long waves, λ, incident from the open sea is approximately equal to four times the length, L, of the bays) is

$$\sigma^2 = \frac{4\pi^2}{g}\frac{Ld}{h}\frac{T_s}{T_bT_p^2}.$$

Here g, the acceleration due to gravity; d, the distance between the two bays; h, the depth of water; T_s, the period of free oscillation of bay water, viz. $(4L/\sqrt{gh})f$, f being the mouth-correction factor; T_p, the period of the long waves coming from the open sea; T_b, the period of beats in the amplitudes of the seiches; c, the velocity of a long wave, viz. $\sqrt{gh}$. In addition some discussions have been made on the respective relations between the coupling coefficient σ and each of the quatities T_p, $\psi(= T_s/T_p)$ and d, based upon the data of the hydraulic model experiments.

1. Introduction

It is a well-known fact that when two pendulums connected with a spring oscillate, a beat phenomenon occurs. The amplitude of oscillation of each pendulum changes periodically due to fluctuation of energy between them through the spring. Therefore, it is not unreasonable to conjecture that an analogous phenomenon may occur in the case of bays considered as "liquid pendulums." Thus, when two bays exist side by side, the mode of oscillation of a single bay may be affected by the presence of the other closely situated bay, and a beat phenomenon may be expected to occur in its amplitude of oscillation.

A good example of this phenomenon is afforded by the case of seiches in two adjacent bays, Koaziro Bay and Moroiso Bay (Moroiso-Aburatubo Bay), situated in Miura Peninsula at the mouth of Tokyo Bay. Figure 1(a) shows a map of these bays, and Fig. 1(b) shows a result of simultaneous observations of seiches at Koaziro and Aburatubo, situated near the heads of the respective bays (NISINA, 1930). Looking at Fig. 1(b), we can recognize that, as Nisina has pointed out, the seiches in both bays are not only regular, but also the variations of their amplitudes are such that, while the one becomes weak, the other becomes strong, and vice versa. Concerning this pheno-menon, NAKANO (1932) once gave a theoretical explanation, considering two adjacent rectangular bays of the same size and form as well as of equal depth by supposing that a kind of coupling takes place between the two bays through a part of water flowing across the mouth-line of each bay. In the present paper, an extension of the previous theory has been made, and its validity has been proved by a series of hydraulic model experiments.

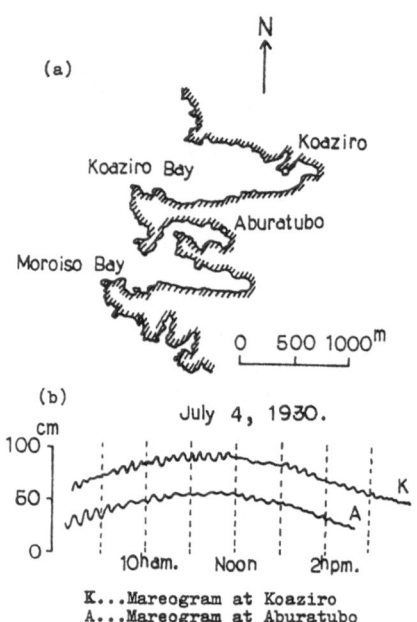

Fig. 1. (a) Koaziro Bay and Moroiso Bay. (b) Mareograms recorded simultaneously at Koaziro and Aburatubo (after N. NISINA, 1930).

2. Theoretical Considerations

First, consider a rectangular bay of length L, breadth b and of uniform depth h. Let the x-axis be taken along the axis of the bay, the z-axis vertically upwards, and the origin of the co-ordinates on the still water level at the mouth of the bay. Now, suppose that a train of long waves proceeds from the open sea toward the bay as well as the neighbouring coasts, and that there is a periodic rise and fall of the sea surface which can be expressed by $Z \cos{(at + \alpha)}$ at the mouth of the bay. The equations of motion and continuity to be solved in this case are, in the usual notation,

$$\frac{\partial^2 \xi}{\partial t^2} = c^2 \frac{\partial^2 \xi}{\partial x^2} \quad (c = \sqrt{gh}) \tag{1}$$

and

$$\zeta = -h \frac{\partial \xi}{\partial x}, \tag{2}$$

respectively, together with the boundary conditions:

$$\left. \begin{array}{l} (\zeta)_{x=0} = Z\cos(at + \alpha), \\ (\xi)_{x=L} = 0. \end{array} \right\} \tag{3}$$

The solutions of (1) and (2), satisfying (3), are

$$\xi = -Z \frac{T_p c}{2\pi h \cos\dfrac{2\pi L}{T_p c}} \cdot \sin\frac{2\pi}{T_p c}(x - L) \cdot \cos\left(\frac{2\pi}{T_p}t + \alpha\right), \tag{4}$$

$$\zeta = \frac{Z}{\cos\dfrac{2\pi L}{T_p c}} \cdot \cos\frac{2\pi}{T_p c}(x - L) \cdot \cos\left(\frac{2\pi}{T_p}t + \alpha\right), \tag{5}$$

where $T_p \, (= 2\pi/a)$ denotes the period of the long waves coming from the open sea, and $T_p c \, (= \lambda)$ is their wave length. Equations (4) and (5) represent the stationary oscillations or seiches occurring in the rectangular bay under consideration, and these equations correspond, respectively, to

$$\xi_i = B\cos lx \cdot \sin{(at + \alpha)} \tag{N1}$$

and

$$\zeta_i = Bhl \sin lx \cdot \sin{(at + \alpha)} \tag{N2}$$

in the previous theory.

Now, we write, for convenience, ξ_i and ζ_i for ξ and ζ, respectively, in Eqs. (4) and (5), then the kinetic and potential energies in the bay are expressed by

$$\text{K.E.,} \; \mathcal{T}_i = \frac{1}{2}\rho bh \int_0^L \xi_i^2 \, dx = \frac{1}{4}\rho bhL \cdot \frac{1}{\sin^2\dfrac{2\pi L}{T_p c}} \cdot \left(1 - \frac{T_p c}{4\pi L}\sin\frac{4\pi L}{T_p c}\right) \cdot \xi_0^2, \tag{6}$$

$$\text{P.E.,}\ \mathscr{V}_i = \frac{1}{2}\rho bg \int_0^L \zeta_i^2\, dx = \frac{1}{4}\rho bgh^2 l^2 L \cdot \frac{1}{\sin^2\dfrac{2\pi L}{T_p c}} \cdot \left(1 + \frac{T_p c}{4\pi L}\sin\frac{4\pi L}{T_p c}\right)\cdot \xi_0^2, \qquad (7)$$

where ρ is the density of water, l is $2\pi/\lambda$ viz. $(2\pi/T_p c)$, and ξ_0 is the value of ξ_i at $x = 0$ viz. at the mouth of the bay.

Again, the kinetic energy, $\mathscr{T}_a$, outside the bay can be expressed by

$$\mathscr{T}_a = Q\cdot\mathscr{T}_i = \frac{1}{4}\rho bhL \cdot \frac{1}{\sin^2\dfrac{2\pi L}{T_p c}} \cdot \left(1 - \frac{T_p c}{4\pi L}\sin\frac{4\pi L}{T_p c}\right)\cdot Q\xi_0^2, \qquad (8)$$

where Q is a non-dimensional quantity (Honda et al., 1908). The potential energy, $\mathscr{V}_a$, outside the bay is small in comparison with $\mathscr{V}_i$, so it can be neglected for the first approximation. Therefore, considering that the regions inside and outside the bay constitute together a mechanical system, the total kinetic and potential energies of the system are given by

$$\mathscr{T} = \mathscr{T}_i + \mathscr{T}_a = \frac{\rho bhL}{4\sin^2\dfrac{2\pi L}{T_p c}} \cdot \left(1 - \frac{T_p c}{4\pi L}\sin\frac{4\pi L}{T_p c}\right)\cdot (1 + Q)\cdot \xi_0^2, \qquad (9)$$

$$\mathscr{V} = \mathscr{V}_i + \mathscr{V}_a = \frac{\rho bgh^2 l^2 L}{4\sin^2\dfrac{2\pi L}{T_p c}} \cdot \left(1 + \frac{T_p c}{4\pi L}\sin\frac{4\pi L}{T_p c}\right)\cdot \xi_0^2. \qquad (10)$$

Next, consider the case where two rectangular bays of same size and form as well as of equal depth exist side by side with mutual distance $d(= EF)$ as shown in Fig. 2. For simplicity suppose that the depth of the neighbouring ocean is constant and is equal to the depth of the bays, h, and denote the quantities corresponding to the two

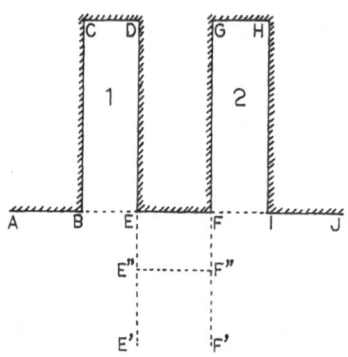

Fig. 2.   Two adjacent rectangular bays.

bays by suffixes 1 and 2, respectively. Further, put $bh\xi_{10} = X_{10}$, $bh\xi_{20} = X_{20}$, then we have

$$
\left.
\begin{aligned}
\mathcal{T}_1 &= \frac{\rho L}{4bh\sin^2\dfrac{2\pi L}{T_p c}} \cdot \left(1 - \frac{T_p c}{4\pi L}\sin\frac{4\pi L}{T_p c}\right) \cdot (1 + Q) \cdot \dot{X}_{10}^2, \\[2ex]
\mathcal{V}_1 &= \frac{\rho g l^2 L}{4b\sin^2\dfrac{2\pi L}{T_p c}} \cdot \left(1 + \frac{T_p c}{4\pi L}\sin\frac{4\pi L}{T_p c}\right) X_{10}^2, \\[2ex]
\mathcal{T}_2 &= \frac{\rho L}{4bh\sin^2\dfrac{2\pi L}{T_p c}} \cdot \left(1 - \frac{T_p c}{4\pi L}\sin\frac{4\pi L}{T_p c}\right) \cdot (1 + Q) \cdot \dot{X}_{20}^2, \\[2ex]
\mathcal{V}_2 &= \frac{\rho g l^2 L}{4b\sin^2\dfrac{2\pi L}{T_p c}} \cdot \left(1 + \frac{T_p c}{4\pi L}\sin\frac{4\pi L}{T_p c}\right) \cdot X_{20}^2.
\end{aligned}
\right\}
\tag{11}
$$

Now, we shall consider the mechanism of coupling between the two bays in question. The coupling will be made through the water in the vicinity of the coast EF in Fig. 2. For example, in the case when the water surfaces at CD and GH fall simultaneously, a sensible rise of water level will occur in the region EFF'E' due to the water flowing out the bays across BE and FI, and, on the contrary, if the water surfaces at CD and GH rise simultaneously, a sensible fall of water level will occur in this region. This *extra* rise and fall, so to speak, of water level near the coast EF plays the same role as the said spring connecting the two pendulums. To simplify the problem, we shall introduce a model as follows. First, consider an imaginary vertical wall fixed at E″F″ which is at some finite distance, b for example, from the coast EF, and suppose that two pistons devoid of inertia are placed at EE″ and FF″, and that the coupling action in question is restricted within the region EFF″E″. Further suppose that the vertical displacement of the water surface over this region is uniform. Now, when a quantity X_0 of water flows across BE or FI, suppose that a certain portion of it, say $\sigma \cdot X_0$ ($0 < \sigma < 1$), participates in the coupling, in other words, a quantity σX_0 of water flows into or out from the region EFF″E″, moving the piston EE″ or FF″, resulting in a uniform extra rise or fall of water surface in that region. The quantity σ is a constant proper to the mechanical system under consideration, and is provisionally called the "coupling coefficient." Thus, if we denote the elevation of water surface occurring in the region EFF″E″ by z, we have

$$
bdz = -\sigma(X_{10} + X_{20}). \tag{12}
$$

Hence, the potential energy $\mathcal{V}_{12}$ due to the elevation or depression of water surface in this region is expressed by

$$\mathscr{V}_{12} = \int_0^z bd\rho gz \, dz = \frac{\rho g\sigma^2}{2bd}(X_{10} + X_{20})^2.$$ (13)

Again, since the kinetic energy has a quadratic form in the velocities, we may assume that the kinetic energy $\mathscr{T}_{12}$ in this region can be written in the form $\mathscr{T}_{12} = (\mu_1/2)\dot{X}_{10}^2 + (\mu_2/2)\dot{X}_{20}^2 + \mu\dot{X}_{10}\dot{X}_{20}$, where μ_1, μ_2 and μ are to be considered as functions of σ, and in the present case where the two bays are quite equal, $\mu_1 = \mu_2(\equiv \mu')$. Accordingly,

$$\mathscr{T}_{12} = \frac{\mu'}{2}(\dot{X}_{10}^2 + \dot{X}_{20}^2) + \mu\dot{X}_{10}\dot{X}_{20}.$$ (14)

Hence the total kinetic and potential energies of the coupled system under consideration are as follows:

$$\text{K.E., } \mathscr{T} = \mathscr{T}_1 + \mathscr{T}_2 + \mathscr{T}_{12} = \frac{\rho L}{4bh\sin^2\dfrac{2\pi L}{T_p c}} \cdot \left(1 - \frac{T_p c}{4\pi L}\sin\frac{4\pi L}{T_p c}\right).$$

$$(1 + Q)(\dot{X}_{10}^2 + \dot{X}_{20}^2) + \frac{\mu'}{2}(\dot{X}_{10}^2 + \dot{X}_{20}^2) + \mu\dot{X}_{10}\dot{X}_{20},$$ (15)

$$\text{P.E., } \mathscr{V} = \mathscr{V}_1 + \mathscr{V}_2 + \mathscr{V}_{12} = \frac{\rho gl^2 L}{4b\sin^2\dfrac{2\pi L}{T_p c}} \cdot \left(1 + \frac{T_p c}{4\pi L}\sin\frac{4\pi L}{T_p c}\right)(X_{10}^2 + X_{20}^2)$$

$$+ \frac{\rho g\sigma^2}{2bd}(X_{10} + X_{20})^2.$$ (16)

Suppose that the present system is an isolated one, namely, that its total energy is constant. Then the Lagrangian equation $d/dt(\partial\mathscr{T}/\partial\dot{X}_{\kappa 0}) - \partial\mathscr{T}/\partial X_{\kappa 0} = -\partial\mathscr{V}/\partial X_{\kappa 0}$ ($\kappa = 1,2$) gives the following simultaneous equations,

$$\left.\begin{array}{l} M'\ddot{X}_{10} + \mu\ddot{X}_{20} + N'X_{10} + \nu X_{20} = 0, \\ M'\ddot{X}_{20} + \mu\ddot{X}_{10} + N'X_{20} + \nu X_{10} = 0, \end{array}\right\}$$ (17)

where

$$\left.\begin{array}{l} M' \equiv M + \mu' \equiv \dfrac{\rho L}{2bh\sin^2\dfrac{2\pi L}{T_p c}} \cdot \left(1 - \dfrac{T_p c}{4\pi L}\sin\dfrac{4\pi L}{T_p c}\right) \cdot (1 + Q) + \mu', \\[6mm] \nu = \dfrac{\rho g}{bd}\sigma^2, \\[6mm] N' \equiv N + \nu \equiv \dfrac{\rho gl^2 L}{2b\sin^2\dfrac{2\pi L}{T_p c}} \cdot \left(1 + \dfrac{T_p c}{4\pi L}\sin\dfrac{4\pi L}{T_p c}\right) + \dfrac{\rho g}{bd}\sigma^2. \end{array}\right\}$$ (18)

To solve (17), put $X_{10} = G_1 e^{int}$, $X_{20} = G_2 e^{int}$, then we have

$$\left.\begin{array}{c} (N' - M'n^2)G_1 + (v - \mu n^2)G_2 = 0, \\ (v - \mu n^2)G_1 + (N' - M'n^2)G_2 = 0. \end{array}\right\} \qquad (19)$$

Therefore,
$$\begin{vmatrix} N' - M'n^2, & v - \mu n^2 \\ v - \mu n^2, & N' - M'n^2 \end{vmatrix} = (N' - M'n^2 + v - \mu n^2) \cdot$$

$$(N' - M'n^2 - v + \mu n^2) = 0, \qquad (20)$$

which gives two values of n^2.

Now, suppose that the coupling is very weak, in other words, consider μ'/M, v/N, etc. to be small, and neglect their squares. Further, let T_s be the period of the fundamental oscillation of each bay in the case when there was no coupling under consideration and so the two bays oscillate quite independently. Then

$$T_s = \frac{4L}{\sqrt{gh}}(1 + Q)^{\frac{1}{2}} \equiv \frac{4L}{\sqrt{gh}} \cdot f. \qquad (21)$$

Hence

$$\frac{N}{M} = \frac{ghl^2}{(1 + Q)} \cdot \frac{1 + \dfrac{T_p c}{4\pi L}\sin\dfrac{4\pi L}{T_p c}}{1 - \dfrac{T_p c}{4\pi L}\sin\dfrac{4\pi L}{T_p c}},$$

or putting

$$\frac{2\pi}{T_s} = n_0,$$

$$\frac{1 + \dfrac{T_p c}{4\pi L}\sin\dfrac{4\pi L}{T_p c}}{1 - \dfrac{T_p c}{4\pi L}\sin\dfrac{4\pi L}{T_p c}} = \gamma,$$

we have

$$\frac{N}{M} = n_0^2 \gamma.$$

Accordingly, the two values of n are:

$$\left.\begin{array}{l} n_1 = n_0\gamma^{\frac{1}{2}}\left(1 + \dfrac{v}{N} - \dfrac{\mu' + \mu}{2M}\right) \equiv n_0\gamma^{\frac{1}{2}}(1 + \delta_1), \\[3mm] n_2 = n_0\gamma^{\frac{1}{2}}\left(1 - \dfrac{\mu' - \mu}{2M}\right) \equiv n_0\gamma^{\frac{1}{2}}(1 + \delta_2). \end{array}\right\} \qquad (22)$$

Thus, if G_1', G_1'', G_2', G_2'', are four constants,

$$\left.\begin{array}{l} X_{10} = G_1'\, e^{in_1 t} + G_1''\, e^{in_2 t}, \\ X_{20} = G_2'\, e^{in_1 t} + G_2''\, e^{in_2 t}, \end{array}\right\} \tag{23}$$

where G_2' and G_2'' are not arbitrary, but are to be determined by G_2' and G_1'', respectively, from (19). Namely, putting n_1 and n_2 for n in (19),

$$\left.\begin{array}{l} \left(\dfrac{G_2}{G_1}\right)_{n=n_1} = \dfrac{M'n_1^2 - N'}{v - \mu n_1^2} = \dfrac{Mn_0^2\gamma\left(\dfrac{2v}{N} - \dfrac{\mu}{M}\right) - v}{v - \mu n_0^2\gamma\left(1 + \dfrac{2v}{N} - \dfrac{\mu' + \mu}{M}\right)} \equiv \kappa, \\[3em] \left(\dfrac{G_2}{G_1}\right)_{n=n_2} = \dfrac{v - \mu n_2^2}{M'n_2^2 - N'} = \dfrac{v - \mu n_0^2\gamma\left(1 - \dfrac{\mu' - \mu}{M}\right)}{\mu n_0^2\gamma - v} \equiv \kappa'. \end{array}\right\} \tag{24}$$

But, since it can be shown that $\kappa + \kappa' = 0$ or $\kappa' = -\kappa$, we have

$$(G_2)_{n=n_1} = \kappa(G_1)_{n=n_1}, \; (G_2)_{n=n_2} = -\kappa(G_1)_{n=n_2}.$$

Accordingly, the solutions become:

$$\left.\begin{array}{l} X_{10} = G_1^{(1)}\, e^{i(n_1 t + \varepsilon_1)} + G_1^{(2)}\, e^{i(n_2 t + \varepsilon_2)}, \\ X_{20} = \kappa G_1^{(1)}\, e^{i(n_1 t + \varepsilon_1)} - \kappa G_1^{(2)}\, e^{i(n_2 t + \varepsilon_2)}. \end{array}\right\} \tag{25}$$

Taking the real parts, the following solutions are obtained:

$$\left.\begin{array}{l} X_{10} = G_1^{(1)}\cos(n_1 t + \varepsilon_1) + G_1^{(2)}\cos(n_2 t + \varepsilon_2), \\ X_{20} = \kappa G_1^{(1)}\cos(n_1 t + \varepsilon_1) - \kappa G_1^{(2)}\cos(n_2 t + \varepsilon_2). \end{array}\right\} \tag{26}$$

These are the required solutions for the horizontal motion of water at the mouths of the bays for the case when the coupling action is supposed to be small. Since n_1 and n_2 are nearly equal, we can say that both X_{10} and X_{20} are made up of two oscillatory motions with nearly equal frequencies. The same can be said for the case of the vertical motion of water surface which is connected with the horizontal motion by the equation of continuity. Therefore, we may say that a train of seiches exhibiting the appearance of beat can occur in a pair of bays lying side by side.

We shall now consider the case where seiches exist at first in only one of the two bays. For simplicity, we assign the following initial conditions

$$(X_{10})_{t=0} = 1, \; (\dot{X}_{10})_{t=0} = 0, \; (X_{20})_{t=0} = 0, \; (\dot{X}_{20})_{t=0} = 0.$$

Then we have

$$G_1^{(1)} = G_1^{(2)} = \tfrac{1}{2}, \; \varepsilon_1 = \varepsilon_2 = 0.$$

Hence,

$$X_{10} = \cos n_0 \gamma^{\frac{1}{2}}\left(1 + \frac{\delta_1 + \delta_2}{2}\right)t \cdot \cos n_0 \gamma^{\frac{1}{2}}\left(\frac{\delta_2 - \delta_1}{2}\right)t,$$

$$X_{20} = \kappa \sin n_0 \gamma^{\frac{1}{2}}\left(1 + \frac{\delta_1 + \delta_2}{2}\right)t \cdot \sin n_0 \gamma^{\frac{1}{2}}\left(\frac{\delta_2 - \delta_1}{2}\right)t. \tag{27}$$

Moreover, since we assume that the two bays are equal in size and form, and that the energy of the coupled system under consideration is constant, it follows that $\kappa = 1$ in the above equation. Again, since δ_1 and δ_2 are small,

$$n_0 \gamma^{\frac{1}{2}}\left(1 + \frac{\delta_1 + \delta_2}{2}\right) \gg n_0 \gamma^{\frac{1}{2}}\left(\frac{\delta_2 - \delta_1}{2}\right).$$

Therefore, putting

$$\frac{2\pi}{n_0 \gamma^{\frac{1}{2}}\left(1 + \dfrac{\delta_1 + \delta_2}{2}\right)} = \frac{T_s}{\gamma^{\frac{1}{2}}\left(1 + \dfrac{\delta_1 + \gamma_2}{2}\right)} = T,$$

$$\frac{2\pi}{n_0 \gamma^{\frac{1}{2}}\dfrac{|\delta_2 - \delta_1|}{2}} = \frac{2T_s}{\gamma^{\frac{1}{2}}|\delta_2 - \delta_1|} = T_b, \tag{28}$$

we see that both X_{10} and X_{20} represent oscillations with period T and amplitudes varying slowly with period T_b. Further, we can see from Eq. (27) that, when X_{10} attains its maximum amplitude, X_{20} is minimum (zero), and vice versa. The results of the observation by NISINA (1930) shown in Fig. 1(b) can thus be explained as the principal feature of the oscillations.

In what follows we shall consider the formulation of the "coupling coefficient" σ. As mentioned above, since in the present case we may regard κ as 1, we have:

$$Mn_0^2\gamma\left(\frac{2v}{N} - \frac{\mu}{M}\right) - v = v - \mu n_0^2 \gamma\left(1 + \frac{2v}{N} - \frac{\mu' + \mu}{M}\right). \tag{29}$$

From this equation and Eqs. (22), we have

$$\mu = 0,$$

$$\mu' = 2\left(1 - \frac{n_2}{n_0}\gamma^{-\frac{1}{2}}\right) \cdot M,$$

$$v = \frac{n_1 - n_2}{n_0} \cdot \gamma^{-\frac{1}{2}} \cdot N. \tag{30}$$

Therefore, referring to Eq. (18), we obtain:

$$\sigma^2 = \frac{bd}{\rho g} v = \frac{bd}{\rho g}\left(\frac{n_1 - n_2}{n_0}\right)\frac{N}{\gamma^{\frac{1}{2}}}$$

$$= \left(\frac{n_1 - n_2}{n_0}\right)\frac{l^2 Ld}{2}\frac{1}{\sin^2\frac{2\pi L}{T_p c}}\left\{1 - \left(\frac{T_p c}{4\pi L}\sin\frac{4\pi L}{T_p c}\right)^2\right\}^{\frac{1}{2}};$$

or, using the relations

$$\frac{n_1 - n_2}{n_0} = \delta_1 - \delta_2 = \frac{2T_s}{T_b}, \quad l^2 = \left(\frac{2\pi}{\lambda}\right)^2 = \frac{4\pi^2}{T_p^2 gh},$$

we have

$$\sigma^2 = \frac{4\pi^2}{g}\frac{Ld}{h}\frac{T_s}{T_b T_p^2}\frac{1}{\sin^2\frac{2\pi L}{T_p c}}\left\{1 - \left(\frac{T_p c}{4\pi L}\sin\frac{4\pi L}{T_p c}\right)^2\right\}^{\frac{1}{2}}. \qquad (31)$$

This is the required expression for σ. Equation (31) corresponds to the equation

$$\sigma^2 = \frac{4\pi^2}{g}\frac{Ld}{h}\frac{T_s}{T_b T_p^2}$$

in the original theory which is applicable to a special case where the length of the long waves, λ, incident from the open sea is approximately equal to four times the length, L, of each of the bays.

3. Experiments

In order to ascertain the validity of the above-stated theory, a series of hydraulic model experiments was carried out.

3.1 Experimental apparatus

Figure 3 shows the general configuration of the experimental apparatus, and Fig. 4 shows the dimensions of each part of the experimental tank. At one end of the tank is a plunger P to generate waves by which the period and height of the waves can be altered. "A" is a wave-absorbing device made of wood shavings wrapped with nets of small mesh and fastened to the inside wall of the tank by iron frames "I". "F" is a wave-filtering device of similar structure as "A". The object of this device is to absorb shorter waves among the waves emitted by the plunger and thus to obtain a train of regular sinusoidal waves to the degree as possible. The model bays are made of wood, and near the head of each one is located a servo-limnimeter sensor "G".

Fig. 3.   General view of the experimental apparatus.

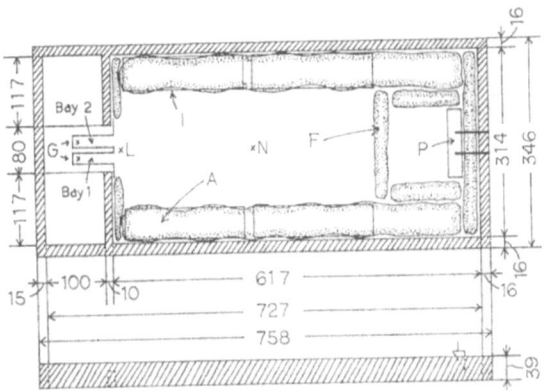

Fig. 4.   Dimensions of each part of the experimental tank (unit: cm). P, plunger; F, wave filter; A, wave-absorbing device; I, iron frame; G, limnimeter; N, L, wave measuring points.

3.2   Long wave condition
3.2.1   Long wave condition for incoming waves

In order that the waves generated by the plunger and transmitted to the bays be long waves, the following condition must be satisfied

$$\frac{h}{\lambda} \text{ (relative depth)} < \frac{1}{20}. \tag{32}$$

The period of the waves, or that of the plunger's up-and-down motion, T_p, must also satisfy the relation

$$\frac{\lambda}{T_p} = \sqrt{gh}.\qquad(33)$$

Therefore,

$$\frac{h}{T_p\sqrt{gh}} < \frac{1}{20} \quad \text{or} \quad T_p > \sqrt{\frac{400}{g}}\cdot\sqrt{h} \quad (\equiv T_{p\,min}),\qquad(34)$$

which gives the minimum value of T_p necessary to satisfy the long wave condition for a given value of h. The above relation is diagrammatically shown in Fig. 5.

3.2.2   Long wave condition for the seiches in the bay model

The period of the seiches (fundamental oscillation) in a bay is approximately given by

$$T_0 = \frac{4L}{\sqrt{gh}} \text{ (Merian's formula)}\qquad(35)$$

and the period T_s corrected according to the bay mouth by

$$T_s = T_0\cdot f = T_0(1 + Q)^{\frac{1}{2}} = \frac{4L}{\sqrt{gh}}\left\{1 + \frac{2b}{\pi L}\left(0.9228 - \ln\frac{\pi b}{4L}\right)\right\}^{\frac{1}{2}}.\qquad(36)$$

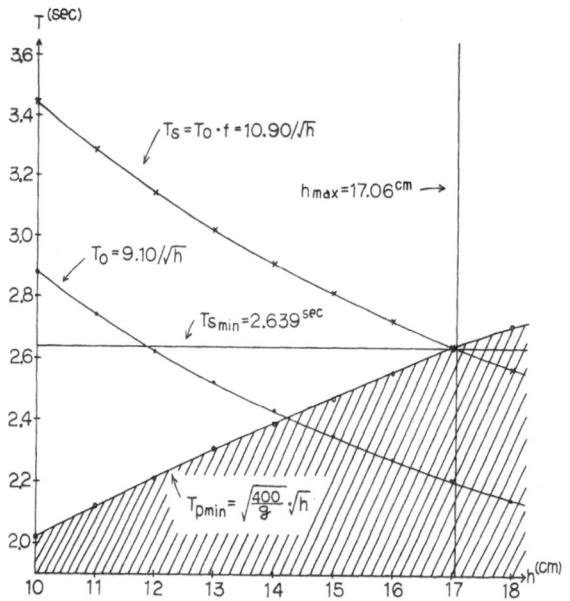

Fig. 5.   Respective relations between h and each of T_o, T_s, T_p. h, depth of water; T_o, theoretical period of seiches in a bay (approximate value calculated by Merian's formula); T_s, ditto (corrected as for the bay mouth). (Note: Hatched portion belongs to intermediate waves.)

The bay model used in the present experiment is of such a size as $L = 71.2$ cm, $b = 20.0$ cm, (height, $H = 25.0$ cm), so in the present case,

$$T_0 = 9.10/\sqrt{h} \text{ (sec)},$$
$$f = 1.198,$$
$$T_s = 10.90/\sqrt{h} \text{ (sec)}. \tag{37}$$

The graphs of these equations are also shown in Fig. 5. Since the wavelength λ' corresponding to the period T_s of the seiches in a bay is $4Lf$, in order that the waves of the seiches occurring in the bay model may be regarded as long waves, the following condition must be satisfied:

$$\frac{h}{\lambda'} = \frac{h}{4Lf} < \frac{1}{20}, \text{ or } h < \frac{Lf}{5} (\equiv h_{max}). \tag{38}$$

In the present case we have $h_{max} = 17.06$ cm. The value of T_s corresponding to this value of h_{max} should be designated as $T_{s\ min}$, and Eq. (37) gives $T_{s\ min} = 2.639$ sec.

Again, in the case when T_p differs from T_s, oscillations with the same period as T_p should occur in the bay generally, so the long wave condition for that case is the same as in Case (a).

To recapitulate—in the present experiment, if we choose the depth of water to be shallower than about 17 cm, and the period of the incoming waves to be longer than about 2.7 sec, then we may consider that the long wave condition is satisfied both inside and outside the bay model (see Fig. 5).

3.3   Experiment on the effect of the wave-absorbing device

Before proceeding to the experiment proper, a preliminary experiment was made in order to examine the effect of the wave-absorbing device.

First, after removing all the wave-absorbing devices from the tank, waves were generated in different ways such as, for example, by moving the plunger, or by one stroke with a plate, and the oscillations of water surface were measured with limnimeters placed at two points, "N" and "L", in the tank. Then the same procedure was repeated after the tank was equipped with the wave-absorbing device. An example of records thus obtained is shown in Fig. 6, where the plunger's motion was stopped after generating ten successive waves of 3.0 sec period. As seen from this figure, in the case when the tank was equipped with the wave-absorbing device, the water oscillation ceases as soon as the last wave generated by the plunger arrives at the measuring point, while, on the contrary, in the case with no wave-absorbing device, the water oscillation continues fairly long even after the plunger motion was stopped. Judging from the above result, the present wave-absorbing device seems to be very effective, especially for waves such as those with periods of about 3 sec generated by the plunger.

3.4   Experimental procedure and results

The experiment proper was carried out in the following way in conformity with the initial condition mentioned in the above theory. First, by sending waves to Bay 2 by operating the plunger, seiches were generated in Bay 2, while the mouth of Bay 1

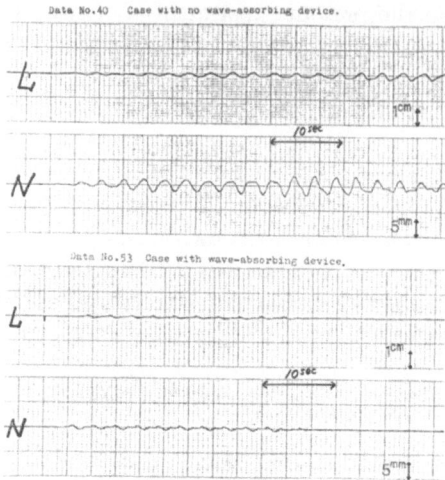

Fig. 6. An example of records showing the effect of the wave-absorbing device. Period of the plunger motion, $T_p = 3.0$ sec.

remained closed. After the seiches in Bay 2 attained a steady state, the mouth of Bay 1 was quickly opened. Figure 7 shows some examples of records of the seiches thus obtained. Looking at these records, we can see that the phenomenon of beat actually occurs, and that the variations of amplitudes of the seiches in the two bays are opposite in phase. Further, looking at these records more closely, we can recognize the following. In the case when the period of the incoming waves T_p is larger than that of the proper oscillation of the bay T_s, in other words, if $\psi\ (= T_s/T_p)$ is relatively small, the beat phenomenon occurs conspicuously, and the beat period T_b is short, and, as T_p decreases, T_b increases. This tendency continues until T_b becomes very large, when no beat phenomenon occurs in either bay. But, when T_p becomes still smaller, the phenomenon of beat with long period again appears, and as T_p decreases, the beat period T_b also decreases.

Figure 8 is an example of records showing the behaviour of oscillations of the two bays when the water depth h is changed (T_p remaining nearly the same). Looking at this figure, we can recognize that as the water depth h increases, in other words, as ψ decreases, the beat period T_b becomes shorter and shorter. Figure 9 gives another example of records showing the behaviour of oscillations of the two bays when the water depth h is changed. (In this case the value of ψ is nearly constant.) As seen in this figure, the beat period T_b remains nearly constant. Thus it appears that, if ψ does not change, T_b does not change, even if the water depth h changes.

Figure 10 gives some examples of records showing the behaviour of oscillations of the two bays when the distance, d, between them is changed. Looking at this figure, we notice the following: In the case when ψ is less than a certain value, say ψ_1, which is nearly equal to 1, the beat period T_b increases as the distance d increases, while, on the other hand, in the case when ψ is larger than ψ_1, T_b tends to decrease as d increases. These relations are schematically shown in Fig. 11.

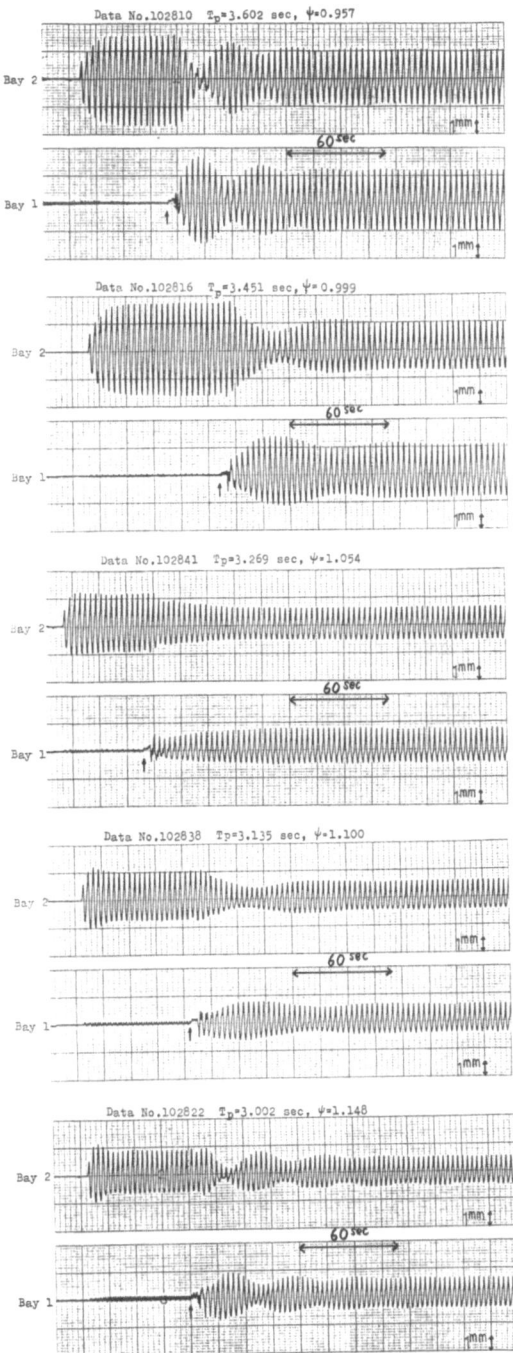

Fig. 7.   Behaviour of oscillations of the two bays when the period, T_p, of the incoming waves is changed. Note: the arrow in each figure indicates the instant at which the mouth of Bay 1 was opened.

354   M. NAKANO and N. FUJIMOTO

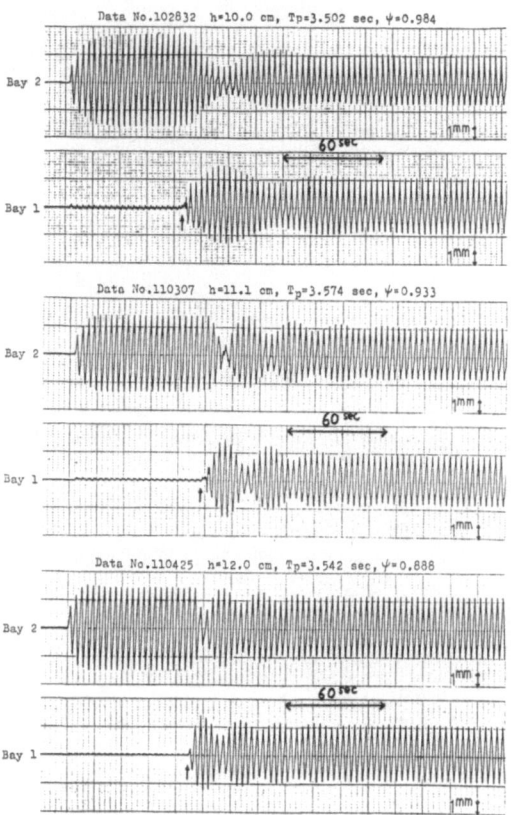

Fig. 8. Behaviour of oscillations of the two bays when the depth of water h is changed (T_p remaining nearly the same). Note: the arrow in each figure indicates the instant at which the mouth of Bay 1 was opened.

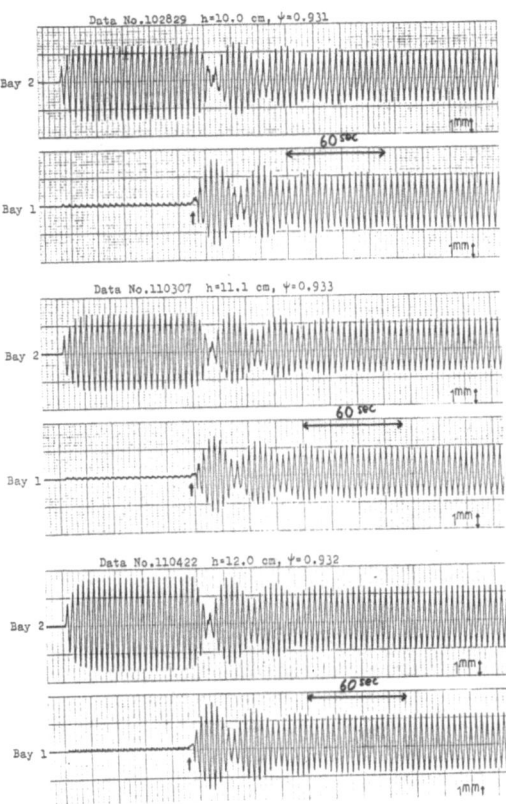

Fig. 9.   Behaviour of oscillations of the two bays when the depth of water *h* is changed (ψ remaining nearly the same). Note: the arrow in each figure indicates the instant at which the mouth of Bay 1 was opened.

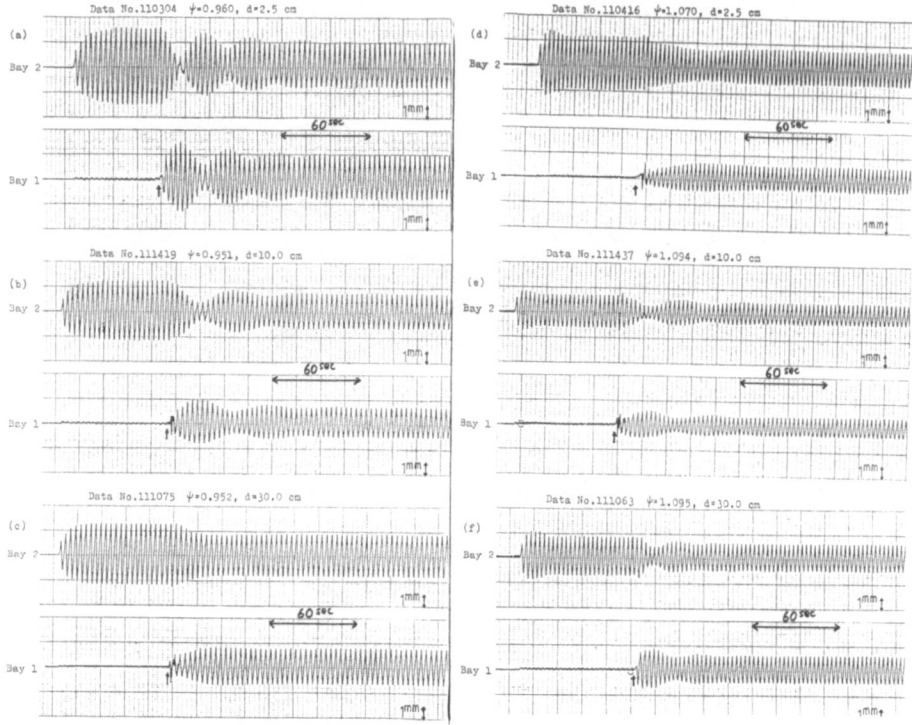

Fig. 10. Behaviour of oscillations of the two bays when the distance, d, between them is changed. Note: the arrow in each figure indicates the instant at which the mouth of Bay 1 was opened.

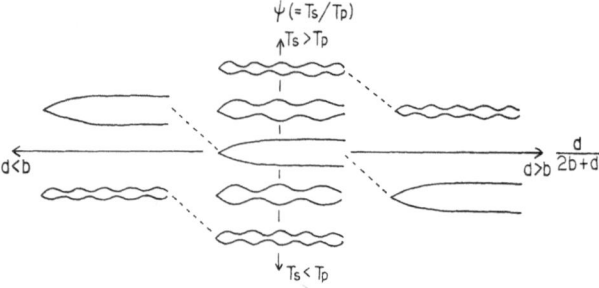

Fig. 11. Characteristic changes of the amplitude curves of the oscillation in Bay 1 when T_p, h and d are changed (schematic diagram).

In what follows the relations among the coupling coefficient σ and other quantities will be noted. Figure 12 shows the relation between σ and T_p, and Fig. 13, that between σ and ψ. From the latter figure we can see that, for the given coupled bays, the value of σ remains constant if the value of ψ remains constant, regardless of the values of each of the quantities h, T_s and T_p.

Finally, Fig. 14 shows the relation between σ and ψ in the case when the distance d between the two bays is taken as variable.

4. Conclusion

It has been confirmed experimentally that the phenomenon of beat does occur for the situation of two rectangular bays forming a coupled system, and the theory developed has been proved adequate for explaining the phenomenon at least as a first approximation.

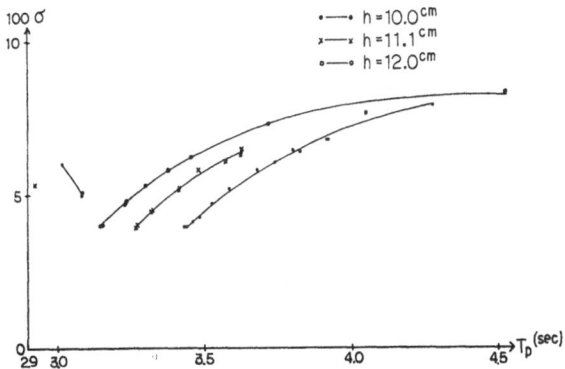

Fig. 12.   Relation between the coupling coefficient σ and the period of the incoming waves T_p ($d = 2.5$ cm).

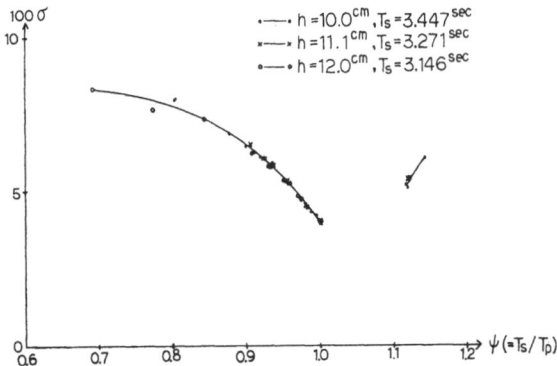

Fig. 13.   Relation between the coupling coefficient σ and ψ ($= T_s/T_p$) ($d = 2.5$ cm).

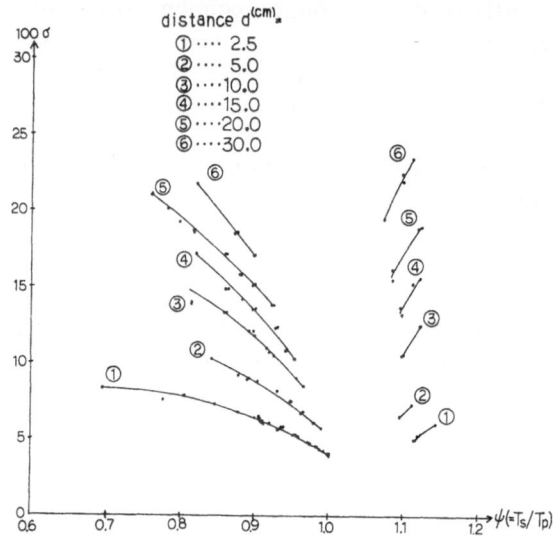

Fig. 14. Relation between the coupling coefficient σ and ψ ($= T_s/T_p$) in the case when the distance d between the two bays is taken as a parameter.

The authors wish to express their sincere thanks to Messrs. Tatsuo Miyata, Masao Kemuyama and Hiroki Kawabata for their helpful suggestions and advice.

REFERENCES

Honda, K., T. Terada, Y. Yoshida, and D. Isitani, An investigation on the secondary undulations of oceanic tides, *J. Coll. Sci., Imp. Univ. Tokyo*, **24**, 59, 1908.
Nakano, M., Die Seiches in gekoppeltes System formenden Buchten, *Geophys. Mag.*, **5**, 163–170, 1932.
Nisina, N., Moroiso, Aburatubo oyobi Koaziro no "Seiche" Kansoku, *Disin.*, **2**(12), 1930.

Tsunamis—Their Science and Engineering, edited by K. Iida and T. Iwasaki, 359–385.

The Excitation of Harbors by Tsunamis

F. Raichlen,* T. G. Lepelletier,** † and C. K. Tam***

*W. M. Keck Laboratory of Hydraulics and Water Resources,
California Institute of Technology,
Pasadena, California, U.S.A.
**Setec Travaux Publics, Paris, France
†Formerly Graduate Student in Civil Engineering,
W. M. Keck Laboratory of Hydraulics and Water Resources,
California Institute of Technology, Pasadena, California, U.S.A.
***W. M. Keck Laboratory of Hydraulics and Water Resources,
California Institute of Technology, Pasadena, California, U.S.A.

(Received September 16, 1981)

This study is based on a nonlinear-dispersive-dissipative numerical model which determines the response of a harbor to transient incident waves. The response of Hilo Harbor, Hawaii, has been studied evaluating the importance of the planform and the offshore bathymetry on the amplification and attenuation characteristics of the harbor. It has been found that, to a first approximation, a linear-dissipative numerical model describes the response of this harbor reasonably well, and indicates extreme effects due to the shape of the harbor and the offshore bathymetry. An attempt has been made to use Hilo Harbor as a "transducer" to infer the spectrum of the incident tsunami from tide gage measurements inside the harbor. This has not been entirely satisfactory in this case due to the extreme frequency sensitivity of this harbor.

1. Introduction

The effect of tsunamis on harbors is very much dependent on the local configuration of the harbor and the nearby offshore bathymetry. This has been realized for many years; nevertheless, it is interesting to present an example which dramatically demonstrates the selective dynamics of a harbor. In Fig. 1, tide gage records are presented for the tsunamis which resulted from the Chilean earthquake of 1960 and from the Alaskan earthquake of 1964 at Honolulu, Hawaii and at Mokuoloe Island, Hawaii. These two locations are on or very near the island of Oahu and are approximately 50 km apart. It is quite evident from Fig. 1 that the response is very dependent on location, since tsunamis from different earthquake sources produce similar tide gage records at the same location and different records for nearby locations for the same event. It is the influence of the local features, the transient nature of the incident waves, and the harbor dynamics which are the subject of this discussion.

Over the years there have been numerous studies dealing with the response of harbors to incident waves. The majority of the theoretical investigations have approached the problem using linear inviscid theories. Among these, one of the earlier

F. Raichlen *et al.*

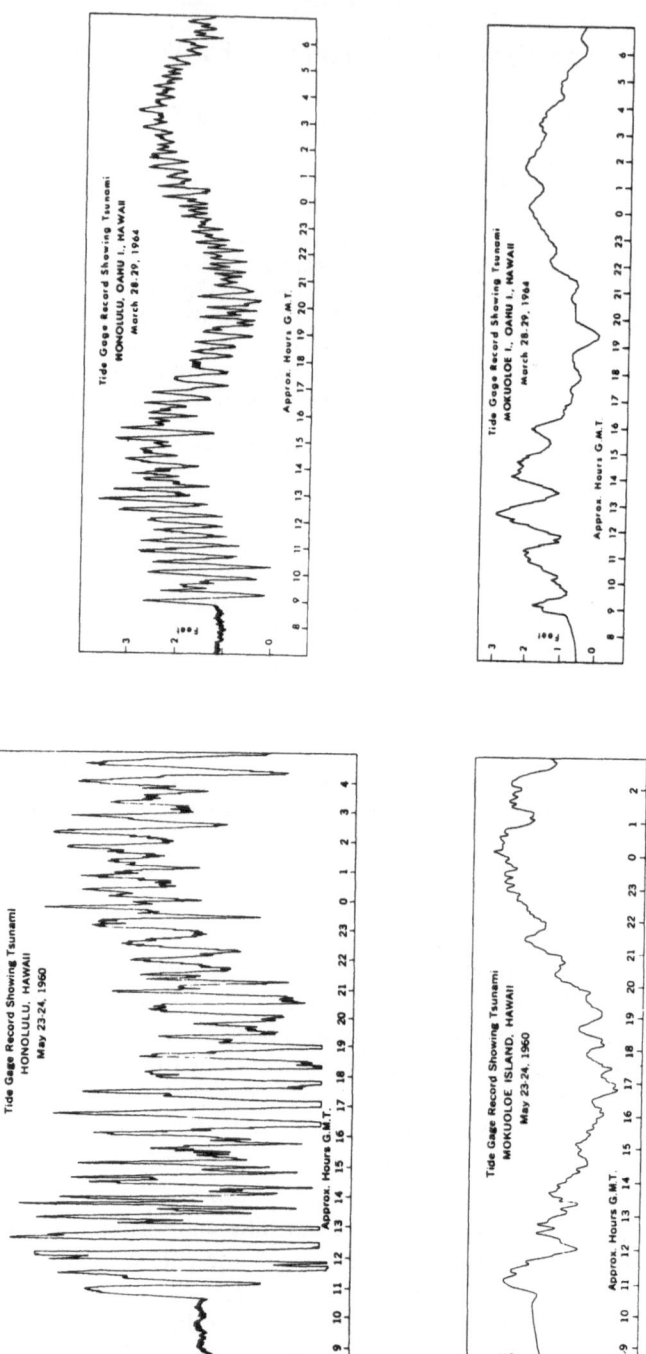

Fig. 1. Tide gage record at Honolulu and Mokuoloe Island, Oahu I., Hawaii, for the 1960 Chile Tsunami and the 1964 Alaska Tsunami.

investigations by MILES and MUNK (1961) introduced the effect of the radiation of wave energy from the harbor mouth to the open sea thus limiting the magnitude of the amplification of waves in a harbor excited at resonance. MILES and MUNK (1961), IPPEN and RAICHLEN (1962), IPPEN and GODA (1963), HWANG and TUCK (1970), LEE (1971), and CHEN and MEI (1974) have studied different aspects of this problem. In addition, various survey articles have appeared which summarize these and other works such as: RAICHLEN (1966), WILSON (1972), MILES (1974), RAICHLEN (1976, 1979); the interested reader is directed to these articles for a more complete discussion of the subject.

In a recent study LEPELLETIER (1980) investigated both theoretically and experimentally the response of arbitrary shaped harbors with variable depths to transient incident waves considering the effect on the resultant excitation of nonlinear incident waves, the nonlinear response of a harbor, and several sources of energy dissipation. The most important source of dissipation included in the analysis related to the effect of flow separation at the harbor entrance. The problem was formulated using a nonlinear-dispersive theory for shallow water waves incorporating dissipation and solving the equations numerically using finite element techniques.

In the investigation which is reported herein, the example chosen for the application of the analysis was the response of Hilo Harbor (and Hilo Bay), Hawaii to transient waves. This harbor has been exposed to several large tsunamis in the past with devastating effects and yet the full reason for its aggravated response has not been firmly established. A map of the harbor and the offshore area are presented in Fig. 2

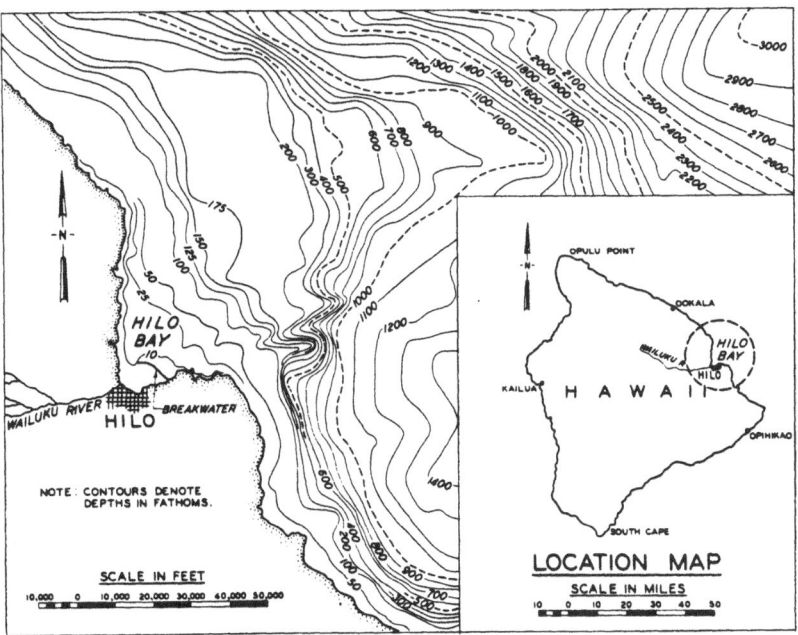

Fig. 2.   Location of Hilo Harbor, Hawaii, and bathymetry near Hilo Bay.

showing three important features: first, the large change in depth offshore with depths varying from about 8.8 m (29 ft) near the tide gage station inside the harbor to a maximum of about 5,500 m (18,000 ft) which occurs about 50 km (170,000 ft) offshore, second, the existing breakwater, and third, the converging planform of the bay. Three-dimensional representations of the harbor and a portion of the offshore-harbor region are presented in Fig. 3 where the vertical scales of the two shown are exaggerated compared to the horizontal; 50 times in the upper part and 10 times in the lower part of the figure. The harbor is shown without the breakwater in place. The importance of the large change in depth and the contorted nature of the offshore bathymetry are evident in this figure.

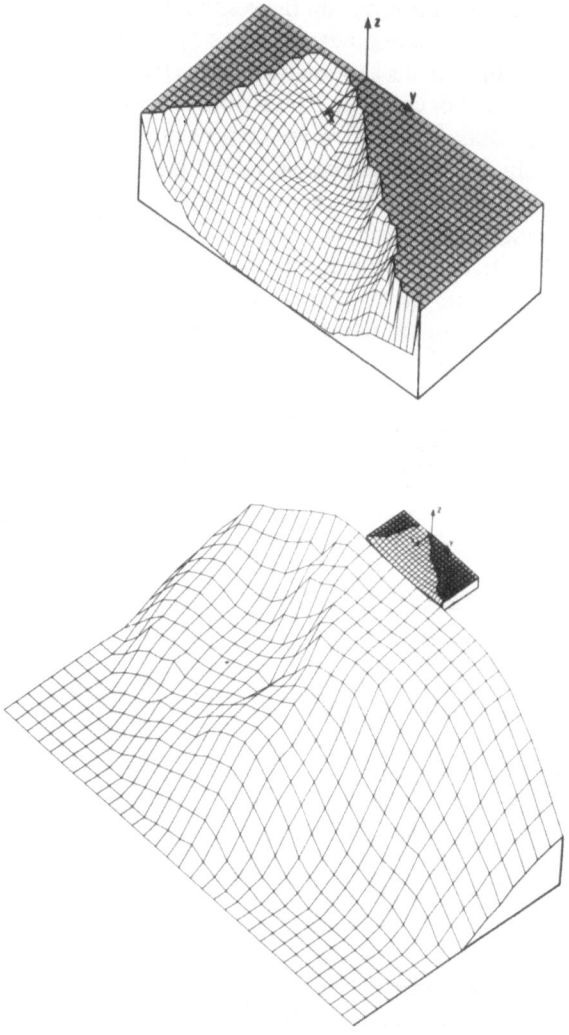

Fig. 3. Bathymetry of (a) Hilo Bay, and (b) the offshore region.

This paper primarily discusses the application of the analytical method developed by LEPELLETIER (1980) to the example of Hilo Harbor to investigate the importance of various features of the coastal region in its response to transient incident waves. The numerical approach utilizing finite element techniques will be summarized in Section 2. In Section 3 the analytically determined response of Hilo Harbor will be discussed.

2. Theoretical Considerations

The numerical model of LEPELLETIER (1980) is a potential function formulation of the nonlinear-dispersive-dissipative long wave equations which includes the effects of variable depth and variable planform. This approach is based on a method proposed by WU (1979) for the inviscid case which allows the continuity equation and the momentum equation to be combined into one equation using a pseudo-velocity potential. The unknown potential of the equation is determined using the appropriate boundary and initial conditions by finite element techniques. In this section the theoretical approach and the numerical method will be summarized; the interested reader is directed to LEPELLETIER (1980) for a complete discussion.

The harbor configuration and the coordinate system are defined in Fig. 4; the analysis is restricted to the fluid domain bounded by the semicircle Γ_R and the curve EDF. The problem consists of determining the waves in the harbor induced by

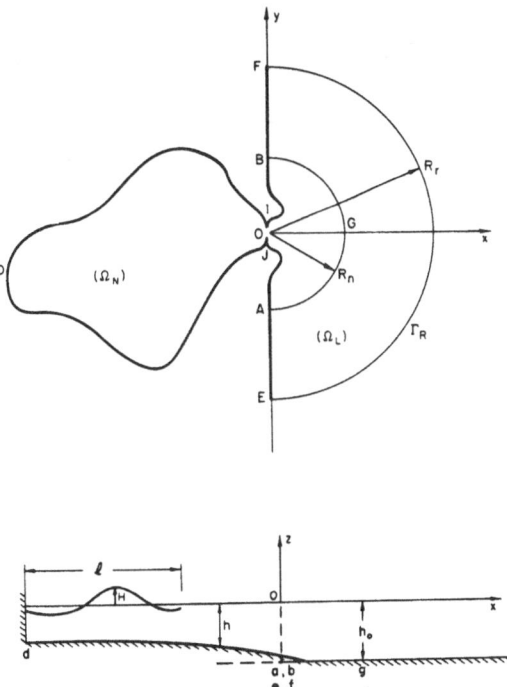

Fig. 4.   Definition sketch of an arbitrary shaped harbor.

transient incident plane waves with known characteristics propagating perpendicular to the coastline.

i) The harbor (or inner) region, denoted by Ω_N, is the fluid domain bounded in planform by the curve AGBDA, and along the bottom by the curve agbda. The sources of energy dissipation considered in the general treatment are: laminar boundary friction at the bottom and the water surface, quadratic flow separation losses across narrow passages inside the harbor region, and energy losses due to leakage underneath the lateral vertical boundaries. (The latter is a loss which may be associated with laboratory experiments or in an approximate way can represent non-perfectly reflecting boundaries in a prototype harbor.) The equations used in the harbor region are the weakly nonlinear dispersive and dissipative long wave equations.

For the sake of clarity in the presentation and in the notation, only one narrow gap with separation loss (in region Ω_N) is considered in the subsequent analysis. This gap is represented by the segment IJ in Fig. 4. To facilitate treatment of the equations, the harbor region Ω_N is subdivided into two non-overlapping subregions Ω_N^1 and Ω_N^2 in each of which the solution is differentiable. Region Ω_N^2 is, by definition, the fluid domain bounded by the curve IJDI in Fig. 4 and subregion Ω_N^1 refers to the fluid domain bounded by the curve IBGAJI.

ii) The outer region, denoted by Ω_L is the fluid domain bounded by the curve EAGBF and the semicircle Γ_R. The water depth is assumed to be constant in this region and is denoted by h_0; the portions of the coastline BF and AE are assumed to be straight and perfectly reflective. The effects of viscous dissipation, convective non-linearities, and dispersion are neglected in this region. As a consequence, the waves are considered to result from the linear superposition of the known incident and reflected waves and the radiated wave emanating from the harbor mouth. Finally, proper boundary conditions are applied on the semicircle, Γ_R, to allow smooth transmission of the radiated wave through it. A matching procedure must be applied to connect the two regions. This is done by imposing continuity of the flow rate and the wave amplitude across the boundary AGB.

In the remainder of this section, the physical variables are expressed in the following dimensionless form:

$$x = \frac{x^*}{l} \qquad y = \frac{y^*}{l} \qquad h = \frac{h^*}{h_0} \qquad t = \frac{t^*\sqrt{gh_0}}{l}$$

$$z = \frac{z^*}{h_0} \qquad \Phi = \frac{h_0}{H}\frac{\Phi^*}{l\sqrt{gh_0}} \qquad u = \frac{h_0}{H}\frac{u^*}{\sqrt{gh_0}} \qquad \eta = \frac{\eta^*}{H}$$

where h_0 denotes the still water depth outside the harbor region, g is the acceleration of gravity, H is a characteristic wave height, l is a characteristic wavelength, and the starred symbols refer to the dimensional variables: t^* is the time, x^* and y^* are the coordinates in the horizontal plane, η^* is the wave elevation, Φ^* denotes the depth-averaged velocity potential function, and u^* is the depth-averaged velocity vector. Henceforth, all the equations will be dimensionless unless specifically stated otherwise, and all dimensionless terms are of the order of unity.

The basic unknowns to be solved for are the pseudo-potential functions $\Phi^1(x, y, t)$, and $\Phi^2(x, y, t)$ defined in regions Ω_N^1 and Ω_N^2, respectively, and the radiation potential function $\psi(x, y, t)$ defined by:

$$\Phi_L = \Phi_I + \psi,$$

where Φ_I denotes the known potential function associated with the incident-reflected wave system, i.e., the wave system which would exist if the coastline extended straight from E to F in Fig. 3.

The considerations stated above lead to a definition of the solution as: Find the functions $\Phi^1(x, y, t)$, $\Phi^2(x, y, t)$, and $\psi(x, y, t)$ differentiable in the domains Ω_N^1, Ω_N^2, and Ω_L, respectively, and in the time interval $0 \le t \le \tau$ such that:

$$\Phi_{tt}^i - \nabla \cdot (h\nabla\Phi^i) + \frac{\gamma_s}{h}\Phi_t^i - \beta\left[\frac{h}{2}\nabla \cdot (h\nabla\Phi_{tt}^i) - \frac{h^2}{6}\nabla^2\Phi_{tt}^i\right] + \beta\nabla \cdot$$

$$\left\{\left(\frac{h}{6}\Phi_{tt}^i + \frac{h}{3}\nabla h \cdot \nabla\Phi^i\right)\nabla h\right\} = -\alpha[\nabla\Phi^i \cdot \nabla\Phi_t^i + \nabla \cdot (\Phi_t^i\nabla\Phi^i)] \quad \text{in } \Omega_N^i, \quad i = 1, 2 \quad (1)$$

$$\psi_{tt} - \nabla^2\psi = 0 \quad \text{in } \Omega_L \quad (2)$$

with the following boundary conditions:

$$\frac{\partial\Phi^1}{\partial n} = -\frac{\varepsilon}{h}\Phi_t^1 \quad \text{on JA, IB}$$

$$\frac{\partial\Phi^2}{\partial n} = -\frac{\varepsilon}{h}\Phi_t^2 \quad \text{on IDJ}$$

$$\frac{\partial\psi}{\partial n} = 0 \quad \text{on AE, BF}$$

$$\frac{\partial\psi}{\partial n} = -\psi_t - \frac{1}{2R_r}\psi \quad \text{on } \Gamma_R$$

$$\frac{\partial\Phi^1}{\partial n} = -\frac{\partial\Phi^2}{\partial n} = -\left(\frac{2}{\alpha f_e}\right)^{1/2}|\Phi_t^1 - \Phi_t^2|^{1/2}\,\text{sign}\,(\Phi_t^1 - \Phi_t^2) \quad \text{on IJ}$$

$$\frac{\partial\Phi^1}{\partial n} = \lambda_*(\Phi^1 - \Phi_I - \psi)$$

$$\text{on AGB}$$

$$\frac{\partial\psi}{\partial n} = -\lambda_*(\Phi^1 - \Phi_I - \psi) - \frac{\partial\Phi_I}{\partial n}. \quad (3)$$

And the initial conditions:

$$\Phi^i(x, y, 0) = 0 \quad \text{in } \Omega_N^i, i = 1, 2$$

$$\psi(x, y, 0) = 0. \quad \text{in } \Omega_L \quad (4)$$

The dimensionless parameters appearing in these equations are defined as follows: the nonlinear parameter, $\alpha = H/h_0$; the dispersion parameter, $\beta = (h_0/l)^2$; the laminar boundary friction parameter, $\gamma_s = (v\sigma/2)^{1/2}(1 + C)(l/h_0\sqrt{gh_0})$, where σ is a characteristic frequency of the wave motion and C is the surface contamination factor (MILES,

1967); the entrance friction factor, f_e; the leakage parameter ε, which can, to a first approximation, also be related to a reflection coefficient $\bar{r}$, is defined as $\varepsilon = (1 - \bar{r})/(1 + \bar{r})$; the contact stiffness parameter λ_*, which ensures continuity of the flow rate and wave amplitude between regions Ω_N and Ω_L provided it is chosen large enough (i.e. $\lambda_* \geq 10^6$).

Once $\Phi(x, y, t)$ is known, the wave elevation $\eta(x, y, t)$ and the depth-averaged velocity vector $\mathbf{u}(x, y, t)$ can be derived simply from Φ at the lowest order as:

$$\eta = - \Phi_t + (\alpha, \beta, \gamma_s) \tag{5}$$

$$\mathbf{u} = \nabla\Phi + O(\beta). \tag{6}$$

For normally incident waves the potential function $\Phi_I(x, y, t)$ can be derived, to a first approximation, from the incident-reflected wave height at the coastline as:

$$\Phi_I(x, y, t) = - F_I(t + x) - F_I(t - x) \tag{7}$$

where

$$F_I(\theta) = \frac{1}{2} \int_0^\theta \eta_I(\tau) d\tau. \tag{8}$$

The weak formulation is derived from the strong form by multiplying both sides of Eqs. (1) and (2) by a trial (or test) function and by integrating each equation in its respective domain, using Green's identity for all the integrals which involve spatial second derivatives. Then the boundary conditions Eq. (3) are substituted into the line integrals resulting from the use of Green's identity. From the Galerkin formulation finite element discretization, the following matrix system results:

$$M\ddot{d} + C\dot{d} + Kd = f_I(t) + g_1(d, \dot{d}) + g_2(\dot{d}) \tag{9}$$

$$d(0) = \dot{d}(0) = 0 \tag{10}$$

in which the vector d includes all the unknown quantities $\Phi_i^1(t)$, $\Phi_i^2(t)$, and $\psi_i(t)$ at the nodes in the entire fluid domain. The matrices M, C, and K are symmetric positive except in the case of variable depth, where matrices M and K become unsymmetric. The quantity $f_I(t)$ is a known "force" vector associated with the incident-reflected wave data, g_1 includes the nonlinear convection terms, and g_2 accounts for the quadratic head loss across the segment IJ. All the integrations were performed using the 2×2 point Gauss Quadrature rule.

A time integration algorithm was developed to solve the nonlinear second order differential equation, Eq. (9), and it has features similar to the "implicit-explicit operator splitting" technique (HUGHES *et al.*, 1978). All linear terms are treated implicitly, and some of the nonlinear terms are treated explicitly, using a predictor-multicorrector algorithm, the others are treated implicitly. This results in a simple matrix system Ax = b for each iterative step at a given time step, which can be solved by the standard technique of LDU decomposition.

3.   Results and Discussion of Results

As discussed previously, the major objective of the study reported herein has been to investigate the dynamic response of Hilo Harbor, Hawaii to incident tsunamis. In pursuing that aspect of the problem also it was of interest to see whether it would be possible to work backwards from the tide gage records for several tsunamis to define the spectrum of a tsunami in the open sea.

To demonstrate the applicability of the nonlinear-dispersive-dissipative long wave model discussed in Section 2, several experimental and theoretical wave records are shown in Fig. 5 for the response of a trapezoidal constant depth harbor to transient waves where the abscissa is normalized time, and the ordinate is the ratio of the wave amplitude to the depth. The wave record shown at the top of the figure is labelled $\sigma L/\sqrt{gh} = 0$ where σ is the circular wave frequency, L is the length of the trapezoidal

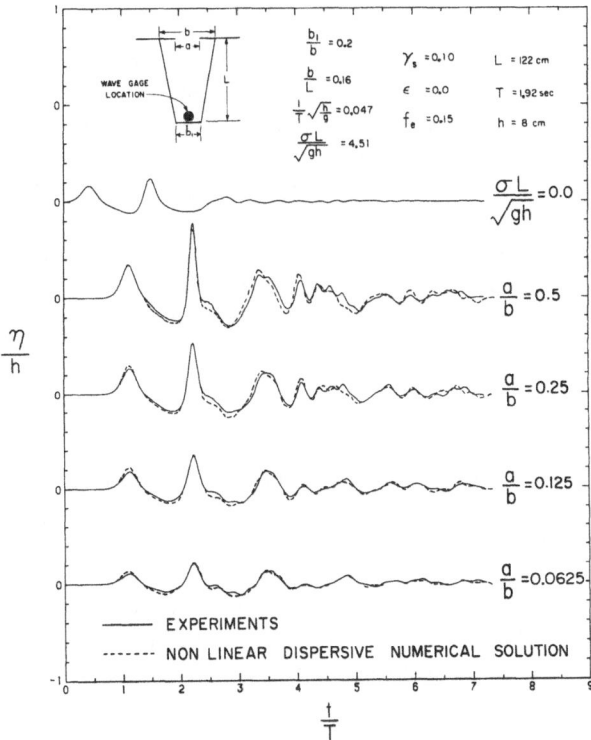

Fig. 5.   Transient wave records at the backwall for a trapezoidal harbor with partially closed entrance, comparison between experiments and nonlinear solutions, $L = 122$ cm, $T = 1.92$ s, $h = 8$ cm, $b = 20$ cm, b_1 = 4 cm.

harbor, and h is the depth. Thus, this corresponds to the incident wave, since $L = 0$ indicates that the harbor entrance is closed. The other water surface profiles are for different entrance openings varying from 6% open to 50% open. The entrance dissipation used in Eq. (3) takes the coefficient $f_e = 1.15$. The magnitude of this coefficient determined experimentally was found to be relatively independent of a nondimensional parameter similar to the Strouhal Number when this parameter is large, i.e., the case of tsunami excitation. The agreement between the experiments and the numerical theory appears to be relatively good both with regard to the amplification (or attenuation) of the incident wave by the harbor and the details of the amplitude-time history. In addition, it is interesting how the effects of separation loss at the entrance of the harbor limits the amplification; indeed, it is only after the relative entrance width has been reduced to between 6% to 12% that dissipation becomes more important than energy trapping. (In additon to this, other work reported by LEPELLETIER (1980) lends support to the applicability of the previously described numerical theory to the case of Hilo Harbor, Hawaii.)

In Fig. 6 tide gage records measured at the tide station, Pier 3 at Hilo, Hawaii, show tsunamis from the Alaskan earthquake of March 27, 1964, the Tokachi-Oki earthquake of May 16, 1968, and the Hawaiian earthquake of November 29, 1975. These are the three tsunami records that will ultimately be used to demonstrate the feasibility (or infeasibility) of obtaining the open-sea tsunami spectrum from tide gage records. These three tsunamis represent distant and near records at the site, and the different character of the near and distant tsunamis can be seen easily in the spectra presented in Fig. 7 for these three tide gage records; in each case the tide has been removed. The abscissa of each spectrum is the frequency in hr^{-1} and the ordinate is the energy density expressed in $cm^2 hr$; thus, the area under the spectrum is the mean square of the record in cm^2. The root mean squares of the 1964, 1968, and 1975 tsunamis at this site as obtained from the spectra are: 35.7 cm, 5.2 cm, and 7.5 cm, respectively. The site of the measurements can be seen in the inset of Fig. 7 at node 9 located shoreward of the existing breakwater. One interesting feature of the spectra is that for the tsunamis from the distant earthquakes, energy is concentrated near the frequencies of $2 \, hr^{-1}$ and $3 \, hr^{-1}$, whereas for the tsunami from the nearby Hawaiian earthquake the important frequencies appear to be about $2 \, hr^{-1}$, $3 \, hr^{-1}$, $4 \, hr^{-1}$, and $4.8 \, hr^{-1}$. Thus, the two tsunamis which have travelled farther appear to have energy concentrated at lower frequencies. Of course, the earthquake magnitudes are quite different for these three events which also must have an important bearing on the distribution of energy with frequency. (It should be noted that some care must be used in interpreting the higher frequencies in the spectra due to the dynamics of the tide gage.)

The finite element grid which is used to obtain the numerical solution discussed in Section 2 is presented in Fig. 8; the region inside the harbor is more easily seen in the inset of Fig. 7. The region is divided into two parts: the bay and the portion of the outside region in which there is a depth variation, and an outer constant depth zone. The dividing line between the two is the curve AGB. The boundary condition on Γ_R defined by Eq. (3) is applied on the outermost boundary in Figure 8 to minimize the effects of "reflection" from the boundaries of the numerical model.

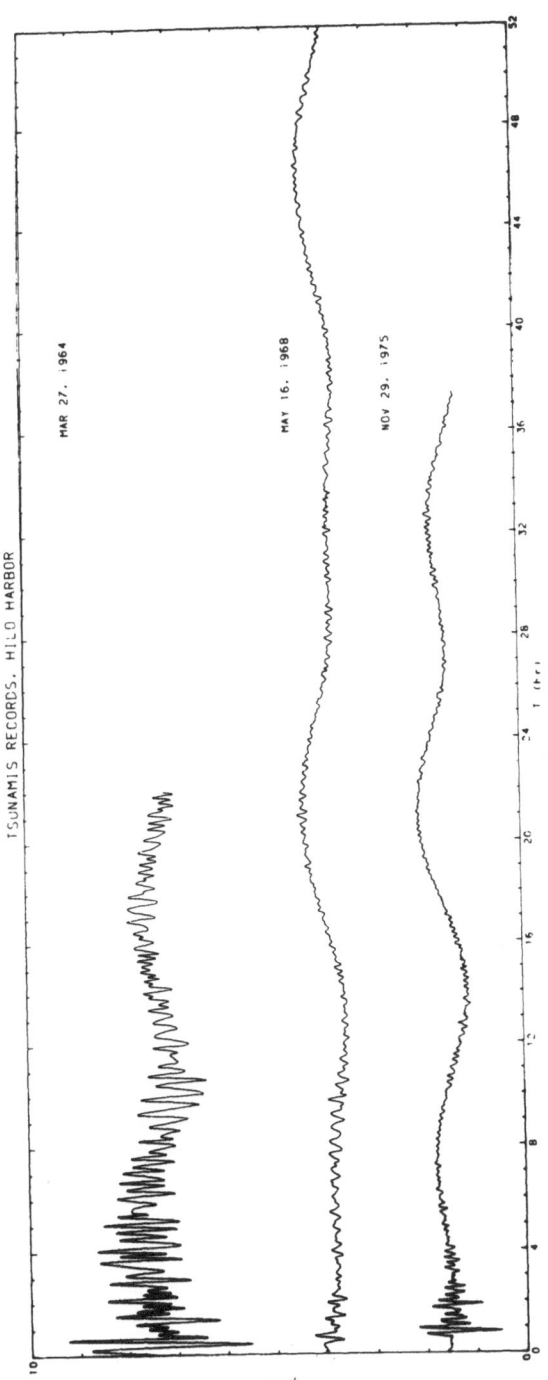

Fig. 6.   Tide gage records at Hilo Harbor for several tsunamis.

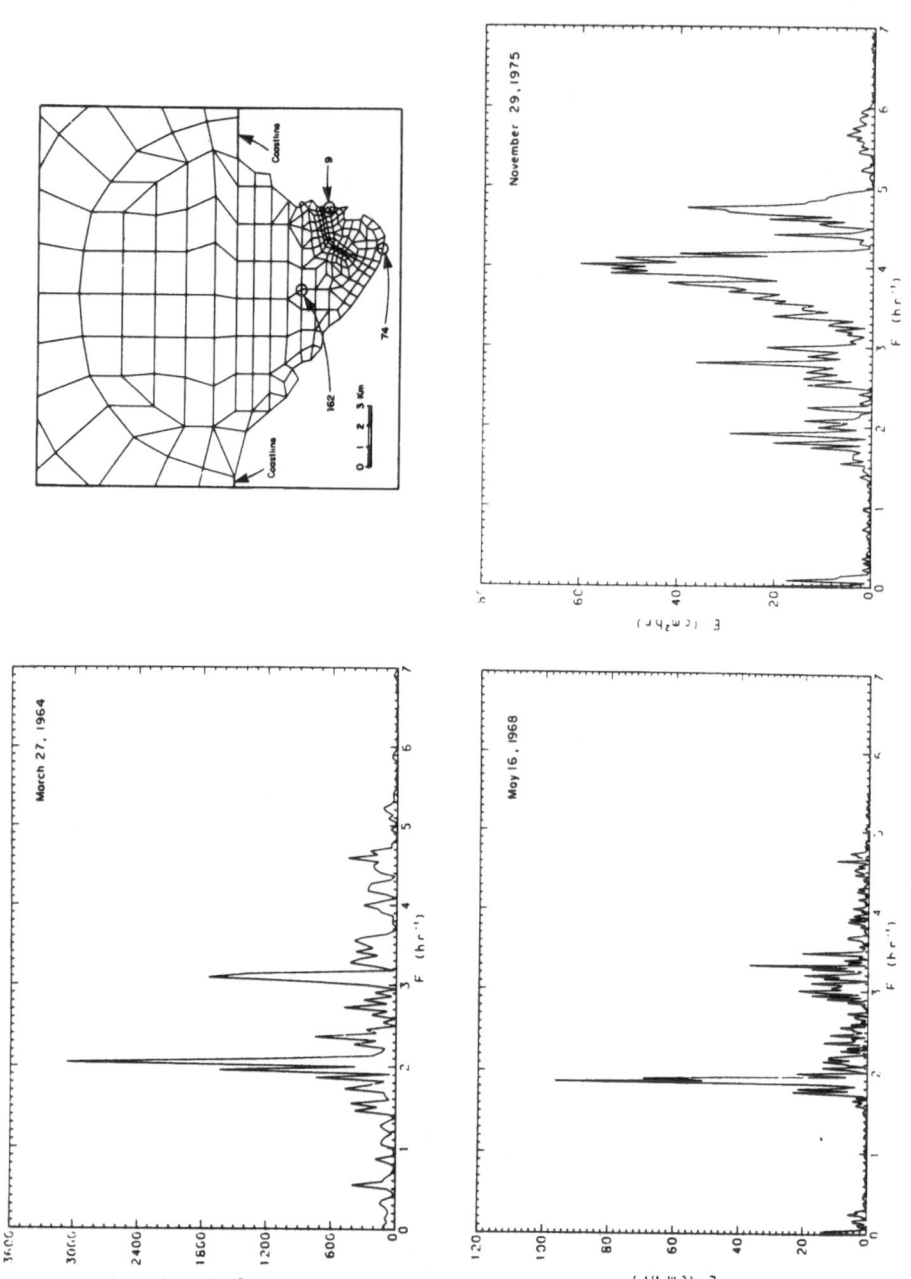

Fig. 7. Energy spectra for three tsunamis at location 9.

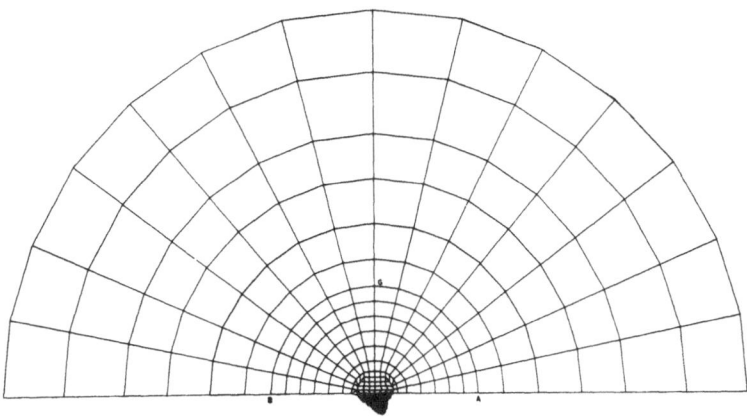

Fig. 8.   Finite element grid for numerical model of Hilo Harbor.

There are 400 elements and 444 nodes in this model with the size of the elements varying depending upon the location within the bay. Generally, the size of the element increases in a seaward direction. The elements are sized so that there will be at least eight nodal points covering the distance of a characteristic wavelength. Figure 8 shows that the bay has a convergent planform and Fig. 3 shows the depth within the bay decreases relatively uniformly from about 100 m at the entrance to a minimum depth for this model of about 1.5 m. Approximately 40 km offshore of the entrance, the depth increases to a depth of 3,840 m which is taken to be the constant depth region outside of AGB in Fig. 8. The width of the entrance of the bay where it connects to the open sea is 15.42 km and the length of the bay measured perpendicular to this section is 7.43 km.

In the first portion of this investigation a comparison was made between results obtained from the nonlinear-dispersive-dissipative theory with a linear theory which included the effects of dissipation. (In all of the analyses conducted the dissipation is that associated with losses due to flow separation at the existing breakwater gap. The breakwater was considered to be impervious and not overtopped by wave activity which indeed may not be the case.) It was planned that if the linear theory was determined to be adequate, a transfer function could be developed from the transient analytical approach.

The time marching method requires an incident transient wave to be used as input to the finite element program; therefore, an incident wave was constructed which consisted of 39 harmonic components each with the same amplitude but different frequencies defined as:

$$\eta(t) = -a \sum_{i=1}^{39} \sin \{2\pi[(i \Delta f)t + \theta_i]\} \tag{11}$$

where a is the amplitude of the harmonic components taken relatively arbitrarily as 0.007 m, t is the time in hours, Δf is the chosen frequency interval in hr^{-1}, and θ_i is the normalized phase angle associated with the i^{tb} harmonic wave. The phase angle associated with each component given in Eq. (11) is shown in Table 1.

Table 1. Phase angle associated with each harmonic.

j	θ_j	θ_{10+j}	θ_{20+j}	θ_{30+j}
1	.42	.36	.30	.21
2	.27	.92	.89	.22
3	.61	.38	.49	.56
4	.53	.45	.57	.59
5	.03	.68	.38	.71
6	.22	.10	.07	.64
7	.57	.78	.29	.66
8	.85	.92	.93	.27
9	.46	.51	.57	.13
10	.58	.97	.80	—

With the values of θ_i chosen in an arbitrary manner by a random number generator, the maximum wave height for the incident wave is about 0.154 m. Since it is necessary that $\eta(0) = 0$, and it is desired that the first extremum of η be positive, at most the first 38 values of θ_i can be arbitrarily chosen. The highest frequency component chosen for the exciting wave was $39 \, \Delta f = 6.06 \, \text{hr}^{-1}$. This was chosen with some knowledge *a priori* of the response characteristics of Hilo Bay so that adequate coverage of the range of frequency intervals was realized. The duration of the input was determined (6.922 hr) so that the bay would reach an approximately steady state condition in that chosen time.

In Fig. 9 the response of Hilo Bay is presented at five locations, due to the incident wave defined by Eq. (11) and Table 1, as determined by a linear theory with dissipation and by a nonlinear, dispersive, dissipative theory. In both cases, the incident wave was identical and is shown in Fig. 9 at the top of each sequence of water surface amplitudes. The location of each record is shown in the inset in the figure with location 9 at the tide gage station behind the breakwater. It should be noted, due to computational costs, the results of the nonlinear-dispersive-dissipative theory was obtained only to a duration of 4.5 hr compared to 6.92 hr for the linear theory. It appears that the general trend of the results at each location is similar from the two theories, but the detailed aspects of the oscillation appear somewhat different at least at two of the locations. However, at location 9, the results compare favorably. Therefore, to investigate the response characteristics of Hilo Harbor at locations 9, 74, and 162, a linear theory predicts results that are approximately the same as the more complete, but more costly, nonlinear-dispersive approach. Thus, this will be used herein, nevertheless, the possible failure of this type of analytical approach must be kept in mind.

It is of interest using this linear numerical model to ask questions about the response of Hilo Bay under several different conditions: without the breakwater, with the breakwater but without entrance dissipation, and with the breakwater and with entrance dissipation. The resultant water surface time histories for these three cases are shown in Fig. 10 and are defined in the figure. The first (and obvious) observation is the extremely large amplification apparent at each location relative to the incident waves. Comparing first the two cases with the breakwater but with and without entrance

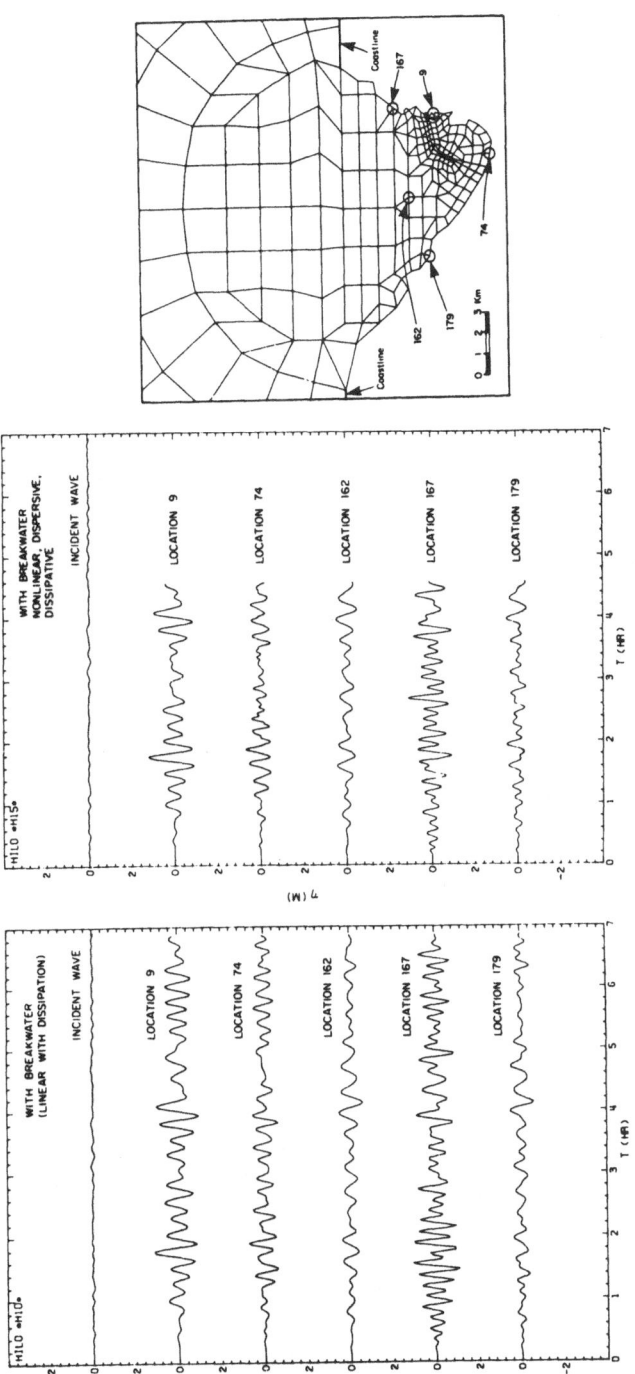

Fig. 9. Water surface-time histories at five locations in Hilo Bay using a linear dissipative model and a nonlinear dispersive-dissipative model.

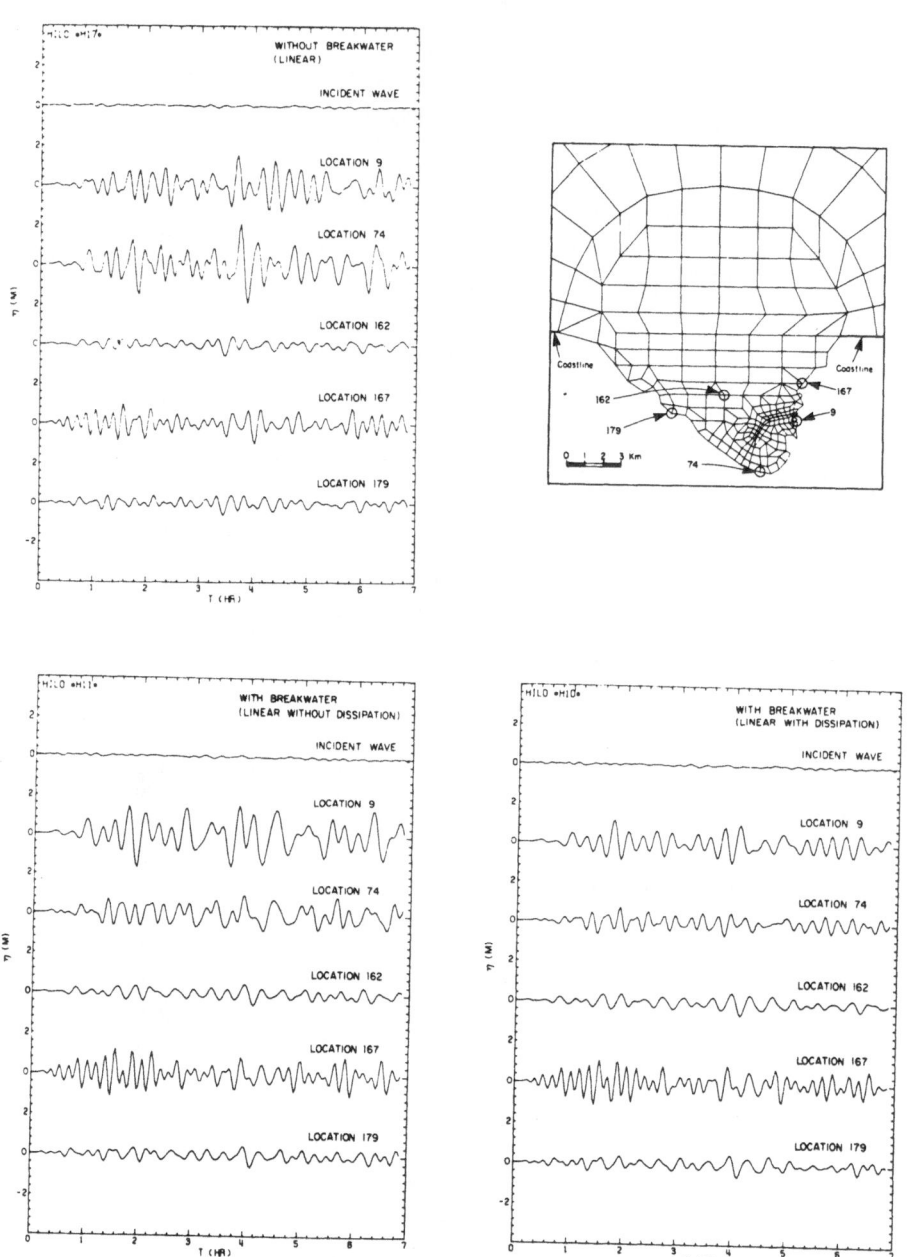

Fig. 10.   Water surface-time histories for several conditions.

dissipation, a significant difference in amplification is noted at locations 9 and 74 in the region of Hilo Bay behind the breakwater. However, for the other three locations seaward of the breakwater, large differences are not as apparent; this is as would be expected. Comparing the cases with and without the breakwater with the former without dissipation, the trapping of energy by the breakwater at location 9 is apparent. However, it appears that at location 74 at the coast, the wave amplitudes without the breakwater are worse than for the case with the breakwater; the reason for this will be discussed later. It is obvious that different locations are sensitive to different frequencies.

These effects are more readily seen when referring to the response curves (or transfer function) at various locations. The response curve is determined by evaluating the energy spectrum of the water surface time histories as shown in Fig. 10 and comparing the energy content in the various frequency bands to those in the incident spectrum as:

$$R^2(f) = \frac{S_R(f)}{S_I(f)} \tag{12}$$

where $S_R(f)$ is the spectral energy content centered at the frequency f as determined at the location of interest, $S_I(f)$ is the spectral energy content in the incident wave centered at the frequency of interest and $R(f)$ is the amplitude response function. Response curves are presented in Fig. 11 for the three conditions mentioned and treated in Fig. 10 for location 9. In each case, the response curve is compared to the case of Hilo Bay *with* a breakwater *with* dissipation. A fourth case is presented here where the depth offshore at the entrance to the bay is constant and equal to that at the entrance.

First the response at location 9 without the breakwater is compared to that with the breakwater with dissipation. Differences in the amplification are significant with the major effect of the breakwater apparently modifying the response near $f = 4 \text{ hr}^{-1}$. The magnitude of the amplification is somewhat startling compared to most harbors. When dissipation is removed from the configuration with the breakwater the balance between energy trapping and dissipation can be observed. In fact, it appears that the lower frequencies are most affected by the introduction of dissipation at the entrance. No doubt this is because the mode shape for these lower modes provide a larger velocity at the entrance than at higher frequencies. The third portion of the figure compares two cases with the breakwater and with dissipation: the realistic case with the larger offshore depth and that with a depth offshore which is the same as that at the entrance to the bay. It appears that a significant effect on the amplification may be attributed to the larger change in depth between the offshore regions with a depth of 3840 m and the site of the tide gage in a depth of about 8.8 m. It should be noted that just considering Green's Law the amplification would be about 4.6, and this does not include the effect of bathymetric variations offshore. Therefore, it appears that dissipation at the breakwater entrance and the large depth variation are quite important in defining the response of Hilo Harbor.

The same comparisons are made at locations 74 and 162 and are presented in Figs. 12 and 13, respectively. With attention first given to location 74 the breakwater appears

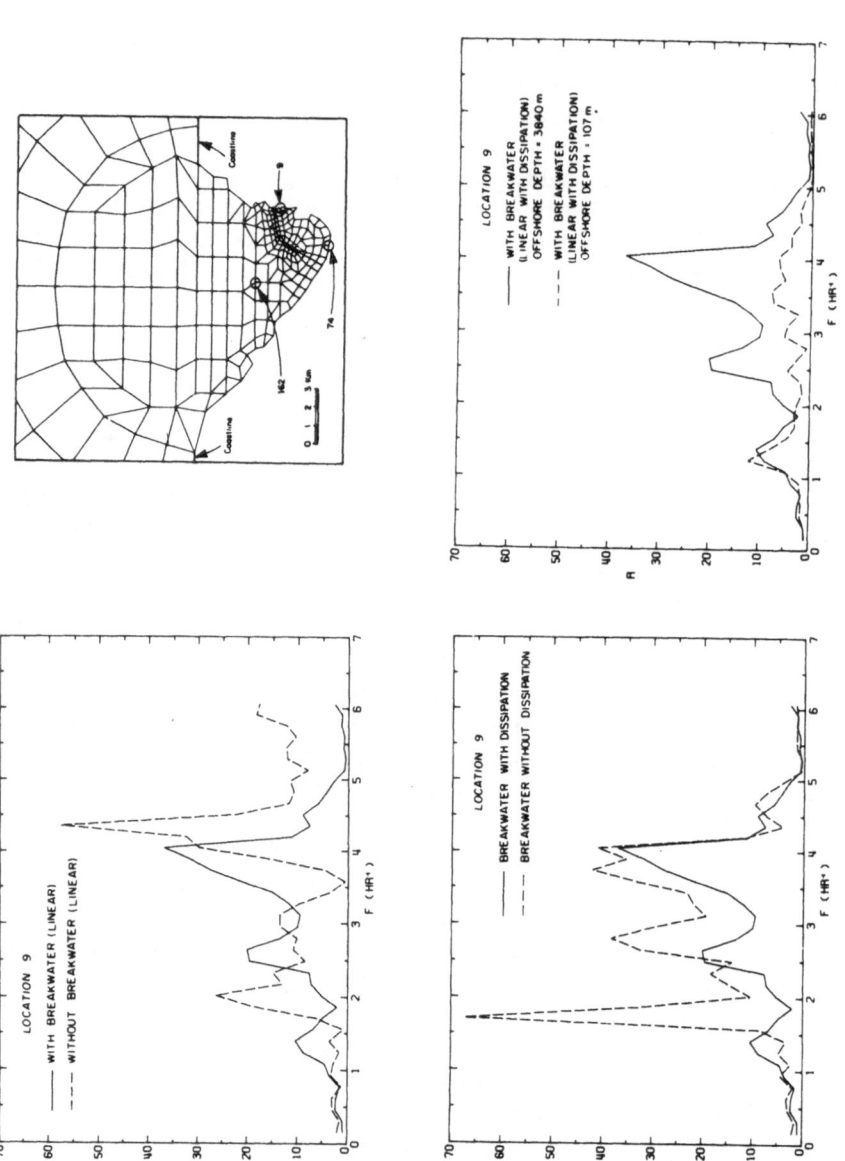

Fig. 11.   Response curves at location 9 for various conditions.

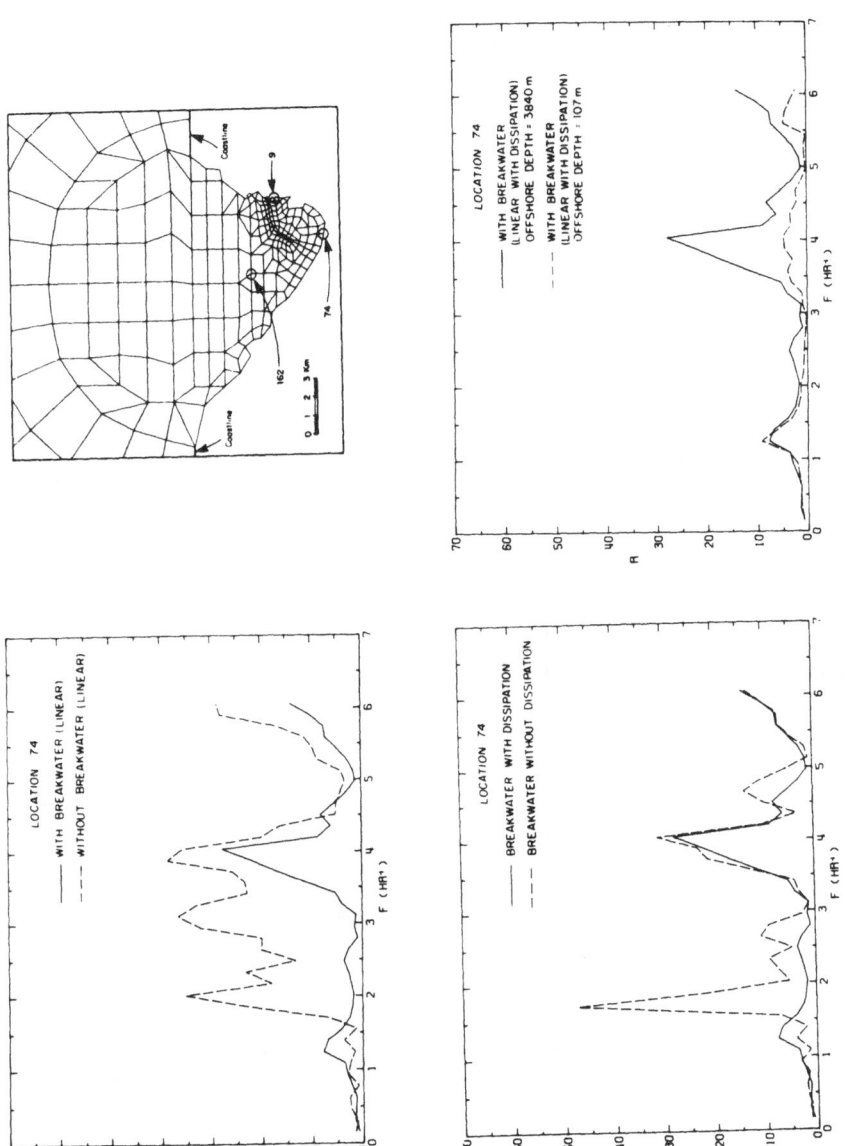

Fig. 12.   Response curves at location 74 for various conditions.

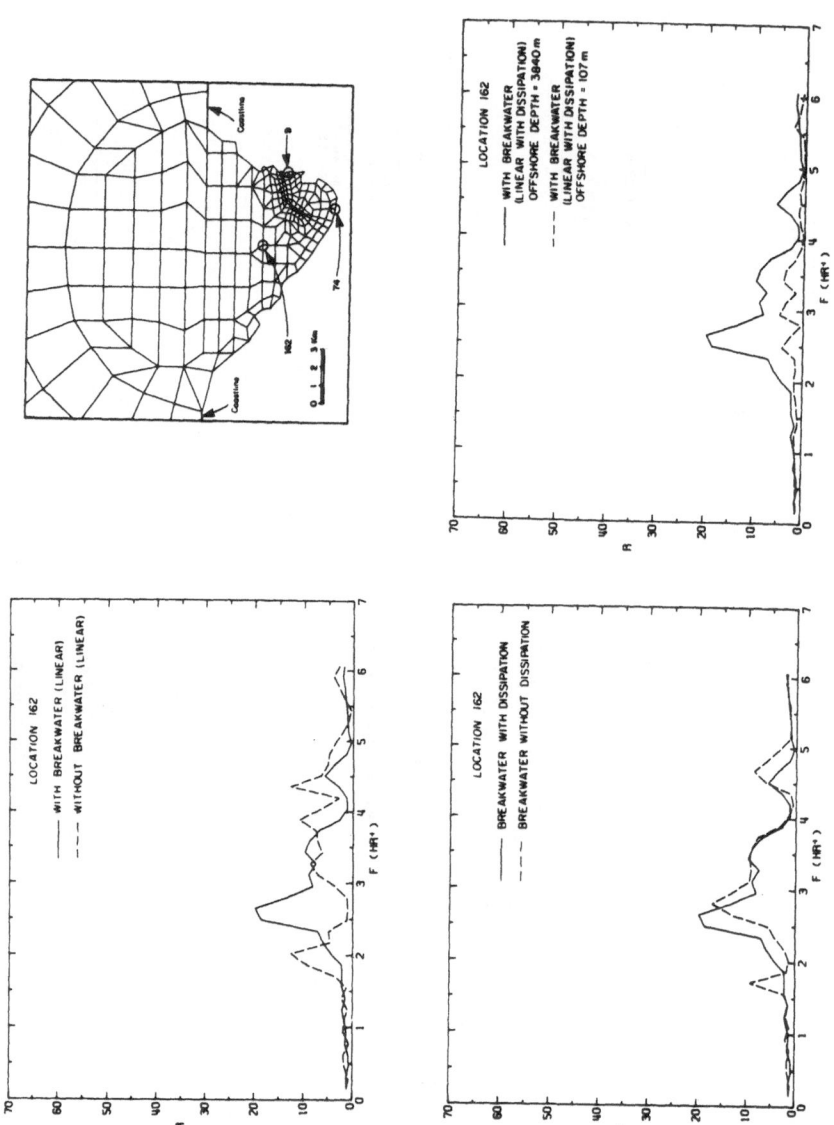

Fig. 13. Response curves at location 162 for various conditions.

to play a major role in modifying the response in the bay. Dissipation is again important mainly at the lowest mode as probably would be expected. At this location, the effect of the deep water offshore is apparent primarily at the peak corresponding to $f = 4 \, \text{hr}^{-1}$. At location 162, which from the inset is seen to be near the center of the bay seaward of the breakwater, the dissipation associated with the breakwater does not appear to have a major effect on the response. Of course, there is a larger difference in the response at that location with and without the breakwater. This also applies to the case with the large depth variation. Therefore, it appears that the breakwater has a major effect on the response at several locations within the harbor. In addition, the large offshore depth also has a major effect on the response. Both of these are not surprising nor are they peculiar to Hilo Harbor. What is perhaps surprising and certainly relates to the devastating attack that Hilo has received from past tsunamis, is the extreme values of amplification realized within the bay for certain frequencies due to these effects.

In Fig. 14 the mode shape is presented for $f = 4 \, \text{hr}^{-1}$ ($T = 15$ min) which is a frequency of major response at the tide gage station (location 9). This was obtained by exciting the numerical model with a sinusoidal input with a period of 15 min. In the lower part of Fig. 14 contours of constant amplitude are shown for a time which is

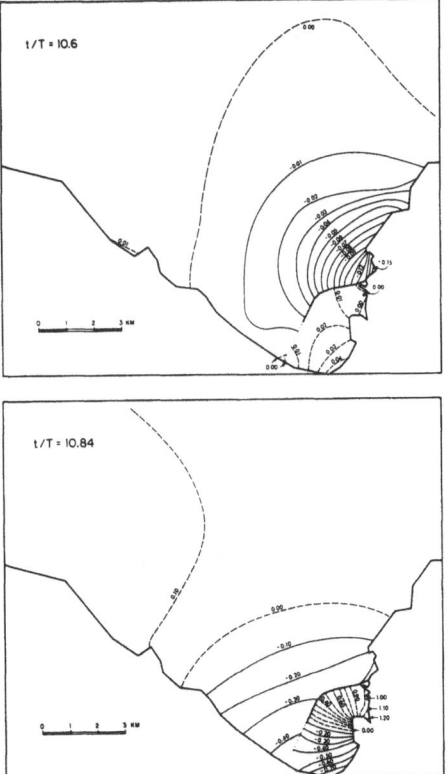

Fig. 14.   Mode shape for sinusoidal excitation at $f = 4 \, \text{hr}^{-1}$.

2.71 hr after the beginning of the excitation. The amplitudes are normalized with respect to that at the tide gage station. In this figure, solid lines correspond to negative amplitudes and dashed lines to positive amplitudes. It is apparent that the wave amplitudes offshore are quite small with amplitudes near location 9 near a maximum. The response with and without entrance dissipation at this frequency shown in Figs. 11, 12, and 13 tend to be supported by this mode shape where velocities across the breakwater entrance must be relatively small. The discontinuity in amplitude across the entrance is due to the boundary condition imposed (see Eq. (3)). The upper part of this figure demonstrates the standing wave nature of this oscillation with the response essentially zero in the harbor for a time about one-quarter of a period before the time of the conditions for the lower figure.

The unique response of a harbor to transient waves provides the opportunity to attempt to determine the spectrum of the incident tsunami from measurements obtained inside the harbor. One must realize that this is taking a very complicated problem and obtaining a solution which is no doubt over-simplified. However, this demonstrates certain problems which may arise in treating a harbor as a "transducer." Among other simplifications, the response is treated linearly and the application of the transfer function to the input spectrum as shown in Eq. (12) follows the usual linear approach. However, the introduction of nonlinear energy dissipation at the entrance, as shown in the boundary conditions of Eq. (3), actually negates this type of approach. In addition, for such long waves as tsunamis the breakwater at Hilo Harbor would probably be relatively transparent. Thus, the treatment in the numerical model of the breakwater as an impermeable barrier must be in error, and referring to Fig. 11 the response at the tide gage station (location 9) probably should lie between the case with and without a breakwater. There are several other aspects of the problem which affect the linear approach taken here, such as the assumption of normal wave incidence, and the lack of runup at the boundaries, among others, but they will not be pursued further here. In actual fact, the most accurate way to approach this would be to treat the problem using the nonlinear-dispersive-dissipative numerical model, but the problem of working backward through the model and the expense of computing are formidable. Therefore, the linear approach mentioned was used realizing that the final results may be somewhat approximate.

In Fig. 15 results are presented using this linear approach. In the upper part of Fig. 15 the spectrum obtained from the tide gage record at Hilo Harbor for the tsunami which was caused by the Alaskan earthquake of March 27, 1964, is shown. The ordinate is proportional to the energy content per frequency bandwidth expressed in units of $cm^2 hr$ and the abscissa the frequency in hr^{-1}. The tidal component seen in the record of Figure 6 has been removed by a numerical filter. Certain concentrations of energy are apparent in this portion of the figure. The middle part of Fig. 15 is the response curve for that location in Hilo Harbor treating the problem linearly with quadratic energy dissipation at the harbor entrance. The lower part of Fig. 15 shows the suggested form of the spectrum of the incident wave obtained from the measured spectrum and the transfer function. The overall energy content is significantly reduced, and large concentrations of energy are apparent at frequencies of $2\ hr^{-1}$ and $5.25\ hr^{-1}$ where minimum response occurs. Thus, one obvious difficulty in applying this

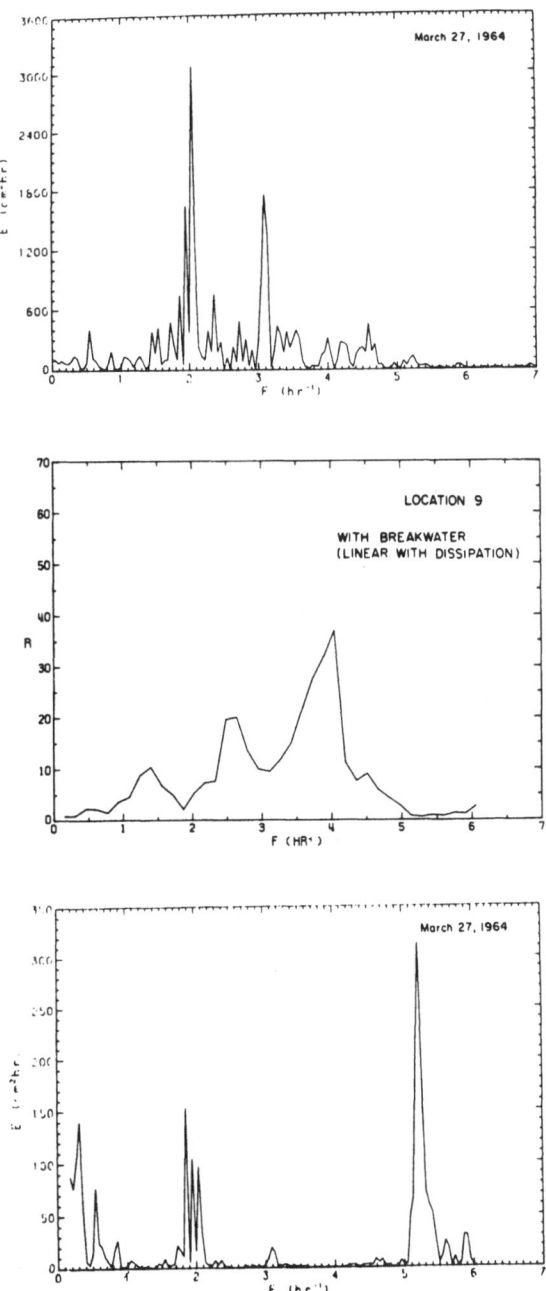

Fig. 15. Proposed spectrum of incident tsunami March 27, 1964.

approach to a harbor with such an extreme response (maxima of the order of ten to forty) is that the energy in regions of minimum response may be accentuated more than appears to be reasonable. The accuracy of these spectral estimates may be questioned since minor changes in the response curves could result in important changes in the derived spectrum.

In Fig. 15 similar results are presented for the tsunami at Hilo Harbor due to the Tokachi Oki earthquake of May 16, 1968, originating in Japan. It appears that there are similar features in the energy content from this tsunami and the Alaskan tsunami of 1964 although certainly the energy content is much less in this case than for the wave which resulted from the Alaskan earthquake. Applying a similar approach using the same transfer function, the energy spectrum which results is shown in the bottom part of Fig. 16. Again this shows energy concentrations near $f = 2 \, \mathrm{hr}^{-1}$ and $5.25 \, \mathrm{hr}^{-1}$, regions of minimum response.

Perhaps the most interesting case is the tsunami which resulted from the Hawaiian earthquake of November 29, 1975. This was centered off the island of Hawaii and very close to Hilo Bay (about 73 km away along the coast). Comparing the measured spectrum and the transfer function in Fig. 17 the similarity between the two is striking. It is apparent that the tsunami measured inside Hilo Harbor due to this earthquake must be very dependent upon the harbor response. The incident wave spectrum shown in the lowest part of this figure supports this observation and gain indicates an "amplification" due to the minimum response function at frequencies of $f = 2 \, \mathrm{hr}^{-1}$ and between $5 \, \mathrm{hr}^{-1}$ and $6 \, \mathrm{hr}^{-1}$. Note the reduction in magnitude of the spectrum due to the large amplification associated with the response.

4. Conclusions

Certain major conclusions can be drawn from this study. It appears that the numerical procedure developed and reported by LEPELLETIER (1980) at the very least as a first approximation can predict the response of a complicated harbor such as Hilo Harbor. Indeed, even for this case, the nonlinear aspects of such a response may not be too important in delineating certain features such as the importance of the offshore depth change and the planform shape of the harbor and its depth variation on the amplification process. It appears that a linear theory can be applied to this type of problem at least to give results which are a good first approximation. Energy dissipation due to a breakwater entrance is extremely important in limiting the response, especially for lower modes of oscillation. The dissipation causes a shift in the frequency of the modes of oscillation indicating the presence of certain nonlinear effects associated with quadratic dissipation; therefore, the approach outlined here of developing a transfer function from the input and output of a numerical model at best must be considered approximate. The application of this method to a measured tsunami to use the harbor as a "transducer" to infer the incident wave system probably must still remain questionable from this investigation. It is apparent that in order to approach the problem in this manner it is necessary that the response of the harbor *not* be as sensitive to the incoming wave energy as is Hilo Harbor. Therefore, any conclusions which may be drawn about the character of the incident wave spectrum

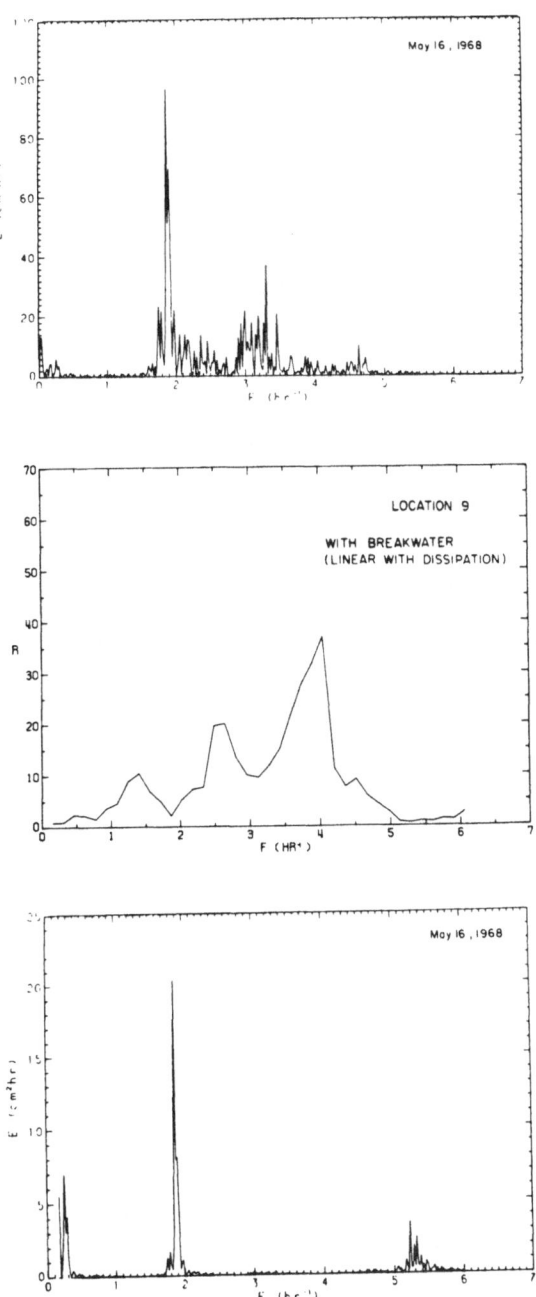

Fig. 16.   Proposed spectrum of incident tsunami May 16, 1968.

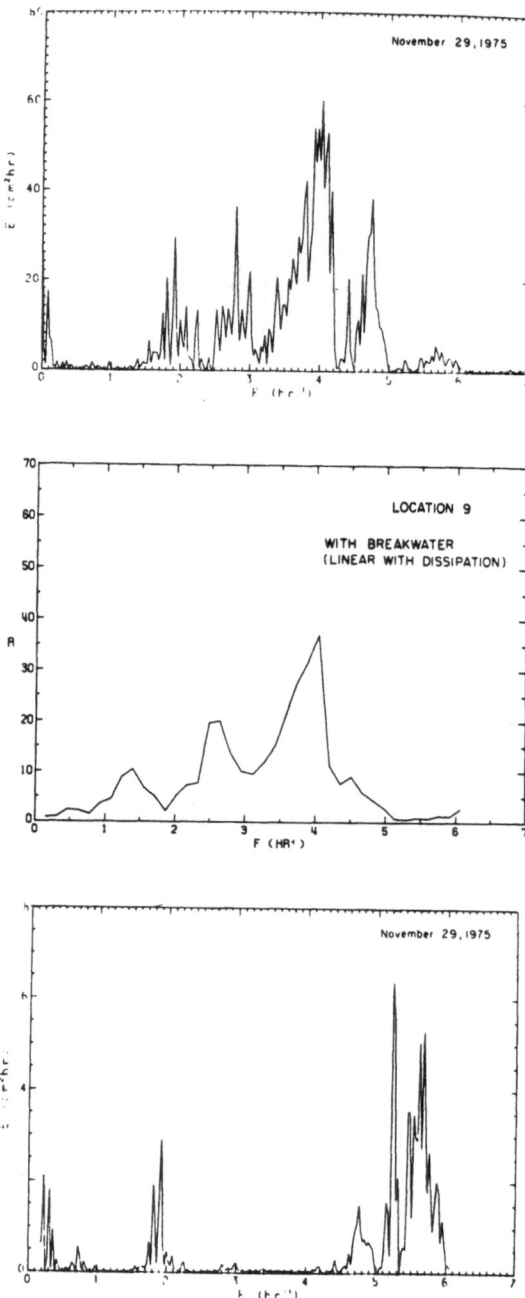

Fig. 17.   Proposed spectrum of incident tsunami November 29, 1975.

obtained by the method outlined using Hilo Harbor as the response mechanism must be left to the interpretation of the reader.

The results reported herein result from research sponsored by the National Science Foundation Grant Nos. ENV72–03587 and ENV77–20499.

REFERENCES

CHEN, H. S. and C. C. MEI, Oscillations and wave forces in an offshore harbor, Report 190, R. M. Parsons Laboratory for Water Resources and Hydraulics, MIT, 1974.

HUGHES, T. J. R., K. S. PISTER, and R. L. TAYLOR, Implicit explicit finite elements in nonlinear transient analysis, *Proc.* Fenomech '78 Stuttgart West Germany, August 30-September 1, 1978.

HWANG, L. S. and E. O. TUCK, On the oscillation of harbors of arbitrary shape, *J. Fluid Mechanics*, **2**, 447–464, 1970.

IPPEN, A. T. and F. RAICHLEN, Wave induced oscillations in harbors: the problem of coupling of highly reflective basins, Report No. 69, Hydrodynamics Laboratory, Massachusetts Institute of Technology, May 1962.

IPPEN, A. T. and Y. GODA, Wave induced oscillations in harbors: the solution of a rectangular harbor connected to the open sea, Report No. 59, Hydrodynamics Laboratory, Massachusetts Institute of Technology, 1963.

LEE, J. J., Wave induced oscillations in harbors of arbitrary geometry, *J. Fluid Mechanics*, **45**, 375–394, 1971.

LEPELLETIER, T. G., Tsunamis—harbor oscillations induced by nonlinear transient long waves, Report No. KH-R-41, W. M. Keck Laboratory of Hydraulics and Water Resources, California Institute of Technology, October 1980.

MILES, J. W., Surface wave damping in closed basins, *Proc. Royal Society*, **A 297**, 1967.

MILES, J. W., Harbor seiching, *Ann. Rev. Fluid Mechanics*, **6**, 17–36, 1974.

MILES, J. W. and W. H. MUNK, Harbor paradox, *Proc. ASCE, J. Waterways, Harbors Division*, **87**, 111–180, 1961.

RAICHLEN, F., *Estuary and Coastline Hydrodynamics*, edited by A. T. Ippen, McGraw-Hill, 281–340, 1966.

RAICHLEN, F., Coastal wave hydrodynamics—theory and engineering applications, Lecture notes for Summer courses at MIT, 1976.

RAICHLEN, F., Tsunamis, Proc. National Science Foundation Workshop, Organized and edited by L. S. Hwang and Y. K. Lu, 1979.

WILSON, B. W., Seiches, in *Advances in Hydroscience*, Academy Press, N. Y., Vol. 8, 1972.

WU, T. Y., Tsunamis, Proc. National Science Foundation Workship, Organized and edited by L. S. Hwang and Y. K. Lu, 1979.

SEA WALLS AND BREAKWATERS

Tsunamis—Their Science and Engineering, edited by K. Iida and T. Iwasaki, 389–396.

Tsunami Countermeasures in Fishing Villages along the Sanriku Coast, Japan

Tatsuma Fukuchi and Koji Mitsuhashi

*Fishing Port Department, Fisheries Agency, Ministry of Agriculture,
Forestry and Fisheries, Government of Japan*

(Received September 12, 1981; Revised January 5, 1982)

Many fishing villages are scattered along the fringe of the intricate coast of the Sanriku district. These fishing villages have been developed because of the very rich fishing resources existing offshore, even though most of them have experienced several times heavy loss of life and property caused by large tsunamis since ancient times.

After the great disaster of the Sanriku tsunami in 1933, people in villages sustaining damage were recommended to move to higher ground. But many people are now again living in the dangerous low land near the shore because of an increase of population and shortage of safe land space.

For this reason, tsunami walls now surround most villages on the coast. A tsunami wall presently under construction is so huge that its inside is planned as a road. In the future, artificial elevated ground will be constructed in some villages for increasing their safety against tsunamis.

1. Introduction

In Japan, most of the areas which have been repeatedly attacked by tsunamis are probably places where fishery has been extensively carried on. The Sanriku coast is one such area. "Sanriku" is the name of the northern part of the main island, "Honshu", of Japan which faces the Pacific Ocean. The coast is well-known for big tsunami waves which have been observed many times since ancient times. During the last hundred years, the coast has experienced three large tsunamis occurring in 1896, 1933 and 1960. In Japan, the last one is called the Chilean tsunami because it originated in Chile.

In this report, several types of tsunami protective seawalls (tsunami walls) are introduced which are erected in fishing villages of Iwate Prefecture on the Sanriku coast of Japan. We also describe the reasons why such structures were constructed in these villages.

2. The Circumstances Surrounding Fishing Villages on the Sanriku Coast

Two ocean currents pass near Japan. One is the warm current, called "Kuroshio" in Japan, and the other is the cold current, "Oyashio". Owing to these two currents, the sea around Japan has rich fishery resources. In particular, the offshore of the Sanriku coast is one of the richest fishing grounds in the world because of the confluence of the

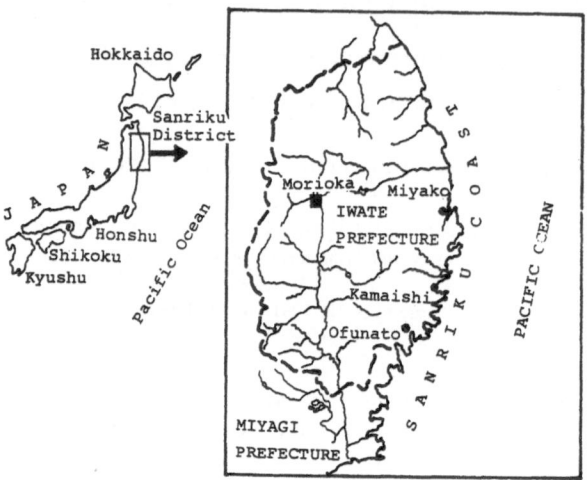

Fig. 1.   Location of the Sanriku coast.

main flows of these two currents.

In addition, this coast has many landlocked calm bays to serve as good berthing for fishing boats. This is the reason why people on the Sanriku coast have depended on fishery for their livlihood and developed many fishing villages on the fringe of these bays since ancient times.

Recently, progress in marine culture techniques in Japan has brought a shift of fishing activities towards sea farm breeding from fish catching. This tendency is also seen on the Sanriku coast.

The greater part of cultured marine products is composed of voluminous ones such as shellfish and seaweeds. As a result, the total amount of fishery products landed on fishing ports along the coast has been rapidly increasing, and a shortage of fishing port facilities and land space for handling and processing these fresh products has become serious in the ports fronting many villages on the Sanriku coast. Photo 1 shows such a case where fishermen and their families are handling cultured seaweed in the limited space on an old breakwater, caused by the shortage of usable land area.

On the coast, generally speaking, it is very expensive to enlarge a fishing port to provide port facilities and land space, because the coast forms a steep hills or cliffs directly from seashore, and its seabottom also forms a steep slope. For this reason, usable land space is very valuable on the coast.

3.   Past Implementation of Tsunami Countermeasures in Fishing Villages on the Sanriku Coast

After a heavy loss of life and property by the great Sanriku Tsunami occurring in 1933, the Government of Japan recommended the damaged villages prepare new housing sites on higher ground with the assistance of a government subsidy. Thereafter, new housing sites were created in many villages, but the space for these sites was not

Photo 1.   Landing of Seaweed (called "Wakame" in Japanese).

Photo 2.   Typical fishing village, (Ryoishi), on the Sanriku coast.

sufficient to accept all the people who wanted to relocate because the scale of the new space depended on the above mentioned geographical conditions. Furthermore, the increase of households in recent years and the desire of fishermen to live close to their home ports caused many people to inhabit the dangerous low land near the shore. Because of this situation, construction of tsunami walls was required.

Only a few villages had completed construction of tsunami walls before the attack of the 1960 Chilean tsunami, while in other places, the construction of tsunami walls had been slowed down or suspended due to financial difficulties in those days. The 1960 Chilean tsunami brought a very heavy loss along the Sanriku coast, because most of the villages on the coast had an insufficient scale of tsunami walls.

After the disaster, the Japanese Government enacted a special law to subsidize about 80% of the construction cost for restoring damaged places and for constructing new tsunami walls against tsunamis whose wave height would be less than or equal to that of the 1960 Chilean tsunami. In 1966, all construction work based on the special law was completed. The total length of these walls reached 14.5km in 28 fishing villages of Iwate Prefecture on the coast. Figure 2 is a typical tsunami wall constructed in a village. This type of structure is aimed at protecting its hinterland from tsunamis and, at the same time, providing port facilities for promoting fishery in a limited land area.

For supporting daily activities and assisting smooth evacuation, many steel gates were installed on the tsunami walls. The distance between gates is about 150 m to 200 m. Today, the responsible authorities are planning to replace these steel gates one by one with aluminium gates, because the later are easy to handle in case of an emergency and do not need painting for prevention against corrosion. Photo 3 shows an example of an aluminium gate.

4.   Present day Implementation of Tsunami Countermeasures

Soon after the completion of the Chilean tsunami countermeasures, the second stage tsunami countermeasure improvement plan was started. It aimed at protecting villages against larger scale tsunamis than that of the 1960 Chilean tsunamis. The crown height of these second stage tsunami walls were decided based on records and data on the 1896 and 1933 Sanriku great tsunamis. In some cases, the planned crown height of tsunami walls after improvement are over 10 m from Datum Level (D. L.: almost the same as the lowest low water level in Japan) and the highest one is about 16 m from D. L., while, for reference, the highest crown height of the Chilean tsunamis was 7 m from D. L.

In the case of such large walls, some were planned for improvement in two steps as shown in Fig. 3, where a step tsunami wall is under construction. A high and big wall such as in Fig. 3 is not very good for daily life in a village because it prevents circulation of fresh air and creates a large shadow. Additionally, it does not blend with the landscape in and around the village.

One of the effective measures to diminish the height of a tsunami wall is the construction of tsunami breakwaters. Such breakwaters have already been constructed in Ofunato Bay, and the mitigating effect of these was verified when the Tokachi-oki tsunami attacked in 1968. However, as a result of the recent development of cultural

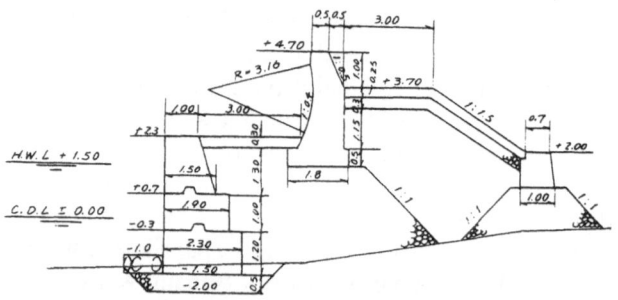

Fig. 2.   Dual purpose structure (composite type tsunami wall).

Photo 3. Aluminium gate (by the courtesy of Nippon Light Steel Co. Ltd.).

Photo 4. Contrast of old part and improved part of a tsunami wall.

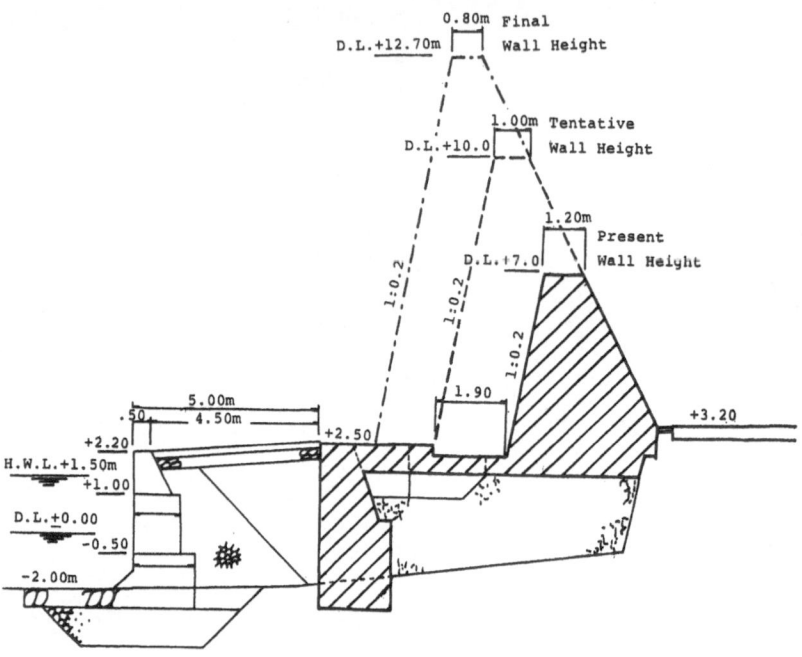

Fig. 3. Stage improvement plan of a tsunami wall.

fisheries on the Sanriku coast, the construction of breakwaters at the mouth or inside of a fishery-oriented bay is resisted by fishermen, because such structures may change the water circulation in the bay. In addition, favorable sites for constructing tsunami breakwaters from the viewpoint of geographical conditions, are limited along the Sanriku Coast. Judging from this situation, the construction of tsunami breakwaters will become more difficult in the near future although it is a very sophisticated method of mitigating tsunami waves. This situation has compelled us to make plans to protect fishing villages against tsunamis not offshore but on the shore. This means that we have to construct large expensive tsunami protective walls along the coast.

5.   Future Implementation of Tsunami Countermeasures

In the future, as a third stage implementation, multi-purpose tsunami walls should be constructed for effective use in order to relieve the shortage of usable land. Figure 4 is one such example now under construction. The inside of this wall is utilized for a road to replace the old road. Judging from the shortage of land in many villages on the coast such tsunami walls will increase in number in order to accommodate fishing boats, to store fishing gear and other goods, and to provide parking space. If the surface of a tsunami wall is made in the appropriate shape, it can be used as a safe working place for drying fish, seaweed and nets, and for other activities.

As a final stage, artificial land area should be created by using reinforced concrete columns and slabs for recovering the sunlight, breeze and a pleasant view. This will

Photo 5. Cultural fisheries field.

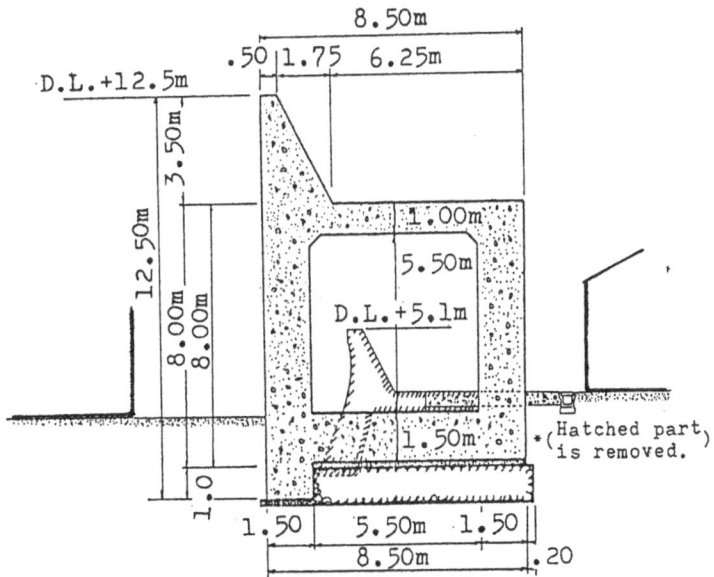

Fig. 4. Multi-purpose tsunami wall.

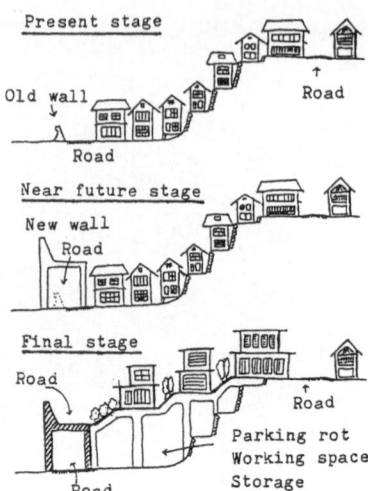

Fig. 5.   Sketch of a village improvement plan.

improve the living conditions in villages where houses will be constructed on artificial ground.

As a conclusion, we should continue efforts to improve and construct large tsunami walls for increasing the safety of villages against tsunamis. However, it should not be believed that these tsunami walls will be a perfect countermeasure against all future tsunamis. Therefore, other measures such as a tsunami alarm system, emergency access, and high ground refuge space should be developed in parallel to the construction of tsunami walls.

The data and figures used in this report are presented by the courtesy of Mr. Noboru Oshima, Director, Fishing Port Division of Iwate Prefectural Government, and Mr. Sunao Sakai, Deputy Director, Disaster Prevention and Coastal Protection Division, Fishing Port Department, Fisheries Agency, Government of Japan.

Tsunamis—Their Science and Engineering, edited by K. Iida and T. Iwasaki, 397–407.

Design and Construction of Ohfunato Tsunami Protection Breakwater

Teruji MATSUMOTO* and Yuhzo SUZUKI**

*Director-General, The Second District Port Construction
Bureau, Ministry of Transport, Yokohama, Japan
**Deputy-Head, Yokohama Research and Design Office,
The Second District Port Construction Bureau, Ministry of Transport, Yokohama, Japan

(Received September 3, 1981; Revised December 3, 1981)

The Second Port Construction Bureau, Ministry of Transport, constructed a tsunami protection breakwater in Ohfunato Bay for the first time in Japan. Until then, seawalls had been constructed as the tsunami protection works. However, the construction of a breakwater held more benefits than seawalls in case of Ohfunato Bay. This paper explains its design and work. The maximum water-depth of the construction site was 35 meters. There was not so deep-sea breakwater as this at the time in Japan and now is neither. Many new ideas were adopted into its design and the construction work was executed by surmounting numerous difficulties. Since completion, several large earthquakes have occured and a few tsunamis have attacked. The peak-cut effect of the breakwater was demonstrated. We are now observing the acceleration of the caissons during earthquakes, tsunami height and the settlement of the breakwater after such events.

1. Introduction

The Sanriku Coast where ria coast prevailed had sustained severe damage several times by large tsunamis caused by nearby earthquakes before the Chilean tsunami attacked. But the tsunami protection works had been executed slowly until then. With the disaster caused by the Chilean tsunami as a turning point, the government enacted the "Law for Special Tsunami Protection Measures" and began to execute tsunami protection works according to a plan for the prevention of the recurrence of future such disasters. The Ohfunato Tsunami Protection Breakwater was one of the largest works among them. The breakwater was constructed by the Second District Port Construction Bureau, Ministry of Transport. The construction work started in 1963, three years after the Chilean tsunami, and the breakwater was completed in March, 1967. The total cost amounted to 1.9 billion yen at that time. This was the first time that a breakwater was constructed as a tsunami protection work.

2. Merits of the Construction of the Breakwater and its Effects

Up to that time, seawalls had been constructed along the seashore as the main tsunami protection works. However, we examined and found that the construction of a

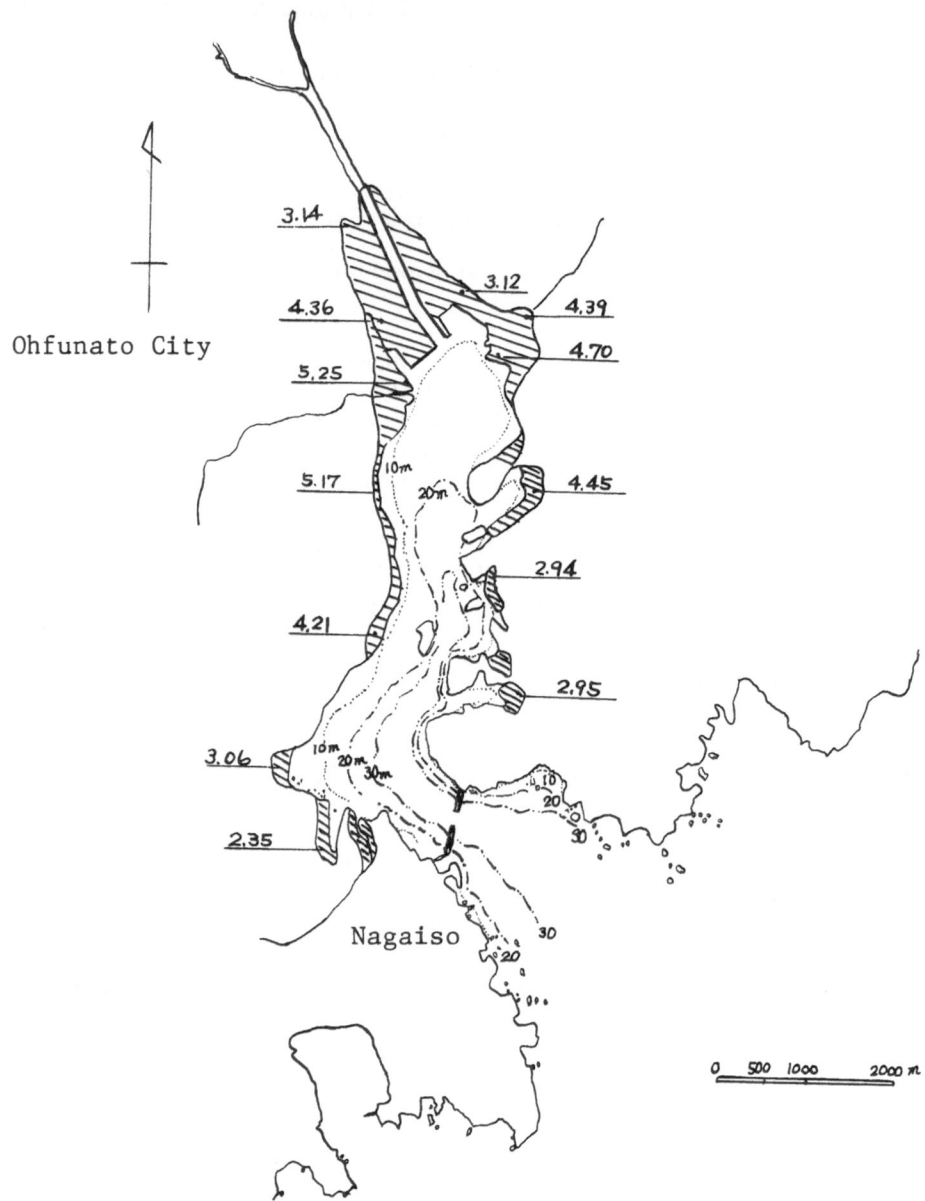

Fig. 1. The map of Ohfunato Bay. ▨ The inundated area by the Chilean tsunami. <u>4.00</u> The number shows the tsunami height by the Chilean tsunami.

breakwater would be better as the tsunami protection works than seawalls in the case of Ohfunato Bay. There were several reasons. The first was that seawalls might act as an obstacle against future development of the Ohfunato Port. If large and high seawalls were to be constructed along the shore, they would separate the port from the city. Furtheremore, Ohfunato City is lacking in flat land because the mountains there approach almost up to the bay. If seawalls were constructed, not only the port could not function sufficiently, but also the city would be prevented from development. A second reason was that we had to preserve the assets of the bay, such as many ships, nurseries and timber, when a big tsunami attacked the bay. These couldn't be protected by seawalls, but only by a breakwater. The third reason is that the shape of the bay is narrow and long.

Figure 1 shows the area inundated by the Chilean tsunami. It would be necessary to construct long seawalls as a tsunami protection works. The construction cost would be extremely high. On the other hand the entrance of the bay is narrow, and it is suited to reduce tsunamis flow-in in conjunction with a breakwater. The fourth reason was that several beneficial effects could be expected by the construction of a breakwater. The most important effect was the peakcut effect by the breakwater which was estimated to be 1.9–2.6 meters by calculation for the same scale as the Chilean tsunami and other large tsunamis caused by nearby earthquakes. Furthermore, the breakwater would reduce the velocity of the tsunami flowing in toword the protected area, except in the vicinity of the entrance, and it would delay the tsunami arrival time to the city and the port. There were problems to consider, such as the volume of permeating water through the rubble mounds, resonant oscillation and influence on other places outside and in the vicinity of the breakwater. We examined these problems and found them not to be a cause of worry.

Several places were examined as the construction site. The site at Nagaiso had good foundation. The sea bed consists of rocks. And we would be able to obtain a great deal of rocks for the rubble mounds from the site. Therefore, this site was selected on the basis of easy availability of rocks compared to other sites.

3. Structure Design of the Breakwater

Figures 2 and 3 show the longitudinal section of the breakwater. Its total length was 737 meters. The harbour entrance channel was secured 16.3 meters in depth and 200 meters in width so that 100,000 D.W.T. class ships would be able to sail through. So, the two visible portions of breakwater was constructed for a total of 537 meters and the submerged portion of the gap was constructed for 200 meters. The water depth of the construction site ranges from 12 meters to 35 meters. The maximum height of the breakwater was 40 meters. Next we examined the stability of the breakwater.

Figure 4 shows some cross sections of the breakwater. Sections A–A and C–C are Caisson-type composite breakwaters. Caissons were installed on rubble mounds consisting of weathered rocks ranging from 10 to 50 kilograms each. Rocks weighing more than 300 kilograms a piece protect the slope of the mound against wave force. The weight of the largest caisson filled with sands and coping concrete was about 5123

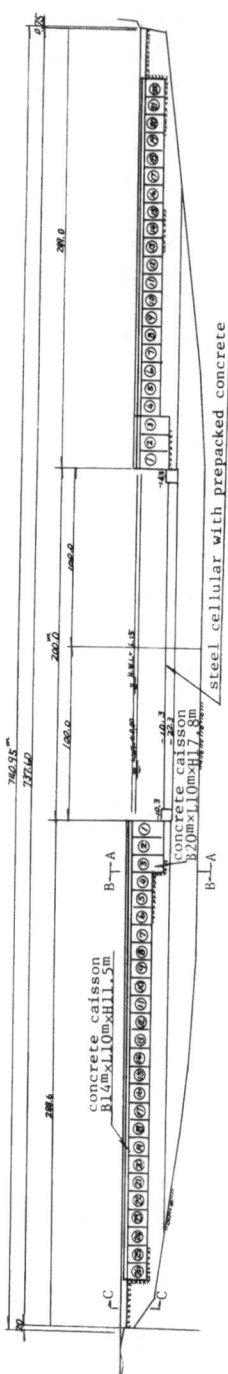

Fig. 2.   The longitudinal section of the Ohfunato Tsunami Protection Breakwater.

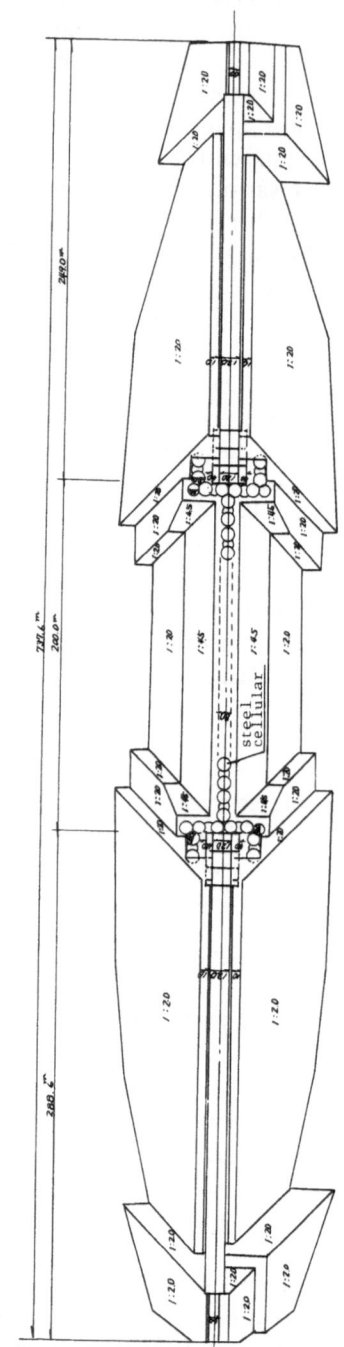

Fig. 3.   The plane section of the Ohfunato Tsunami Protection Breakwater.

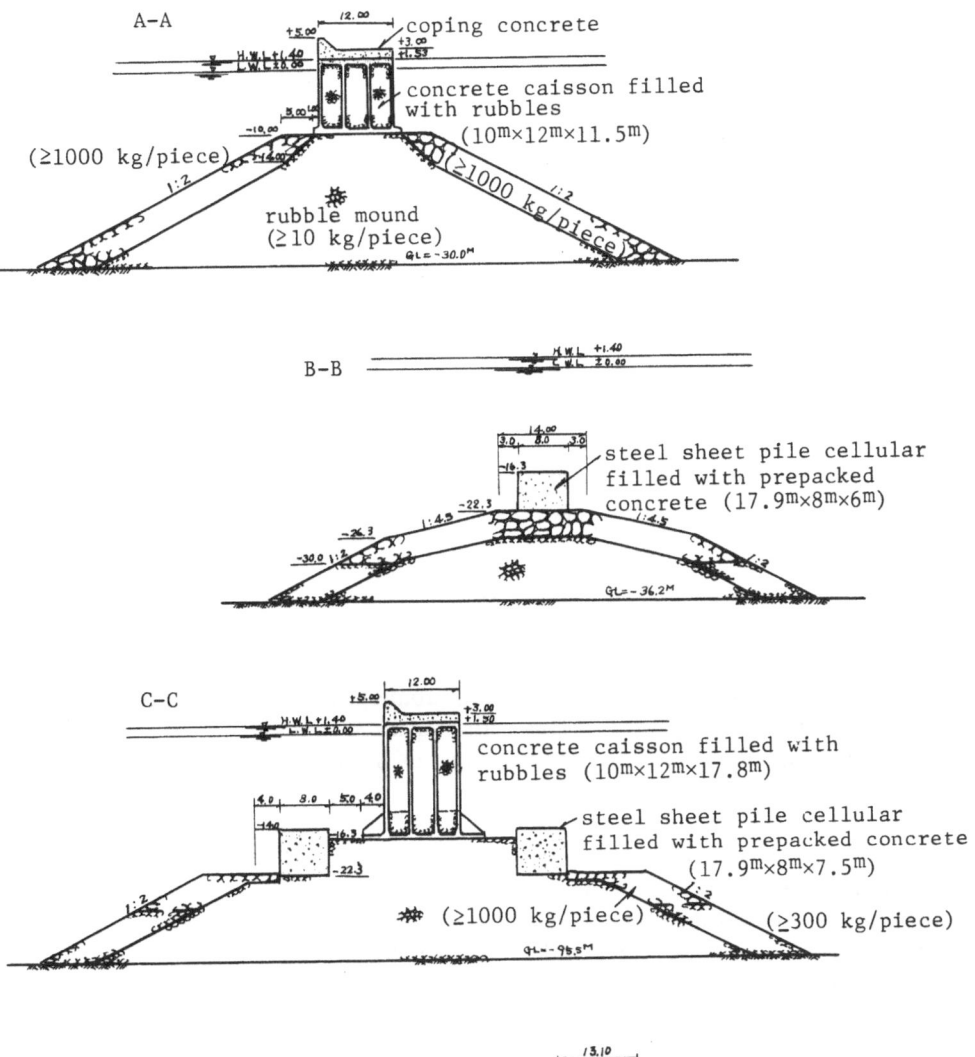

Fig. 4. Standard cross section of the breakwater.

Table 1. Dimensions of the upright section of the breakwater.

	Size of caissons or cell	Caisson weight in air	Upright structure weight in air	Upright structure weight in sea
Concrete caisson for deepwater section (Section C-C)	Length × Breadth × Height $10^m × 12^m × 17.8^m$	1420.4 tons	5123 tons	2656 tons
Concrete caisson for shallow-water section (Section A-A)	$10^m × 12^m × 11.5^m$	908.5 tons	3420 tons	1773 tons
Steel sheet pile cell filled with prepacked concrete (Section B-B)	$17.9^m × 8^m × 6^m$	—	1690 tons	942 tons

tonnage in air (see Table 1). Section B–B is a submerged steel-cellular type composite breakwater. The steel-cellular caissons with prepacked concrete were used to protect the top of the mound from return flow of a tsunami. Section D–D is transitional part of the breakwater and is rubble mound breakwater.

As for the stability of the upright section, the sliding and the overturning of the structure and the bearing capacity of the foundation on the top of the mound were calculated. External forces to the breakwater were buoyancy and wave force. For earthquakes, inertial force and dynamic water pressure were considered instead of wave force. The resistance force is the deadweight of the structure and the shearing resistance force of the mounds. Breakwaters are in general designed against wave force and against seismic force. But this breakwater had to be designed not only against those forces, but also against tsunami force. We considered the tsunami force and hydrostatic pressure due to the difference in water level between both sides of the breakwater during tsunami. Tsunami force was calculated as elliptic trochoidal wave. As for seismic design, we adopted modified seismic coefficient method because of the height of the breakwater being 40 meters. The dimensions of the caissons were determined by tsunami force because the design wave height was 3.5 meters and not so high. As for the stability of the rubble mounds, the examination of slip was made on all planer slip planes and circular slip planes, by considering apparent seismic coefficient in water. The gradient of the mound slope was determined by seismic force. The section at harbour entrance part was determined so as to have sufficient stability against a current force during tsunami flowin. Thus the dimensions of the steel-cellular caissons

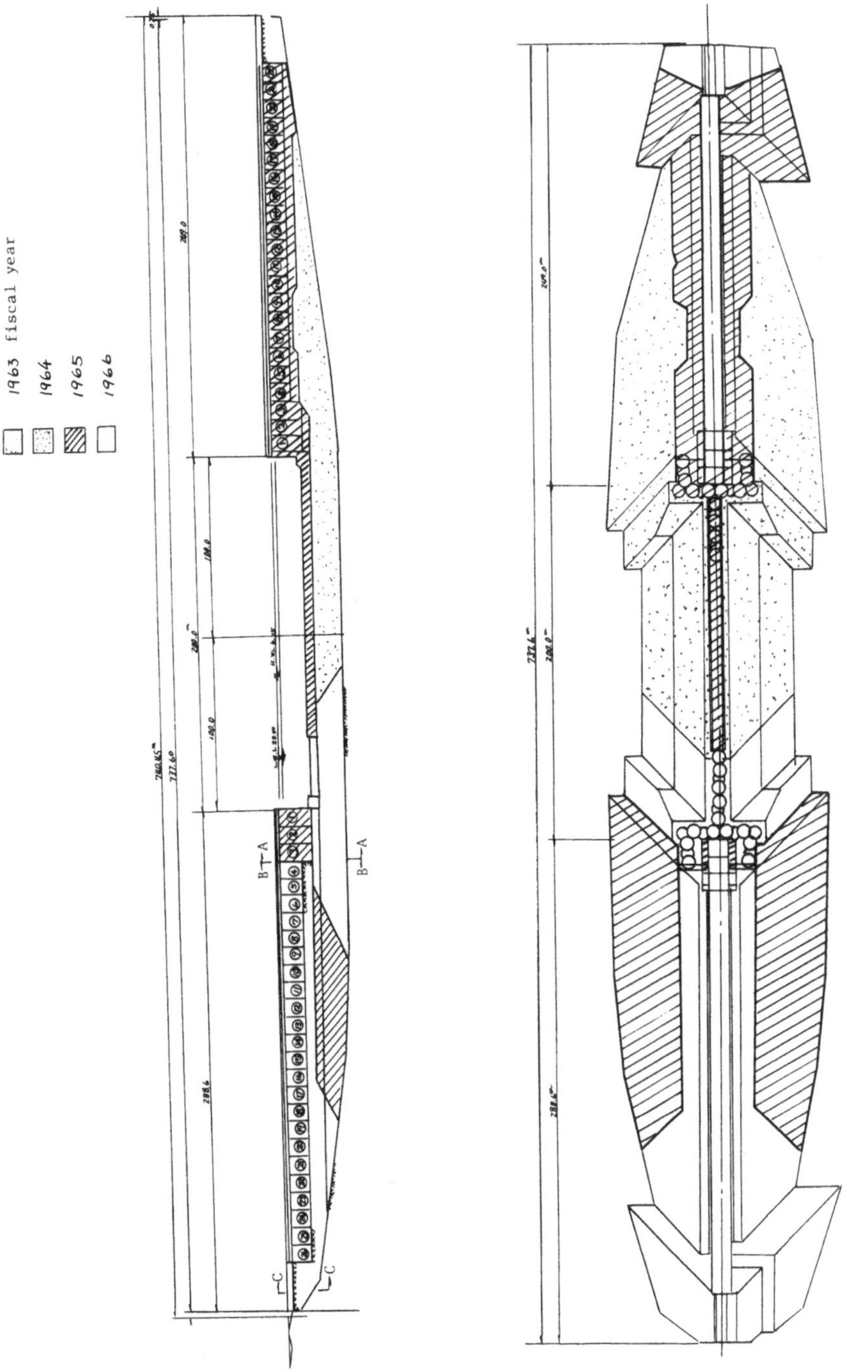

Fig. 5. Construction year of Ohfunato Tsunami Protection Breakwater.

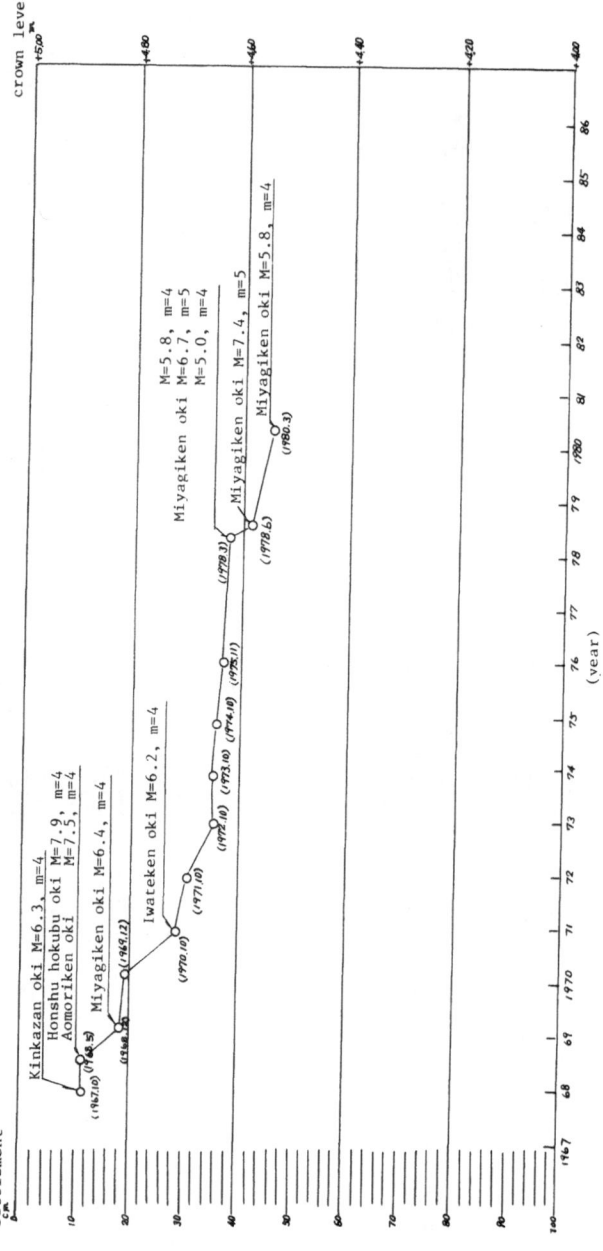

Fig. 6.   Settlement observation diagram of the breakwater at certain points.

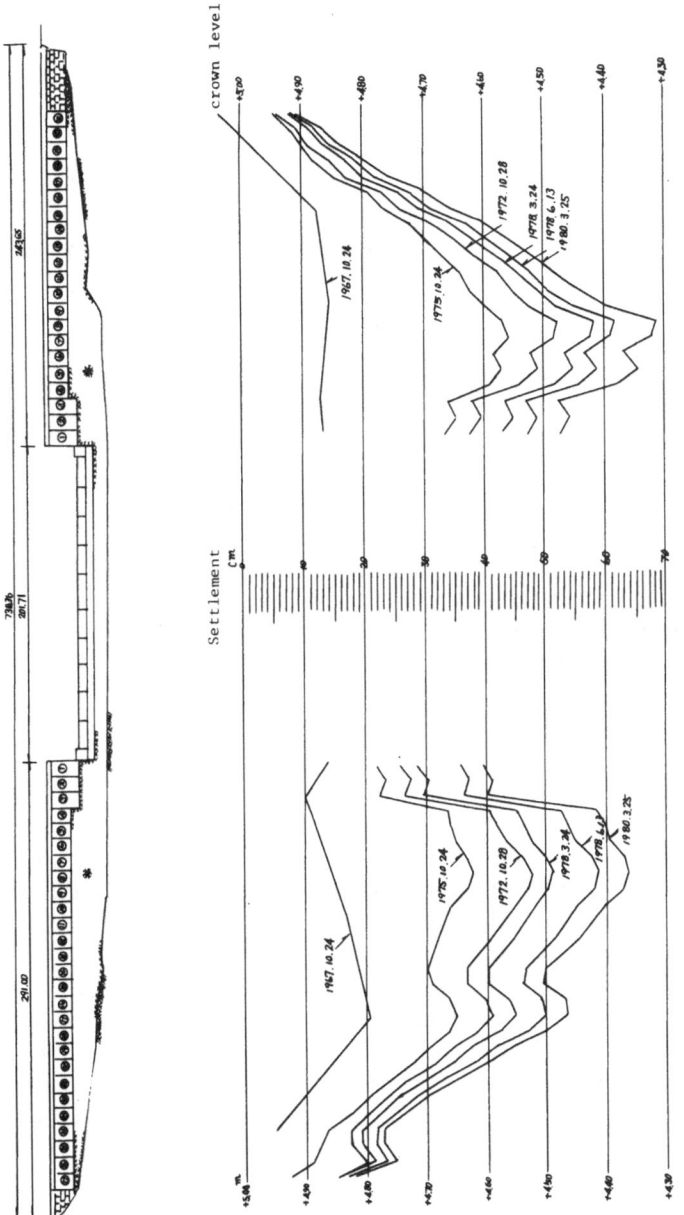

Fig. 7. Settlement observation diagram of the breakwater along the longitudinal section.

were determined. The sea bottom at the construction area was very firm, because of very stiff clay-slate stones outcropping. Therefore we did not need to examine the stability of the foundation.

4.   Construction Work of the Breakwater

Figure 5 shows the construction procedure for the breakwater. First, the submerged breakwater and the rubble mounds were constructed. Rubble was quarried from a mountain in the vicinity of the construction area. It was carried by stone carriers with grab buckets and was dumped into the water. The total amount of the rubble utilized in the construction, which was the main material of the breakwater, was about 1.13 million cubic meters. After the top of the mound was leveled by divers, the steel cellular caissons were installed by floating crane, and then prepacked concrete was filled into the cells as quickly as possible. After the completion of the submerged section, we constructed the deeper sections. The concrete caissons, which were used as the upright structure of the breakwater, were fabricated at the concrete caisson yard of Miyako Port. That was because there was no caisson yard in Ohfunato Port. Miyako Port is about 120 kilometers far from Ohfunato Port by sea. In all, 48 caissons, were towed from Miyako to Ohfunato Port in about 22 hours by tug boats. Six of these caissons, which were used at the deeper-sea area, were modified with additional concrete placing work for 7.8 meters in height at the temporary placing yard after being towed. This was due to capacity limitation of the caisson yard. Thus the breakwater was constructed by surmounting many difficulties such as the construction of the huge rubble mound and the diving works in the deep-sea.

5.   The Breakwater after Completion

Fourteen years have passed since the breakwater was constructed. By now the breakwater has been attacked several times by earthquakes and tsunamis. Especially, when the Tokachi offshore tsunami attacked in Dec. 1968, which was soon after completed construction, the deviation of the water level was observed to be 1.2 meters by tide guage stations installed inside and outside of the breakwater. But the deviation without the breakwater was estimated to be about 2.2 meters by calculation. The peakcut effect of the breakwater was recognized to be about 1 meter and we were confident that the breakwater was very effective for protection against tsunamis.

Figure 6 shows the cumulative settlement of the breakwater. The top height of the breakwater was 5 meters. The maximum settlement of the breakwater has amounted to about 45 centimeters. This settlement was mainly caused by nearly large earthquakes, you can see.

Figure 7 shows settlement along the longitudinal section. The section which has thicker mounds has larger settlement. This is because the rubble mound is settling by compression every time a large earthquakes occurs. We are now examining the total settlement and its influence to the function of the breakwater.

Recently we have started a project to construct a tsunami breakwater at Kamaishi Port which is located in the neighbourhood of Ohfunato Port. This breakwater is to be constructed at the sea area with 65 meters in maximum water depth. This will be one of the biggest breakwaters in Japan, and also in the world. We are examining the stability of the breakwater by referring to the tsunami and seismic observation at the Ohfunato breakwater. We can say that the construction of the Ohfunato Tsunami Protection Breakwater was one of the monumental works in Japan not only as a tsunami protection works, but also as a large-scale civil work in the deep-sea.

REFERENCES

ITO, T., K. TANIMOTO, and T. KIHARA, Digital Computation on the effect of breakwaters against long-period waves (4th Report), Report of Port and Harbour Research Institute, Vol. 7, No. 4, pp. 55–83, 1968 (in Japanese).

IWATE PREFECTURE, Reconstruction report from the disaster by the Chilean Tsunami, pp. 192–202, Mar., 1969 (in Japanese).

MIYAKO PORT WORK OFFICE, The Second District Port Construction Bureau; The construction work report about Ohfunato Tsunami Protection Breakwater—Chapter of the plan and research, Mar., 1965; Chapter of the construction, Aug., 1967 (in Japanese).

Tsunamis—Their Science and Engineering, edited by K. Iida and T. Iwasaki, 409–421.

A Hybrid Simulation System Developed for Model Tests of Tsunamis in a Harbor

Toshio IWASAKI

Department of Civil Engineering, Faculty of Engineering,
Tohoku University, Sendai, Japan

(Received September 10, 1981; Revised December 16, 1981)

A scale model of Kamaishi and Ryoishi bay is described with a discussion of scale ratios, inclusive region and generating system of tsunami waves. Numerical procedures to obtain design tsunamis are discussed with particular focus on mesh sizes. The tsunami input point for a model basin should be adjusted by the bathymetric difference between model and prototype, which also necessitates modification of the wave height by Green's law. Then incidental component is computed using a method of characteristics. Measured wave forms are compared with those of the numerical models and results show that reflected components cannot be eliminated in the scale model. Since the major portion of the oscillations have passed along before the arrival time of the waves again reflected at the bay heads, evaluation of effects of breakwater to reduce inundation heights can be successfully achieved. Some ideas on refinement of the system are presented.

1. Introduction

Recent progress in numerical computation has made possible analytical assuming of tsunamis along coast line. However, the method is still unreliable because the vertical velocity component, the effect of surface curvature, and frictional effects are dominant near the coastline. Although runup and drawdown have been successfully modeled in recent years, the results are still not completely reliable.

Engineering evaluation of structures against tsunami disaster necessitates much quantitative data. Such data may be obtained by physical model experiments, as is usally done for problems on short waves or tides, since the effects of topography and non-linearity can be included in scale models. However, tsunami inputs have not been employed yet. Instead, sinusoidal, cnoidal or solitary waves have commonly been used. Since a tsunami is not a permanent wave, but transient with a power spectrum, inputs should be specified with design tsunami for engineering structures. This has not been done before, and is the object of this study.

Firstly, a tsunami model basin and generating system of tsunami are described. Numerical procedures to obtain design tsunamis and tsunami inputs at the entrance of the model basin are then presented and the results of a simulation and evaluation of a breakwater are illustrated. Finally, other possible applications of physical modeling and limitations of the system are discussed with an idea for refinement of the system.

2.   Tsunami Model Basin and Tsunami Generating System

Figure 1 shows a scale model of the area which includes Kamaishi Bay and Ryoishi Bay. The problem was to invesigate the hydraulic behavior of tsunamis in both bays in relation with construction of a tsunami breakwater at the mouth of Kamaishi Bay. The choice of the vertical scale was decided by requirement of plausible accuracy in the wave height measurement. This gave a scale of 1/120, which might be minimum since 1 mm in the model was equivalent to 12 cm in the prototype. The depth of the outer region to the bays could not be simulated for the actual bathymetry of the continental slope and was therefore set to a uniform depth of 100 m in the prototype scale, which was slightly deeper than the depth at the bay mouth. This illustrates one of the problems with physical models. The continental slope is 1/100 from the baymouth to 12.65 km offshore, followed by a slope of 1/45 until 25 km offshore.

Horizontal length scale was taken as 1/600. HIGUCHI (1961) presented a formula relating roughness coefficients between prototype and model.

$$n_r = (h_r)^{2/3}/x_r^{1/3}. \tag{1}$$

In the above equation, n_r, h_r, and x_r are the ratio of roughness coefficients, vertical and horizontal scales of the prototype against those of the model, respectively. Thus if $h_r = 120$, $x_r = 600$ and roughness coefficient of the prototype 0.03, a roughness coefficient of the model is to be 0.010 which can be realized by coating of the model surface by cement mortar.

Due to the relative positions of wave sources to the bays, the flow direction of the 1933 Sanriku Tsunami was estimated to be parallel with the latitudinal line which is

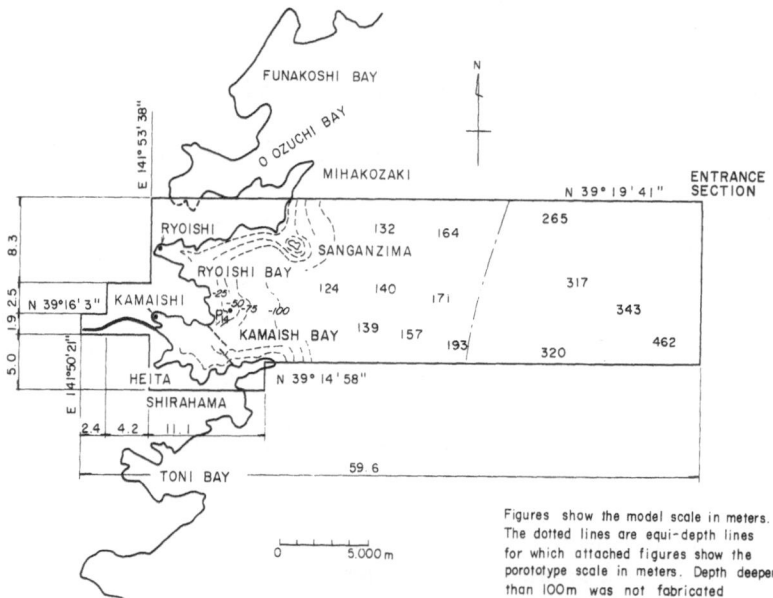

Fig. 1.   Area contained in the scale model (horizontal scale 1 : 600, vertical scale 1 : 120).

parallel to the axis of the area of two bays. The flow direction of the 1896 Sanriku Tsunami, which will be described below, was not parallel with the above one as seen in Fig. 3. However, a flow direction which coincides with the area axis is more dangerous.

The water flow of a tsunami coming from the above direction might be bounded by the capes which seperate Ryoishi Bay from O-Ozuchi Bay to the north and Kamaishi Bay from Toni Bay to the south. Thus, the area in the model basin was limited to that included by the two capes.

Figure 2 shows the tsunami wave generating system. The flow discharge is supplied from the tank into the model basin by an 600 mm axial flow pump and is returned from the model basin to the tank by the difference of water levels. Inflow to the basin is controlled by the diverging flow through a 600 mm butterfly valve, and the return flow is controlled by regulating the openings of a 350 mm rotovalve and a 500 mm rotovalve. Equating the inflow rate with the outflow rate produces no discharge into the model basin. A positive discharge is generated by closing the rotovalves from the equilibrium setting, while a nagative discharge is produced by opening the valves. Regulation of the rotovalves is accomplished by horizontal movement of pistons which is converted to rotation of rotovalves around the vertical axis through worm gears. On-off switching of the oil supply into either side of the piston chambers is electro-magnetic controlled so that the difference between the programmed and actual water level at the entrance of the model basin is minimal.

3. Design Tsunamis

Design tsunamis were specified by large tsunamis which actually occurred in the past. For Kamaishi Bay, the 1896 Sanriku Tsunami and the 1933 Sanriku Tsunami caused the most serious disasters. The basic equations are the two-dimensional,

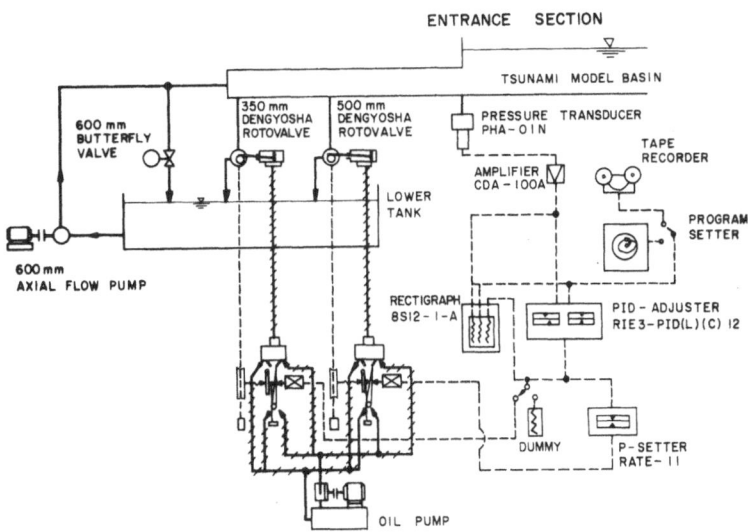

Fig. 2.   Schematic diagram of tsunami generator.

vertically integrated shallow water equations with transmission conditions for the open boundary and with zero velocity of the normal component for shorelines with a hypothetical depth specified in order to avoid drying up at the drawdown stage. A finite difference method has been used for the numerical computation since 1972 and the modeling of wave sources has already been investigated. (IWASAKI, 1974, 1976) Although fault models produce small tsunamis, it was pointed out that fault models are most reliable since these are deduced by seismic evidence with the use of the theory of elastic deformation in a semi-finite medium (YONEKURA and ANDO, 1973).

Thus efforts to refine the computational procedure have continued. This was done by adopting a telescoping technique from 10 km in the outer sea region to a very fine size down to 10/27 km in the bay regions, connected by a ratio of 1/3 in each step.

Figure 3 shows the computed area which includes the Hokkaido and Tohoku area, with the Japan Sea Trench from 140°E to 146°E in meridian and from 37°N to 43°N in latitude. Bathymetric contours are shown by light curves. The grid size is 10 km and the computed points are denoted J1 to J50 in latitude and K1 to K70 in

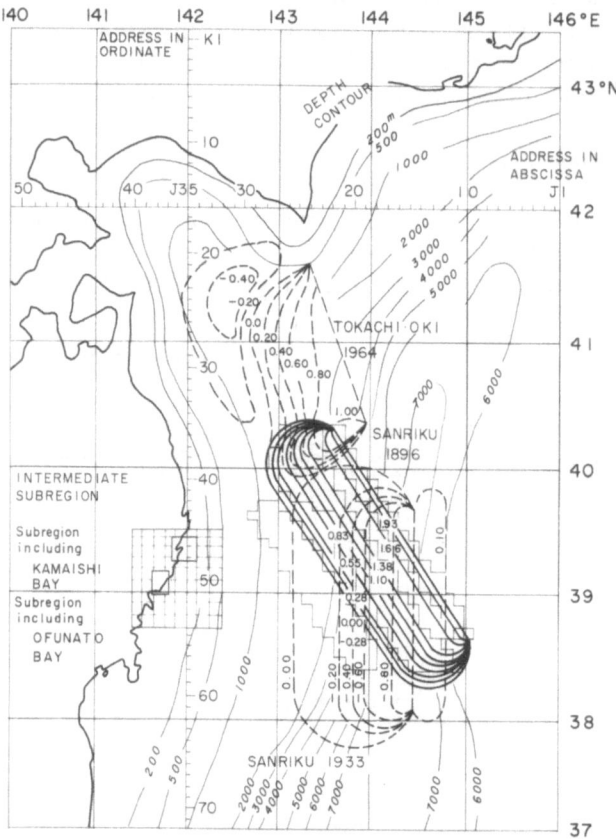

Fig. 3. Computed region with tsunami wave sources. (Numbers attached to the sources are assumed distributions of vertical displacement of sea bed.)

meridian. The intermediate subregion is shown in Fig. 3 where the grid size is 10/3 km. The subregion including Kamaishi Bay and the one including Ofunato Bay where the step size is 10/9 km are included. Relations between these subregions are shown in Fig. 4, and the final regions are the Kamaishi Bay Region and the Ofunato Bay Region where the grid size is 10/27 km.

Besides the two tsunamis mentioned above, the 1964 Tokachi-Oki Tsunami was analyzed for a verification test of the numerical procedure using marigrams, since the scale was moderate, without overshooting by huge undulations. For this tsunami, the model presented by ABE (1974) was used. In Fig. 3, wave sources are shown by dotted heavy lines with assumed distributions of vertical displacement of the sea bottom. Table 1 shows the dimensions of the wave sources.

The computed maximum wave height, ζ_m, was compared with observed trace records, ζ_b, by deducing a ratio k_0 between them. As seen in Figs. 3 and 4, outputs of the maximum wave height were obtained at several points in each subregion which enabled correlation of the ratio k_0 with the grid size. Figure 5 shows the results obtained, in which the abscissa is the mesh size of the subregion of output points and the ordinate is k_0. The ratio k_0 is about 10 to 20 for points in the grid size of 10 km as reported already in 1976. However, the ratio reduces as the grid size becomes finer, until nearly equalling unity for the finest grid size. Thus attention should be paid to grid size when one wishes to test the reliability of numerical results with actual traces.

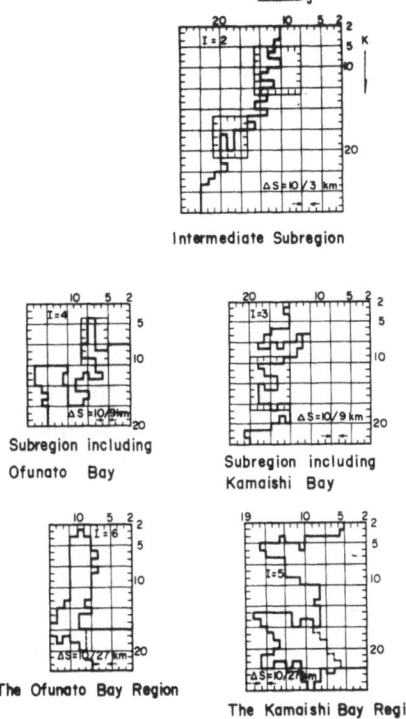

Fig. 4.   Subregions.

Table 1. Wave source parameters of design tsunami.

	1896 Sanriku Tsunami	1933 Sanriku Tsunami	1964 Tokachi-Oki Tsunami
Length of the longer Axis (km)	270*	300*	230**
Length of the shorter Axis (km)	110*	120*	110**
Mean vertical displacement of sea bed of wave source (m)	1.35	0.74	4.98
Parameters of fault model			
Length (km)	230	185	150
Width (km)	80	100	100
Area (km²)	1.84×10^4	1.85×10^4	1.50×10^4
Dislocation (m)	3.65†	1.65	4.10
Inclination of fault plane against horizontal	30°	45°	20°
Angle of fault line against meridian line (clockwise)	332°	0°	336°
Slip angle of faulting	90°	90°	
Type of faulting	Adverse dip slip with low angle	Normal dip slip	Adverse dip slip with low angle accompanied by strike slip component
Total displacement volume of water (cm³)	3.1×10^{16}	2.1×10^{16}	1.4×10^{16}
Effective moment M_0 (dyne · cm)	4.8×10^{28}	2.2×10^{28}	2.8×10^{28}
Rupture velocity (km/sec)		3.5	3.5
Time constant (sec)	100	10	

*HATORI, 1974.
**ABE, 1974.
†Estimated by a method proposed by KANAMORI (1972).

Figure 6 shows a comparison of wave forms at Nagasaki just outside of the Ofunato Breakwater and at the head of Ofunato Bay for the 1968 Off-Tokachi Tsunami. Figure 7 also shows a comparison of computed wave forms with visual observations. The discrepancy is not so great which verifies use of results of the computations for design tsunamis.

4. Tsunami Inputs at the Entrance of Model Basin

In the one-dimensional x-t plane, there are two sets of curves. C_1 and C_2, called characteristics, which are the solution curves of the linear long wave equations

$$C_1 : dx/dt = c_0$$
$$C_2 : dx/dt = -c_0. \tag{2}$$

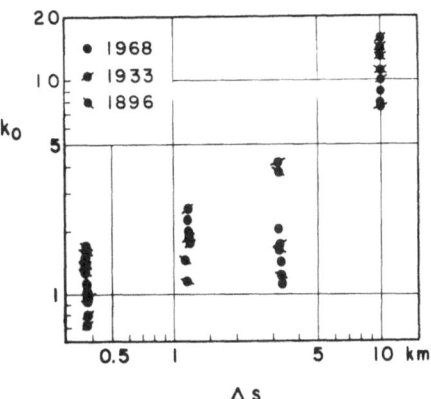

Fig. 5. Ratio of computed maximum wave height and observed trace record k_0 against mesh size of subregion of output points.

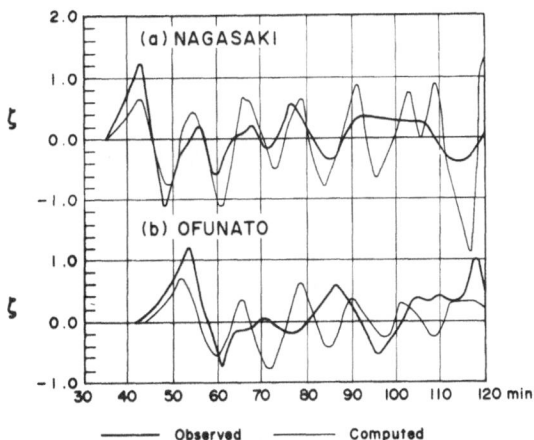

Fig. 6. Comparison of observed and computed time histories for 1968 Tokachi-Oki Tsunami. ($t = 0$ is the instant of outbreak of the earthquake.)

Also, the following relations hold,

$$\zeta_A + q_A/c_0 = \zeta_B + q_B/c_0 = Z^+ \qquad \text{along curve } C_1$$
$$\zeta_A - q_A/c_0 = \zeta_C - q_C/c_0 = Z^- \qquad \text{along curve } C_2 \tag{3}$$

in which ζ, q, and c are the surface elevation above a level without a tsunami, discharge across a cross section of unit breadth, and long wave celerity, respectively. The suffixes A, B, and C refer to values at an arbitrary point, at a point one step offshore, and at a point one step onshore, respectively. Suffix 0 means average value around the region including A, B, and C (Fig. 8).

These relations are used to find the tsunami input at the entrance of the model basin. Since the length of the part with depth equivalent to 100 m in the prototype is

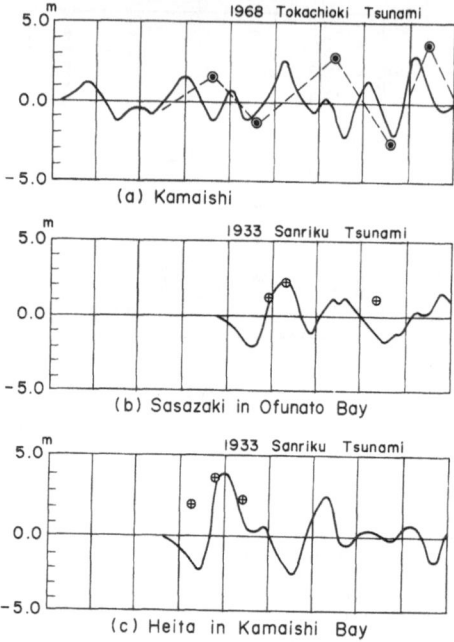

Fig. 7.   Computed time histories with visual observations.

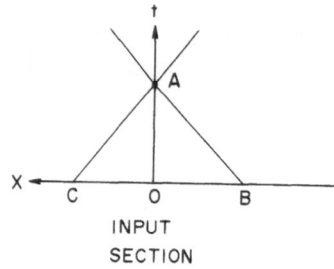

Fig. 8.   Lines of characteristics.

41.9 m, the travel time of long waves through this part is 14.66 sec in the model or 803 sec in the prototype, during which tsunamis may propagate through 25.41 km in the actual ocean since the depths are 226 m at a point 12.65 km offshore and 510 m at a point 25.41 km offshore from the entrance of the bays. Thus the above-mentioned 25.41 km offshore point was considered as equivalent with the entrance. Then Z^+ was computed as the sum of ζ and q/c_0. The water level at the entrance of the wave basin was given by $\zeta_B = 1/2\, Z^+$ and was used in the input program to regulate the water level at the entrance. This is of the character of a progressive wave component, aince in Z^+ the reflected wave component is absent. The input derived above is called a progressive wave. In contrast, if the water level at the entrance was regulated to follow the computed water level ζ, then from Eq. (3) we have $q_A = 0$ and the section works as a solid boundary. Thus such an input is called a standing wave input. It is obvious that progressive waves are to be generated as tsunami inputs.

Figures 9(a) and (b) are two such kinds of input waves for the 1896 Sanriku Tsunami and the 1933 Sanriku Tsunami, respectively, obtained by numerical analysis for the actual bathymetry. The difference between the two curves is caused by reflected components from the coastline. This is recognized since 25 min after the initiation of waves, while the wave front has travelled back again from the cape, P_{14}, shown in Fig. 1. Outstanding oscillations are confined before this time duration and the degree of reflection is not so great. This provides a strong basis for the credibility of the experiments.

5. Results of Simulation and Evaluation of Breakwater

Figures 10 and 11 show experimental results for the 1896 and the 1933 Sanriku Tsunamis, respectively. The locations of measurement points are shown in Fig. 1. The point P_{14} is on the nose of the cape which seperates Kamaishi Bay and Ryoishi Bay. Kamaishi and Ryoishi are small ports situated in the inner most part of the bays just in front of the quaywalls. Full lines are results of experiments for cases without the

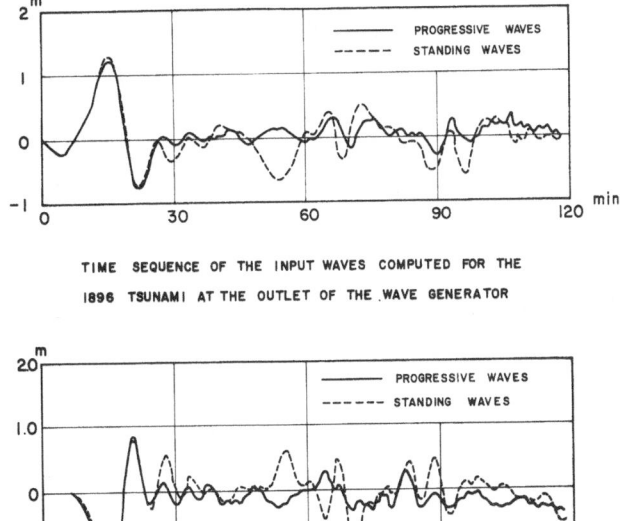

Fig. 9. Time history of input waves at the outlet of the tsunami generator. (a) for the 1896 Sanriku Tsunami, (b) for the 1933 Sanriku Tsunami.

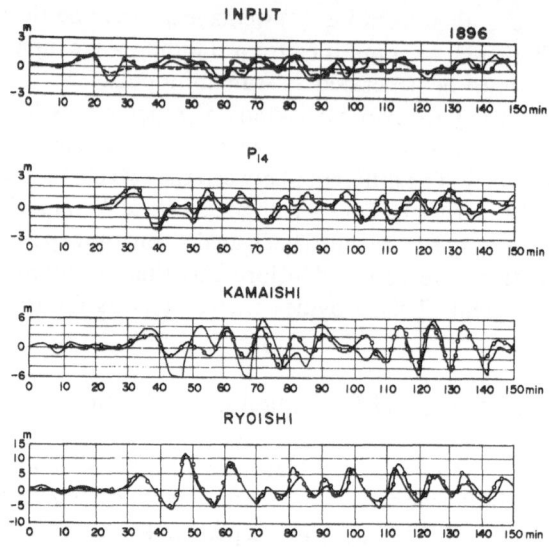

Fig. 10.   Experimental results for the 1896 Sanriku Tsunami.

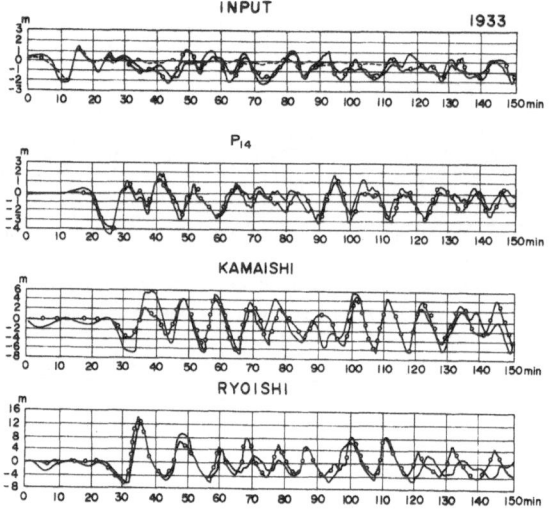

Fig. 11.   Experimental results for the 1933 Sanriku Tsunami.

Kamaishi Breakwater, full lines with circles are those for cases with Kamaishi Breakwater and dotted lines shown in INPUT are the progressive waves shown in Fig. 9.

    Since the PID-adjuster was set at the maximum rate of proportionality without rating of integration and differentiation, there was some time lag and gain loss of simulated waves relative to the programs. Vertical scaling was adjusted so that the maximum wave heights roughly coincided with inundation heights. However, the discrepancy between the full lines and the dotted lines in input curves is not so small. The reason for the discrepancy is believed to be as follows.

1)   The dotted lines show progressive components, but the full lines include retrogressive components.

2)   Time lag and gain loss owing the generating system built up a distortion of the wave form at an early stage before waves reflected from the bay model reached the entrance.

3)   Three kinds of reflected waves arrived. One was a part reflected from the cape, P_{14}, for which travel time was 14 min, the others were from Kamaishi Bay and Ryoishi Bay with travel times of 20 min and 17 min, respectively. These made up wave groups with three peaks which appeared roughly in every 30 min.

A clear correspondence is observed between the input curves and curves at P_{14}. Without the breakwater, the wave height was on the same order of that at the entrance, which means the reflected part at the cape was small. Wave energy mostly entered into the two bays. However, with the breakwater, much of the energy which had flowed into Kamaishi Bay was stopped and the wave height was augmented about twice. This means, moreover, that energy would be reflected at the cape. Examination of curves for Ryoishi verifies the above interpretation. Wave heights were not altered by construction of the breakwater in the neighboring bay.

A remarkable modification of wave configurations took place in both bays. Kamaishi Bay has two main natural frequencies. One is 20 min, for which the loop is at the bay head and the node is at the bay mouth. The other is 10 min for which loops are at Kamaishi and at Shirahama. Ryoishi Bay also has two frequencies, one of which is 16 min with loop and node at the head and the mouth, respectively, and the other is 13 min with loop at the head and node at a cross section 3 km off where the bay breadth narrows abruptly.

Thus, the tsunami energy which is confined in the first one or two waves of input is to be taken into account. This means that in Fig. 10, up to 50 min for P_{14}, and up to 57 min for Kamaishi and for Ryoishi are essential parts of the tsunami. Also, in Fig. 11, up to 37 min for P_{14} and up to 45 min for Kamaishi and for Ryoishi are time readings which are to be taken into consideration as experimental results.

In order to evaluate effects of the breakwater against tsunamis, maximum inundation heights should be quantitatively discussed. However, locality greatly and unevenly affects tsunami distributions. Table 2 shows trace heights reported by several organizations. They are remarkably inconsistent. Stating the location of the survey is definitely important. However, such descriptions are usually unavailable. Thus as a basis of structure design, inundation heights given by authorities from the engineering standpoint are taken as follows.

1)   For the 1896 Sanriku Tsunami, the maximum was T. P. 7.9 m at Kamaishi and 13.0 m at Ryoishi, when the mean water level was T. P. 0.64 m.

2)   For the 1933 Sanriku Tsunami, the maximum was T. P. 5.4 m at Kamaishi and T. P. 6.4 m at Ryoishi, when the mean water level was T. P. 0.0 m.

Meanwhile reasonable experimental results are given in Figs. 9 and 10 such as,

1)   For the 1896 Sanriku Tsunami, maximum heights before 57 min are 5.1 m at

Table 2.   Reported maximum traces by various sources.

	1896 Sanriku Tsunami		1933 Sanriku Tsunami	
	Ryoishi	Kamaishi	Ryoishi	Kamaishi
Haruo Matsuo	11.7^m	6.0^m	9.5^m	4.4^m
Iahimoto; Earthquakes and their Investigations	11.2	6.3	10.4	3.7
Ministry of Interior	6.7	7.9	5.5	4.1
Iwate Prefecture	13.7	7.6	9.8	4.6
Bureau of Meteorology	10.6	5.4	6.4	5.4

Kamaishi and 11.4 m at Ryoishi. If a correction is done by adjusting the dislocation of the fault plane and the sea level by the tsunami, the multiplication factor should be (7.9 − 0.64) + 5.1 = 1.42 for the 1896 Sanriku Tsunami. Then the maximum height at Ryoishi is 11.4 × 1.42 + 0.64 = T. P. 16.8 m.

2) For the 1933 Sanriku Tsunami, maximum heights before 45 min are 6.3 m at Kamaishi and 13.0 m at Ryoishi. If a correction is considered so that the maximum level at Kamaishi coincides with the design height, the multiplication factor should be 5.4 ÷ 6.3 = 0.86. Then the maximum height at Ryoishi is 13.0 × 0.86 = 11.2 m.

3) With construction of a breakwater at the mouth of Kamaishi Bay, the maximum height at Kamaishi will become 2.5 × 1.42 + 0.64 = T.P. 4.2 m for the 1896 Tsunami and 2.4 × 0.86 = 2.1 m for the 1933 Tsunami.

4) Ryoishi may not be affected by breakwater construction at Kamaishi. The crown height of sea walls constructed now along the shoreline of Kamaishi Port is T. P. 4.0 m. So, by construction of the breakwater, inundation by design tsunamis can be mostly prevented by incorporation of sea walls.

6. Conclusions

Without the recognition that tsunamis are not permanent but transient, design of engineering works have founded on insufficient basis. In this paper, the concept of design tsunami is proposed with corroboration by numerical analysis to obtain offshore wave forms. Physical model experiments were undertaken using these forms by extraction of progressive components which made possible evaluation of effects of breakwater construction for diminishing wave heights at bay heads.

Much other scientific information such as velocity field, vortex formation, eddy viscosity and so on can be obtained by this hybrid modeling, and will be useful in refining numerical models in turn.

However, since this was the first step of such a modeling, a deficiency was clearly found. Although, the intention was to eliminate reflection in the bay model from input waves, the automatic regulator system did not work completely and a gain loss and time lag were recognized. Fortunately, the length of the approach channel was long enough to delay the arrival of the reflected waves again at the bay heads until almost the whole part of the outstanding forced oscillations by tsunamis had passed by. Future refinement to eliminate this deficiency might be worked out firstly by adjusting the PID-operation and then by preliminary transformation of input waves, taking into account of characteristics of the automatic regulator. Accordingly, the system described in this paper is useful for bays where the basic period of free oscillations is sufficiently short and for tsunamis for which the duration of the main part of the oscillations is also short in comparison with the re-arrival time of waves reflected at the bay mouth.

The writer wishes to express his sincere gratitude to the Faculty of Engineering, Tohoku University, and to the Ministry of Education for continuing support by permitting usage of space and funds for this investigation, especially for administrative establishment of the Experimental Station of Tsunami Engineering in 1981. He also acknowleges, the Ministry of Transportation and Iwate Prefecture for contracting Research Projects on Defense Works for Tsunamis in Kamaishi Harbor. The Association of the Advancement of Construction Engineering assisted in these contracts and donated a part of the experimental facilities.

He also withes to express his thanks to members of the Station for their enthusiastic work on the experiments. In particular, Dr. Akira Mano, Lecturer of Tohoku University, has been engaging in the numerical and physical studies. Mr. Eiji Sato, Engineer of Tohoku University, also has been supporting the investigations.

REFERENCES

ABE, K., Kinematics of faulting of earthquake, *Science*, **44**–**3**, 139–145, 1974 (in Japanese).

HATORI, T., Wave sources of tsunamis in Pacific Ocean along north-eastern Japan, *Earthquake*, **27**, 321–337, 1974 (in Japanese).

HIGUCHI, H., Hydraulic model experiment on the oscillation of water level in Sakai Channel, *Coastal Eng. Japan*, **4**, 35–45, 1961.

IWASAKI, T., Computer aid for optimum design of tsunami waves, Proc. 14th Coastal Engineering Conference, pp. 642–659, 1974.

IWASAKI, T., Numerical models of huge tsunamis off the Sanriku Coast, Proc. 15th Coastal Engineering Conference, pp. 1044–1059, 1976.

KANAMORI, H., Mechanism of tsunami earthquakes, *Phys. Earth Planet. Inter*, **6**, 346–359, 1972.

YONEKURA, A. and M. ANDO, Crustal deformations accompanied by huge earthquakes along submarine ditches and topography, *Science*, **43**, 92–101, 1973.

Tsunamis—Their Science and Engineering, edited by K. Iida and T. Iwasaki, 423–435.

On the Hydraulic Aspects of Tsunami Breakwaters in Japan

Katsutoshi Tanimoto

Port and Harbour Research Institute, Ministry of Transport, Yokosuka, Japan

(Received September 8, 1981; Revised December 25, 1981)

In the present paper, previous studies of numerical computation on the effect of the Ofunato tsunami breakwater are first reviewed. The principle is to solve numerically the linearized long wave equations with the introduction of a quadratic term for head loss at the breakwater opening. The head loss coefficient is assumed to be 1.5.

The method is applied to the 1968 Tokachi-oki earthquake tsunami. Good agreement between the computed and the observed water level variation at the innermost point of the bay confirms the practical validity of the method. The computation is extended to the original bay when the breakwater had not been constructed. It is demonstrated that the breakwater would have reduced the tsunami height by one half for the 1968 tsunami.

Another aspect is the stability of the breakwater against tsunami forces. A method to calculate tsunami forces on a vertical wall is proposed on the basis of laboratory experiments. The stability of the existing vertical walled Ofunato tsunami breakwater is re-examined according to the present method. The result indicates that the Ofunato tsunami breakwater is stable with respect to sliding if the incident tsunami height is less than about 6 meters.

1. Introduction

As the coasts of Japan have suffered from many tsunamis, various counter-measures have been adopted in the coastal areas frequently attacked by great tsunamis. The most conventional countermeasure work is to construct seawalls along the waterline high enough to prevent the overflow of tsunamis. Some of these seawalls have a creast height of 16 meters above the mean sea level. The construction of high seawalls along the waterline of a port area, however, causes hindrance and inconvenience for daily port operation. Reduction of tsunami height by means of offshore breakwaters provides a solution for the protection of port areas from tsunami attacks.

After the Chilean earthquake tsunami in 1960, the construction of breakwaters as a countermeasure against tsunamis began to be considered. The Chilean tsunami caused severe damages along the Pacific coasts of Japan. Directly after the damages, a special governmental council was established to examine works for prevention of disaster due to possible future tsunamis with heights equal to those of the Chilean tsunami. For port areas, the construction of breakwaters which would reduce the height of tsunami without offering serious obstruction to daily port operation was investigated as a permanent countermeasure against tsunamis. Finally, the council approved the construction of tsunami breakwaters for the four ports of Ofunato (Iwate Pref.), Hachinohe (Aomori Pref.), Onagawa (Miyagi Pref.) and Mori (Wakayama Pref.). The locations of these ports are indicated in Fig. 1.

The breakwaters of Ofunato and Hachinohe were constructed directly by the Ministry of Transport, since both are of large scale and contained complex engineering problems which had to be solved for the construction. In connection with the planning of the Ofunato tsunami breakwater, the Port and Harbour Research Institute, Ministry of Transport, investigated the effect of breakwaters against tsunamis by means of numerical computation. In the present paper, these previos studies by ITO *et al.* (1968) and ITO (1970) are reviewed to demonstrate the effectiveness of the breakwater against tsunamis. The stability of vertical wall against tsunami forces is also re-examined on the basis of recent laboratory test results.

2. The Effect of Ofunato Tsunami Breakwater

The Ofunato tsunami breakwater was constructed in the period from 1963 to 1968. The breakwater site, near to the bay mouth, is 38 meters deep (maximum) and 738 meters wide. The open section was reduced to a width of 200 meters and a depth of 16.3 meters below the chart datum.

The principle of numerical analysis is to solve step by step the difference equations converted from the differential equations of motion and continuity for long waves. Higher order terms for inertia and bottom friction are omitted in the basic equations, but the following quadratic term for head loss is considered at the breakwater opening,

$$\Delta \zeta_l = \frac{f}{2g} u_b^2 \tag{1}$$

where, $\Delta \zeta_l$ is the head loss at the breakwater opening, u_b is the velocity through the breakwater opening, f is the head loss coefficient, and g is the acceleration of gravity. The head loss coefficient is simply assumed to be 1.5, as a sum of the coefficients of sudden contraction and expansion of 0.5 and 1.0, respectively.

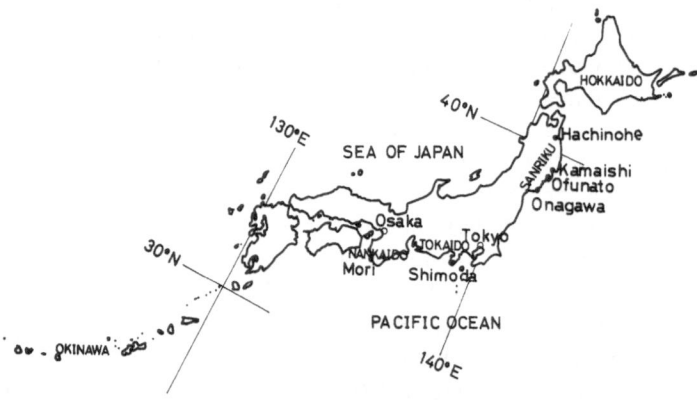

Fig.1.   Location map of Japan.

Figure 2 displays the grid system for Ofunato bay. The grid intervals Δx and Δy are both set equal 280 meters in the bay area. The breakwater opening has a width of 200 meters and a water depth of 16.9 meters below the mean sea level. In the grid system, the width of the breakwater opening is expanded to 280 meters with an equivalent water depth of 12 meters ($= 16.9 \times 200/280$), so that no modification of the continuity equation is necessary for the calculation of water level at both sides of the breakwater opening. This distortion of the breakwater opening gives no effect on the result in the present method, since the flow sectional area through the breakwater opening is kept equal for both of the prototype and the numerical model.

The outer sea is replaced by a channel of constant depth and width with a larger grid interval. This imaginary outer sea is only used for supplying the incoming tsunami into the bay area. Special equations are applied at the bay mouth because of the difference in the grid intervals there.

The incident tsunami is treated as a progressive wave train travelling through water which is originally at rest. The computation is started at the time when the front of tsunami arrives at the bay mouth. The time interval Δt is selected as 10 seconds.

As a boundary condition, the velocity component normal to the breakwater and shorelines including the sides of the imaginary outer sea is assumed to be zero. Another boundary is placed at the offshore end of the imaginary outer sea, where the velocity variation due to succeeding tsunami is given as a function of time. This offshore boundary is so located that any reflected waves from the breakwater or shorelines will not affect the phenomena in the bay area, after being re-reflected from the offshore boundary, before the end of computation. Recently, TANIMOTO et al. (1975) introduce a non-reflective offshore boundary for numerical computation of short-period waves in order to eliminate this imaginary outer sea.

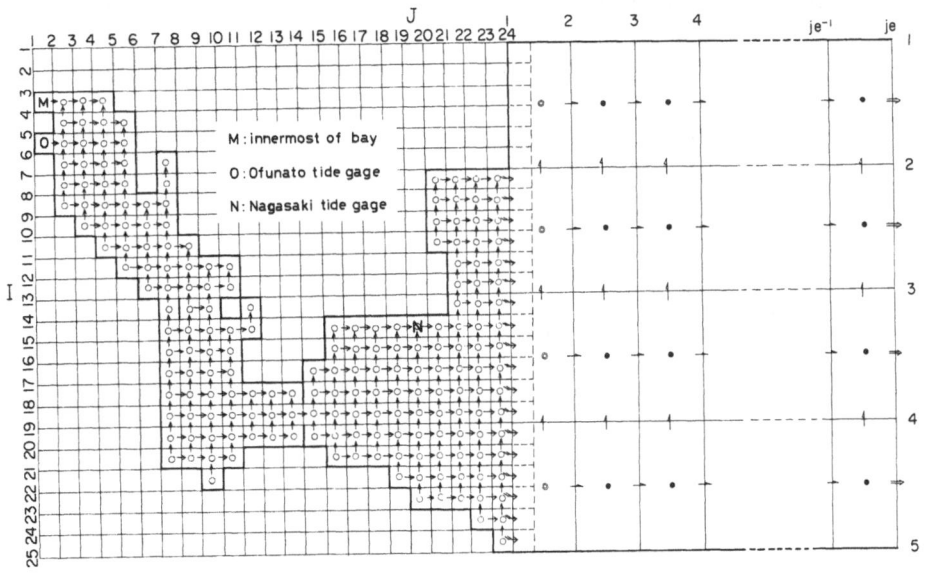

Fig.2.   Grid system for Ofunato Bay.

Firstly, the effect of the tsunami breakwaer was investigated for a hypothetical incident tsunami of sinusoidal wave form. Figure 3 exhibits the highest water level from the transient to the stationary states at the innermost point of Ofunato bay for an incident tsunami with an amplitude of 0.5 meter in the outer sea. The results evidently show the effect of the tsunami breakwater. The resonance of the original bay without the breakwater at the natural period of 37 minutes completely disappears after the break-water construction and the highest water level of 7 meters is reduced to approximately one meter. Although new resonances are found at the periods of around 16 and 8 minutes, the highest water levels are still less than or at most equal to those in the original bay.

After the completion of breakwater construction, the Tokachi-oki earthquake tsunami of 16 May, 1968, provided the first opportunity to demonstrate the effectiveness of the new breakwater. Actual records of water level variation were obtained at two tide gage stations, one at Ofunato located at the innermost point of the bay, and the other at Nagasaki located outside of the breakwater. In the same year, computations for this tsunami were made to examine the numerical analysis method and to confirm the effect of the tsunami breakwater.

In the analysis, the incident tsunami profile has to first be determined. The record of the outer tide gage station cannot be regarded as a progressive wave profile because of the influence of reflection from the boundaries. The incident tsunami profile in the imaginary outer sea for computation is determined so that the computed water level variation at the outer tide gage station agrees as closely as possible with the observed record. As for actual procedure, the record during the first 124 minutes is selected for discussion, and the water level variation at the outer tide gage is analyzed by Fourier sine series, having a fundamental period of 248 minutes. Then, only the first 30 components are taken into consideration. For the incident tsunami profile, the amplitude of each component is devided by the corresponding amplifying ratio which is defined as the average ratio of computed amplitude at the outer tide gage station to that of the incident sinusoidal wave during the first 124 minutes.

Figure 4 shows the observed and computed variations of water level at the outer

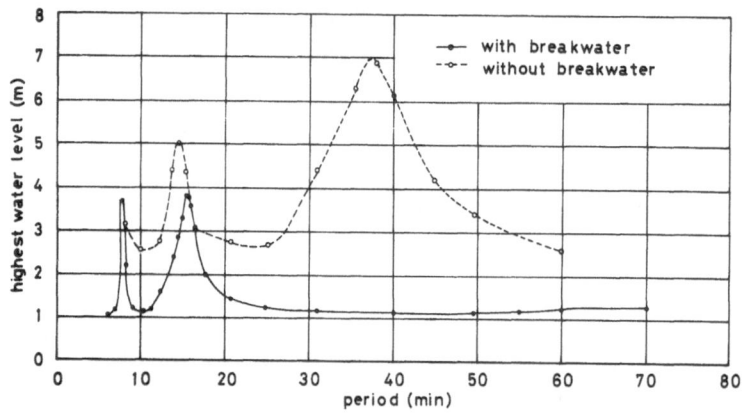

Fig. 3.   Response diagram of Ofunato Bay.

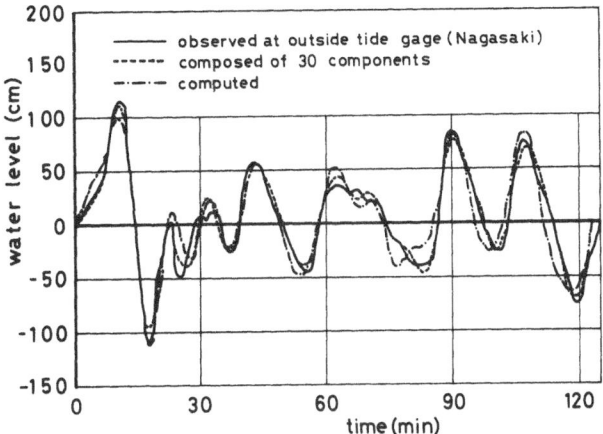

Fig. 4. Comparison of the computation with the observation at the outside tide gage station.

tide gage together with the approximate water level variation composed of 30 components from the observed tsunami. Similarly, Fig. 5 shows a comparison between the
computation and the observation at the inner tide gage station. The water level
variation computed assuming $f = 0$ is also presented in the same figure in order to
demonstrate the importance of head loss in evaluating the effect of the breakwater
against tsunami. The good agreement of the computed water level, assuming $f = 1.5$,
with the observed water level confirms the practical validity of the present numerical
method.

The effect of the existing tsunami breakwater is easily investigated by applying the
same incident tsunami to the bay without the breakwater. Figure 6 displays the

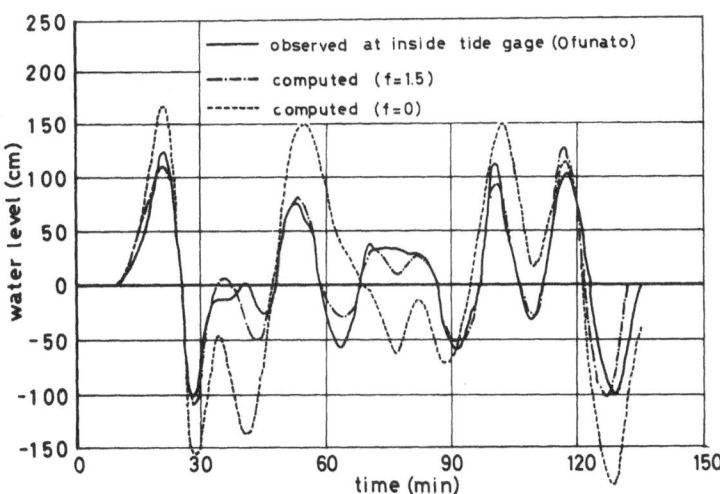

Fig. 5. Comparison of the computation with the observation at the inside tide gage station.

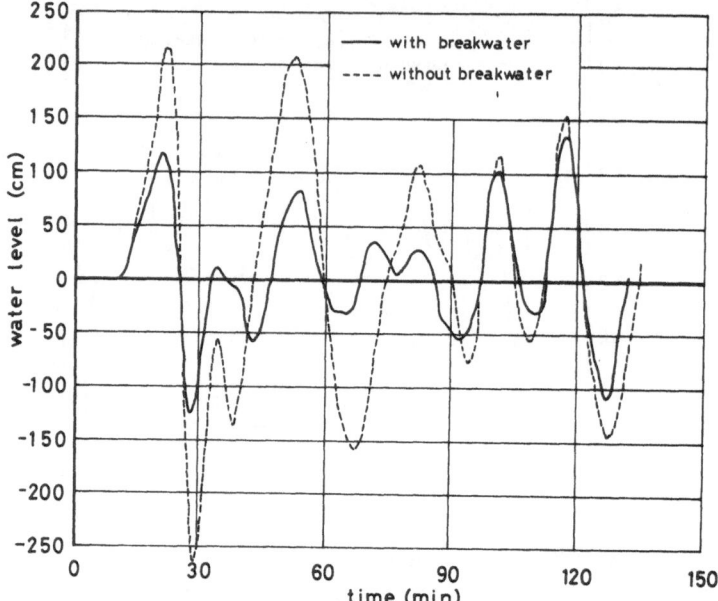

Fig. 6.   Computed water level variation at the innermost.

computed water level variations at the innermost point of the bay with and without breakwater. It can be seen that the breakwater considerably reduces the peak of the tsunami. Figure 7 shows the distribution of the water level at the first peak. The highest level after the breakwater construction is around 0.9 to 1.2 meters at the bay mouth, 0.5 meter directly inside the breakwater and about 1.1 to 1.2 meters at the innermost point. Without the breakwater, the highest water level reaches more than 2 meters at the innermost point. It is also seen that the breakwater has no serious influence on the water level elevation near the bay mouth.

It can be concluded from these computations that the tsunami breakwater in Ofunato bay produced the desired effect. The highest water level due to the 1968 tsunami would have reached almost twice the actual elevation if no breakwater had been constructed.

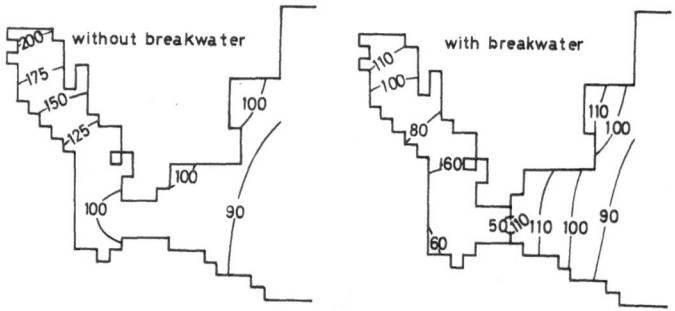

Fig. 7   Distribution of computed water level at the first peak of the 1968 tsunami.

3. Stability of a Vertical Wall Against Tsunami Forces

The 1968 Tokachi-oki earthquake tsunami attacked the Pacific coasts of Hok-kaido and the northeastern area of the Japan main island. The collapse of the Kawaragi breakwater in Hachinohe port is listed among damages due to this tsunami. Figure 8 shows the plan of Hachinohe port at the time of the 1968 tsunami. The kawaragi breakwater, which is a composite type with caissons having a width of 4.5 meters, had been situated at the end of a series of breakwaters in the direction of the tsunami propagation and collapsed over the total length of 338 meters. It was concluded after the field investigation that the main cause of the collapse was the significant difference of water levels between the outside and the inside of the breakwater which was caused by the phase difference of the tsunami effected by the breakwaters (BUREAU OF PORTS AND HARBOURS et al., 1968). The collapse of this breakwater due to a tsunami is instructive for the design of future tsunami breakwaters.

In Japan, it is a standard practice that the design wave forces on a vertical wall due to storm waves are calculated by Goda's formula (GODA, 1974; BUREAU OF PORTS AND HARBOURS et al., 1980), but no standard method to estimate tsunami forces on a vertical wall has been established. For this reason, the Port and Harbour Research Institute carried out a model experiment of wave forces on a vertical wall including long-period waves in connection with the project of a new tsunami breakwater at Kamaishi bay.

The experiments were made in a long wave channel of which the length is 160 meters, the depth is 1.5 meters, and the width is 1.0 meter. The experimental water depth h at the breakwater site on a bottom of 1/100 slope was 60 centimeters. The thickness of the rubble mound, and consequently the bottom depth of the vertical wall, h', and the crest depth of the armoring units of the rubble mound, d, were changed to produce three cases, keeping other dimensions as constant. Wave conditions are indicated in Table 1, in which T is the wave period, L is the wavelength at the depth h, and H is the incident wave height. Wave pressures were measured with transducers fixed on the vertical wall, and the time variation of the total wave pressure acting on the vertical wall was analyzed.

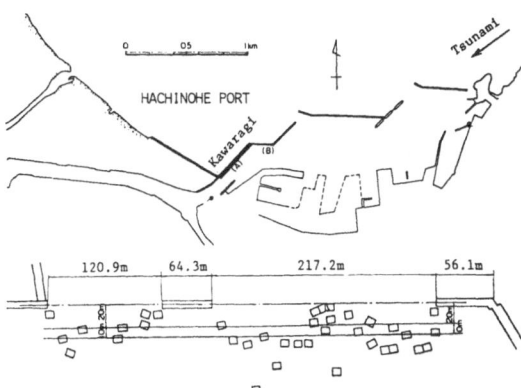

Fig. 8.   Collapse of Kawaragi breakwater in Hachinohe Port due to the 1968 tsunami.

Table 1. Experimental wave conditions.

T (s)	$\dfrac{h}{L}$	H (cm)	$\dfrac{H}{h}$
1.4	0.222	4.0–15.4	0.067–0.257
2.0	0.138	3.6–34.0	0.060–0.567
3.6	0.0709	4.0–32.0	0.067–0.533
5.6	0.0448	4.0–40.0	0.067–0.667
14.0	0.0177	4.0–32.0	0.067–0.533

Figure 9 shows the variation of $\bar{p}/w_0 H$, where $\bar{p}$ is the mean wave pressure intensity averaged over the height of the vertical wall for maximum horizontal wave force during the time variation, and w_0 is the specific weight of water. The abscissa is the relative water depth h/L. For the conditions of relatively short-period waves, the wave pressure intensity is greatly changed by the crest depth of the armoring units, d, and the wave height H. In particular, very powerful breaking wave forces are caused when high waves act on a vertical wall placed on a high rubble mound, as pointed out by TANIMOYO *et al.* (1981) in a different study. As the wave period becomes longer and the relative water depth becomes smaller, the mode of wave action on a vertical wall becomes hydrostatic, even if high waves act on the vertical wall placed on a high rubble mound. The results presented in Fig. 9 demonstrate an evident tendency that the mean wave pressure intensity $\bar{p}$ converges to about 1.0 w_0H, as the relative water depth becomes smaller. A similar tendency is observed in the uplift pressure which acts upwardly beneath the bottom of a vertical wall.

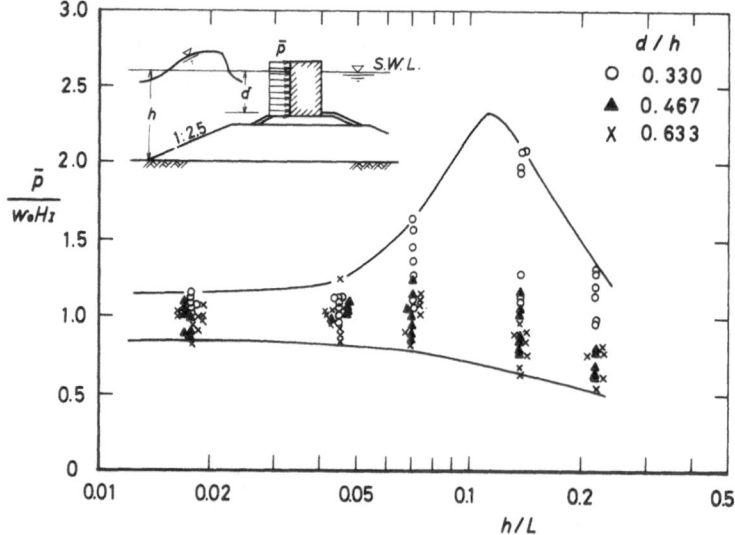

Fig. 9. Mean wave pressure intensity on a vertical wall.

Based on these experimental results, and considering Goda's formula for storm waves, a wave pressure distribution to calculate tsunami forces on a vertical wall in relatively deep water offshore, so that no breaking tsunami like a bore acts on a wall, is proposed as shown in Fig. 10. In this Figure,

$$\eta^* = 1.5\,H \tag{2}$$

$$p = p_u = 1.1\,w_0\,H \tag{3}$$

where, η^* is the height above the still water level at which the wave pressure intensity is zero, p is the wave pressure intensity which acts uniformly on the vertical wall below the still water level, and p_u is the intensity of toe uplift pressure.

The proposed method is equivalent to the extreme relation of Goda's formula to long waves, when the term α_2 representing the effect of a rubble mound on wave forces in Goda's formula is omitted.

According to the present method, the total horizontal wave force, P, uplift force, U, and the overturning moments around the heel of the caisson, M_P and M_U, are expressed as follows:

$$P = \left\{ 1 + \left(1 - \frac{h_c^*}{3H} \right) \frac{h_c^*}{h'} \right\} ph' \tag{4}$$

$$M_P = \left\{ \frac{1}{2} + \left(1 - \frac{h_c^*}{3H} \right) \frac{h_c^*}{h'} + \left(\frac{1}{2} - \frac{4h_c^*}{9H} \right) \left(\frac{h_c^*}{h'} \right)^2 \right\} ph'^2 \tag{5}$$

$$U = \frac{1}{2} p_u B \tag{6}$$

$$M_U = \frac{1}{3} p_u B^2 \tag{7}$$

where,

$$h_c^* = \min\{\eta^*, h_c\}$$

$$\min\{a, b\}: \text{ smaller of } a \text{ or } b \tag{8}$$

h_c is the crest height of a vertical wall above the still water level, and B is the width of vertical wall.

Experiments were done only for waves of normal incidence to the vertical wall, but

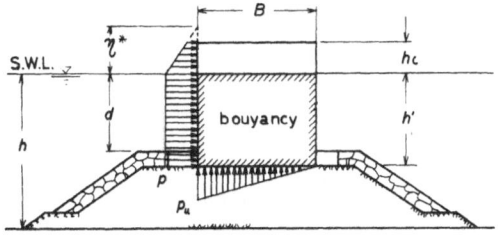

Fig. 10.   Wave pressure distribution due to non-breaking long-period wave.

the calculation method of tsunami forces may be applied to obliqely incident waves without any correction for the incident wave angle. Once tsunami forces are calculated, the stability of a vertical wall against sliding and overturning can be examined according to the coventional method.

As an example of application, the stability of the Kawaragi breakwater for the 1968 tsunami will be examined in order to confirm the proposed method. Although the water depth at the breakwater site is very shallow, from 4.5 to 6.5 meters below the datum level, no breaking tsunami had been observed for the 1968 tsunami. According to the report of the field investigation for the 1968 tsunami (BUREAU OF PORTS AND HARBOURS et al., 1968), the water levels outside and inside of the breakwater were + 3.35 meters and − 1.00 meter above the datum level, respectively, and the astronomical tide level can be estimated to be nearly equal to the datum level at the time of collapse. Therefore, it is necessary to consider the effect of a lowering water level behind the vertical wall from the astronomical tide level in this case. The pressure due to this lowering of water level behind the vertical wall can be assumed completely hydrostatic.

The greatest difficulty is in the estimation of the incident tsunami height, H. Suppose that H is nearly equal to the water level above the astronomical tide in front of the vertical wall, considering the superposition of the reflected waves from the boundaries. Then,

$$H = 3.35 \ (m).$$

When the deepest section of the Kawaragi breakwater is considered for the examination, the following dimensions are given (Fig. 11):

$$h_c = 2.85 \ (m), \ h' = 4.00 \ (m), \ B = 4.50 \ (m).$$

The weight of the vertical wall in the air is 64.6 tw/m, where the abbreviation "tw" stands for "tons weight". The weight in the water, W_0, and the arm length around the heel of the vertical wall, x_{W_0}, are obtained as follows:

$$W_0 = 64.6 - 1.03 \times \frac{1}{2}(4.00 + 3.00) \times 4.50 = 48.4 \ (tw/m)$$

$$x_{W_0} = 2.24 \ (m).$$

Tsunami forces P and U are calculated by Eqs. (4), (6),

$$P = \left\{ 1 + \left(1 - \frac{2.85}{3 \times 3.35} \right) \left(\frac{2.85}{4.00} \right) \right\} \times 3.80 \times 4.00 = 22.9 \ (tw/m)$$

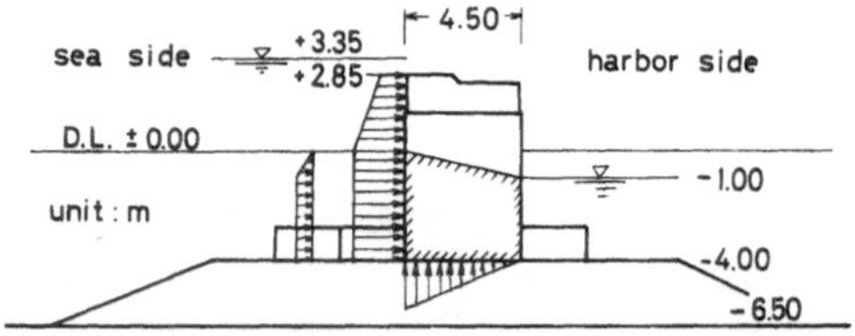

Fig. 11. Cross section of Kawaragi breakwater (−6.5m section).

$$U = \frac{1}{2}3.80 \times 4.50 = 8.54 \ (\text{tw}/\text{m}).$$

The hydrostatic horizontal force due to the lowering of water level behind the vertical wall, P_s, is:

$$P_s = 1.03\left(\frac{1}{2}1.00^2 + 1.00 \times 3.00\right) = 3.61 \ (\text{tw}/\text{m}).$$

The safety factors, S_{sliding}, against sliding and, $S_{\text{overturning}}$, against overturning are examined as follows:

$$S_{\text{sliding}} = \frac{0.6\,(64.6 - 8.5)}{22.9 + 3.6} = 0.90$$

$$S_{\text{overturning}} = \frac{108.6 - 25.6}{65.1 + 6.4} = 1.16.$$

Here, the coefficient of friction between the concrete slab and the rubble mound is taken to be 0.6.

A safety factor of less than 1.0 indicates the possibility of failure of the vertical wall. It is concluded that the proposed method explains the fact of the collapse of the Kawaragi breakwater, since the safety factor against sliding of 0.9 is less than 1.0.

It is interesting to re-examine the stability of the existing Ofunato tsunami breakwater according to the present method. In general, sliding of a vertical wall is more critical than overturning. The critical incident tsunami height, H_c, causing sliding will be calculated, assuming that the water level behind the vertical wall happens to agree with the astronomical tide. Then, H_c is expressed by the following relation:

$$H_c = \frac{\mu W_0 + 1.1 w_0 h_c^{*2}/3}{\left(1 + \dfrac{h_c^*}{h'} + \dfrac{\mu}{2}\dfrac{B}{h'}\right) \times 1.1 \, w_0 h'} \tag{9}$$

where, μ is the coefficient of friction.

A high water level + 1.40 meters is considered for the astronomical tide, because the higher tide is more critical than the lower tide. Then, the dimensions of the vertical wall for the shallower section are given as follows (MATSUMOTO and SUZUKI, 1981):

$$h_c = 3.6 \ (\text{m}), \ h' = 11.4 \ (\text{m}), \ B = 12.0 \ (\text{m})$$

and,

$$W_0 = 202.0 \ (\text{tw}/\text{m}).$$

The following critical height is estimated by Eq. (9):

$$H_c = 5.98 \ (\text{m}).$$

Similarly, the critical height is estimated to be 6.29 meters for the deeper section. This implies that the vertical wall of the Ofunato tsunami breakwater will be stable with respect to sliding if the incident tsunami height is less than about 6 meters.

The present examination reconfirms the original design, since the vertical wall had been designed to be stable for a combination of hydrostatic pressure due to 3 meters difference of water elevation between the sides of the breakwater and the pressure of progressive waves by the elliptic trochoidal wave theory with a height of 6 meters.

4. Concluding Remarks

Protection of coastal areas from great tsunamis is not easy, because the tsunami runup is quite high and the protection works are quite costly when the frequency of extreme tsunami is taken into consideration. Neverthless the severity of great tsunami damages removes any doubt against the necessity of tsunami protection works. Port areas with dense populations and busy daily operation are one of the most difficult places for providing good protection against tsunamis.

In the present paper, an earlier study on the numerical computation of tsunami in a harbor is reviewed to demonstrate the effectiveness of the breakwater on the reduction of tsunami height. It is concluded that the construction of a tsunami breakwater is a desirable solution, if the configuration and the bathymetry of the harbor is favorable. The Ofunato tsunami breakwater is one of the most successful examples.

Recent progress in the numerical analysis of tsunami propagation is particularly notable as shown in several papers presented at the International Tsunami Symposium 1981. The new projects of the Kamaishi tsunami breakwater on the Sanriku coast and the Shimoda breakwater on the Izu peninsula are results of tsunami prediction schemes in which these advanced numerical techniques have been applied. Numerical analysis as well as hydraulic model tests will further aid us in making adequate plans for tsunami protection works for port areas.

Another aspect which is very important for design is the stability of a breakwater against tsunami forces. In the present paper, a method to calculate non-breaking tsunami forces on a vertical wall is proposed. The application to the existing Ofunato tsunami breakwater indicates that the vertical wall will be stable against sliding, if the incident tsunami height is less than about 6 meters.

Associated with the new project of the Kamaishi tsunami breakwater, intensive studies for the design problems against tsunamis as well as storm waves and earthquakes have been carried out by the Port and Harbour Research Institute and the Second District Port Construction Bureau, Ministry of Transport. The proposed method to calculate tsunami forces on a vertical wall is one of these results. The stability of the opening section of the new tsunami breakwater is a future subject which has to be examined for the construction.

The author wishes to acknowledge Dr. Y. Ito for permission to use the computed results on the effect of a breakwater against tsunamis. Acknowledgement is also expressed to Dr. Y. Goda, Head of Hydraulic Engineering Division, PHRI, for his encouragement for the presentation of this paper.

REFERENCES

BUREAU OF PORTS AND HARBOURS, PORT AND HARBOUR RESEARCH INSTITUTE, THE SECOND DISTRICT PORT CONSTRUCTION BUREAU, MINISTRY OF TRANSPORT, AND PORT AND HARBOUR DIVISION, HOKKAIDO DEVELOPMENT BUREAU, HOKKAIDO DEVELOPEMENT AGENCY, The investigation of the tsunami caused by the 1968 Tokachi-oki Earthquake, pp. 211–281, 1968 (in Japanese).

BUREAU OF PORTS AND HARBOURS, PORT AND HARBOUR RESEARCH INSTITUTE, MINISTRY OF TRANSPORT, Technical standards for port and harbour facilities in Japan, 317 p. 1980.

GODA, Y., New wave pressure formula for composite breakwaters, Proceedings of the Fourteenth Coastal Engineering Conference, pp. 109–112, 1974.

ITO, Y., On the effect of tsunami-breakwater, *Coastal Engineering in Japan*, Vol. 13, pp. 89–102, Japan Soc. Civil Engrs., 1970.

ITO, Y., K. TANIMOTO, and T. KIHARA, Digital computation on the effect of breakwaters against long-period waves (4th Report), *Rep. of Port and Harbour Res. Inst.*, Vol. 7, No. 4, pp. 55–83, 1968 (in Japanese).

MATSUMOTO, T. and Y. SUZUKI, Design and construction of Ofunato tsunami protection breakwater, Abstracts of the International Tsunami Symposium, pp. 109–112, 1981.

TANIMOTO, K., K. KOBUNE, and K. KOMATSU, Numerical analysis of wave propagation in harbours of arbitrary shape, *Rep. of Port and Harbour Res. Inst.*, Vol. 14, No. 3, pp. 35–58, 1975 (in Japanese).

TANIMOTO, K., S. TAKAHASHI, and T. KITATANI, Experimental study of impact breaking wave forces on a vertical-wall caisson of composite breakwater, *Rep. of Port and Harbour Res. Inst.*, Vol. 20, No. 2, pp. 3–39, 1981 (in Japanese).

TSUNAMI RUN-UP

Tsunamis—Their Science and Engineering, edited by K. Iida and T. Iwasaki, 439–451.

Numerical Simulation of Tsunami Propagations and Run-up

Chiaki Goto and Nobuo Shuto

Department of Civil Engineering, Tohoku University, Sendai, Japan

(Received August 31, 1981; Revised December 22, 1981)

Three problems are discussed in relation to the required accuracy in a numerical simulation of tsunamis. The spatial mesh size in the computation should be selected in relation not only to the hydraulic characteristics of the tsunami and the topographical properties of the site, but also to the boundary condition at the wave front used in the computation. With a steep surface at the wave front, oscillations coherent only with the numerical computation appear and become a hindrance in the simulation. An artificial diffusion term is introduced to suppress this oscillation. Finally, the relative importance of terms in the fundamental equations are examined in an example of a practical application, after problems in the numerical computation have thus been solved.

1. Introduction

In recent years, numerical simulations have been developed and used to compute the behaviour of tsunamis in shallow water and on land. The results of these numerical computations are often referred to in practical designs of defense works against tsunamis. However, several problems remain to be solved before the accuracy of the numerical results can be improved.

In a numerical simulation in which a tsunami is followed from its source to the final run-up, the spatial mesh sizes are varied from coarse in the open sea to fine on land. The run-up height of the tsunami depends to some extent on the mesh size on land. Therefore, to obtain a correct result, a condition must be taken into account for the selection of an adequately sized spatial mesh, in addition to the CFL condition. This additional condition is established in the present paper by a comparison of the numerical results obtained with different expressions of the boundary condition at the wave front to an analytical solution.

If a wave front becomes nearly vertical, another problem occurs. With a sudden change in wave charactaeristics at the front, oscillations appear with a wave-length strongly dependent on the spatial mesh size, and increase in amplitude so as to lead to a break-down of the computation. This is found only in numerical computations and not in hydraulic experiments. A well-designed artificial diffusion term will successfully suppress this oscillation. At the same time, however, the slope of the wave front is made slightly gentler than that expected. To correct this weakness, a dispersion term is artificially introduced. With these diffusion and dispersion terms, both of which are only

439

artificially introduced and are made effective only in the neighborhood of the wave front, numerical computations can be continued to yield the expected results. The CFL condition is modified by the introduction of these terms.

In shallow water and on land, nonlinear equations should be used in the computation whereas linear equations are accurate enough in the deep sea. In order to understand what the necessary terms are in shallow sea and on land, a tsunami in Ofunato Bay, Iwate Prefecture, Japan is computed.

2.   Effect of Spatial Mesh Size on Run-up Height

One of the problems in tsunami computations on land is the boundary condition at the wave front. Water runs up on dry land and runs down to leave a dry bed. Even when there is no breaking at the wave front, the movement of water on the slope can not be easily expressed. There are several methods to overcome this difficulty. Among them, three methods are selected and compared in the present paper, and their applicability to a two-dimensional problem of practical importance is considered.

AIDA (1977) assumes that the Froude number at the wave front takes a constant value of 0.5. HOUSTON and BUTLER (1979) assume a weir formula. In both assumptions, the water particle velocity at the wave front is uniquely correlated with the water depth there. In the condition of Houston and Butler, the Froude number is not always equal to 0.5. In the present work, two values of the Froude number, 0.5 and 2.0, are assumed and called Aida's condition I and II respectively.

IWASAKI and MANO (1979) use a different method. A continuous topography is approximated by a series of discontinuous, horizontal steps. The velocity of water particles flowing on to a dry step is evaluated from the hydraulic characteristics one step behind the wave front.

Both Aida and Iwasaki and Mano use equations in the Eulerian description, which is based upon information obtained at spatially fixed points.

The present authors, GOTO and SHUTO (1977, 1979) have developed another method expressed in the Lagrangian description, by which the boundary condition at the wave front is easily satisfied with no special assumption. The equations themselves are constructed for water particles. Therefore, if we follow the movement of the water particles which constitute the shoreline at the initial instant, then the wave front is automatically determined.

In order to examine and compare the three methods, we consider a simple one-dimensional topography of a uniform slope connected with a horizontal channel of constant water depth. The theoretical run-up height of a sinusoidal wave train has been obtained (J. B. KELLER and H. B. KELLER, 1964; SHUTO, 1972) provided that perfect standing waves are formed without breaking. The ratio of the run-up height, R, measured above the still water level to the incident wave height, H, is given by

$$\frac{R}{H} = \left[J_0^2\left(4\pi\frac{l}{L}\right) + J_1^2\left(4\pi\frac{l}{L}\right) \right]^{-\frac{1}{2}} \tag{1}$$

where L is the wave length of incident waves, l the horizontal distance between the toe of the slope and the shoreline, and J_0 and J_1 the Bessel functions of the first kind of order 0 and 1.

The second, third and fourth columns in Table 1 show the length of spatial meshes, slopes and wave periods used in the computation for the comparison. The water depth is 50 m and wave height is 1 m. With longer wave periods, all the numerical results coincide with the theoretical result regardless of the boundary condition. Shallow water theory in the Eulerian description is used, combined with Aida's and Iwasaki and Mano's conditions. Linear theory in the Lagrangian description is used for the present work. If the maximum run-up height alone is of concern, no significant difference is found between linear and nonlinear theories.

The sixth, seventh, eighth and ninth columns in Table 1 give the computed maximum run-up height divided by the theoretical results, obtained by Aida I ($F_r = 0.5$), Aida II ($F_r = 2.0$), Iwasaki and Mano and the authors. The computation in each case was continued over eight wave periods and the stationary values attained are tabulated. Values marked by an asterisk correspond to cases in which numerically-induced Gibbs oscillations were observed.

Table 1. Comparison of numerical results obtained with different methods to the analytical results.

Run	Spatial mesh Δx (m)	Slope α	Wave period T (s)	$\dfrac{\Delta x}{\alpha g T^2} \times 10^4$	Accuracy			
					Aida[1] ($F_r = 0.5$)	Aida[1] ($F_r = 2.0$)	Iwasaki and Mano[3]	Authors[6,7]
1	25	1/10	300	2.83	1.00	1.00	1.00	0.98
2		1/10	600	0.709	1.00	1.00	1.00	1.00
3		1/25	300	7.09	0.94	0.93	0.91	0.99
4		1/25	600	1.77	1.02	1.02	0.97	0.99
5		1/50	300	14.2	0.92	0.92	0.84	0.99
6		1/50	600	3.54	1.04	1.04	0.99	0.99
7		1/100	300	28.3	0.46*	0.45*	0.55*	1.00
8		1/100	600	7.09	0.97	0.97	0.96	0.99
9	50	1/10	300	5.67	0.96	0.96	0.96	0.97
10		1/10	600	1.42	0.99	1.00	0.99	0.99
11		1/25	300	14.2	1.00	1.00	0.96	0.98
12		1/25	600	3.54	0.98	0.98	0.97	0.98
13		1/50	300	28.3	0.84	0.86	0.84	0.97
14		1/50	600	7.09	1.03	1.03	0.98	0.98
15		1/100	300	56.7	0.48*	0.45*	0.47*	1.00
16		1/100	600	14.2	0.91	0.92	0.89	0.99
17	100	1/10	300	11.3	0.96	0.96	0.96	0.95
18		1/10	600	2.83	0.99	0.99	0.99	0.99
19		1/25	300	28.3	0.95	0.95	0.94	0.97
20		1/25	600	7.09	1.01	1.01	0.97	0.98
21		1/50	300	56.7	0.83	0.84	0.74	0.99
22		1/50	600	14.2	1.00	1.00	0.94	0.96
23		1/100	300	113.0	0.48*	0.49*	0.40*	1.00
24		1/100	600	28.3	0.84	0.84	0.82	0.99

Figures 1(a), (b), and (c) compare wave profiles of numerical results with those of theoretical ones for the maximum run-up and run-down. The slope is 1/50 and the spatial mesh size is 50 m. The solid and dotted lines are the results obtained with linear theory in the Lagrangian description for wave period of 600 s and 300 s, respectively, while the white and black circles are the numerical results.

Shallow water theory combined with Aida's and Iwasaki and Mano's wave front conditions yields wave profiles with almost the same results as those obtained by the linear theory in the Lagrangian description for the maximum run-up. However, when the maximum run-down occurs, shallow water theory gives higher water surface. As far as the maximum run-up height is concerned, no difference results from assuming, in Aida's method, that F_r is either 0.5 or 2.0.

For a gentler slope and a wave train of shorter period, the numerical results obtained with shallow water theory are worse. A finer spatial mesh can improve the situation. Accordingly, the accuracy of the numerical computations may be judged from the slope α, wave period T, and mesh size Δx. Figure 2 shows the results. Ratios of computed values to theoretical ones are plotted against a dimensionless parameter, $\Delta x/\alpha g T^2$. Although the method proposed by the authors gives the best result, it is not convenient in practical applications. Convergence or divergence of water particles on land inevitably occurs in two-dimensional computations. Then, no run-up height can be

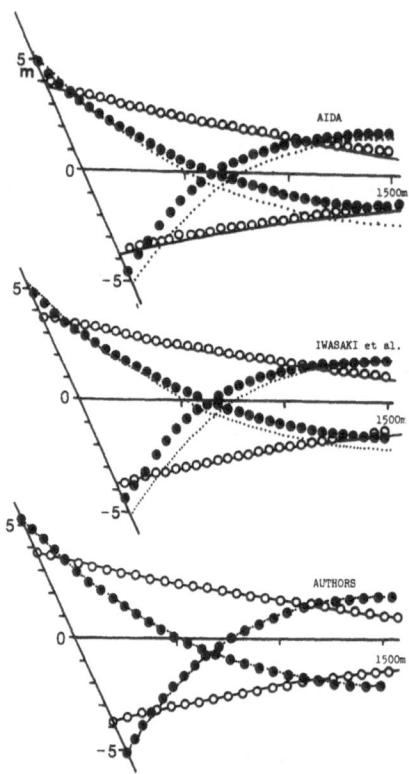

Fig. 1.   Comparison of wave profiles.

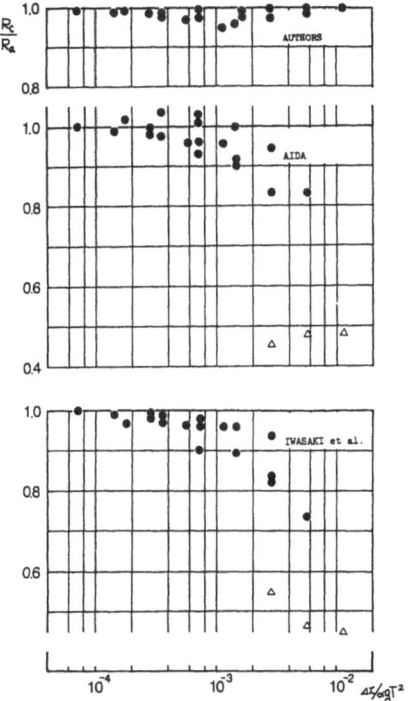

Fig. 2.   Dependence of the numerical results on the size of spatial mesh.

predicted at places where no water particle arrives. On the other hand, if the equations are taken in the Eulerian description and used with any of the conditions proposed by Aida or Iwasaki and Mano, the run-up height may easily be computed with a possible error of less than 5% at any designated point, provided that the spatial mesh size is selected to satisfy $\Delta x/\alpha g T^2 < 4.0 \times 10^{-4}$.

3.   Numerical Scheme at a Nearly Vertical Wave Front

As is shown in Table 1, tsunamis of short wave period have a bore-like wave front on gently sloping land. If the leap-frog scheme is applied to the computations of the tsunamis, Gibbs oscillations generated at the sharply discontinuous front will constitute problems difficult to handle.

Figure 3 shows examples of these oscillations. The bottom slope is 1/200 and the period of the incoming waves is 300 s. The spatial mesh size is 12.5 m in Case A and 25 m in Case B. Iwasaki and Mano's front condition is used. Both figures give three consecutive wave profiles, beginning with the profile at $t = 1,000$s when the maximum run-up of the first wave occurs. It is clearly noticed that oscillations have a wave length twice the size of the spatial mesh. In the computations, either no such oscillation was obtained at the wave front of the first incoming wave, or else it was negligibly small.

In order to clarify the cause of the oscillations, numerical experiments were carried

C. Goto and N. Shuto

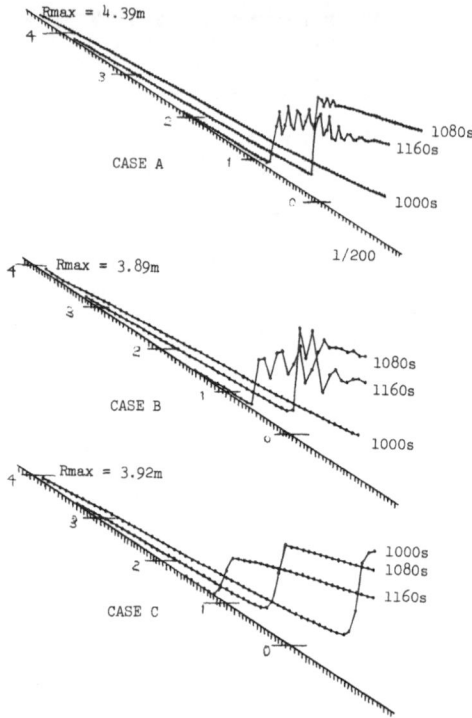

Fig. 3. Oscillations induced by numerical computations, (A) with $\Delta x = 12.5$ m, (B) with $\Delta x = 25.0$ m, and (C) wave profiles without oscillations obtained with the artificial diffusion term.

out for a simple bore on a horizontal bed, by changing the flow conditions ahead of and behind the vertical wave front. Figures 4(a), (b), and (c) show the results. In each figure, two values of the Froude number are given. The first one is for the flow behind the wave front and the other for the flow in front of it.

On the left-hand side of the figures, thick lines give the analytical results which show no deformation with time while thin lines connecting small black circles are the conputed results. Oscillations always occur near the wave front, but they are small. If enlarged, they are given in the right-hand side of Fig. 4 with the broken lines showing the path of the bore front. They propagate downstream (upstream) if the Froude number behind the wave front is smaller (larger) than unity. In these examples, the maximum wave height of the oscillation is, at maximum, less than one-tenth of the original bore height in Case C.

It is deduced that if the water particle velocity behind the wave front is sufficiently strong, and that ahead of the wave front is also sufficiently strong but in the reverse direction, small oscillations which inevitably occur due to the digitized numerical procedure are accumulated near the front and increase in amplitude even if they are very small at their initiation.

Elimination of these undesirable oscillations will be accomplished by the introduction of a diffusion term which forces the oscillations to propagate so as to

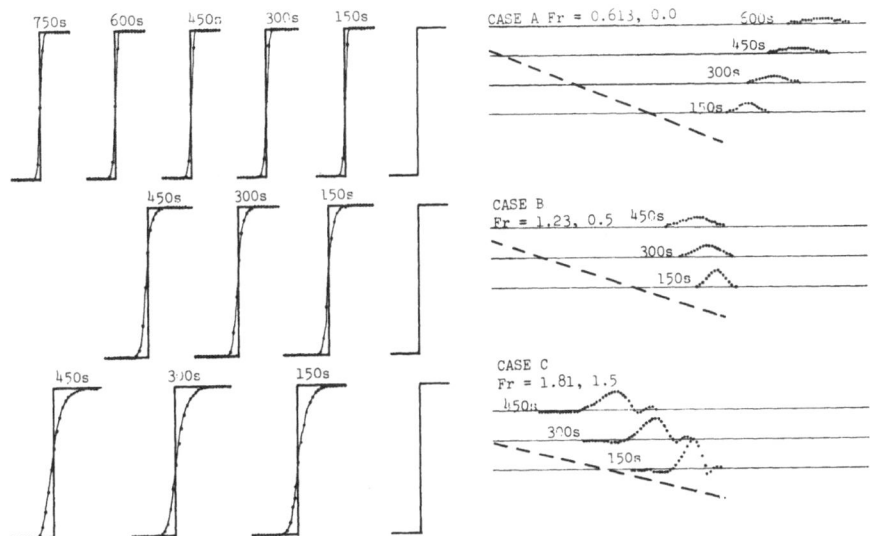

Fig. 4.   Numerical examples of a bore on a horizontal bed, computed with the S.L.F. scheme for different Froude numbers.

become of no significant magnitude. The diffusion term proposed here is

$$\beta_1(\Delta x)^3 \sqrt{\frac{g}{D}} \left|\frac{\partial^2 \eta}{\partial x^2}\right| \frac{\partial^2 \eta}{\partial x^2} \tag{2}$$

where β_1 is given by

$$\beta_1 = 0 \text{ for } \Delta x \left|\frac{\partial^2 \eta}{\partial x^2}\right| - \gamma \left|\frac{\partial \eta}{\partial x}\right| \leqq 0$$

$$\tag{3}$$

$$\beta_1 = \text{Const. for } \Delta x \left|\frac{\partial^2 \eta}{\partial x^2}\right| - \gamma \left|\frac{\partial \eta}{\partial x}\right| > 0.$$

If we select an adequate value for γ, the diffusion term has a strong effect only in the area where the oscillations are apt to be generated and has no contribution at all otherwise.

In Fig. 3(c), an example of the computation with the diffusion term is shown. Oscillations are completely suppressed and at the same time no significant reduction in the bore height results. However, the effects of the diffusion term need to be checked in more detail. With the diffusion term, the stability condition of the FDM is modified to

$$1 \geqq (|u| + \sqrt{gD})\frac{\Delta t}{\Delta x} + \beta_1(\Delta x)^2 \frac{1}{D}\left|\frac{\partial^2 \eta}{\partial x^2}\right|. \tag{4}$$

Another effect expected is "over-smoothing" which makes the slope of the water surface at the wave front gentler. In Fig. 3(c), the wave profile for $t = 1,160$ s seems to suggest this effect. In order to examine this effect due to the diffusion term, a hydraulic jump is computed and compared in Figs. 5(a), (b), and (c). Figure 5(a) shows the results

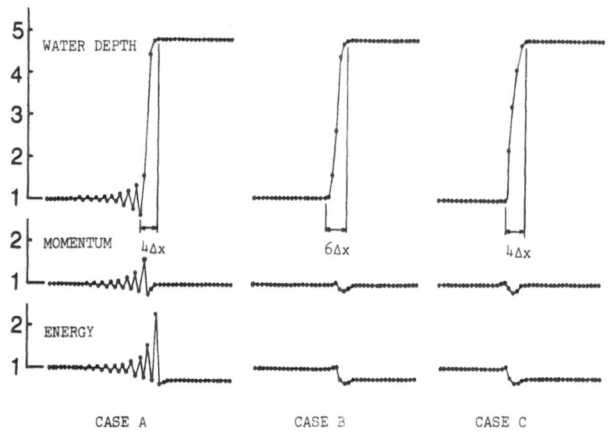

Fig. 5. Numerical examples of a hydraulic jump; (A) with no artificial terms, (B) with the diffusion term and (C) with the diffusion and viscosity terms.

without the diffusion term. Oscillations exist in profiles of water surface elevation, momentum and energy. The transition from shallow water to deep water occurs within a length of $4\Delta x$. In Case B, the included diffusion term suppresses the oscillations but the zone of transition is broadened from $4\Delta x$ to $6\Delta x$.

In order to reduce the length of the transition zone, an artificial viscosity similar to that introduced by von Neumann and Richtmyer (1950) is effective. The equation of conservation of energy in the transition zone is expressed by

$$\frac{\partial}{\partial x}(uDE) = \left(\frac{2}{\pi}\beta_2 \Delta x\right)^2 uD\frac{\partial^3 E}{\partial x^3} \tag{5}$$

where E is given by $u^2/2 + gD$, and β_2 is a constant. The right-hand side is the term introduced and its effect in reducing the length of transition zone can be understood from an analytical solution. Solving Eq. (5) with the boundary conditions,

$$\left. \begin{array}{ll} E = E_1 & \text{at} \quad x = -\beta_2 \Delta x \\ E = E_2 & \text{at} \quad x = \beta_2 \Delta x \end{array} \right\} \tag{6}$$

then the solution is given by

$$E = \frac{1}{2}(E_1 + E_2) - \frac{1}{2}(E_1 - E_2)\sin\left(\frac{\pi}{2} \cdot \frac{x}{\beta_2 \Delta x}\right) \tag{7}$$

and is also shown in Fig. 6. The solution connects two regions continuously and the length of the transition zone is of the order of $\beta_2 \Delta x$. Therefore, with an adequately selected value of β_2, this term is easily controlled to reduce the "over-smoothing" introduced by the diffusion term.

Figure 5(c) shows an example computed for the hydraulic jump by setting $\beta_1 = 1.0$ and $\beta_2 = 0.8$. The transition zone is improved so as to be of the order of $4\Delta x$.

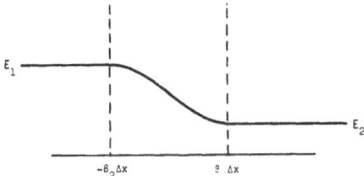

Fig. 6.   Continuous profile in the transition zone, obtained with Eq. (7).

4.   Magnitude of Terms in Tsunami Computations

A set of nonlinear equations is used in the followings to clarify the relative importance of the pressure, convection, bottom friction and dispersion terms. Conditions and artificial terms discussed in the preceding sections are included in the numerical scheme.

The equations are

$$\frac{\partial \eta}{\partial t} + \frac{\partial M}{\partial x} + \frac{\partial N}{\partial y} = 0$$

$$\frac{\partial M}{\partial t} + \frac{\partial}{\partial x}\left(\frac{M^2}{D}\right) + \frac{\partial}{\partial y}\left(\frac{MN}{D}\right) + gD\frac{\partial \eta}{\partial x} + \frac{gn^2}{D^{7/3}}M\sqrt{M^2 + N^2} =$$

$$\frac{gD^3}{6}\left\{\frac{\partial^3 \eta}{\partial x^3} + \frac{\partial^3 \eta}{\partial x \partial y^2}\right\} - \frac{gD^2}{2}\left\{\frac{\partial^2}{\partial x^2}\left(D\frac{\partial \eta}{\partial x}\right) + \frac{\partial^2}{\partial x \partial y}\left(D\frac{\partial \eta}{\partial y}\right)\right\}$$

$$\frac{\partial N}{\partial t} + \frac{\partial}{\partial x}\left(\frac{MN}{D}\right) + \frac{\partial}{\partial y}\left(\frac{N^2}{D}\right) + gD\frac{\partial \eta}{\partial y} + \frac{gn^2}{D^{7/3}}N\sqrt{M^2 + N^2} =$$

$$\frac{gD}{6}\left\{\frac{\partial^3 \eta}{\partial x^2 \partial y} + \frac{\partial^3 \eta}{\partial y^3}\right\} - \frac{gD^2}{2}\left\{\frac{\partial^2}{\partial x \partial y}\left(D\frac{\partial \eta}{\partial x}\right) + \frac{\partial^2}{\partial y^2}\left(D\frac{\partial \eta}{\partial y}\right)\right\} \qquad (8)$$

where M and N are the discharges in the x- and y-directions, g the gravity acceleration, D the total water depth and η the water surface elevation above the still water level. A leap-frog method is used with mesh intervals, $\Delta x = \Delta y = 50$m and $\Delta t = 1.5$s. The stability condition for Eq. (8) which includes the dispersion term is slightly more restricted than usual and is given by

$$(|u| + \sqrt{gD})\frac{\Delta t}{\Delta x} \leq 1 \text{ and } \Delta x \geq \frac{\sqrt{2}}{3}D. \qquad (9)$$

According to this condition the dispersion term is taken into the computations only in the area where $D < 30$m is satisfied. In what follows, Manning's roughness is equal to 0.025 everywhere, and Iwasaki and Mano's front condition is used.

The Meiji Great Sanriku Tsunami in 1896 is computed for Ofunato Bay, Iwate Prefecture, Japan. Input wave profiles at the seaward boundary of the computation are given by

$$\eta = -0.75 \sin\left(\frac{2\pi}{T}t\right) \quad \text{for} \quad 0 \leqq t \leqq \frac{T}{2}$$

$$= -3.00 \sin\left(\frac{2\pi}{T}t\right) \quad \text{for} \quad \frac{T}{2} \leqq t \leqq T \tag{10}$$

$$= -0.75 \sin\left(\frac{2\pi}{T}t\right) \quad \text{for} \quad T \leqq t$$

which is determined by considering Aida's (1977) results, and T is 600s.

Figure 7 shows the contours in meters of Ofunato Bay. The thick line indicates the shoreline and the thin smooth curves are isodepths. Thin stepped lines bound the dry land included in the computation.

Figure 8 shows the distribution of the highest water surface elevation. Ofunato Bay is L-shaped. A tsunami turns to the right by an angle of nearly 90° after entering the bay. The wave period of the tsunami in the computation is 600s. It is shorter than the period of natural frequency of the bay which is about 40 min. The maximum height in the bay is 5.02m and appears at Point A which faces directly to the open ocean. At Point B, the head of the bay, the run-up height is about 2m.

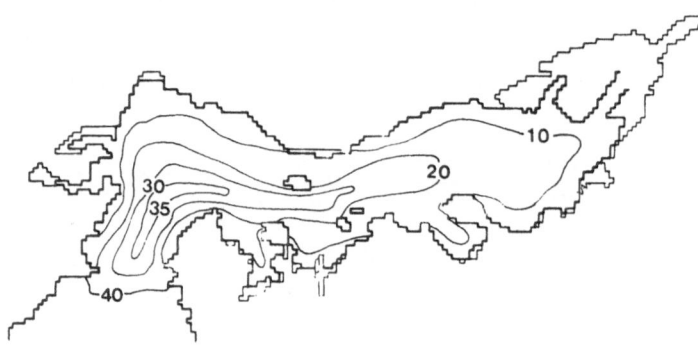

Fig. 7.   Contours in Ofunato Bay.

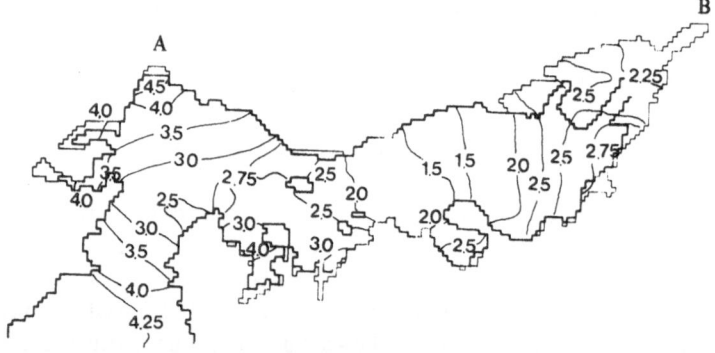

Fig. 8.   Distribution of the highest water surface elevation.

Figure 9 shows the distribution of the highest water particle velocity. At narrow portions of the bay and on land, the velocity becomes higher. It rises to more than 5.0 m/s when the tsunami reduces.

Figures 10 through 13 compare the convection, pressure, friction and dispersion terms with the non-steady acceleration term. Maximums of the resultant values are computed and compared at every point. The pressure term is of the same order of magnitude as the non-steady acceleration term almost everywhere in the bay. At narrow portions of the bay and at the area where the tsunami runs up high, the convection term becomes very significant. This is typically noticed at Point A where the run-up is highest. The magnitude of the convection term at Point A is more than twice the non-steady acceleration term. Judging from this fact, the convection term is very important in the computation of tsunami run-up. The magnitude of the friction term is not very great but is not negligible. The dispersion term has a negligible contribution.

5. Conclusions

The FDM applied to shallow water theory including bottom friction gives satisfactory results for the run-up of tsunamis. The leap-frog method is normally used.

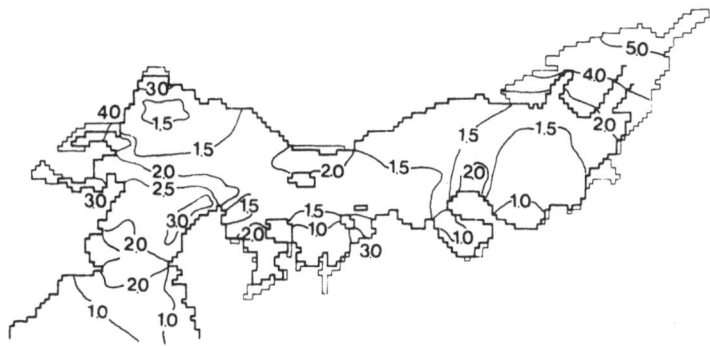

Fig. 9.   Distribution of the highest water particle velocity.

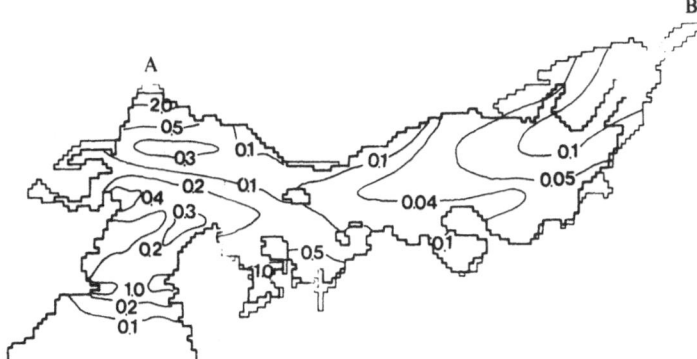

Fig. 10.   Distribution of the ratio of the magnitude of the convection term to that of the non-steady acceleration term.

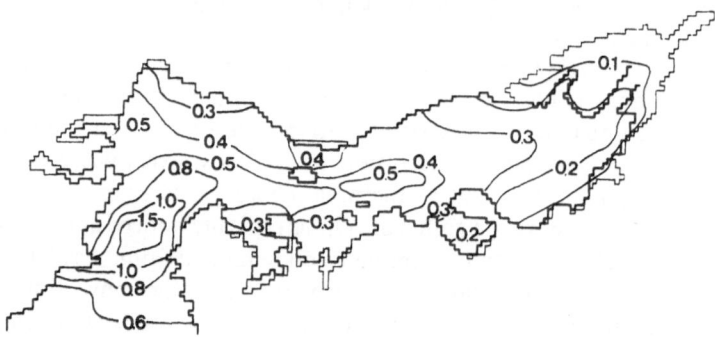

Fig. 11. Distribution of the ratio of the magnitude of the pressure term to that of the non-steady acceleration term.

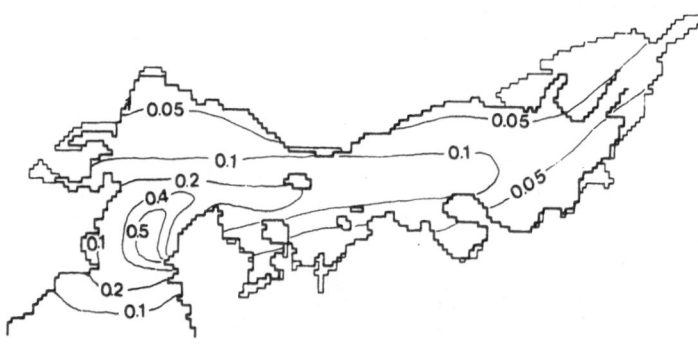

Fig. 12. Distribution of the ratio of the magnitude of the friction term to that of the non-steady acceleration term.

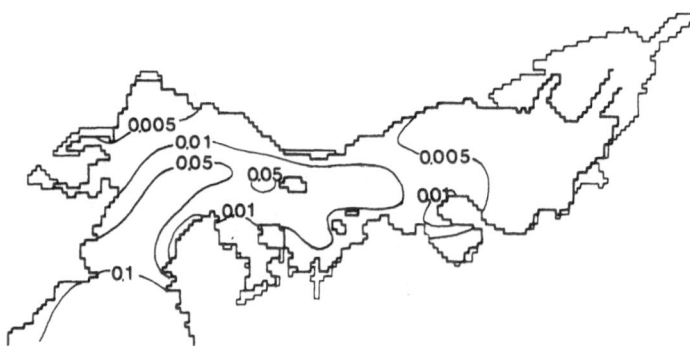

Fig. 13. Distribution of the ratio of the magnitude of the dispersion term to that of the non-steady acceleration term.

Even when there are no breaking waves, the boundary condition at the wave front has an important influence on the accuracy of the numerical results. Although equations described in the Lagrangian description give accurate results with no ambiguity of the boundary condition, they are not convenient in practical applications. Aida's and/or Iwasaki and Mano's conditions which are vague from a physical point of view are successfully applied to the problem, provided that the size of spatial mesh be determined to satisfy $\Delta x/\alpha g T^2 < 4.0 \times 10^{-4}$.

When waves break at their front, oscillations of wave length twice the spatial mesh size often appear and lead to a break down of the numerical scheme. In order to suppress these oscillations, a diffusion term is introduced, with two adjustable constants, β_1 and γ. The constant β_1 determines the strength of the diffusion term while γ determines the region where the diffusion term works. The diffusion term is shown to function very well but at the same time it induces "over-smoothing" of the wave front. An artificial viscosity is introduced to counteract the "over-smoothing". It contains a constant β_2, by which we can freely control the length of the transition zone.

With all these considerations taken into account, a computation is carried out for a tsunami in Ofunato Bay. It is concluded that dispersion is not important in this case. No dispersion may be required except in the case of tsunamis in gently sloping shallow water or in rivers.

The present report was partially supported by a research grant from the Ministry of Education.

REFERENCES

AIDA, I., Numerical experiments for inundation of tsunamis, *Bull. Earthq. Res. Inst.*, **53**, 441–460, 1977 (in Japanese).

GOTO, C. and N. SHUTO, Numerical simulation of tsunami run-ups, Coastal Eng. in Japan, JSCE, Vol. 21, pp. 13–20, 1978.

GOTO, C. and N. SHUTO, Two-dimensional numerical computation of nonlinear tsunami run-ups, Proc. of 26th Conf. on Coastal Eng., JSCE, pp. 56–60, 1979 (in Japanese).

HOUSTON, J. R. and H. L. BUTLER, A numerical model for tsunami inundation, U. S. Army Engineer Waterways Experiment Station, Tech. Report HL–79–2, 1979.

IWASAKI, T. and A. MANO, Two-dimensional numerical computation of tsunami run-ups in the Eulerian description, Proc. of 26th Conf. on Coastal Eng., JSCE, pp. 70–74, 1979 (in Japanese).

KELLER, J. B. and H. B. KELLER, Water wave run-up on a beach, Res. Report No. NONR-3828(00), Office of Naval Res., Rept. of Navy, 1964.

SHUTO, N., Standing wave in front of a sloping dike, Proc. 13th Conf. on Coastal Eng., pp. 1629–1647, ASCE, 1972.

VON NEUMANN, J. and R. D. RICHTMYER, A method for the numerical calculation of hydrodynamic shocks, *J. Appl. Phys.* **21**, 232–238, 1950.

Tsunamis—Their Science and Engineering, edited by K. Iida and T. Iwasaki, 453–466.
Copyright © 1983 by Terra Scientific Publishing Company (TERRAPUB), Tokyo.

Tsunami Run-Up and Back-Wash on a Dry Bed

Kan Kok Chu* and Tetsuo Abe**

Graduate Student, Asian Institute of Technology, Bangkok, Thailand
**Associate Professor, Asian Institute of Technology, Bangkok, Thailand*

(Received August 3, 1981)

A numerical analysis is performed by the characteristic method to describe the behaviour of a long-period wave such as a tsunami wave running up on a beach. Three particular problems to be analysed in this study are the wave motion near the shoreline, the subsequent run-up and the back-wash effect on the run-up. It is shown that the run-up process on the dry bed is concerned with the bore front condition, but the final run-up height depends on the wave characteristics behind the bore front and the water body specified by the initial wave profile. The back-wash which is considered as a return flow decreases the run-up height of the following wave.

1. Introduction

Estimates of the run-up of waves are very important for shore protection works. A tsunami wave which arrives at the shoreline as a bore causes considerable concern. The bore run-up with a high speed associated with the subsequent back-wash may cause serious damage to coastal structures and human properties. Therefore the case that a wave travels toward a uniform impermeable slope ended by a horizontal bottom is studied with the intention of giving a qualitative description.

The method of characteristics which is based on the nonlinear shallow water wave equations has been used for a long time for the theoretical investigations on the problems of wave deformation, breaking, bore inception and propagation, and wave run-up over a beach. Principle works were done by STOKER (1948) and others. It has been noted that once the vertical acceleration is neglected, a bore is predicted sooner than it will be observed.

KELLER *et al.* (1960), Ho and MEYER (1962), and FREEMAN and LE MÉHAUTÉ (1964) have shown that if the bottom friction is neglected, the bore will suddenly collapse as it reaches the shoreline and the run-up is equal to the square of the velocity at the shoreline divided by two times the gravitational acceleration. This solution shows a mathematical singularity at the shoreline with zero depth. It is not a good approximation of practical interest since the run-up is independent of the beach slope and the wave profile. Moreover, FREEMAN and LE MÉHAUTÉ (1964) derived a solution of the run-up by including the bottom friction. The leading edge of the wave is cut short due to the effect of bottom friction, and the leading wave element appears as a "borelike" wave with a vertical wall of water at the front. In reality, it is difficult to estimate the value of the bottom friction for an unsteady wave motion; the friction factors which consist of trees, houses and the debris that may be carried by the wave should be taken into account.

KISHI and SAEKI (1966), CAMFIELD and STREET (1967), and MILLER (1968) performed experimental studied by using a solitary wave. The results of the above investigators indicate that for flatter slopes the run-up height appears equal to or less than the wave height at the shoreline. For steeper slopes, the run-up height increases as the slope increases. And MILLER (1968) describes that the bore run-up is in the form of a thin, fast moving, greatly elongated wedge.

Under certain conditions, a long period wave will decompose into a train of waves when it passes into a shallower water depth. In case the shore is impacted by such waves, a back-wash which occurs continually after a preceding wave run-up should be considered. The effect of the back-wash decreases the run-up height of the following waves, and such as effect is predominant if two consecutive wave crests are not far apart. KEMP and PLINSTON (1974) obtained a solution for the velocities by assuming a sinusoidal movement for the up-rush and back-wash on a sloping beach. In which the up-rush and back-wash were considered indentical in nature but of opposite sign. HIBBERD and PEREGRINE (1978) derived a numerical solution which indicates that a bore is formed on the landward side when the back-wash is retarded by the slower moving water. This bore only lasts for a short peri and is then changed to wave motions.

Numerical solutions for the case of a solitary wave run-up are derived in the present study by using a Fortran Program run on a IBM 370 computer system. Results are compared with the experimental data given by KISHI and SAEKI (1966), and the theoretical solutions according to FREEMAN and LE MÉHAUTÉ (1964). However, the wave run-up becomes complicated if the influence of a back-wash is taken into account. An approach which is different from that of Kemp and Plinston is proposed herein to treat the back-wash as a return flow under gravitational motion. With some simplications, the model presented is also applied to the sinusoidal wave in order to gain insight on the wave run-up accompanied with a back-wash.

2. Numerical Model

2.1 Governing and characteristic equations

The wave motion is assumed to be a two-dimensional, in compressible, invisid fluid motion, and the vertical velocity component is negligibly small. Then the two governing equations are:

Continuity:
$$\frac{\partial \zeta^*}{\partial t^*} + \frac{\partial}{\partial x^*}\{(h^* + \zeta^*)U^*\} = 0 \tag{1}$$

Momentum:
$$\frac{\partial U^*}{\partial t^*} + U^*\frac{\partial U^*}{\partial x^*} + g^*\frac{\partial \zeta^*}{\partial x^*} = -f\frac{U^*|U^*|}{h^* + \zeta^*}. \tag{2}$$

The superscript, *, refers to a variable with dimension. As shown in Fig. 1, where h^* is the water depth from still water level, ζ^*, the free surface elevation, x^*, the horizontal coordinate positive toward the shore. t^*, g^*, and U^* are the time, gravi-

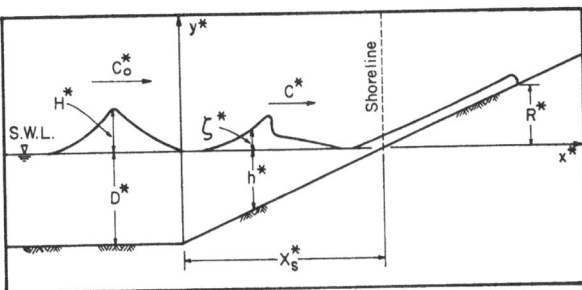

Fig. 1.   Definition sketch.

tational acceleration and mean horizontal particle velocity. The friction coefficient f in Eq. (2) can be related to the Chezy coefficient C_h^* as $f = g^*/C_h^{*2}$ or related to the Manning roughness n^* as $f = n^{*2}g^*/(h^* + \zeta^*)^{1/3}$. The bottom friction has a small value on the seaward side but has a non-negligible influence near the shoreline and on the dry bed.

Choosing $l_0^*, D_0^*, C_0^* (= \sqrt{g^* D_0^*}$, where D_0^* is equal to the water depth at the origin, D^*) and $t_0^* (= D_0^*/C_0^*)$ as the standard values of the horizontal distance, vertical distance, wave celerity and time respectively, the following non-dimensional notations are introduced:

$$x = \frac{x^*}{l_0^*}, \quad h = \frac{h^*}{D^*}, \quad \zeta = \frac{\zeta^*}{D^*}, \quad t = \frac{t^*}{t_0^*}, \quad U = \frac{U^*}{C_0^*}, \quad C = \frac{C^*}{C_0^*}.$$

The wave celerity C^* is equal to $\sqrt{g^*(h^* + \zeta^*)}$, by setting $l_0^* = D^*$, Eqs. (1) and (2) are rearranged into non-dimensional forms as

$$2\frac{\partial C}{\partial t} + 2U\frac{\partial C}{\partial x} + C\frac{\partial U}{\partial x} = 0 \tag{3}$$

$$\frac{\partial U}{\partial t} + U\frac{\partial U}{\partial x} + 2C\frac{\partial C}{\partial x} + S = -f\frac{U|U|}{(h + \zeta)} \tag{4}$$

where $S = -\partial h/\partial x$ is the bottom slope.

The characteristic equations which are used in the numerical model are derived from the equations of continuity and momentum for the long waves with finite amplitudes. The forward (or positive) characteristic equation is defind as

$$U + 2C - Gt = \text{constant, along the curve } dx/dt = U + C. \tag{5}$$

And the backward (or negative) characteristic equation is defined as

$$U - 2C - Gt = \text{constant, along the curve } dx/dt = U - C \tag{6}$$

where $G = -S - fU|U|/(h + \zeta)$.

2.2 Spilling breaker and bore propagation

There is breaker inception as two forward characteristics cross each other at a point. The breaker, which starts to break near the crest and travels behind the leading characteristic, is defined here as a spilling breaker. A bore occurs when the leading element is taken over by following wave elements. Physically, the above mentioned wave profiles both have a vertical appearance at a point (Fig. 2), and the propagation of a spilling breaker is treated with the same mathematical approach as that of a bore. At the intersection, two different sets of the values of flow velocity and wave celerity exist at the same time namely, (U_u^*, C_u^*) and (U_d^*, C_d^*) where subscripts u refers to the high side and d refers to the low side of the bore. The bore proceeds over a water layer with or without a velocity. The values of U_d^*, C_d^* can be calculated by ordinary process, and the values of U_u^*, C_u^* and the wave speed W^* are determined by using continuity and momentum equations for a shockwave. The governing equations are written as

Continuity:
$$q^* = h_u^*(U_u^* - W^*) = h_d^*(U_d^* - W^*) \tag{7}$$

Momentum:
$$\frac{1}{2}g^*(h_u^{*2} - h_d^{*2}) = q^*(U_d^* - U_u^*). \tag{8}$$

By introducing the water depths $h_u^* = C_u^{*2}/g^*$, $h_d^* = C_d^{*2}/g^*$ and $W = W^*/C_0^*$ into the above equations, the non-dimensional expressions are derived as

$$C_u^2 U_u - C_d^2 U_d = W(C_u^2 - C_d^2) \tag{9}$$

$$C_u^4 - C_d^4 = 2C_d^2(W - U_d)(U_u - U_d) \tag{10}$$

With known values of U_d, C_d, there are still three unknowns in two equations, and this can be solved by introducing one more equation based on the foreward characteristic. Details are referable to the work of Freeman and Le Méhauté (1964).

2.3   Initial condition and seaward boundary condition

Coastal waters are assumed to be still as an initial condition. It means that a wave proceeds toward the shore and the water particle velocity at the tip of the wave is equal to zero ($U_d = 0$).

The seaward boundary condition is given by the water level specified by the wave

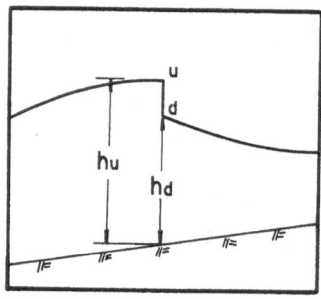
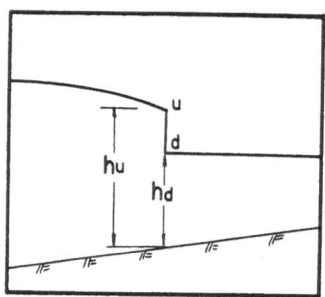

a. Spilling Breaker   b. Bore

Fig. 2.   Definition sketch of spilling breaker and bore.

profile of a solitary wave or a sinusoidal wave, where the flow velocity $U = C\zeta/h$ is used as a first degree of approximation. On the other hand, the corresponding velocity $U = 2(C - \sqrt{h})$ can be determined by the characteristic equation, when an incident wave element is not affected by a wave element coming down, where the celerity is given as $C = \sqrt{h + \zeta}$.

2.4 Shoreline boundary conditions

Two shoreline boundary conditons are adopted: the initial shoreline and the moving shoreline. An initial shoreline is given at the intersection of the still water level with the beach. When a wave arrives at the shoreline, the front condition is assumed constant for the wave run-up. Immediately after the wave attains the maximum run-up height, it runs down, then the receding edge of the back-wash flow is defined as a moving shoreline.

2.5 Bore run-up

As a bore reaches the shoreline, normal procedure for the bore propagation is no longer applied since there is a zero water depth. FREEMAN and LE MÉHAUTÉ (1964) derived the run-up on a dry bed from the leading wave element which is considered as the water front of a rarefaction wave instead of a bore. The solution is obtained by the integration of the leading characteristic, which is,

$$R = \frac{U_s^2}{2} \frac{(1 + A)(1 + 2A)}{1 + f/A^2 S}. \tag{11}$$

Such an estimation depends on the values of the leading edge at the shoreline, where U_s is the bore velocity, f, the friction coefficient, A, a constant corresponding to Froude number, i.e. $A = F_r^{-1}$, and S is the bottom slope.

FREEMAN and LE MÉHAUTÉ (1964) considered A less than 1/2. KISHI and SAEKI (1966) determined A by measuring the depth and the velocity of the leading element at the shoreline, the results show that the velues of A are nearly constant for a given slope and are independent of the initial wave height, and A tends to increase with S. In present model, the wave rushing up the dry bed is considered keeping a bore shape. With regard to the backward characteristic equation the Froude number can be shown equal to two.

$$U_u - 2C_u = U_d - 2C_d + G_d(t_u - t_d) \tag{12}$$

where $G_d = - S - fU_d|U_d|/C_d^2$.

At the shoreline C_d equals to zero since there is zero water depth, consequently, the friction term in G_d is shown to be infinitive. And the velocity U_d decreases to zero under a very large friction. Moreover the bore inception takes place in one location at the same time, hence, $t_u = t_d$. Substituting the above relations into Eq. (12), it implies that $F_r = U_u/C_u = 2$ (IWASAKI and TOGASHI (1969) also used this relation).

As shown in Fig. 3, the characteristics of points P and E are known, where P is the local bore position, E lies on the characteristic curve next to the bore line. Point Q at the shoreline is to be computed. Point O is a neighbouring characteristic that will catch up

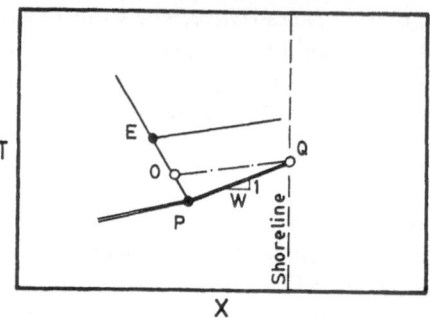

Fig. 3.   Bore run-up.

with the bore front, and its values can be obtained by interpolation between P and E. According to the foreward characteristic equation.

$$U_q + 2C_q = U_0 + 2C_0 + G_0(t_q - t_0).$$

As a front condition, a specified Froude number which is theoretically derived as 2 is introduced into the above equation.
Hence,

$$U_q = \frac{1}{(1 + 2/F_{r,q})}\{U_0 + 2C_0 + G_0(t_q - t_0)\}. \tag{13}$$

The same procedure enables the run-up calculation on the dry bed. It should be noted that beyond the shoreline the flow characteristics behind the bore front will affect the run-up, i.e., more wave elements catch up with the leading front, giving a higher run-up. Until the wave elements behind the bore front become almost at rest, then, the fluid motion is considered as a jet of water rushing up with a certain speed. This speed, in place of U_s, is incorporated into Eq. (11) to obtain the maximum run-up.

2.6   Back-wash

Back-wash is assumed to be a return flow under the gravitational motion immediately after the maximum run-up of a wave. The non-dimensional flow velocity is expressed as

$$U_b = -\sin\alpha\,(\Delta t) \tag{14}$$

where U_b refers to the back-wash velocity, α is the angle between the beach and the horizon, Δt, the time increment starting from the time at the maximum run-up of the preceding wave. The following wave run-up accompanied with a back-wash is considered similar to the case of a tidal bore intruding into a river flow.

3.   Application to a Solitary Wave

3.1   Wave profile

Boussinesq's solution for the solitary wave is used for the input at the seaward boundary condition, the origin of x axis is taken at the toe of the sloping bottom.

$$\zeta = H \operatorname{sech}^2 \left\{ \sqrt{\frac{3H}{4}} \left(x - Ct + \frac{L}{2} \right) \right\} \tag{15}$$

$$C = \sqrt{1 + H} \tag{16}$$

$$U = \frac{C\zeta}{(1 + \zeta)}. \tag{17}$$

The solitary wave is a limiting case of the cnoidal wave and it has an infinite wave length. However, the wave length, L, which is taken at not less than ten times of the depth, can be considered as a good approximation.

Figure 4 shows that the breaking solitary wave forms a bore when it proceeds toward shore. Beyond the shoreline, the height of the bore decreases and a maximum run-up takes place.

3.2 Froude number

As shown in Fig. 5, before the bore inception the Froude number is smaller than unity just like the tsunami wave advancing across the river (IWASAKI et al. (1978)), then, the value increases especially near the shoreline. And the surge region becomes wider as the wave height increases, this is due to the fact that a higher wave breaks sooner in the shoaling water. Moreover, Fig. 6 indicates the Froude number at a section near the shoreline, most of the values fall between 2 and 3.

3.3 Wave run-up
3.3.1 Maximum run-up height

Numerical results of the wave run-up on beach slopes 1/15, 1/20, and 1/30 are

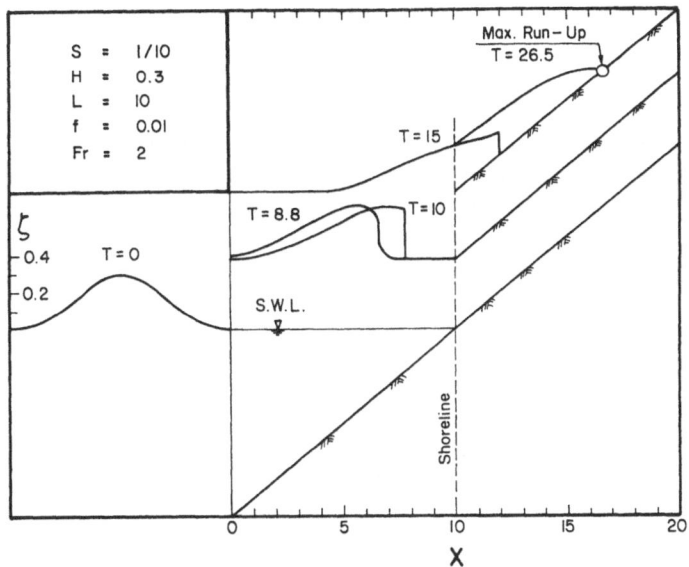

Fig. 4. Time-history of the solitary waves climbing over a beach.

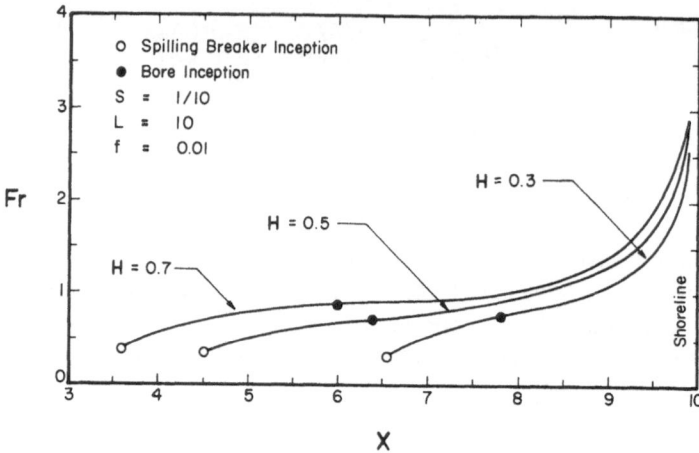

Fig. 5.   The Variation of froude number toward shore.

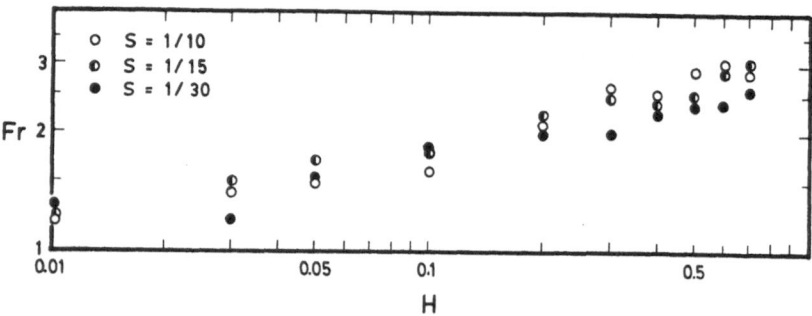

Fig. 6.   Froude number near shoreline ($X = 0.99 X_s$, X_s is the shoreline distance from the origin).

shown in Fig. 7 together with the experimental data given by KISHI and SAEKI (1966) and the theoretical solutions according to FREEMAN and LE MÉHAUTÉ (1964). The results are slightly different from one another in terms of only small wave heights for the mild bottom slope, and the values agree well as a whole. Present results which predict a higher run-up for the smaller incident wave heights are good to meet our assumption. As a wave breaking nearer to the shoreline, a larger amount of energy is conserved. Hence, more wave elements overtake the leading front on the dry bed, thus the run-up is increased. Referring to Fig. 8, it is also demonstrated that the run-up height increases as the slope increases.

3.3.2   Run-up duration

A constant front condition is assumed for the wave run-up. The cases for $F_r = 1.5$, 2.0, 2.5, and 3.5 were calculated under an incidient wave height $H = 0.3$ and $S = 1/10$. The results are plotted in Fig. 9 which indicates that the final run-up heights are almost the same. It is specified that the run-up depends on the shape of the wave rather than the bore front velocity, and the wave front condition only affects the process until the run-up reaches the final stage.

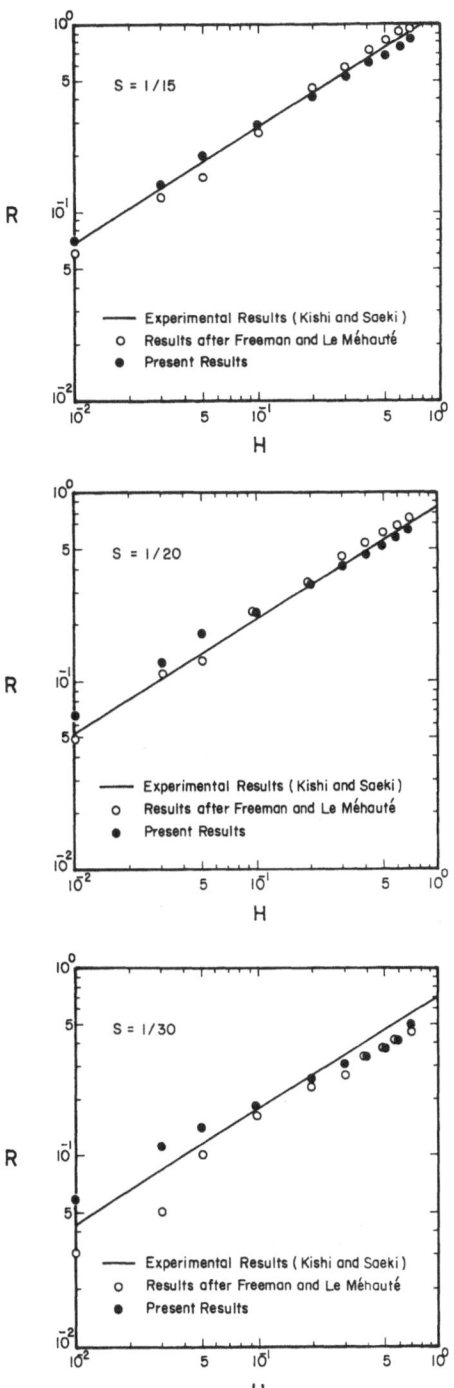

Fig. 7. Run-up of solitary waves on slopes 1/15, 1/20, and 1/30 ($L = 10, f = 0.01, F_r = 2$).

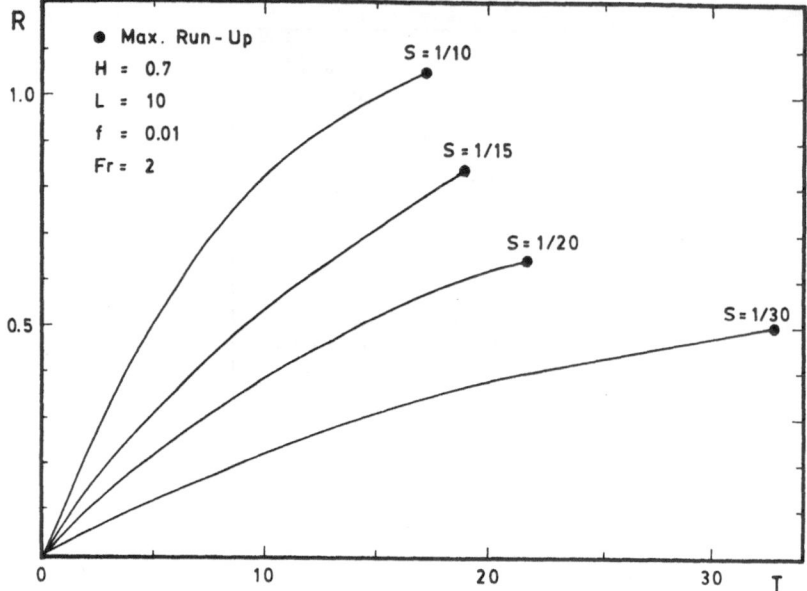

Fig. 8.   Run-up varies with bottom slope ($T = 0$ at shoreline).

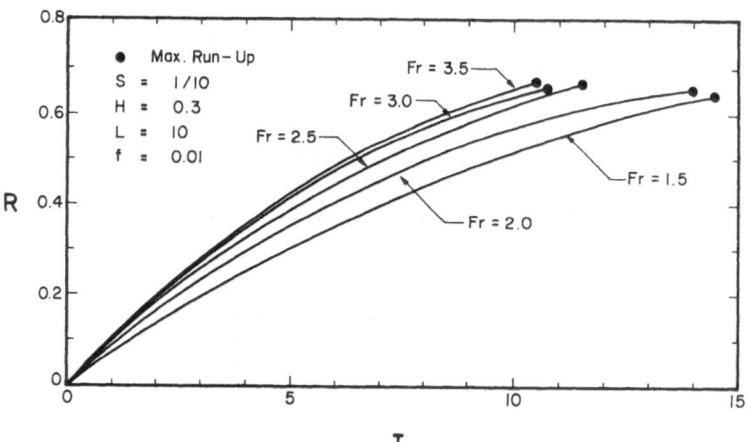

Fig. 9.   Run-up duration curves from shoreline to maximum height.

3.3.3   Bottom friction

The friction coefficient for sea bottom can be verified by field investigation, i.e., PUTNAM and JOHSON (1949) obtained $f = 0.01$ under the ripple formation. But the onshore friction coefficient is difficult to be determined since it depends on the beach configuration, beach materials, structures and vegetations, and these factors are expected to produce a large resistance against a tsunami wave. Arbitrary values of f are selected equal to 0.01, 0.05, and 0.1 respectively for the run-up calculation. Results obtained from Eq. (11) and present results are plotted in Fig. 10. Both of the predictions show that the bottom friction has a counter effect on the wave run-up. But two sets of values are remarkably different from each other under high friction and small wave height, it is because the effect of wave characteristics behind the front has been neglected by using Eq. (11). And Eq. (11) is considered valid for small bottom friction. For high bottom friction, we cannot come at any conclusion since practical information is not available even now. Therefore the value 0.01 is chosen for f in the present study.

4.   Application to a Sinusoidal Wave

4.1   Single sine wave with positive front

The time-history of the sinusoidal wave on the run-up indicates that the process is similar to that of a solitary wave (Fig. 11). But the maximum value of the run-up height is different from that of the solitary wave with the same initial wave height.

4.2   Twin sine wave and back-wash

The purpose of the twin sine wave calculation is to see the effect of a back-wash on the second wave run-up. This example was done with the following simplications:

1)   Input at the seaward boundary condition is divided into two parts but with consecutive time interval.

2)   Still coastal water condition is assumed for the second wave before the back-wash takes place.

According to the results shown in Fig. 12, the effect of the back-wash is to decrease the run-up. The second bore height at its final stage indicates a higher value compared with the one for the first wave, and its profile still keeps the appearance of a bore. Moreover a

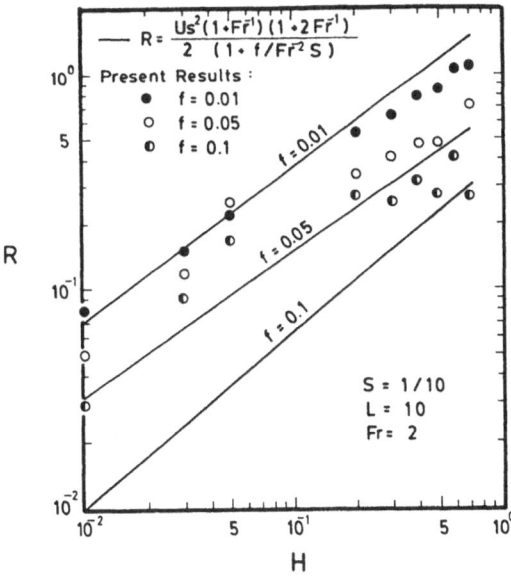

Fig. 10.   Run-up varies with bottom friction.

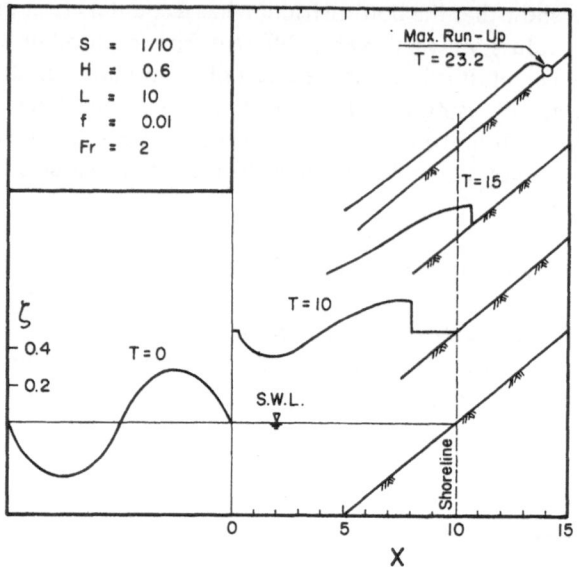

Fig. 11.   Time-history of the sinusoidal wave climbing over a beach.

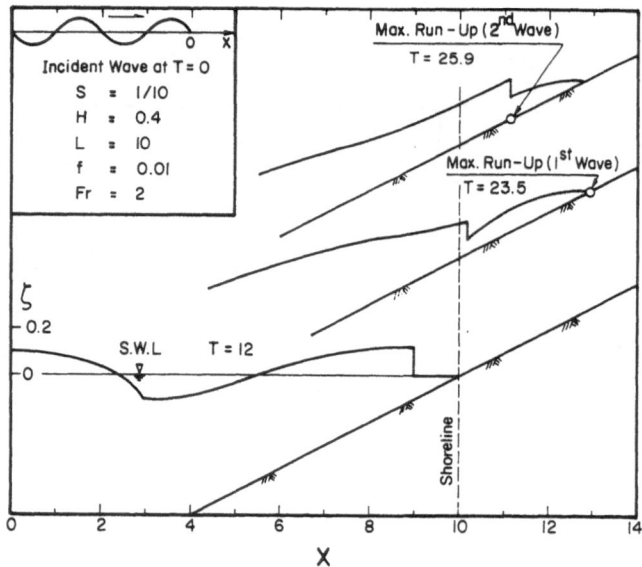

Fig. 12.   Wave run-up accompanied with back-wash.

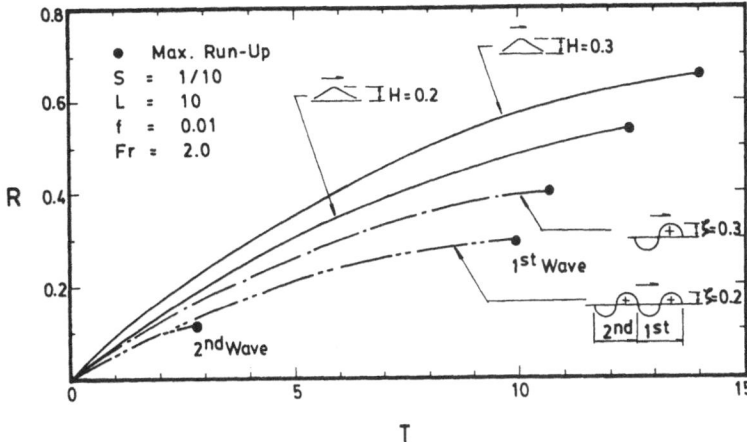

Fig. 13.   Run-up-duration of solitary waves and sinusoidal waves ($T = 0$ at shoreline).

comparison between the run-up of solitary waves and sinusoidal waves is shown in Fig. 13, and the differences are obviously due to the dissimilar wave profiles.

5.   Conclusion

A numerical method for the calculation of a bore run-up on the dry bed has been developed. The final run-up height depends on the water body specified by the initial wave profile rather than the bore front condition. However, the run-up process which will play an important role in the studies on the prevention of disaster due to tsunamis is affected by the bore front condition.

The back-wash which is considered as a return flow under gravitational motion has been taken into account for the wave run-up. The effect of the back-wash is to reduce the subsequent run-up. It is, however, recalled that a still basin conditon is also assumed for the second wave. In practice, the movement of the coastal waters which is induced by the preceeding wave should not be ignored. Hence a more precise method remains to be developed.

The authors wish to express their sincere gratitude to Professor T. Iwasaki for his cogent suggestions during his visit to AIT in January, 1981.

REFERENCES

CAMFIELD, F. E. and R. L. STREET, An investigation of the deformation and breaking of solitary waves, Tech. Report No. 81, DDC AD 664 249, Dept. of Civil Eng., Stanfore University, California, 1967.

FREEMAN, J. C. and B. LE MÉHAUTÉ, Wave breakers on a beach and surges on a dry bed, *J. Hydraulics Div.*, ASCE, No. 3834, HY2 187–216, 1964.

HIBBERD, S. and D. H. PEREGRINE, Surf and run-up on a beach: A Uniform Bore, *J. Fluid Mech.*, **95**, Part 2, 323–345, 1979.

Ho, D. V. and R. E. MEYER, Climb of a bore on a beach, Part 1, Uniform beach slope, *J. Fluid Mech.*, **14**, 305–318, 1962.

IWASAKI, T. and H. TOGASHI, Front condition of tsunami run-up, Proc. 16th Conf. Coastal Eng. Japan, pp. 359–364, 1969 (in Japanese).

IWASAKI, T., T. ABE, and K. HASHIMOTO, Tsunami run-up in a river, Proc. 25th Conf. Coastal Eng. Japan, 137–140, 1978 (in Japanese).

KELLER, H. B., D. A. LEVINE, and G. B. WHITHAM, Motion of a bore on a sloping beach, *J. Fluid Mech.*, 7, 302–316, 1960.

KEMP, P. H. and D. T. PLINSTON, Internal velocities in the uprush and backwash zone, Proc. 14th Conf. Coastal Eng., 575–585, 1974.

KISHI, T. and H. SAEKI, The shoaling, breaking and run-up of the solitary wave on impermeable rough slopes, Proc. 10th Conf. Coastal Eng., 322–348, 1966.

LE MÉHAUTÉ, B., On surge on a dry bed and wave run-up, United States-Japan Cooperative Scientific Research Seminars on Tsunami Run-Up, Sapporo, Japan, Apr., 1–18, 1965.

MILLER, R. L., Experimental determination of run-up of undular and fully developed bores, *J. Geophys. Res.*, 73, 4497–4510, 1968.

PUTNAM, J. A. and J. W. JOHNSON, The dissipation of wave energy by bottom friction, *Trans. Am. Geophys. Union*, 30, No. 1, 67–74, 1949.

STOKER, J. J., The formation of breakers and bores, *Comm. Appl. Math.*, 1, 1–87, 1948.

List of Symbols

A	= coefficient for the wave front
C^*, C_0^*	= wave celerity
C_h^*	= Chezy coefficient
D^*	= water depth at the origin
F_r	= Froude number
f	= friction coefficient
g^*	= gravitational acceleration
H^*	= incident wave height
h^*	= water depth from still water level
L^*	= wave length
n^*	= Manning roughness
R^*	= run-up height from still water level
S	= bottom slope
t^*, t_0^*	= time
U^*	= horizontal particle velocity
U_b^*	= back-wash velocity
U_s^*	= velocity at shoreline
W^*	= bore speed
x_s^*	= horizontal distance between shoreline and the toe of bottom slope
x^*	= horizontal coordinate positive toward the shore
y^*	= vertical corrdinate positive upward from still water level
ζ^*	= free surface elevation from still water level
α	= beach angle with horizon
$*$	= superscript refers to a dimensional quantity

Dimensionless quantities are defined as follows:

C	$= C^*/C_0^*$
H	$= H^*/D^*$
h	$= h^*/D^*$
t(or T)	$= t^*/t_0^*$
U	$= U^*/C_0^*$
x	$= x^*/D^*$
ζ	$= \zeta^*/D^*$

Tsunamis—Their Science and Engineering, edited by K. Iida and T. Iwasaki, 467–478.

Breaking and Run-Up of Long Waves

M. S. KIRKGÖZ

Department of Civil Engineering
Cukurova University, Adana, Turkey

(Received September 10, 1981)

Breaking and run-up of long-period oscillatory waves propagating on a beach sloping at 1/10, are studied theoretically. Nonlinear shallow-water wave equations are solved by using the method of characteristics. In the numerical scheme the computation is initialized at the transformation point where the wave begins to become asymmetrical about a vertical line through its crest. Solitary wave theory is used to provide data for input values. Although the numerical scheme tends to prduce an early breaking, certain geometrical features of the wave seem to be predicted quite well when compared with the experimental values in the transormation zone. In order to avoid the problematical consequences at the shoreline the classical bore equations are not coupled with the characteristic equations, instead, a geometrical criterion is introduced in the surf zone. An expression for estimating the run-up on roughened beds is presented which gives resonably accurate results when Jonsson's friction formula is used.

1. Introduction

A number of motions in nature such as the tides in the oceans, solitary waves and flood waves in rivers in which the waterdepth to wavelenth ratio is small, and the vertical accelerations of the fluid particles are unimportant, are often considered to be best represented theoretically by the nonliear shallow-water wave equations. This theory which is usually referred to as the theory of long waves gives solutions in which the wave profile changes continuously and becomes increasingly asymmetrical.

It seems that STOKER (1957) was the first to propose the application of the method of characteristics (which is the method of solving the nonlinear long wave equations) to the problem of wave deformation and breaking inception over a beach. Following the pioneering work of Stoker, many authors have attempted to describe the behaviour of long waves breaking in shoaling water (see for example FREEMAN and LE MÉHAUTÉ, 1964; AMEIN, 1966; IWASAKI and TOGASHI, 1970). The application of the nonlinear long-wave theory to the breaking of shoaling oscillatory waves, including the backwash effects, was made by KIRKGOZ (1981).

In this study, the nonlinear long-wave theory is applied to the breaking (by plunging) of long-period oscillatory waves propagating into a still-water of nonuniform depth. In the method of characteristics the initial values for the motion are taken from solitary wave and subsequently some of the geometrical features of the wave are predicted in the breaker zone (description of the wave plunging process and the breaker zone was given by HEDGES and KIRKGOZ, 1981). A theoretical treatment is also given for the wave motion after the shoreline. The motion of wave in the surf zone is not assumed to be like that of an ordinary bore, instead a geometrical criterion is introduced. In this way the shore sigularity problem is avoided. Experimental results relating to plunging breakers (KIRKGOZ, 1978) are incorporated in computational procedures based on long-wave theory, and also used to provide some comparison between what is prediced theoretically and what actually occurs.

2. Nonlinear Shallow-Water Wave Equations on a Sloping Bed

The first order nonlinear shallow-water wave equations are the conservation of mass and momentum:

$$\frac{\partial}{\partial x}(u(d + \eta)) = -\frac{\partial \eta}{\partial t} \tag{1}$$

$$\frac{\partial u}{\partial t} + u\frac{\partial u}{\partial x} = -g\frac{\partial \eta}{\partial x} \tag{2}$$

where $d(x)$ is the depth from still-water-level (SWL), $\eta(x, t)$ is the surface displacement, $u(x, t)$ is the horizontal water-particle velocity, x is horizontal distance, and t is time (see Fig. 1 for definition of symbols).

AIRY (1845) first pointed out that in long waves, different parts travel with different speeds depending on the local water depth. STOKER (1957) confirmed that the speed variations along the length of a long wave would be given by Airy's expression:

$$C = (g(d + \eta))^{\frac{1}{2}}. \tag{3}$$

Some experimental data on breaker crest velocities given by KIRKGOZ (1978), tend to support the use of Eq. (3).

When the method of characteristics is used in the numerical solution of Eqs. (1) and (2), these equations are transformed into a pair of ordinary differential equations called characteristic differential equations which are satisfied along the characteristic curves. On the advancing characteristic curves:

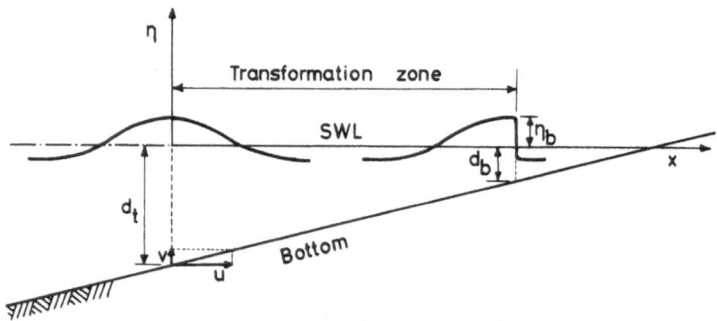

Fig.1.   Definition sketch for symbols and wave transformation zone.

$$u + 2C - g(\mathrm{d}d/\mathrm{d}x)t = \text{constant} \tag{4}$$

$$\mathrm{d}x/\mathrm{d}t = u + C. \tag{5}$$

On the receding characteristic curves:

$$u - 2C - g(\mathrm{d}d/\mathrm{d}x)t = \text{constant} \tag{6}$$

$$\mathrm{d}x/\mathrm{d}t = u - C. \tag{7}$$

The solution of the characteristic equations, Eqs. (4)–(7), may be found once the initial values of x, t, u, and C are known.

3. Theoretical Treatment in the Transformation Zone

The transformation zone (see Fig. 1) covers the region between the point where the wave begins to become asymmetrical about a vertical line through its crest (the transformation point) and the point where the wave front becomes vertical (the breaking point). When the method of characteristics is applied to a wave, paradoxically, it begins to transform and ultimately breaks. Therefore, in this study, it was thought appropriate to initialize computations at the transformation point.

3.1 Numercial scheme

The numerical solution of the nonlinear long-wave equations was carried out in this study by using a grid of characteristics. The basic principle of the method is given below.

It is convenient to introduce the following dimensionless quantities: $\bar{x} = x/d_t$, $\bar{t} = tC_m/d_t$, $\bar{u} = u/C_m$, and $\bar{C} = C/C_m$ in which x is the horizontal distance measured from wave crest at the transformation point, t is the time measured from the instant when wave is at the transformation point, d_t is the still-water depth at the transformation point (Fig. 1), and $C_m = (gd_t)^{\frac{1}{2}}$.

In the characteristic equations the term $\mathrm{d}d/\mathrm{d}x$ merely represents the bed gradient. However, friction and turbulence induced by wave breaking may also be allowed for in this term by writing: $\mathrm{d}d/\mathrm{d}x = G = $ bed slope + friction slope + turbulence slope, or, $G = -S - S_f - S_t$. More details of the determination of the friction slope, S_f, are given below. The turbulence slope, S_t, was not included in the present work as no satisfactory methods as yet exist for its evaluation.

Figure 2 illustrates how the dimensionless variables, $\bar{x}$, $\bar{t}$, $\bar{u}$, and $\bar{C}$ are evaluated within the breaker zone. For example, in order to calculate the values of $\bar{x}_3$, $\bar{t}_3$, $\bar{u}_3$, and $\bar{C}_3$ at point 3, dimensionless form of Eqs. (4)–(7) are
On the advancing characteristic (between points 1 and 3):

$$\bar{u}_1 + 2\bar{C}_1 - G_1\bar{t}_1 = \bar{u}_3 + 2\bar{C}_3 - G_1\bar{t}_3 \tag{8}$$

$$\bar{x}_3 - \bar{x}_1 = (\bar{u}_1 + \bar{C}_1)(\bar{t}_3 - \bar{t}_1). \tag{9}$$

On the receding characteristic (between points 2 and 3):

$$\bar{u}_2 - 2\bar{C}_2 - G_2\bar{t}_2 = \bar{u}_3 - 2\bar{C}_3 - G_2\bar{t}_3 \tag{10}$$

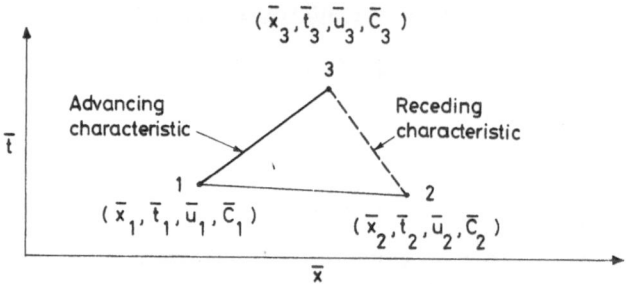

Fig.2. Scheme for numerical computation.

$$\bar{x}_3 - \bar{x}_2 = (\bar{u}_2 - \bar{C}_2)(\bar{t}_3 - \bar{t}_2). \tag{11}$$

In Eqs. (8) and (10) the term G is assumed to be constant between points 1 and 3 and also between points 2 and 3, having the values G_1 and G_2 respectively. From, Eqs. (8)–(11), the four unknown quantities, $\bar{x}_3$, $\bar{t}_3$, $\bar{u}_3$, and $\bar{C}_3$, may be calculated, Once the computation is initiated with known values of $\bar{x}$, $\bar{u}$ and $\bar{C}$ at $\bar{t} = 0$, a complete network of characteristics may be built up using the scheme outlined above (Fig. 3). Depending upon the choice of the number of initial points, as fine a network of characteristics as desired may be obtained. Note, however, that once intersection takes place between the two advancing characteristics, then the number of points which remain for evaluating the next set of points is reduced by one.

Computations involving the characteristic corresponding to the leading wave trough are carried out slightly differently since no intersections with receding characteristics exist. As illustrated in Fig. 3 the computations on the front characteristic may be accomplished by taking the time value of the last point to be equal to that of the penultimate point in a computational step.

3.2 Selection of initial values

As oscillatory waves travel into shallow water their heights increase and crests tend to become separated. If the distance between crests is large enough each wave starts looking and behaving rather like a solitary wave. This resemblance has led many

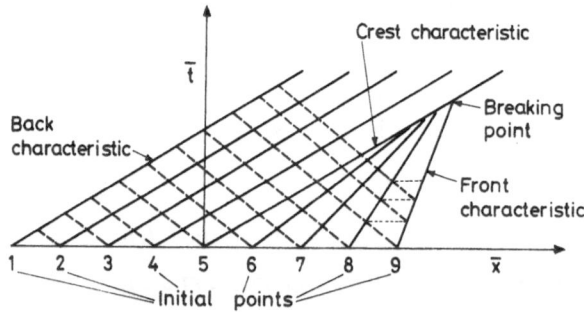

Fig.3. Definition sketch for characteristics network.

authors to suggest that shoaling oscillatory waves close to breaking may be treated as solitary waves. BAGNOLD (1947) pointed out that as the wave period is increased beyond a certain limiting value, T_{eff}, the shoaling wave train tends to split up into a succession of solitary waves. He gave:

$$T_{eff} = (2\pi/M)(d/g)^{\frac{1}{2}} \qquad (12)$$

in which to a first approximation $M = (3H/d)^{\frac{1}{2}}$, H is wave height, and d is water depth. Some experimental evidence in support of the validity of Eq. (12) is given in Fig. 4 in which experimental water particle velocities under the crest of a shoaling oscillatory wave are compared with a number of wave theories (full details of the particle velocity measurements are given elsewhere, KIRKGOZ, 1978). Consequently, in this study, the motion of long-period oscillatory waves has been regarded as similar to that of a similar sized solitary wave. As for the local wave celerities, $\bar{C}$, they were calculated at each point from Eq. (3) using the solitary wave profile.

Above procedure for the treatment of the wave in the transformation zone is carried out up to the time when all advancing characteristics for points in front of the crest have been overtaken by the crest characteristic. At that moment in the computation it is assumed that the wave breaking point has been reached (see Fig. 3).

4. Determination of Roughness Factor

A lot of efforts have been spent in formulating the bottom resistance to waves on horizontal bed (see for example JONSSON, 1966; KAJIURA, 1968; KAMPHUIS, 1975; JONSSON and CARLSEN, 1976). However, the estimation of the friction losses in the breaker zone on a sloping bed seems at present to require further information. Consequently, it is assumed in this study that the information available for waves in constant water depth is also valid for waves breaking in shoaling water.

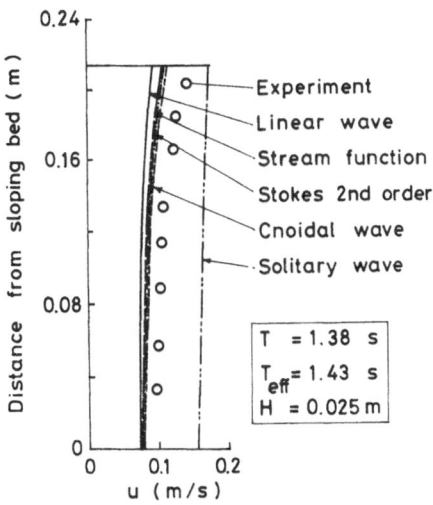

Fig.4. Horizontal particle velocity under the crest.

Using the expression for the local hydraulic radius, $R = d + \eta$, and Eq. (3):

$$R = C^2/g. \tag{13}$$

Using Chézy formula, $u = C_z (RS_f)^{\frac{1}{2}}$, and Eq. (13) the friction solpe, S_f, may then be written:

$$S_f = (g/C_z^2)(u/C)^2 \tag{14}$$

where C_z is the Chézy coefficient which could be related to wave friction factor, f_w, by:

$$C_z = (2g/f_w)^{\frac{1}{2}}. \tag{15}$$

The maximum amplitude Reynolds number, RE, for oscillatory flow is:

$$RE = u_\delta a_\delta/\nu \tag{16}$$

in which u_δ is the maximum orbital velocity just outside the boundary layer and a_δ ($\simeq u_\delta T/2\pi$) is the wave orbital amplitude just outside the boundary layer and ν is the kinematic viscosity of water. It should be noted, however, that u_δ in the present case corresponds to the horizontal particle velocity beneath the crest, since the particle velocity distribution is assumed constant at every vertical section according to the long wave theory. Substituting the values $g = 9.81$ ms^{-2} and $\nu = 1.013 \times 10^{-6}$ m^{2}s^{-1} (at 20°C) into Eq. (16) yields:

$$RE = 0.987 \times 10^6 \, u_\delta a_\delta.$$

The following formulae have been used to determine f_w for different flow conditions; for smooth turbulent flow (Kajiura, 1968):

$$\frac{1}{8.1\sqrt{f_w}} + log\frac{1}{\sqrt{f_w}} = -0.135 + log\sqrt{RE} \tag{17}$$

for rough beds (Jonsson and Carlsen, 1976):

$$\frac{1}{4\sqrt{f_w}} + log\frac{1}{4\sqrt{f_w}} = -0.08 + log\frac{a_\delta}{k_s} \tag{18}$$

where k_s is Nikuradse's sand grain roughness and equivalent values of it for different roughened beds must be found experimentally.

5. Theoretical Treatment in the Surf Zone

The surf zone covers the region between the breaking point and the point where the wave begins to run up the beach (see Fig. 5). In this zone energy dissipation increases due to the highly turbulent nature of the motion which is caused by air entrainment, and the formation of a horizontal roller. Sawaragi and Iwata (1974) estimated that 15–30% of a breaking wave's energy was dissipated by the horizontal roller between the breaking point and the point where the roller disappears. Battjes (1975) also studied the turbulent energy dissipation in the surf zone by relating to the energy dissipation due to breaking. Unfortunately, much still remains to be done before these two mecahnisms of energy dissipation can be adequately described mathematically in a numerical model. Consequently, the present work does not attempt to direcly account for them, though their effect is recognised as part of the complicated process which limits the maximum elevation of the water surface in the surf zone.

The motion of the wave front in the surf zone has commonly been treated as that of a bore. PEREGRINE (1979) pointed out that if the breaking was sufficiently prolonged the motion eventually became quite turbulent and, indeed, on a beach it might be modelled as a turbulent bore.

Following WHITHAM (1958), many investigators have adopted the classical bore equations for describing the front conditions and assumed that the motion immediately behind the front could be described by the long wave theory. These investigators, however, have been faced with the problem that: the bore height was predicted to reduce to zero at the shoreline, and at this point the corresponding bore velocity was infinite (the shore singularity problem). In order to avoid these physically incorrect mathematical predictions, somewhat arbitrary assumptions have been made in the past such as the termination of the computing just before the bore front reaches the shore (AMEIN, 1966) or taking $C = Au$ at the shoreline (A is an arbitraty constant and was chosen 0.2 by FREEMAN and LE MÉHAUTÉ, 1964). In addition, significant geometrical and kinematical differences appear to exist between ordinary bores and wave-induced bores. Consequently, in the present study the bore equations are not used; instead the following procedure is adopted.

5.1 Proposed surf zone treatment

It was experimentally found that at the breaking point on beaches the ratio η_b/d_b (η_b is wave crest elevation from still-water level, and d_b is still-water depth at the breaking point) remained fairly stable at a value of 0.78 (KIRKGOZ, 1978). Since the water depth controlled the maximum elevation of the wave crest at the breaking point, it has been assumed that a similar control would be exerted throughout the surf zone. MIZUGUCHI (1980) also found that on uniformly sloping beaches the height decayed almost linearly in the surf zone.

Knowing the values of η at all depths the following step procedure may be applied to compute $\bar{x}$, $\bar{t}$, $\bar{u}$, and $\bar{C}$ in the surf zone. Referring to Fig. 5, the values of $\bar{x}_1$, $\bar{t}_1$, $\bar{u}_1$, and $\bar{C}_1$ are known at Section 1. The ratio η_b/d_b is also known. Thus the values of $\bar{x}_2$, $\bar{t}_2$, $\bar{u}_2$, and $\bar{C}_2$ may be obtained from the advancing characteristic equation. If Δx is the chosen step distance, then:

$$\bar{x}_2 = \bar{x}_1 + \Delta \bar{x}, \bar{t}_2 = \bar{t}_1 + \Delta \bar{t}, \bar{C}_2 = ((d_2 + \eta_2)/d_1)^{\frac{1}{2}}, \text{ and}$$
$$\bar{u}_2 = \bar{u}_1 + 2\bar{C}_1 + G\Delta \bar{t} - 2\bar{C}_2 \text{ in which}$$
$$\Delta \bar{t} = \Delta \bar{x}/(\bar{u}_1 + \bar{C}_1) \text{ and } G = -S - S_f.$$

The above procedure is terminated at point A in Fig. 5 when the wave front starts climbing the beach. It is then assumed that, because the wave front at point A has a velocity $dx/dt = u_A + C_A$, according to Eq. (5), the leading edge of the run-up bore also has this velocity. That is:

$$u_0 = u_A + C_A. \tag{19}$$

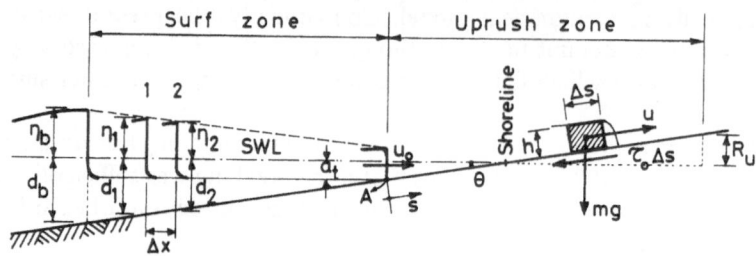

Fig.5.   Definition sketch for surf zone and uprush zone.

Eq. (19) gives a finite value for the velocity of the water at the shoreline, unlike the situation when the bore equations are used.

6.   Theoretical Treatment in the Uprush Zone

When the wave propagation ends at point A in Fig. 5, its motion is replaced by a jet of water which runs up the sloping beach until its kinetic energy is exhausted.

The following expression which has commonly been agreed as giving the maximum vertical rise of water on a smooth, impermeable and dry uniform beach, has also been used in the present study. That is the run-up height from still-water-level:

$$R_u = u_0^2/2g - a_t \tag{20}$$

in which a_t is wave trough amplitude at the transformation point.

During run-up on rough beaches the leading water edge is subjected to both gravity and shear gorces. Assuming that the form of the leading water edge is unchanged in the uprush zone, then, the following derivation may be given for frictional run-up.

From Fig. 5 the equation of motion for the leading edge (considered to be of unit width) is:

$$m\frac{du}{dt} = -mg\sin\theta - \tau_0\,\Delta s \tag{21}$$

in which m is water mass, θ is beach slope angle, τ_0 is shear stress at solid boundary, and s is distance in the uprush zone measured in the direction of run-up. From the shear stress equation, $\tau_0 = \gamma\,RS_f$ (γ is specific weight of water), and the Chézy equation the shear stress, τ_0, is given as:

$$\tau_0 = \rho g u^2/C_z^2 \tag{22}$$

where ρ is density of water. Substituting Eq. (22) into Eq. (21), and using $m = \rho h\Delta s$:

$$\frac{du}{dt} + \frac{g}{hC_z^2}u^2 + g\sin\theta = 0. \tag{23}$$

On integrating Eq. (23), and using the initial condition when $t = 0, u = u_0$, the velocity of the leading edge is:

$$ds/dt = u = (hC_z^2\sin\theta)^{\frac{1}{2}}\tan(P+Q) \tag{24}$$

in which $P = -t \, (g^2 \, \sin\theta/hC_z^2)^{\frac{1}{2}}$ and $Q = \tan^{-1} \, (u_0(1/hC_z^2 \sin\theta)^{\frac{1}{2}})$. On integrating Eq. (24), and using the condition at $t = 0$, $s = 0$, the position of the leading edge is found as:

$$s = (hC_z^2/g) \, \text{Ln} \, (\cos \, (P + Q)/\cos Q).$$

Consequently, the run-up height from still-water-level is:

$$R_u = -\frac{hC_z^2 \sin\theta}{g} \, \text{Ln} \, \cos \left[\tan^{-1} \left(\frac{u_0^2}{hC_z^2 \sin\theta} \right)^{\frac{1}{2}} \right] - a_1. \qquad (25)$$

7.   Results and Discussion

The results are limited to two different waves, on a slope of 1/10, with deep-water steepnesses $(H_0/L_0 =)$ 0.004 and 0.006, whose motions in shoaling water may be regarded as similar to those of similar sized solitary waves. Figure 6 shows a typical network of characteristics and history of wave profile in the breaker zone. The final point on the crest characteristic represents the end of the breaker zone and the beginning of the uprush zone.

Figure 7 gives the variations in the dimensionless wave height, H/H_0, with relative depth, d/L_0, in the transformation zone. As may be seen from the figure, the numerical model agrees reasonably well with the experimental curve provided that $d/L_0 > 0.02$. For lower values of the relative depth, the numerical results diverge from the

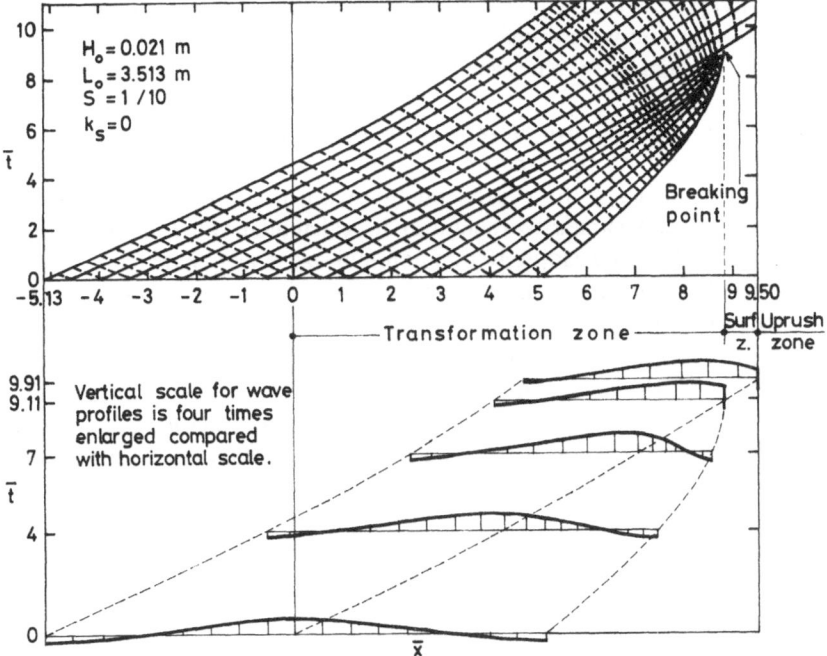

Fig.6.   Network of characteristics and history of wave profile in the breaker zone.

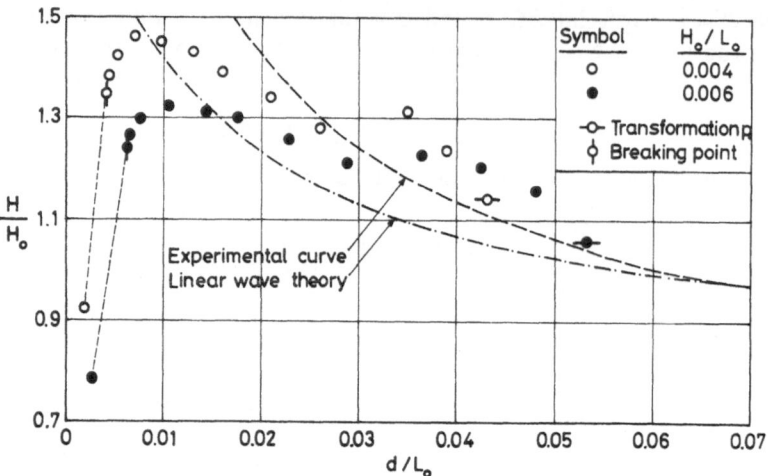

Fig.7.   Wave height variation in the transformation zone.

experimental values but remain in close agreement with those predicted by linear wave theory. It should be noted that the crest elevation of the breaker becomes maximum slightly earlier than the assumed breaking point is reached.

The dimensionless water depths, d_b/H_0, at the breaking point are 0.95 and 1.05 for waves of steepnesses 0.004 and 0.006 respectively. These values are considerably smaller than the experimental value of $d_b/H_0 \simeq 1.5$. This indicates that the method of characteristics gives an early breaking when compared with the experiment. On the other hand, for the same wave conditions the numerical values of the ratio η_b/d_b are found to be 0.90 and 0.76, respectively, and these values seem to reasonably agree with the experimental value of 0.78.

Table 1 shows the relative run-up values, R_u/H_0, predicted by Eqs. (20) and (25), and compare them with the experimental values (smooth beach only) for a beach sloping at 1/10. In using the expression involving friction (Eq. (25)), the Chézy coefficient has been assumed to remain at its calculated value at the beginning of uprush zone. On the rough slope, 10 mm nominal size of aggregate had 8 mm average roughness height. The equivalent roughness height, k_s, was 0.012 m. Eq. (25) which contains friction term is also applicable to smooth surfaces since completely frictionless flow does not occur. It may be seen from Table 1 that the theoretical run-ups predicted by Eq. (20) are considerably greater than experimental values. Equation (25) also gives similar results on smooth beach when Kajiura's friction factor (Eq. (17)) is used. Theoretical run-up results on roughened bed where friction effects are included by using Jonsson's formula (Eq. (18)), seem to agree quite well with the experimental values.

8.   Conclusions

The motion of long-period oscillatory waves in the breaker zone and in the uprush zone is studied theoretically. Initial wave input is taken from the solitary wave theory.

Table 1. Run-up results.

| H_0/L_0 | Relative run-up from still-water-level, R_u/H_0 | | | | |
| | Smooth bed | | Roughened bed | | Experimental values (smooth beach only) |
	Eq. (20)	Eq. (25)	Eq. (20)	Eq. (25)	
0.004	3.33	2.93	2.74	1.15	1.22
0.006	3.12	2.73	2.56	1.06	0.98

The solution of the nonlinear shallow water wave equations by the method of characteristics shows that this method is capable of providing a reasonably accurate description of wave motions over sloping beds provided that energy losses are properly defined and included in the computational procedure. It should be emphasized, however, that the wave breaking point is prematurely predicted. Equation (25) is quite successful in predicting wave run-up on roughened bed when Jonsson's friction formula is used in the numerical scheme to include friction effects.

The computations involved in this study were partly carried out in the Department of Civil Engineering at Southamption University. In this connection, the author is most grateful to Mr. N. B. Webber for giving him an opportunity to work at the Department as a UNESCO fellow during the summer of 1981.

REFERENCES

AIRY, G. B., On tides and waves, *Encyclopaedia Metropolitana*, London, 1845.

AMEIN, M., A method for determining the behaviour of long waves climbing a sloping beach, *J. Geophys. Res.*, **71**, 401–410, 1966.

BAGNOLD, R. A., Sand movement by waves: some small-scale experiments with sand of very low density, *J. ICE*, **27**, 447–469, 1947.

BATTJES, J. A., Modelling of turbulence in the surf zone, in *Symposium on Modelling Techniques, ASCE*, 2, pp. 1050–1061, 1975.

FREEMAN, J. C. and B. LE MÉHAUTÉ, Wave breakers on a beach and surges on a dry bed, *Proc. ASCE*, **90** (HY2), 187–216, 1964.

HEDGES, T. S. and M. S. KIRKGOZ, An experimental study of the transformation zone of plunging breakers, *Coastal Eng.*, **4**, 319–333, 1981.

IWASAKI, T. and H. TOGASHI, On the Shoreline and leading front conditions of tsunami waves in the light of the method of characteristics, *Coastal Eng. in Japan*, **13**, 113–125, 1970.

JONSSON, I. G., Wave boundary layers and friction factors, *Proc. 10th Coastal Eng. Conf.*, **1**, 127–148, 1966.

JONSSON, I. G. and N. A. CARLSEN, Experimental and theoretical investigations in an oscillatory turbulent boundary layer, *J. Hydraulic Res.*, **14**, 45–60, 1976.

KAJIURA, K., A model of the bottom boundary layer in water waves, *Bull. Earthq. Res. Inst.*, **46**, 75–123, 1968.

KAMPHUIS, J. W., Friction factor under osicllatory waves, *Proc. ASCE*, **101** (WW2), 135–144, 1975.

KIRKGOZ, M. S., Breaking waves: their action on slopes and impact on vertical seawalls, *Ph. D. Thesis*, University of Liverpool, 1978.

KIRKGOZ, M. S., A theoretical study of plunging breakers and their run-up, *Coastal Eng.*, **5**, 353–370, 1981.

MIZUGUCHI, M., An heuristic model of wave height distribution in surf zone, *Proc. 17th Coastal Eng. Conf.*, **1**,

278–289, 1980.

PEREGRINE, D. H., Mechanics of breaking waves—a review of Euromech 102, in *Mechanics of Wave-Induced Forces on Cylinders*, edited by T. L. Shaw, pp. 204–214, Pitman, 1979.

SAWARAGI, T. and K. IWATA, Turbulence effect on wave deformation, *Coastal Eng. in Japan*, **17**, 39–49, 1974.

STOKER, J. J., *Water Waves*, Interscience Publishers, New York, 1957.

WHITHAM, G. B., On the propagation of shock waves through regions of nonuniform area or flow, *J. Fluid Mechanics*, **4**, 337–359, 1958.

Tsunamis—Their Science and Engineering, edited by K. Iida and T. Iwasaki, 479–493.

Numerical Analysis of the Run-up of Tsunamis on Dry Bed

Hideo MATSUTOMI

Department of Civil Engineering, Akita University, Akita, Japan

(Received August 31, 1981; Revised February 25, 1982)

A new wave-front condition is proposed to evaluate the time-varying friction factor in tip region of a run-up of tsunami on a dry bed, based upon the flow and topographical characteristics in a way similar to dam-break flows. A method of computation of the run-up of tsunami on a dry bed is also given and the numerical results are compared with experimental results. The comparison shows the method to be very effective. If the friction factor is not selected appropriately, large differences occur between the numerical and experimental results.

1. Introduction

Numerical computations are considered to be an effective method for estimating the run-up of tsunamis on a dry bed. However, several problems remain to be solved. The wave-front condition is one of them. Since a wave-front on a dry bed forms a mathematically singular point, it is difficult to give a wave-front condition in order to continue the computation. A number of wave-front conditions have been proposed (IWASAKI and TOGASHI, 1970; SAKKAS and STRELKOFF, 1973; SHUTO and GOTO, 1977; AIDA, 1977; IWASAKI and MANO, 1979; etc.), but there remains still problems for understanding the mechanism of the flow itself. In addition, there are only a few comparisons of numerical results obtained with these wave-front conditions to experimental results.

Another problem is the friction. When we try to take account of frictional effects in the tip region of a tsunami on land, we usually use a quadratic resistance law with a constant friction factor. The friction factor is empirically determined only on the basis of characteristics such as topography or density of houses without considering the magnitude of the tsunami.

In this study, the author proposes a new wave-front condition similar to dam-break flow, and a method to evaluate a time varying friction factor in the tip region of a run-up of tsunami on a dry bed based upon the flow and topographic conditions.

2. Tip Region of Dam-Break Flow on a Dry Bed

We assume that Whitham's theory (WHITHAM, 1955) can be applied to the dam-break flow problem.

As shown in Fig. 1, Whitham's theory divides the dam-break flow into two parts;

479

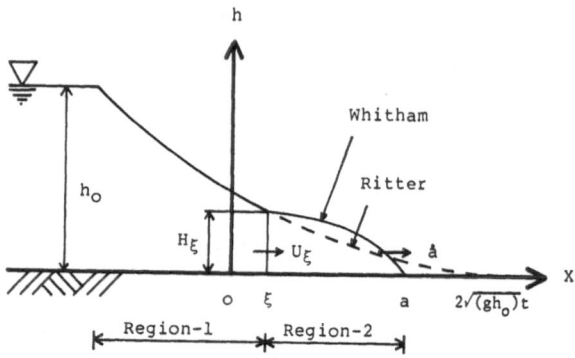

Fig. 1.   Definition sketch of dam-break flow.

Region 1 in which the frictional effect is neglected and Region 2 in which the frictional effect is taken into account. In Region 1, Ritter's theory (RITTER, 1892) is applicable. In Region 2, Whitham assumes that the water pressure is hydrostatic and the water particle velocity is a function of time alone. Then, he derived the basic equation for the wave-front position $a(t)$ as follows.

$$h_0\sqrt{gh_0}\left(1 - \frac{\dot{a}}{2\sqrt{gh_0}}\right)^3 t\ddot{a} = \frac{1}{2}gh_0^2\left(1 - \frac{\dot{a}}{2\sqrt{gh_0}}\right)^4 - f\dot{a}^2\left\{a - \left(\frac{3}{2}\dot{a} - \sqrt{gh_0}\right)t\right\}. \quad (1)$$

The solution is

$$a = \frac{h_0}{f}(0.04862p^3 + 0.02503p^4 + 0.01262p^5 + 0.00635p^6$$

$$+ 0.00319p^7 + 0.00161p^8 + 0.00081p^9 - 0.00167p^{10} \ldots \ldots) \qquad (2)$$

$$t = \frac{1}{f}\sqrt{h_0/g}\,(0.02431p^3 + 0.02163p^4 + 0.01496p^5 + 0.00941p^6$$

$$+ 0.00563p^7 + 0.00161p^8 + 0.00081p^9 \ldots \ldots) \qquad (3)$$

where a is the horizontal distance of the wave-front measured from the dam-gate, h_0 is the initial water depth in the reservoir, f is Whitham's friction factor, t is the time, g is the gravitational acceleration, and $p = [2 - \dot{a}/\sqrt{gh_0}]$ is a dimensionless velocity. Differentiation with respect to time is shown by a dot.

Equations (2) and (3), combined through the parameter p, indicate a trajectory of the wave-front of the dam-break flow. A number of curves are obtained for combinations of various values of h_0 and f.

Figure 2 shows the experimental results of trajectories of the wave-front of dam-break flow in a channel, the bottom of which is covered by a crosspiece roughnesses of height $k = 0.5$ cm with the spatial interval of $s = 5$ cm. The channel used in the experiments is a rectangular glass-walled channel 0.3 m wide, 0.5 m high and 6 m long with a steel horizontal bed. There is a reservoir 0.3 m wide, 0.7 m high and 1.5 m long at the upstream end. Wave-front trajectories are measured with a 16 mm cine-camera and a video system.

In Fig. 2, the broken curves are best-fits to the experimental results which are expressed by Eqs. (2) and (3). The values in brackets are the friction factors adopted for each theoretical curve. From this figure, we can conclude that Whitham's friction factor varies greatly according to the initial water depth h_0 which is one of the flow scale parameters. Therefore, since we would like to predict the position of the wave-front at an arbitrary time with Whitham's theory, it is very important to select the correct value of the friction factor.

Figure 3 shows the relationships between f and h_0 for the crosspiece roughnesses determined by fitting Eqs. (2) and (3) to the experimental results. Broken curves in this figure correspond to the empirical resistance law for crosspiece roughnesses expressed by the Eq. (4), which is a modification of Adachi's resistance law (ADACHI, 1964) for steady flow,

$$f=\left[1.50\log_{10}\left(\frac{s}{k}\right)-1.91+\left\{5.75+0.12\left(\frac{s}{k}\right)^{0.8}\right\}\log_{10}\left(\frac{4h_0}{9\,k}\right)\right]^{-2} \qquad (4)$$

where the range of experiments is $6 \leqq h_0/k \leqq 120$.

The resistance laws of dam-break flows for different bottom conditions are obtained by the same method.

It is possible to predict not only the position of the wave-front but also the wave profile in the tip region if an appropriate friction factor is used. Figure 4 shows an example. In this figure, the dotted line indicates an experimental result and the solid line is a theoretical curve which is expressed by

$$x'=\frac{g}{\ddot{a}}h-fg\left(\frac{\dot{a}}{\ddot{a}}\right)^2\ln\left(1+\frac{1}{f}\frac{\ddot{a}}{\dot{a}^2}h\right) \qquad (5)$$

where h is the local water depth at x', the horizontal distance measured from the wave-front.

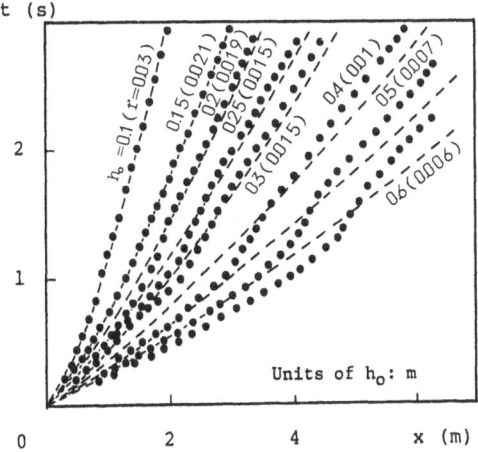

Fig. 2. Trajectories of the wave-front of dam-break flow ($s/k = 10$).

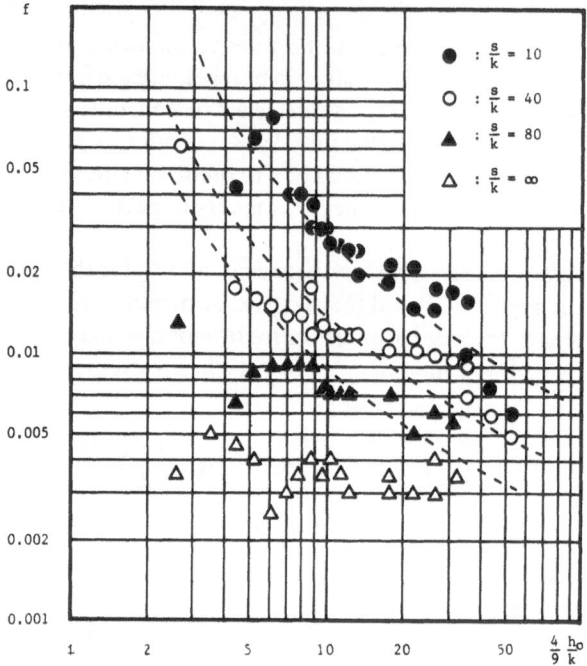

Fig. 3. Approximate curves of Whitham's friction factor.

Equation (5) is derived from the Eq. (12) with the following assumptions (CROSS, 1967): i) the water pressure is hydrostatic, and ii) the water particle velocity is a function of time alone. These are quite the same assumptions used by Whitham for the tip region of a dam-break flow.

Equation (5) is for a channel of constant water depth. In the case of a sloped channel, Eq. (5) should be replaced by

$$x' = \frac{gh}{\ddot{a} + ig} - fg\left(\frac{\dot{a}}{\ddot{a} + ig}\right)^2 \ln\left(1 + \frac{\ddot{a} + ig}{f\dot{a}^2}h\right) \tag{6}$$

where i is the bottom slope. In Fig. 4, the broken line is a theoretical curve given by

$$h = \dot{a}\sqrt{2fx'/g}. \tag{7}$$

This equation is obtained if the pressure term is considered to balance the friction term (CROSS, 1967). From this figure, we can conclude that the contribution of the non-steady acceleration term can't be neglected and is more important than the convective term in the tip region of flow on a dry bed.

We can also evaluate the energy loss in the tip region of dam-break flow on the basis of Whitham's theory. Figure 5 shows an example. In this figure, the broken and solid lines indicate the energy loss E_s due to bottom friction alone and the total energy loss E_t (E_s + energy loss due to breaking, and so on), respectively. They are expressed by

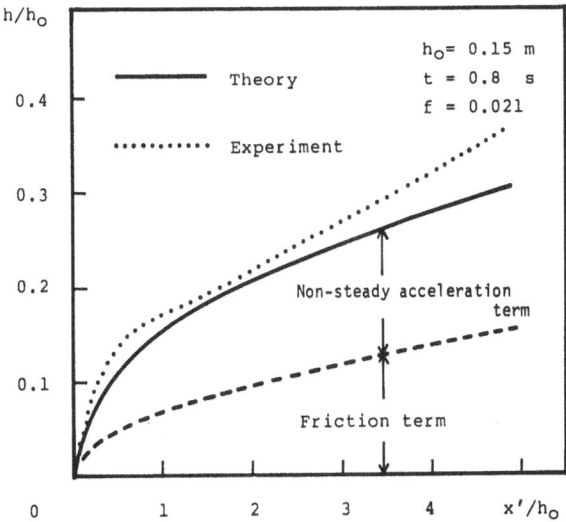

Fig. 4.   Wave profile in the tip region of a dam-break flow ($s/k = 10$).

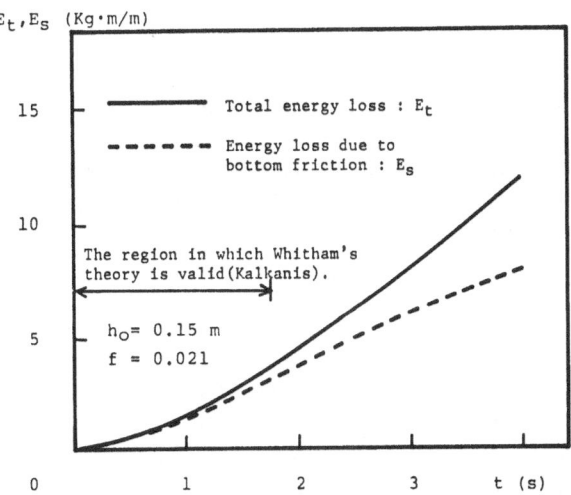

Fig. 5.   Energy loss in the tip region of a dam-break flow on the basis of Whitham's theory ($s/k = 10$).

$$E_s = -\frac{\rho g h_0^3}{f}(0.00152p^7 + 0.000685p^8 + 0.000336p^9 +$$

$$0.000163p^{10} + 0.000098p^{11} + 0.000030p^{12} +$$

$$0.000025p^{13} - 0.000119p^{14} \ldots \ldots) \tag{8}$$

$$E_t = -\frac{1}{16}\rho g h_0^2\sqrt{gh_0}\,p^4\left(1 - \frac{3}{4}p\right)t + \frac{1}{2}\rho g \int_\xi^a h^2 dx \tag{9}$$

where ρ is the density of water and ξ is the horizontal distance of the rear point of Whitham's boundary layer measured from the dam-gate. Figure 5 shows that energy loss due to bottom friction is dominant in the tip region of flow on a dry bed.

3. Wave-Front Condition

Whitham's theory is applied to tsunami problems. In such applications, we can expect that assumptions i) and ii) previously stated are only valid in the region of the very narrow tip of tsunami on a dry bed. In fact, according to the results of a computation done earlier (for example, Iwasaki and Togashi, and so on), we can admit that the region satisfying the assumptions is only of $O(dx)$ near the wave-front, where dx is the spatial mesh size in the computation. Moreover, in order to perform a two-dimensional analysis of tsunami on an arbitrary topography and slope, it is best to assume that the above theory can be applied to a region of $O(dx)$ near the wave-front. Hence, Eqs. (5) and (6) will be applied only to the tip region of length $O(dx)$ of a flow on dry bed (see Appendix).

Values of h, $\dot{a}$, and $\ddot{a}$ at a distance $O(dx)$ from the wave-front can be given, for an arbitrary time, from values computed for regions behind the wave-front, with no regard to the wave-front itself. If the friction factor of the flow in the tip region is estimated, we can calculate x' with Eq. (5) or (6) to determine the position of the wave-front. Therefore, either Eq. (5) or (6) gives a wave-front condition of a tsunami which runs up on a dry bed. The Froude number at a distance of $O(dx)$ from a wave-front can change from nearly zero to infinity.

When the run-down of a tsunami is considered, the same assumptions i) and ii) are mathematically possible but physically impossible. Therefore, Eq. (10) is proposed as a wave-front condition for run-down,

$$\Delta x' = \dot{a} \cdot dt \tag{10}$$

where $\Delta x'$ is the distance over which the wave-front travels in a time interval in computation, dt, and $\dot{a}$ is the water particle velocity at a distance $O(dx)$ from the wave-front. Equation (10) physically means that the water particle velocity is a function of the time alone in the tip region. In a practical computation, the value of $\dot{a}$ can be given from either the present time-step, or from the next-step, or their average. However, no significant difference is found from the choice.

4. Friction Factor in the Tip Region of a Flow on a Dry Bed

In contrast to dam-break flow, which is unsteady but is characterized by the initial water depth h_0 in a reservoir as shown in Fig. 2, a tsunami can't be characterized by such a typical water depth. Therefore, in order to relate tsunami to the dam-break flow problem, it is assumed here that the tip region of a tsunami acts as the tip region of dam-break flow with a time-varying representative water depth h_0.

In the previous section, it was assumed that the length of the region dominated by Eq. (5) or (6) is of $O(dx)$ in the tip region of a flow on a dry bed. In other words, this corresponds to the assumption that the length of Whitham's boundary layer is of $O(dx)$

and the point of distance $O(dx)$ from the wave-front is the rear point of Whitham's boundary layer. Then, the time-varying depth h_0 in the tip region of a tsunami can be determined by the following equation, which is a result of Ritter's theory:

$$h_0 = \frac{H_\varsigma}{4}(Fr_\varsigma + 2)^2 \tag{11}$$

where H_ς and Fr_ς are the water depth and the Froude number. They are given at the same point as the point where h, $\dot{a}$, and $\ddot{a}$ in Eqs. (5) and (6) are given.

It is possible to correlate h_0 to f as shown in Fig. 3. For example, Eq. (4) is the approximate resistance law for the case of the crosspiece roughnesses arranged regularly and densely. Therefore, if we know the resistance laws of dam-break flows for various bottom conditions, we can evaluate the friction factor in the tip region of a tsunami by calculating h_0 with Eq. (11) and putting it into the resistance law. Whitham's theory may be very useful to investigate the resistance laws of dam-break flows for various bottom conditions (see Appendix).

5.   Applications

5.1   Dam-break flow

A dam-break flow is computed with the present theory and the results are compared with experimental results. Both computations and hydraulic experiments were carried out on the condition that i) the dam breaks instantly, ii) the bottom slope condition is two cases which are $i = 0$ and $i = 0.075$, iii) the length of the reservoir is $L_0 = 1.5$ m and iv) crosspiece roughnesses are installed on the bottom. Equation (4) was therefore used as the resistance law in the computation. The basic equations except for the tip region of the flow are the shallow water equations given by

$$\frac{\delta u}{\delta t} + u\frac{\delta u}{\delta x} + g\frac{\delta h}{\delta x} = -ig - f_p\frac{u^2}{h} \tag{12}$$

$$\frac{\delta h}{\delta t} + \frac{\delta}{\delta x}(hu) = 0 \tag{13}$$

where u is the depth-averaged water particle velocity, h is the total depth of water, f_p is a friction factor and the resistance force is defined by $\tau_0 = \rho f_p u^2$.

The condition at the receding wave-front in the reservoir is

$$u = 0, \ c = \sqrt{gh_0} \ \text{on} \ x = -\sqrt{gh_0}\,t \ \text{for} \ t \leqq \frac{L_0}{\sqrt{gh_0}} \tag{14}$$

$$u = 0, \ \text{approximately} \ c = \frac{1}{3}\sqrt{gh_0} + \frac{2}{3}\frac{L_0}{t} \ \text{on} \ x = -L_0 \ \text{for} \ t > \frac{L_0}{\sqrt{gh_0}} \tag{15}$$

where c is the wave celerity. As the initial condition, Ritter's solution was selected for very short time.

Computations were carried out by the method of characteristics. The intervals in the computation, $dx = 0.025$ m and $dt = 0.01$ s, were selected on the basis of the CFL stability condition.

Figure 6 shows the relative importance of each term of the equation of motion (12) in a dam-break flow. In this figure, the solid, chain and dotted lines are the ratios of the friction, convection and pressure terms to the non-steady acceleration term, respectively. This figure suggests the following relation in the tip region of a flow on dry bed.

$$\left| f_p \frac{u^2}{h} \right| > \left| g \frac{\delta h}{\delta x} \right| > \left| \frac{\delta u}{\delta t} \right| > \left| u \frac{\delta u}{\delta x} \right|. \tag{16}$$

Therefore, Whitham's assumptions can be accepted as a first approximation in the tip region of dam-break flow.

Figure 7 shows a comparison between computational and experimental results for the wave-front trajectories. The experimental result is the mean value averaged for five runs. The results of computation in this study method agree very well with the experimental results even for large t. Figure 8 shows the wave profiles in the tip region of a dam-break flow on a dry bed at this time. The agreement between the computed and experimental results is satisfactory in the tip region.

5.2 Run-up of a tsunami

For simplicity, consider the one-dimensional problem of a tsunami incident upon a sloping beach connected with a region of constant water depth. In the computations, Eqs. (12) and (13) neglecting frictional effects were also used as the basic equations except for the tip region of the tsunami on land which was solved by the method of characteristics. Equations (6) and (10) were used as the wave-front conditions on land. The offshore boundary was taken at a point sufficiently far offshore so that reflected waves did not arrive before the end of the computation. The relationship between the water particle velocity and the water surface elevation at this point was given by

$$u_0 = 2\{\sqrt{g(h_* + \eta_0)} - \sqrt{gh_*}\} \tag{17}$$

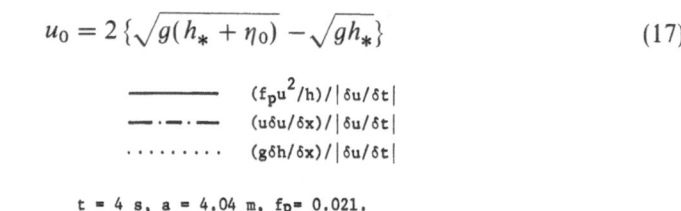

$$\underline{\hspace{2cm}} \quad (f_p u^2/h)/|\delta u/\delta t|$$
$$-\cdot-\cdot- \quad (u\delta u/\delta x)/|\delta u/\delta t|$$
$$\cdots\cdots\cdots \quad (g\delta h/\delta x)/|\delta u/\delta t|$$

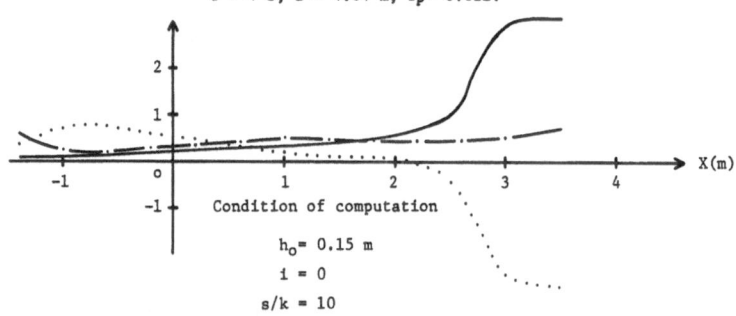

$t = 4$ s, $a = 4.04$ m, $f_p = 0.021$.

Condition of computation

$h_0 = 0.15$ m

$i = 0$

$s/k = 10$

Fig. 6.   Relative importance of each term of the equation of motion expressed by Eq. (12) in a dam-break flow.

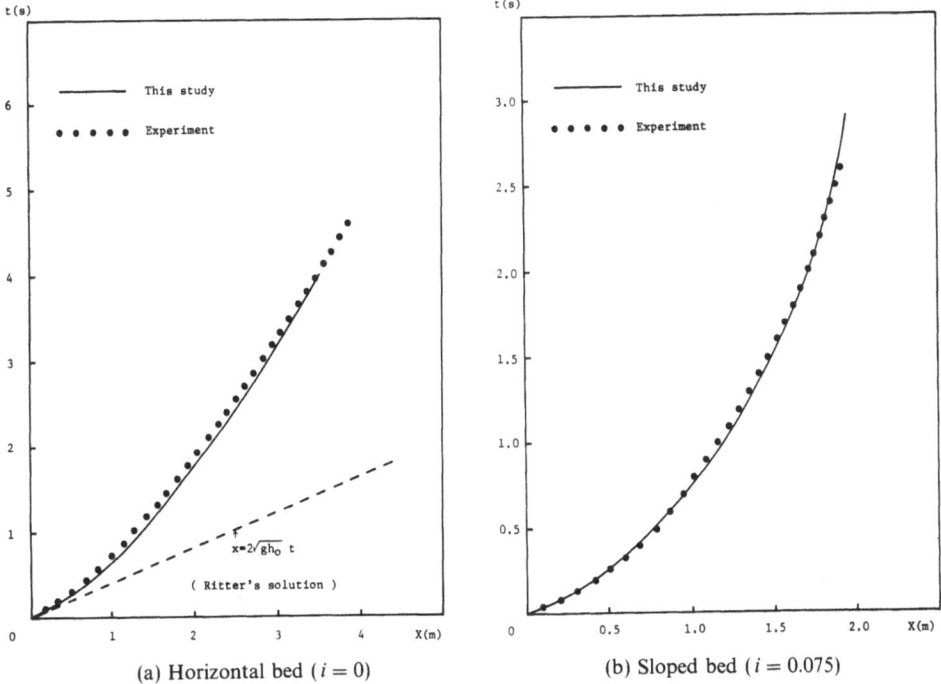

(a) Horizontal bed ($i = 0$)   (b) Sloped bed ($i = 0.075$)

Fig. 7. Comparison between computational and experimental results of the wave-front trajectories ($h_0 = 0.15$ m, $s/k = 10$, $f_p = 0.01$).

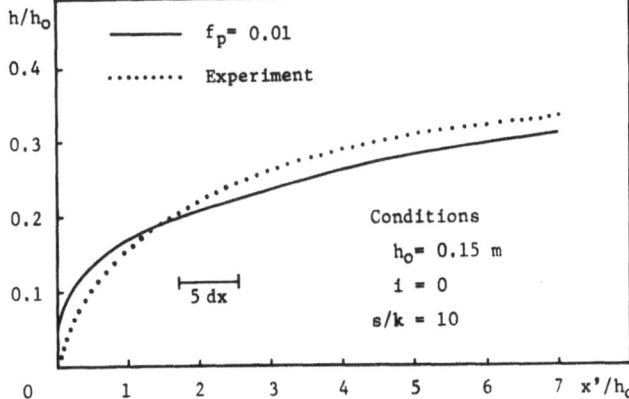

Fig. 8. Comparison between the computed and experimental wave profiles near the wave-front ($a = 3.6$ m).

where u_0 is the water particle velocity, η_0 is the water surface elevation above still water level, and h_* is the still water depth. Equation (17) is the result given by the shallow water theory for waves progressing on water of constant depth.

The intervals $dx = 0.066$ m and $dt = 0.033$ s were selected on the basis of the CFL stability condition. Equation (4) was also used as the resistance law for a constant water depth.

In the hydraulic experiments, the channel used for the dam-break-flows was also used. At an end of the channel, a slope ($i = 0.075$) was installed on one side and a wave generator on the other side. Crosspiece roughnesses were installed only on the slope. Two video systems were used to measure η_0 at the offshore boundary, wave-front trajectories on the dry bed, wave profile in the tip region on the dry bed, and so on.

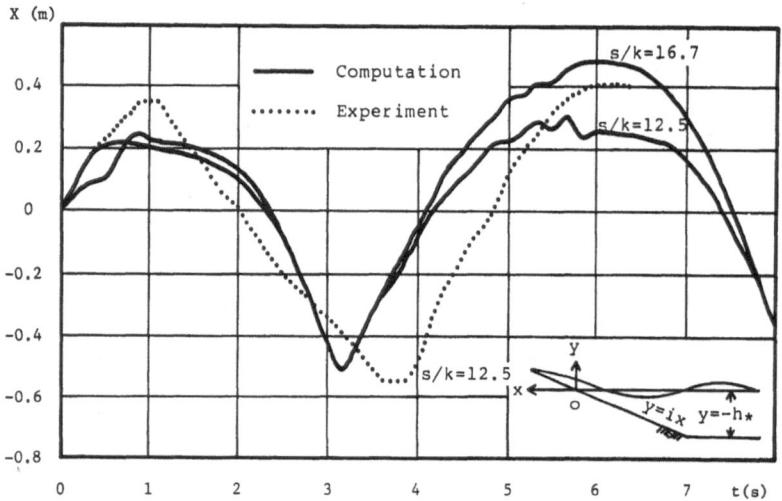

Fig. 9. Time-histories of the wave-front position of the run-up of tsunamis ($h_* = 0.1$ m, $i = 0.075$, $s/k = 12.5$, $s/h_* = 0.5$).

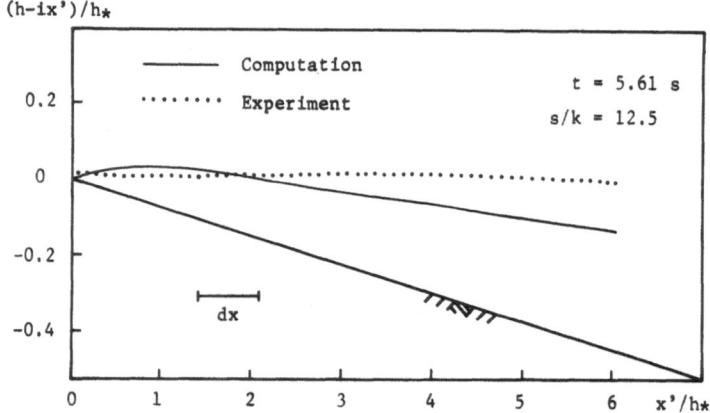

Fig. 10. Comparison between the computed and experimental wave profiles near the wave-front ($h_* = 0.1$ m, $i = 0.075$, $s/k = 12.5$, $s/h_* = 0.5$).

Figure 9 shows the time-histories of the horizontal position of the wave-front of the tsunami on the slope. The dotted line is the experimental result, and the solid lines are the numerical results. In this figure, the computed results show, i) smaller run-up heights both for the first and second wave crests, ii) a shorter wave period, and iii) small oscillations when the tsunami runs up. Concerning i), the reason may be due to the frictional effect at the time when the velocity of the wave-front approaches zero. At this time, the Eq. (4) is out of the range of applicability and gives an overly large friction factor. Table 1 shows the change of the friction factor with respect to time. The friction factor in the tip region increases in value suddenly, from 0.162 at time $t = 5.28$ s to 1.300 at time $t = 5.61$ s. This sudden change shows that Eq. (4) is no longer applicable. If the computation is carried out for the case of $s/k = 16.7$, the run-up height becomes bigger than for the case of $s/k = 12.5$ as shown in Fig. 9, too. From this figure, we can notice that large differences occur between the numerical and experimental results if the friction factor is not selected appropriately.

Concerning ii), two explanations are considered. One is the difficulty in finding the correct position of the wave-fronts in the experiments. The other is the problem of the computation method itself. In the present method, wave-fronts always fluctuate either by running up or down on the sloping beach. But, the wave-fronts could possibly stop

Table 1.   Friction factor, f, in the tip region of the run-up of a tsunami on a dry bed and the length of the boundary layer, $a - \xi$ ($h_* = 0.1$ m, $i = 0.075$, $s/k = 12.5$, $s/h_* = 0.5$).

t(s)	f	ξ/h_*	a/h_*	$(a - \xi)/h_*$
0	—	0	0	0
0.33	0.114	0	0.93	0.93
0.69	0.226	0	1.71	1.71
1.02	—	—	—	—
1.32	—	—	—	—
1.65	—	—	—	—
1.98	—	—	—	—
2.31	—	—	—	—
2.64	—	—	—	—
2.97	—	—	—	—
3.30	0.038	−5.28	−4.20	1.08
3.63	0.038	−3.30	−2.55	0.75
3.96	0.039	−1.98	−0.75	1.23
4.29	0.047	−0.66	0.69	1.35
4.62	0.068	0.66	1.59	0.93
4.95	0.102	1.32	2.21	0.89
5.28	0.162	1.32	2.86	1.54
5.61	1.300	1.98	3.14	1.16
5.94	1.214	1.98	2.61	0.63
6.27	—	—	—	—
6.60	—	—	—	—
6.93	—	—	—	—
7.26	—	—	—	—
7.59	—	—	—	—
7.92	—	—	—	—

moving for a while in the actual phenomena. For example, in Fig. 9 the wave-front does not move near the time $t = 3.7$ s.

For iii), the reason is due to the assumption that only Eq. (5) or (6) is applicable to the region of $O(dx)$ near the wave-front on a dry bed. In this assumption, the position of the rear point of the region suddenly changes. As a result, the friction factor as well as the length of the region suddenly changes, to make the wave-front oscillate. This oscillation is liable to occur at the time when the wave-front velocity on dry bed approaches zero. Figure 10 shows the wave profiles in the tip region of a tsunami on a dry bed. The agreement between the computed and experimental results is satisfactory in the tip region.

6. Conclusions

On the basis of some considerations on dam-break problems, a new wave-front condition was proposed for computation of run-up of tsunami on a dry bed. According to the results of computation done with the wave-front condition, the followings are concluded:

1) Whitham's assumptions can be accepted as a first approximation in the tip region of flow on a dry bed,

2) if we know the resistance laws of dam-break flows for various bottom conditions, it is possible to evaluate the run-up of tsunamis, and

3) if the friction factor in the tip region is not selected appropriately, large differences occur between the numerical and experimental results.

At last, the author would like to point out that the extension of this method to a two-dimensional case is also possible and easy.

The author wishes to express his sincere thanks to Dr. N. Shuto, Professor of the Faculty of Engineering, Tohoku University and Dr. H. Asada, Professor of Mining College, Akita University, for their helpful suggestions and encouragements.

Appendix

A Note on the Definition of the End Point of Whitham's Boundary Layer in the Flow on a Dry Bed

The equation of motion for flows on a dry bed is generally expressed as

$$\frac{\delta u}{\delta t} + u \frac{\delta u}{\delta x} + g \frac{\delta h}{\delta x} = -ig - f \frac{u^2}{h}. \tag{A-1}$$

In this flow, the region dominated by the following equation of motion is described by Whitham's boundary layer (WHITHAM, 1955).

$$\frac{\delta u}{\delta t} + g \frac{\delta h}{\delta x} = -ig - f \frac{u^2}{h}. \tag{A-2}$$

Therefore, the following relations in Whitham's boundary layer must be satisfied.

$$O\left(\frac{\delta u}{\delta t}\right), O\left(g\frac{\delta h}{\delta x}\right), O(ig), O\left(f\frac{u^2}{h}\right) \gg O\left(u\frac{\delta u}{\delta x}\right). \tag{A-3}$$

Now, we introduce the quantities; L, the distance from the wave-front to an arbitrary point in the flow; H, the depth of water at the arbitrary point; U_0, the water particle velocity at $H/L = 1$; and evaluate the order of each term of Eq. (A-1). From the physical cosiderations of the flow, we can put the particle velocity U and $\delta/\delta t$ at the arbitrary point as follows.

$$U \sim \frac{H}{L}U_0 \tag{A-4}$$

$$\frac{\delta}{\delta t} \sim \frac{U}{H} \sim \frac{U_0}{L}. \tag{A-5}$$

Therefore, we can obtain the following approximate relations at an arbitrary point.

$$\left.\begin{aligned}
O\left(\frac{\delta u}{\delta t}\right) &\sim \frac{H}{L^2}U_0^2 \\
O\left(u\frac{\delta u}{\delta x}\right) &\sim \frac{H^2}{L^3}U_0^2 \\
O\left(g\frac{\delta h}{\delta x}\right) &\sim g\frac{H}{L} \\
O(ig) &\sim ig \\
O\left(f\frac{u^2}{h}\right) &\sim f\frac{H}{L^2}U_0^2.
\end{aligned}\right\} \tag{A-6}$$

At the arbitrary point, the relative ratio of each term is

$$\left.\begin{aligned}
O\left(u\frac{\delta u}{\delta x}\bigg/\frac{\delta u}{\delta t}\right) &\sim \frac{H}{L} \\
O\left(g\frac{\delta h}{\delta x}\bigg/\frac{\delta u}{\delta t}\right) &\sim \frac{L}{H}\cdot\frac{gH}{U_0^2} \\
O\left(ig\bigg/u\frac{\delta u}{\delta x}\right) &\sim i\left(\frac{L}{H}\right)^3\cdot\frac{gH}{U_0^2}
\end{aligned}\right\} \tag{A-7}$$

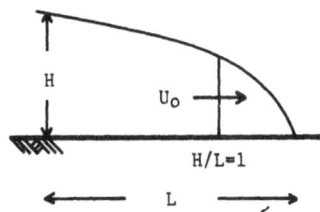

Fig. 11. Definition sketch of hydraulic quantities near the wave-front.

$$O\left(f\frac{u^2}{h}\middle/u\frac{\delta u}{\delta x}\right) \sim f\frac{H}{L}.$$

In Whitham's boundary layer,

$$O\left(u\frac{\delta u}{\delta x}\middle/\frac{\delta u}{\delta t}\right) \sim \frac{H}{L} \ll 1$$

$$O\left(g\frac{\delta h}{\delta x}\middle/\frac{\delta u}{\delta t}\right) \sim \frac{L}{H}\cdot\frac{gH}{U_0^2}$$

$$O\left(ig/u\frac{\delta u}{\delta x}\right) \sim i\left(\frac{L}{H}\right)^3\cdot\frac{gH}{U_0^2} \gg 1 \qquad \text{(A-8)}$$

$$O\left(f\frac{u^2}{h}\middle/u\frac{\delta u}{\delta x}\right) \sim f\frac{H}{L} \gg 1.$$

The rear point of Whitham's boundary layer should satisfy Eq. (A-3), but there are no other limitations. Therefore, if we assume $O(\delta u/\delta t) \sim O(g\,\delta h/\delta x)$ at the end point, we can obtain the following relations.

$$O\left(u\frac{\delta u}{\delta x}\middle/\frac{\delta u}{\delta t}\right) \sim \frac{H_\xi}{L_\xi} \ll 1$$

$$O\left(g\frac{\delta h}{\delta x}\middle/\frac{\delta u}{\delta t}\right) \sim \frac{L_\xi}{H_\xi}\cdot\frac{gH_\xi}{U_0^2} \fallingdotseq \frac{L_\xi}{H_\xi}\cdot\frac{gH_\xi}{\dot{a}^2} \sim O(1)$$

$$O\left(ig/u\frac{\delta u}{\delta x}\right) \sim i\left(\frac{L_\xi}{H_\xi}\right)^3\cdot\frac{gH_\xi}{U_0^2} \fallingdotseq i\left(\frac{L_\xi}{H_\xi}\right)^3\cdot\frac{gH_\xi}{\dot{a}^2} \sim i\left(\frac{L_\xi}{H_\xi}\right)^2 \gg 1 \qquad \text{(A-9)}$$

$$O\left(f\frac{u^2}{h}\middle/u\frac{\delta u}{\delta x}\right) \sim f\frac{L_\xi}{H_\xi} \gg 1.$$

Equation (A-9) means that i) the distance from the wave-front to the end point of Whitham's boundary layer is much greater than the depth of water at the end point, ii) The Froude number is of $O(\sqrt{L_\xi/H_\xi}) > 1$ at the end point, iii) the effect of gravity is negligible when $i \leq (H_\xi/L_\xi)^2$, and iv) the friction term is negligible when $f \leq H_\xi/L_\xi$. The information contained in iii) and iv) is concerned with external factors, not with physical aspects of the flow itself. Hence, on the basis of i) and ii), the point where the following relation is satisfied is defined at the end point of Whitham's boundary layer.

$$\frac{L_\xi}{H_\xi} \fallingdotseq \frac{\dot{a}^2}{gH_\xi} = Fr_\xi^2. \qquad \text{(A-10)}$$

This point surely exists in the flow on a dry bed. And, if we use this definition as the end point of Whitham's boundary layer in the computation, the oscillations appearing in Fig. 9 will not appear. It is also possible to obtain the resistance laws of flows on a dry bed for various bottom conditions by using Eqs. (6) and (A-10) instead of Whitham's theory. Figure 12 shows an example. In this figure, the dotted line is Eq. (4).

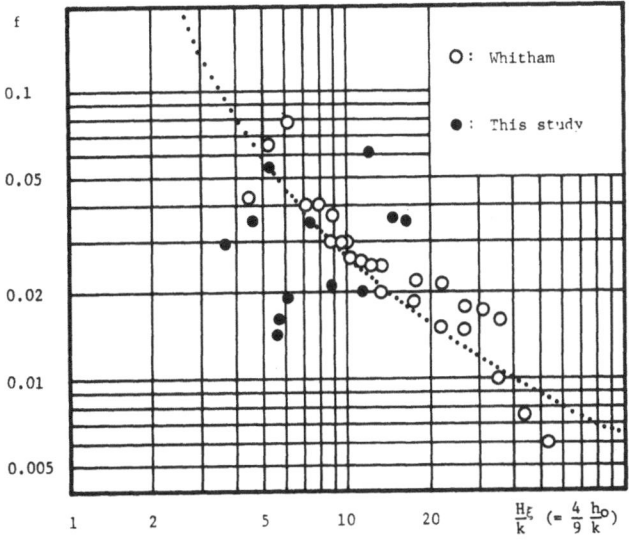

Fig. 12.   Friction factor in the tip region of flow on a dry bed ($i = 0$, $s/k = 10$).

REFERENCES

ADACHI, S., Experimental study on artificial roughness, Trans. of JSCE, No. 104, pp. 33–44, 1964 (in Japanese).

AIDA, I., Numerical experiments for inundation of tsunamis—Susaki and Usa, in Kochi Prefecture—, *Bull. Earthq. Res. Inst.*, **52**, 441–460, 1977 (in Japanese).

CROSS, R. H., Tsunami surge forces, *Proc. of ASCE*, **93**, No. WW4, 201–231, 1967.

IWASAKI, T. and A. MANO, Numerical computations of two dimensional run-up of tsunamis due to the Eulerian description, Proc. 26th Japanese Conf. on Coastal Eng., pp. 70–74, 1979 (in Japanese).

IWASAKI, T. and H. TOGASHI, On the shoreline and leading front conditions of tsunami waves in the light of the method of characteristics, *Coastal Eng. in Japan*, **13**, 113–125, 1970.

RITTER, A., Die Fortpflanzung der Wasserwellen, *Zeitschrift des Vereines deutscher Ingenieure (Berlin)*, **36**, pt. 2, No. 33, 947–954, 1892.

SAKKAS, J. G. and T. STRELKOFF, Dam-break flood in a prismatic dry channel, *Proc. of ASCE*, **99**, No. HY12, 2195–2216, 1973.

SHUTO, N. and C. GOTO, Numerical analysis of run-up of tsunamis, Proc. 24th Japanese Conf. on Coastal Eng., pp. 65–68, 1977 (in Japanese).

WHITHAM, G. B., The effects of hydraulic resistance in the dam-break problem, *Proc. Royal Soc. London, Series A*, **227**, 399–407, 1955.

Tsunamis—Their Science and Engineering, edited by K. Iida and T. Iwasaki, 495–509.

Shoreline Wave Height and Land Run-up Height of Tsunamis on Uniformly Sloping Beaches

Hiroyoshi TOGASHI

Department of Civil Engineering, Faculty of Engineering
Nagasaki University, Nagasaki, Japan

(Received August 31, 1981; Revised May 2, 1982)

This paper addresses the problem of the estimation of the position of maximum tsunami height on a coast according to relationships between incident wave characteristics and beach slope. The phenomena of run-up of tsunamis onto land at the recesses of bays or on flat coasts without significant embayment should be studied as one of the causes of wave height increase or decrease, in addition to the effects of bay water oscillations or energy convergencies and/or divergencies. One of the purposes of this study is to investigate such two-dimensional run-up phenomena.

Therefore, the relationships between shoreline wave height and run-up height at the recesses of a bay or on flat coasts were studied through hydraulic experiments. Rather interesting relationships were obtained which had not previously been noted among incident wave height, shoreline wave height and run-up height on land.

1. Introduction

In a previous paper (TOGASHI and NAKAMURA, 1977), empirical formulas regarding land run-up height of tsunamis were evaluated. KAPLAN's formulas (1955) were employed as representative ones taking wave steepness H/L as a parameter, and the proprieties of these expressions were investigated only for slopes of $S = 1/30$ and $1/60$. Therefore in the present paper, first, results of cases where run-up phenomena are influenced by H/L are discussed using the additional slopes $S = 1/50$ and $1/20$ for periodic waves, and $S = 1/40$ for non-periodic waves. Other cases, where relative shoreline wave heights or run-up heights influenced by both beach slopes and relative depths, are newly investigated for all slopes treated in this paper, and a few empirical formulas are presented. Finally, two examples of formulas applied to historical tsunamis are given.

2. Experimental Apparatus and Waves

As mentioned above, the experiments were of 2 kinds, one of which was carried out with periodic waves using the facilities of Tohoku University. Laboratory wave channels and slopes were made by subdividing a flat water vasin. The surfaces consisted of concrete with four slopes; $S = 1/50 + 1/100, 1/50, 1/30$, and $1/20$. The wave gauge was a registance-type round stainless rod of a 200 mm length, $\phi 3$ mm. Notation for the

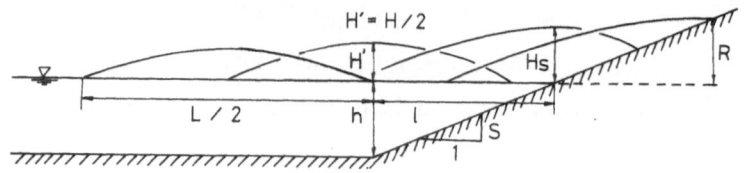

Fig. 1. Definition sketch of periodic wave and symbols.

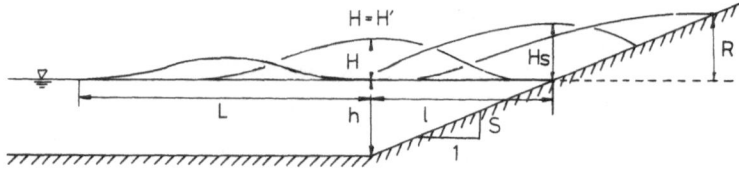

Fig. 2. Definition sketch of non-periodic wave and symbols.

experimental waves in this case is given in Fig. 1. The incident wave profile at the toe of slope is part of a periodic sinusoidal wave, i.e., $\eta = H' \sin (2\pi/T)t, 0 \leq t \leq T/2$, after IWASAKI *et al.* (1970). The other kind of experiment was conducted with solitary-like non-periodic waves using the laboratory equipment of Nagasaki University. A reinforced concrete wave channel 1.0 m wide, 1.0 m deep, and 60.0 m long was equipped at one end with a pneumatic wave generator. The channel was provided with separate steel plate slopes of $S = 1/60$ and $1/40$. The wave gauge was a capacitance-type and the run-up height was measured by eye. The experiment wave notation in this case is given in Fig. 2. The incident wave profile toward the foot of the slope is a non-periodic type of sine squared wave made to resemble a solitary wave as $\eta = H \sin^2(\pi/T)t, 0 \leq t \leq T$, and it produces a single wave on the still water surface.

3. Results and Discussion

3.1 Effects of wave steepness for periodic waves

(1) Figure 3 gives the experimental results for the sloping beach model shown in Fig. 1 and for $S = 1/50$. The solid line is a fitted empirical straight line which can be expressed by

$$\frac{R}{H'} = 16.8 \left(\frac{H}{L}\right)^{0.344}. \tag{1}$$

The dashed line and chain line are averaged curves for clapotis type run-up theory of small amplitude long waves obtained by applying the experimental results in the following manner. First, the wave length shown in Fig. 1 will be defined. It is assumed that for a water depth, h, at the foot of the slope $L/2 = \sqrt{g/h} \times T/2$, where g is the acceleration due to gravity. Accordingly, for U in the theoretical formula of SHUTO (1967),

$$\frac{R}{H'} = \frac{1}{J_0(U)}, \quad J_0: \text{Bessel function of 0th order} \tag{2}$$

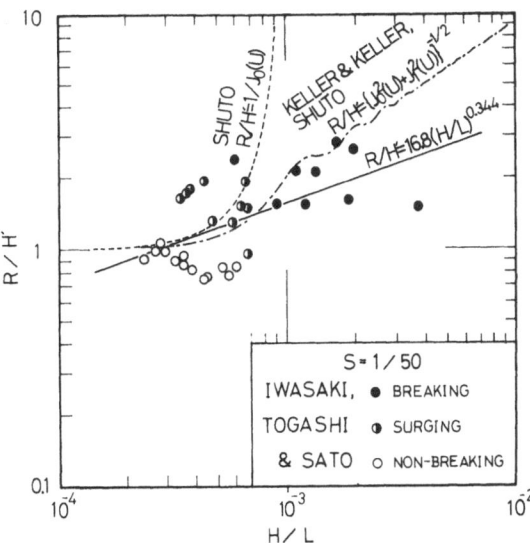

Fig. 3. Relationship between R/H' and H/L ($S = 1/50$).

$$U = \frac{4\pi}{ST}\sqrt{\frac{h}{g}} = \frac{4\pi}{S}\cdot\frac{h}{L}. \tag{3}$$

If the empirical formula for the experimental wave shown in Fig. 4,

$$\frac{h}{L} = 39.0\left(\frac{H}{L}\right)^{1.33} \tag{4}$$

and $S = 1/50$ are substituted into Eq. (3), then

$$U = \frac{4\pi}{S}\cdot\frac{h}{L} = 2.45 \times 10^4\left(\frac{H}{L}\right)^{1.33}. \tag{3'}$$

Thus Eq. (2) can be shown by the dashed line in Fig. 3 as a function of H/L.

Next, the theoretical formulas of KELLER and KELLER (1964) and SHUTO (1972) for the offshore incident wave height, $H_0 = 2a$ (a: amplitude), is given as

$$\frac{R}{H_0} = \frac{1}{\{J_0^2(U) + J_1^2(U)\}^{1/2}}. \qquad \begin{array}{l} J_0: \text{ Bessel function} \\ \quad\ \text{ of 0th order} \\ J_1: \text{ Bessel function} \\ \quad\ \text{ of 1st order} \end{array} \tag{5}$$

In the region of clapotis, the relation to the incident wave height of the experimental wave at the foot of the slope can be considered to be roughly $H' = H_0$. Consequently, Eq. (5) is employed to give the approximation

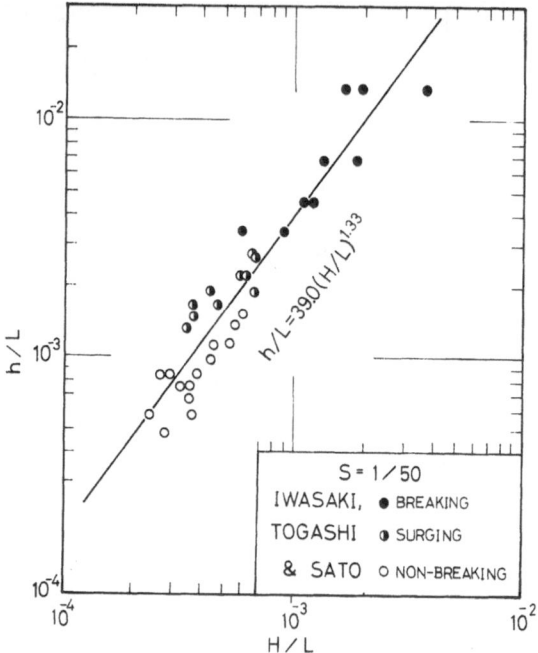

Fig. 4. Experimental wave characteristic ($S = 1/50$).

$$\frac{R}{H'} = \frac{1}{\{J_0^2(U) + J_1^2(U)\}^{1/2}}. \tag{5}'$$

Since Eq. (3)' holds for the case of $S = 1/50$, Eq. (5)' behaves as shown by the chain line in Fig. 3 as a function of H/L. Scatter in the experimental values is so great that nothing exact can be said. However, where H/L is relatively small, that is where the value of U is small, it appears that the two theoretical curves approach the experimental results.

Figure 3 can be rewritten using the shoreline wave height as a medium to Figs. 5 and 6, producing the following equations:

$$\frac{H_s}{H'} = 5.02\left(\frac{H}{L}\right)^{0.145} \tag{6}$$

$$\frac{R}{H_s} = 3.35\left(\frac{H}{L}\right)^{0.199}. \tag{7}$$

In Fig. 7, three experimental lines are collected and compared, while the dashed curve is the average reflection coefficient, r, of the experimental waves obtained from SHUTO's breaking condition (1972). From the viewpoint of only the experimental values in Figs. 3 through 7, it is essential that the following be considered regarding Eqs. (6) and (7). Namely, as H/L decreases, H_s/H' increases due to the shoaling or reflection effect from the foot of the slope to the shoreline and reaches a maximum when $r = 1$, and R/H' also reaches a maximum. However, the land run-up effect, R/H_s, has a

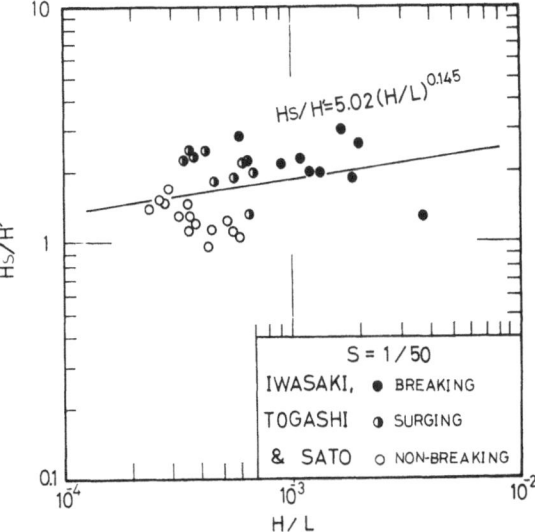

Fig. 5.   Relationship between H_s/H' and H/L ($S = 1/50$).

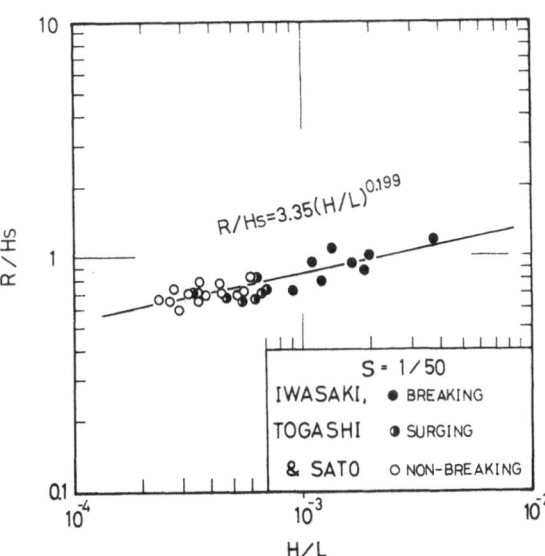

Fig. 6.   Relationship between R/H_s and H/L ($S = 1/50$).

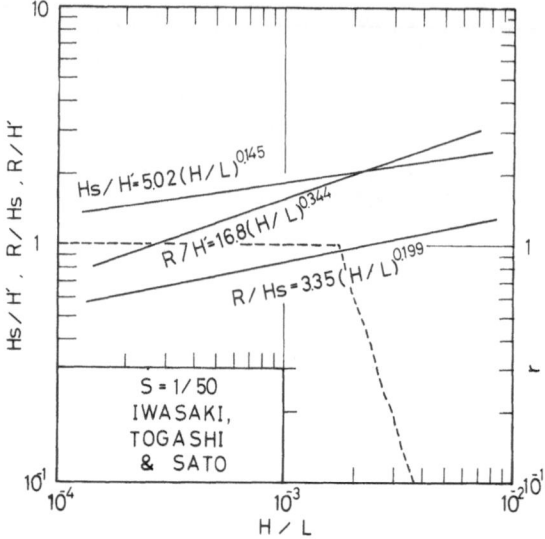

Fig. 7.   Relationships among R/H', H_s/H', R/H_s, r, and H/L ($S = 1/50$).

tendency to decrease uniformly with a decrease of H/L.

Shuto's breaking condition for long waves on a slope is

$$\frac{H'_m}{L} \cdot \frac{1}{S} = \frac{2}{\pi} U^{-1} \{ J_0^2(U) + J_1^2(U) \}^{1/2}. \tag{8}$$

For the same reason as Eq. (5)' was substituted for Eq. (5), here we substitute H' for $H_0 = 2a$ and H'_m for $2a_m$ (a_m is the maximum amplitude of the incident wave at the breaker limit on a slope). Furthermore, the value of U in this case is also provided by Eq. (3)', and accordingly Eq. (8) can be expressed as a function of H/L. Dividing both sides by $(H'/L)/S$ gives

$$r = \left(\frac{H'_m}{L} \cdot \frac{1}{S} \right) \bigg/ \left(\frac{H'}{L} \cdot \frac{1}{S} \right) = \begin{cases} 1, & H < H'_m \\ H'_m/H', & H \geq H'_m \end{cases} \tag{9}$$

as a definition of the reflection coefficient, r, which also becomes a function of H/L, and is illustrated as the broken line in Fig. 7. The quantity r is plotted on the vertical axis.

(2) Figure 8 shows the final experimental results for the slope 1/20, in which three experimental lines shown by the solid lines are gathered and compared to one another, and also compared with r, shown by the dashed line, in the same way as with $S = 1/50$ in Fig. 7. There seem to be no significant differences. However, a comparative study of these two cases follows. Comparing Fig. 7 ($S = 1/50$) and Fig. 8 ($S = 1/20$) for R/H', H_s/H' and R/H_s in the perfect standing wave range, all show a tendency for a similar decrease with decreasing H/L. In particular for R/H', this trend can be affirmed theoretically as well, as shown in Fig. 3. So at least a qualitative tendency for a decrease can be recognized, in contradiction to the tendency for increase obtained by extrapolating the experimental lines of KAPLAN (1955), as pointed out in the authors'

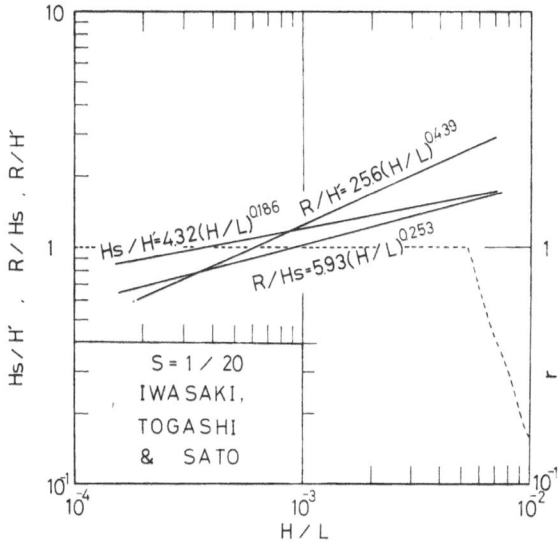

Fig. 8.   Relationships among R/H', H_s/H', R/H_s, r, and H/L ($S = 1/20$).

previous paper (1977). However, it is difficult to detect quantitatively any kind of common overall rule.

3.2   Effects of wave steepness for non-periodic waves

The experimental lines in Fig. 9 are the results for non-periodic waves on a slope S = 1/40 according to the schematic diagram show in Fig. 2. Where the incident wave height at the toe of the slope is treated as a periodic wave, it is always shown with a prime symbol, such as H', while the prime is omitted for non-periodic incident wave

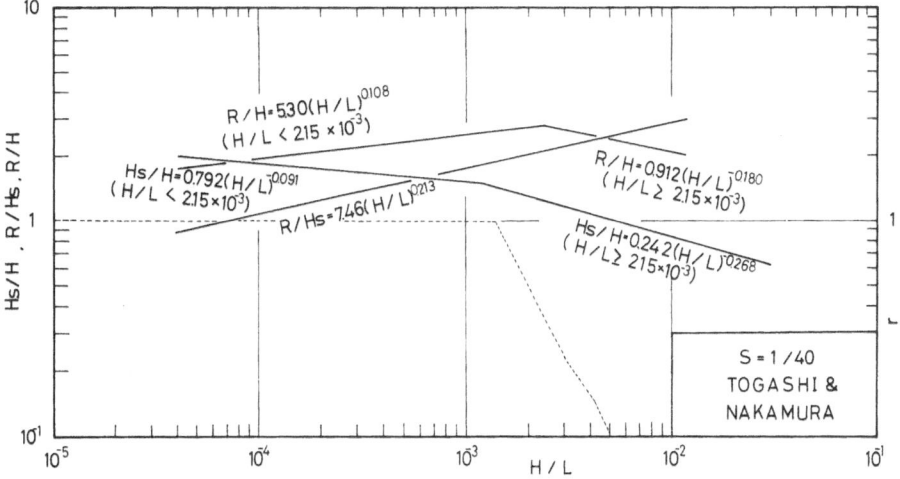

Fig. 9.   Relationships among R/H', H_s/H', R/H_s, r, and H/L ($S = 1/40$).

height H at the toe of the slope. As there is no danger of confusion for the wave steepness (H/L), the prime is omitted regardless of periodicity.

The three experimental equations are shown together by the solid lines in Fig. 9. The dotted line in Fig. 9 is the mean r of all experimental values obtained substituting U into Eqs. (8) and (9) in the same way as for the slope $S = 1/50$. According to Fig. 9 and the other results, generally, for breaking progressive wave ($r < 1$) the energy from wave breaking after shoaling transformation is largely dissipated, and therefore the wave height amplification factor (H_s/H) is small. However, since the land run-up effect (R/H_s) is great for strong surging breakers transformed from bores at the shoreline, R/H reaches a maximum. Both the effects of H_s/H and R/H_s for transitional surging breakers ($r \fallingdotseq 1$) are approximately equal, and for non-breaking standing waves ($r = 1$), they are reverse and land run-up effect (R/H_s) is small and the trend of R/H becomes definite through the amplification effect (H_s/H) by reflection.

3.3   Influences of slope and relative depth

(1) Figure 10 shows the curves for Eqs. (2), (3), and (5)′ as solid curves, contrasted against Kaplan's experimental equations which take H'/h as the parameter. Also shown is what is considered to be the experimental range for each with a group of single and double point chain lines respectively. Kaplan's extended straight lines are shown as a group of broken and dotted lines, and experimental values are plotted for $S = 1/60$ and $1/30$. The two line groups for Kaplan's equations were to be shown by two straight averaged lines. However, since there were no experimental values available, there was

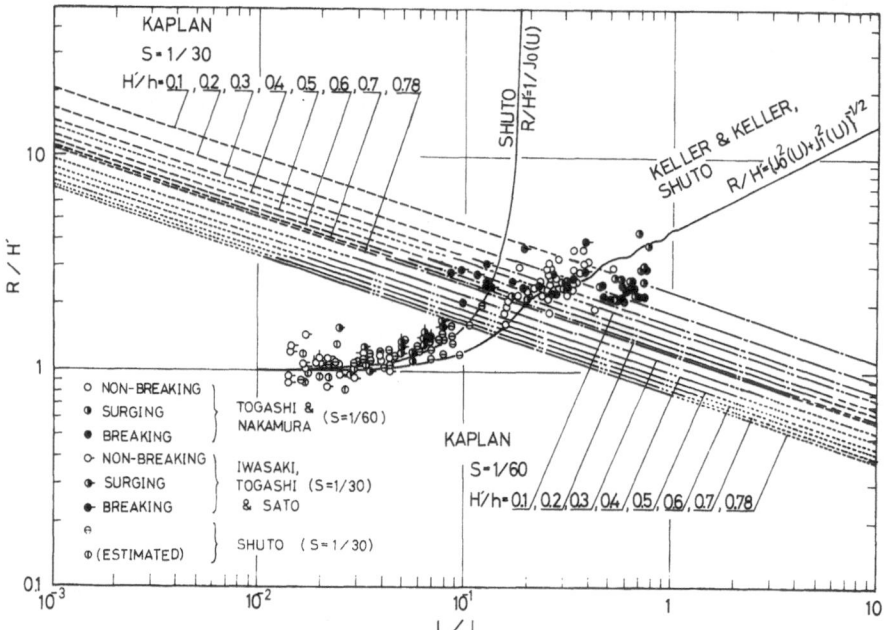

Fig. 10.   Investigation for each of the run-up equations presented in relationship between R/H' and l/L ($S = 1/60$ and $1/30$).

no choice but to use the width as the parameter indicated by the expressions available. However, there is a possibility that the widths may be too large in places. For example, there is some doubt regarding the existence of the double point chain line protruding above the theoretical curve for Eq.(2), where for $S = 1/60$ and $H'/h \geq 0.3$, the distribution is compared with the displayed experimental values. The theoretical curve from Eqs. (2) and (3) is different from the case where the influence of wave steepness is investigated by substituting a mean characteristic Eq. (4) of experimental values for theoretical Eqs. (2) and (3) as such. Regardless of the fact that all three of the experimental values are different, the distribution along the theoretical curve does show a general tendency for qualitative similarity. Thus for the present case, anything above the theoretical curve can simply be considered to be due to non-linear effects of finite wave height. Accordingly, Kaplan's straight line groups, even if extended beyond their intersection with the theoretical curve, still only reach to about $l/L = 10^{-1}$. Where $l/L < 10^{-1}$, the run-up shifts to that of non-breaking clapotis types, and R/H' gradually decreases and approaches to 1. This in actuality corresponds with the rup-up phenomenon of storm surge or tide. Furthermore, in Fig. 10, since within the range of $10^{-1} < l/L < 1$, the resonance effect of clapotis or shoaling effect of progressive waves reaches a maximum, breaking phenomena become more pronounced. This also gives the maximum region for R/H', and the range of $l/L \geq 1$ is ordinarily said to be in the normal range of Green's formula. Accordingly, it may be clear from this, too, that a possible effective range of Kaplan's extrapolated formula can not be found.

(2) In Fig. 11, experimental values for $S = 1/50$, 1/40, and 1/20 are added to the

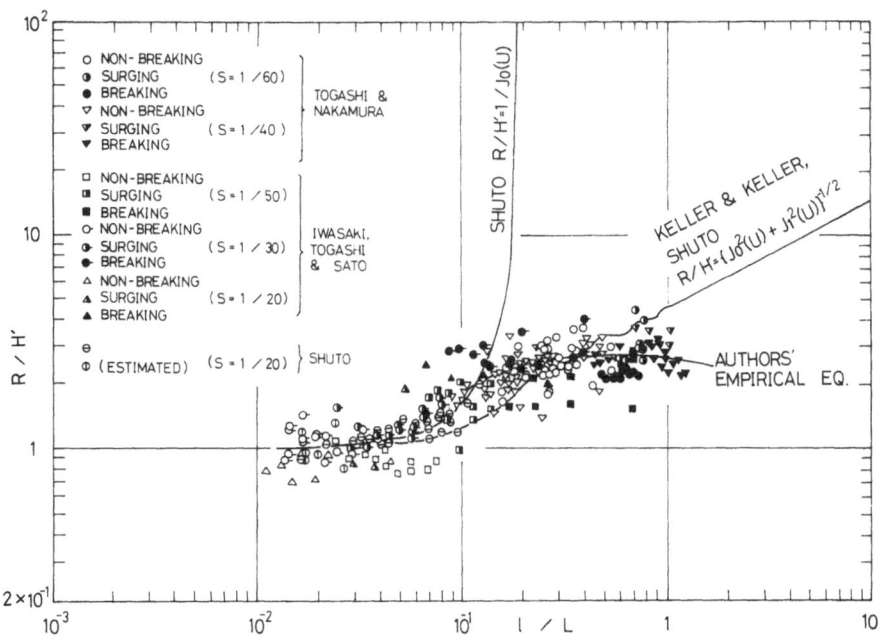

Fig. 11. Investigation for each of the run-up equations proposed in relationship between R/H' and l/L ($S = 1/60$ to 1/20).

experimental values shown in Fig. 10 for $S = 1/60$ and 1/30. Kaplan's experimental lines where the parameter H'/h is uncertain are excluded, and a more suitable experimental curve is drawn and compared with the theoretical ones. However, since this case was already discussed in detail in a previous paper (TOGASHI and NAKAMURA, 1977), the explanation is here omitted. Further, in nearly the same way as above, another relationship between H_s/H' and l/L can be obtained this time as in Fig. 12. Here, both R/H' and H_s/H' for l/L are shown together in a single figure in Fig. 13. In Fig. 13, three similar empirical equations are given for each case of R/H' and H_s/H', and this is because of the investigation as to whether or not there are any differences due to kinds of experimental data and ranges of l/L in making experimental equations, as follows.

The solid lines correspond to the experimental curves shown in Figs. 11 and 12, and are made up only of experimental values for $S = 1/60$ and 1/40 in the region $l/L \geq 10^{-1}$ (TOGASHI and NAKAMURA, 1977). The broken lines are the experimental values for all slopes in the range $l/L \geq 10^{-1}$. The one-point chain lines are obtained for all slopes in the enlarged range $l/L \geq 5 \times 10^{-2}$, this being more than the range of the experimental values shown with the broken lines. However, since there is almost no difference among the three equations, any of them can be used.

Another relationship between R/H_s and l/L is exhibited in Fig. 14, and the three experimental equations of R/H_s for l/L are obtained in Fig. 15 in the same way as those for H_s/H' are obtained in Fig. 13. Of course, these nearly correspond with the results

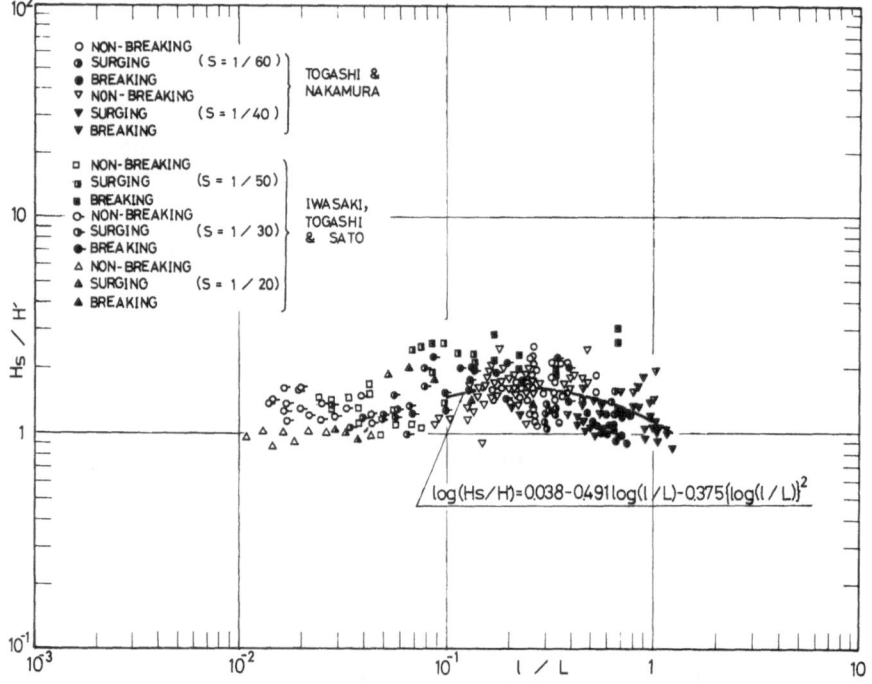

Fig. 12.   Relationship between H_s/H' and l/L ($S = 1/60$ to 1/20).

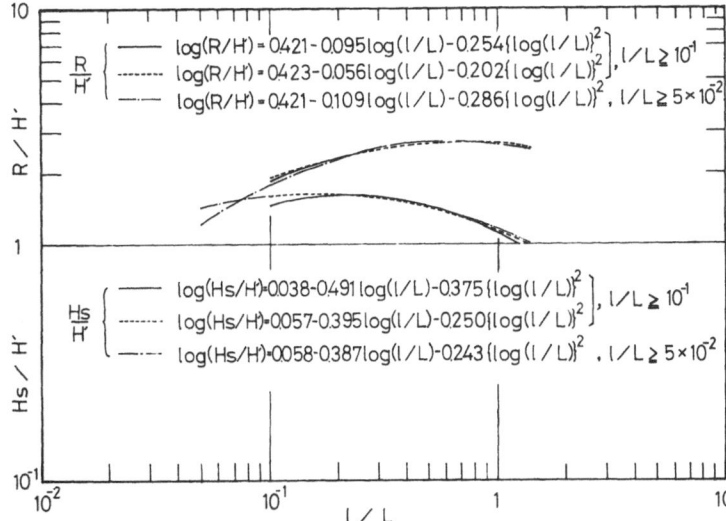

Fig. 13. The author's empirical formulas brought forward in relationships among R/H', H_s/H' and l/L.

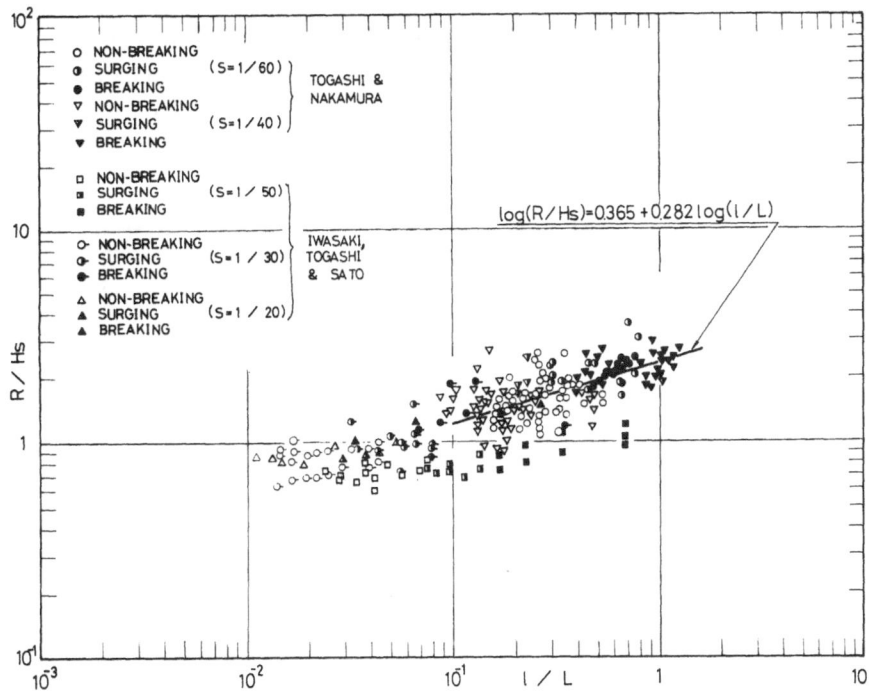

Fig. 14. Relationship between R/H_s and l/L ($S = 1/60$ to $1/20$).

which are given by arithmetical operation from Fig. 13 as $\log(R/H_s) = \log(R/H')$
$- \log(H_s/H')$.

According to Figs. 13 and 15, it is clear that, for an incident wave crest height H' at
the toe of slope, the shoaling factor H_s/H' from the slope toe to the shoreline, the run-
up factor R/H_s from the shoreline to the land surface and the amplification factor R/H'
eventually given all vary as a function of l/L. So, it could be said that a profile of the
mechanism of tsunami run-up is shown in Fig. 13 and 15. It is considered that these
empirical formulas will be useful when we want to estimate some tsunami run-up
heights using only the longitudinal slope of some beach profile of seabed and land
surface.

In application of these formulas to prototype tsunamis in the field, run-up
estimations of reasonably good accuracy can be expected if the water depth at the slope
toe, h, is less than 100 m and the slope, S, is less than 1/100, i.e., $l = h/S < 10$ km. The
hydraulic experiments themselves were conducted assming the scale of the prototype to
be of that order. In the cases where some conditions are out of these ranges or where
there are some extreme changes of topography, such exact values for estimation may
not be obtained.

Finally, two applications of these empirical formulas are given as examples. In
Fig. 16, the run-up heights of the 1854 Ansei-Tokai Tsunami on the coast of Suruga
Bay in Shizuoka Prefecture are displayed (INA CIVIL ENGINEERING CONSULTING CO.,
LTD., 1978). As for H', the values of η_{max} at $l = 1$ km were taken from the results of a
separate numerical computation, and R gives the inundation trace heights of the 1854
Ansei-Tokai Tsunami. Roman characters within open circles (e.g. Ⓐ) are signatures
estimated at locations along the coast. Although the estimated run-up values are fairly
scattered, the results are to a degree convincing, and therefore it could be said that the
empirical formulas explain run-up of actual tsunami relatively well.

In Fig. 17, concerning run-up heights at the furthest recess within Ryoishi Bay
(behind Ryoishi Fishing Port) in Iwate Prefecture, the empirical formulas are

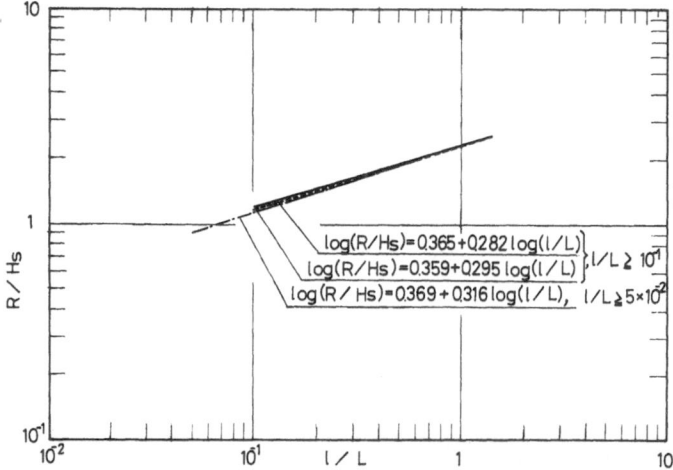

Fig. 15.   The author's empirical formula presented in relationship between R/H_s and l/L.

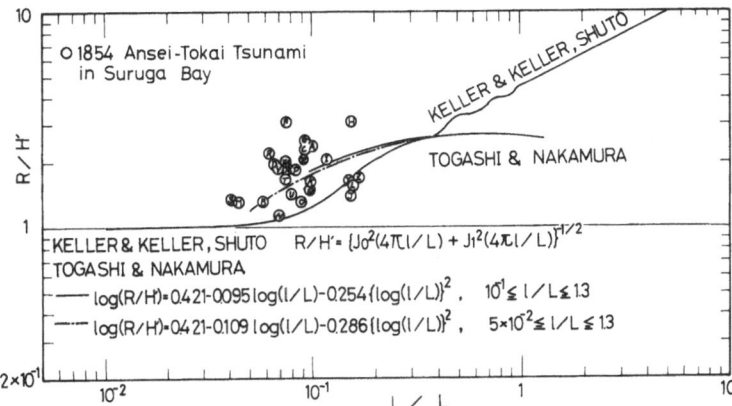

Fig. 16.   Case study of the run-up heights R/H' by the 1854 Ansei-Tokai Tsunami on the coasts of Suruga Bay in Shizuoka Prefecture (after INA CIVIL ENGINEERING CONSULTING CO., LTD., 1978).

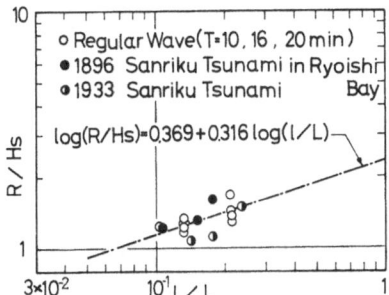

Fig. 17.   Case study based on the experiments of the run-up heights R/H_s by the Sanriku Tsunamis of 1896 and 1933 at recesses within Ryoishi Bay in Iwate Prefecture (Data are used after IWASAKI et al., 1977).

compared with the experimental results by IWASAKI et al. (1977). The length of the bay axis is $l = 4$ km in this case, because the distance from the bay head to the bay mouth, near the line connecting Namiitazaki with Shirasaki, is about 4 km. The wave length, L, is taken as the value at the point where $h = 100$ m, to fit the experimental conditions of Iwasaki et al. The case was confirmed by checking that the values l/L display almost no difference when computed by $\sqrt{h/g}/S\,T$, putting $h = 50$ m as the mean water depth near at the bay mouth, and beach slope $S \doteqdot 1/60$ from the bay mouth to the recess land. Since it is natural for the measured values to be scattered to some degree for the same reason as before, the results can be said to be satisfactory.

As may be seen from the plane topography of Ryoishi Bay which is similar to the letter "W", in some cases oscillations are induced by forming a coupled system with neighboring Kamaishi Bay. In other cases, usual oscillations without coupling occur due to the topography which is now similar to the letter "V", so bay water oscillations at Ryoishi are complicated. Therefore, R/H' or H_s/H' cannot be easily estimated by one simple theoretical or experimental equation as shown here. However, it is expected that the run-up height, R/H_s, from shoreline onto land can be estimated, if there is no extreme changes of land topography.

4. Conclusions and Summary

(1)   Regarding the effects of wave steepness on the relationships of incident wave height to shoreline wave height and land run-up height on sloping beach topographies, they could be experimentally understood only for cases when the beach slope is specified. Even so, by conducting experiments using periodic waves and also solitary-type non-periodic waves, it is difficult to find any general overall unified quantitative tendency. This is still true after making consideration of wave profile and periodicity as characteristics of the incident waves, since not only are the experimental formulas of R/H' and H_s/H' different according to each experimental case for the same slope, but also for each slopes there are great variances according to differences in wave plofile, periodicity and other experimental conditions. Nonetheless, a good result can be obtained for R/H_s, both quantitatively and qualitatively.

(2)   With regard to the effect of beach slope and relative depth upon the relative run-up height R/H', these were first examined and compared with Kaplan's empirical formulas for the slopes $S = 1/60$ and $1/30$, using the theoretical formulas of Keller and Keller, and Shuto. As a result, a partial overlap with Kaplan's experimental formulas is obtained. However, generally not only are there differences within the experimental range, but their extrapolation gives results that are both qualitatively and quantitatively quite different. This suggests that more careful consideration of breaking conditions is necessary.

(3)   Regarding the influence of slope and relative depth on relative run-up height, R/H', on sloping beaches, experimental results are well unified overall, and show a qualitative tendency resembling the theoretical relative run-up formulas of small amplitude long waves for standing waves. However, where effects of non-linearity are great, there are ranges which clearly cannot be explained adequately by the theoretical equations. Thus new emprical equations for R/H' are proposed which are primarily concerned with finite amplitude long waves.

(4)   Similarly to item (3), two more empirical formulas for H_s/H' and R/H_s were presented and the tsunami run-up mechanism was made clearer. These formulas would be useful for a preliminary and rough estimate of tsunami run-up on a simple beach topography.

Finally, there was some discussion for applications to the field, and two examples of applied calculations to actual tsunamis were shown.

The author expresses his gratitude to Research Associate T. Nakamura and Technical Official Y. Hirayama, Department of Civil Engineering, Nagasaki University, for their valuable help in the experiments and in compiling the charts. This work is part of a research program partially supported by a Science Research Grant of the Ministry of Education in Japan, which the author wishes to thank.

REFERENCES

INA CIVIL ENGINEERING CONSULTING CO., LTD., *Report of the Work for Investigation on Tsunami Run-up Simulation Model*, 1978.

IWASAKI, T., H. TOGASHI, and E. SATO, The hydraulic characteristics of tsunamis at shoreline and its run-up, Proc. 17th Japanese Conf. on Coastal Eng., pp. 427–433, 1970 (in Japanese).

IWASAKI, T., A. MANO, and Y. ADACHI, Tsunami run-up heights at the bay recesses, Proc. 24th Japanese Conf. on Coastal Eng., pp. 69–73, 1977 (in Japanese).

KAPLAN, K, Generalized laboratory study of tsunami run-up, Tech. Memo. No. 60, B. E. B., Corps of Engineers, 1955.

KELLER, J. B. and H. B. KELLER, Water wave run-up on a beach, Service Bureau Corporation Research Report, Contract No. NONR-3828(00), Office of Naval Research, Department of the Navy, 1964.

SHUTO, N., Run-up of long waves on a sloping beach, *Coastal Eng. Japan*, **10**, 23–38, 1967.

SHUTO, N., Standing waves in front of a sloping dike, *Coastal Eng. Japan*, **15**, 13–23, 1972.

TOGASHI, H. and T. NAKAMURA, An experimental study of tsunami run-up on uniform slopes, *Coastal Eng. Japan*, **20**, 95–108, 1977.

TOGASHI, H., Study on Tsunami Run-up and Countermeasure, 1981.

Tsunamis—Their Science and Engineering, edited by K. Iida and T. Iwasaki, 511–525.

Effects of Large Obstacles on Tsunami Inundations

Chiaki Goto and Nobuo Shuto

Department of Civil Engineering, Tohoku University,
Sendai, Japan

(Received August 31, 1981; Revised December 22, 1981)

Buildings sufficiently solid not to be washed away may reduce tsunami inundation. Low sea walls, by reflecting part of incoming tsunamis, are also effective. A method is proposed to include both these effects in numerical computations. Modified empirical formulas from steady flow hydraulic experiments are employed. Numerical results obtained for unsteady flow agree fairly well with the hydraulic experiments.

1. Introduction

Several numerical techniques have been developed to analyze the run-up of tsunamis and applied with sufficient accuracy to practical problems. In the computation, however, it is normally assumed either that there are no solid structures capable of withstanding the tsunami, or that the effect of a solid structure can be expressed in terms of a roughness coefficient, such as Manning's n, the value of which is selected without due consideration of the hydraulic characteristics.

The Japanese Sanriku coast, where tsunami defense works have been constructed since 1960, exhibits sea walls along the shoreline high enough to have offered protection against the Chilean Tsunami of 1960 although not against larger tsunamis such as occured in 1896 and 1933. Records show that when the Chilean Tsunami hit Kushiro city, Hokkaido, Japan in 1960 the horizontal distance of tsunami inundation was only 20 m in areas dense with residential structures, but more than 100 m along wide boulevards (Fukushima, 1961). Since the sizes of these obstacles are of the same order of magnitude as the thickness of tsunamis in shallow water and on land, it is necessary to determine their hydraulic effects more precisely to provide detailed information on the behavior of tsunamis passing through them.

The purpose of the present paper is to include, in numerical computation, the effect of low sea walls and solid buildings capable of resisting tsunamis without being washed away, after establishing the laws of resistance through hydraulic experiments. Although tsunamis are highly unsteady, the laws of resistance were determined by using steady flow, because it is not considered possible to carry out hydraulic experiments with unsteady flow with the required accuracy. Once the empirical laws were established, they were modified into a form convenient for numerical computation. Deformations of tsunami profiles due to solid structures were measured in the laboratory using unsteady flow and compared with the numerical results to ascertain the applicability of the numerical scheme established above.

2.   Hydraulic Characteristics of Residential Buildings

2.1   Experimental set-up and procedure

A flume 16 m long, 0.6 m high, and 0.8 m wide was used in the experiments. Buildings were modelled by two kinds of square-sectioned wooden pillars, 75 cm high, 9 cm × 9 cm and 18 cm × 18 cm. Pillars were installed regularly as is shown in Fig. 1. Ratios of areas occupied by the wooden pillars to the total area are given in Table 1. This ratio in Ishinomaki city, Miyagi Prefecture, is 30.6% for the area which was inundated by the Chilean Tsunami in 1960.

The distance, B, between two columns of pillars measured laterally to the flow is given in Table 1, as well as the distance between rows of pillars measured in the longitudinal direction. The number of rows of pillars in the experiments was two, three, four and five.

The discharge was varied from 10.0×10^3 cm^3/s to 40.0×10^3 cm^3/s and was measured with a triangular weir installed downstream. Water levels were usually measured with point gauges and sometimes with servo-type wave gauges when lateral oscillations were induced by a large dischange.

2.2   Division into three regions

The present problem was at first sight considered quite similar to the resistance of piers in rivers. D'Aubuisson (1840) proposed a method to evaluate the loss of energy head due to piers, assuming a discharge coefficient which depends on the plan shape of the piers. Figure 2 shows the discharge coefficient obtained in a similar way to D'Aubuisson for our experiments. The discharge coefficient has no unique value but varies with the ratio of contraction, number of rows of pillars and the Froude number of the flow. The major difference is that piers are arranged in only one row whereas in the present case more than two rows of obstacles are considered.

The authors therefore abandoned the attempt to express the effects of the pillars as a whole. Instead, we divided the whole region into three subregions, according to the major hydraulic phenomena which govern each subregion. The first region is the entry region including the first row of pillars, where the head loss is mainly caused by a

Table 1.

Run	Lateral spacing	Longitudinal spacing	Ratio of area occupied by pillars
A	8.67 cm	7.0 cm	48.6%
B		15.0	36.8
C		22.0	30.4
D	15.0	7.0	39.3
E		15.0	29.8
F		22.0	24.5
G	22.0	7.0	32.4
H		15.0	24.5
I		22.0	20.3
J	6.0	14.0	23.5

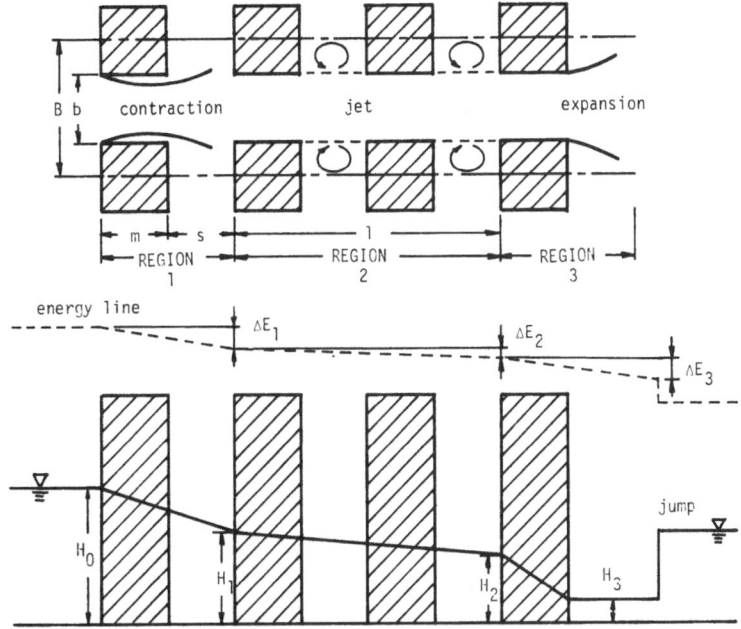

Fig. 1. Schematic diagram of wooden pillars in steady flow and classification of regions.

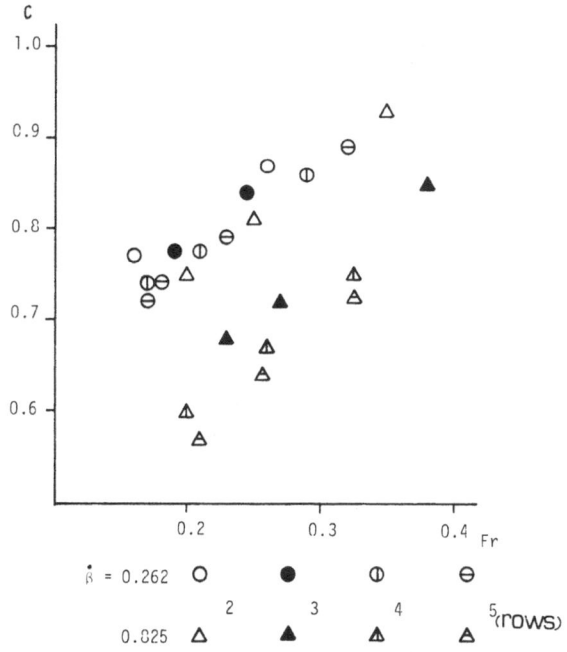

Fig. 2. Discharge coefficient of a group of pillars.

sudden contraction. The third region is the rearmost region, which includes the last row of pillars, where a sudden expansion is of greatest importance. The second region lies between the above two regions. Here, the water flows as if through a narrow channel of roughened walls. The major hydraulic phenomenon in the second region can be expressed in terms of a friction coefficient. The contribution of the second region to the total head loss is, as shown below, fortunately negligibly small in comparison with that of the first region.

2.3   Region I

In this region, the predominant phenomenon is the contraction of the water flow at the entrance, as considered by D'Aubuisson. The continuity equation and the energy equation with a loss term, ΔE_1, are

$$Q = BH_0V_0 = C_c bH_1V_1$$
$$\frac{V_0^2}{2g} + H_0 = \frac{V_1^2}{2g} + H_1 + \Delta E_1 \tag{1}$$

where C_c is the coefficient of contraction. Let ΔE_1 be $K_e V_0^2/2g$. Then the discharge coefficient can be calculated by

$$C = \frac{C_c}{\sqrt{1 + K_e}} = \frac{F_{ro}}{\beta\gamma_0\sqrt{F_{ro}^2 + 2(1 - \gamma_0)}} \tag{2}$$

where β is the ratio of contraction b/B, γ_0 is the ratio of water depth H_1/H_0, and F_{ro} is the Froude number defined by $V_0/\sqrt{g H_0}$.

Figure 3 gives the experimental data of the discharge coefficient. Although the average value of the discharge coefficient is nearly equal to the value of 0.8 given by D'Aubuisson in case of piers of a rectangular plan form, it can clearly be recognized that the discharge coefficient is a function of β and F_{ro}.

Figure 4 shows that the following formula

$$C = 1.46\left(\frac{F_{ro}}{\beta}\right)^{0.7} \text{ for } 0.33 < \beta < 0.55 \text{ and } 0.08 < F_{ro} < 0.26 \tag{3}$$

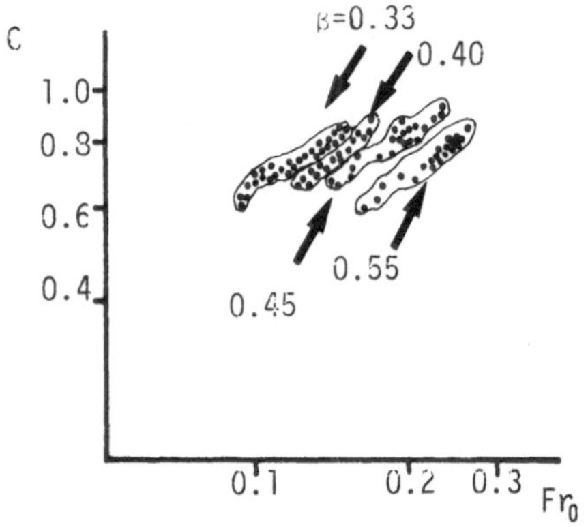

Fig. 3.   Discharge coefficient in region I with a parameter β.

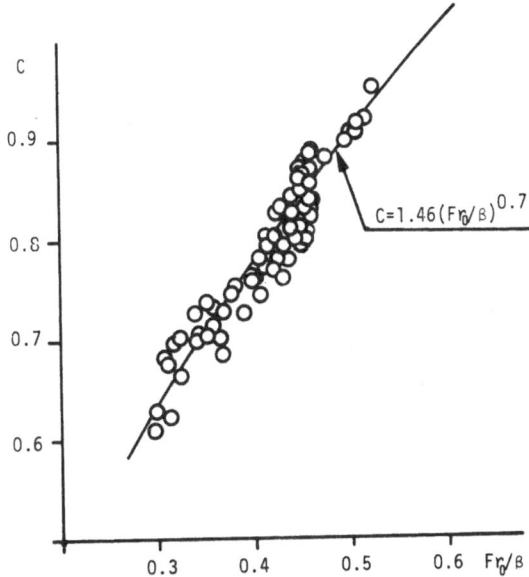

Fig. 4.   Discharge coefficient in region I as a function of F_{r_0} and β.

is sufficiently accurate for practical purposes.

2.4   Region II

The continuity and energy equations are

$$Q = b H_1 V_1 = b H_2 V_2$$

$$\frac{V_1^2}{2g} + H_1 = \frac{V_2^2}{2g} + H_2 + \Delta E_2 \quad . \tag{4}$$

The energy loss in this region can be expressed in analogy to head loss in rough-walled pipes or channel. Therefore, according to the usual expression, the energy loss, ΔE_2, is given by

$$\Delta E_2 = K_a \frac{l}{H_1} \frac{V_1^2}{2g}. \tag{5}$$

Substitution of Eq. (5) into Eq. (4) yields the following formula for the friction coefficient,

$$K_a = \frac{H_1}{l} \left[\left(1 - \frac{1}{\gamma_1^2} \right) + 2(1 - \gamma_1) \frac{1}{F_{r_1}^2} \right] \tag{6}$$

where $\gamma_1 = H_2/H_1$ and $F_{r_1} = V_1/\sqrt{g H_1}$.

The size of the pillars relative to the flow width or their spacing may affect the values of the friction coefficient as in the case of rough-walled pipes or channels. However, as far as our experimental results are concerned, no dependence on these parameters was recognized. An empirical formula

$$K_a = 0.0033\, F_{r_1}^{-5} \text{ for } 0.32 < F_{r_1} < 0.81 \tag{7}$$

was established as shown in Fig. 5.

2.5   Region III

Sudden expansion is of great importance in this region. After region III there occures a hydraulic jump in the case of steady flow, the loss due to which can be evaluated by a familiar formula found in any text book of hydraulics. The hydraulic jump is not included in our consideration.

The equations are

$$Q = b H_2 V_2 = B H_3 V_3$$

$$\frac{V_2^2}{2g} + H_2 = \frac{V_3^2}{2g} + H_3 + \Delta E_3. \tag{8}$$

It is assumed that ΔE_3 can be expressed by $K_p V_2^2/2g$ as a fraction of the velocity head before the expansion. Then, K_p is calculated with

$$K_p = 1 - \frac{\beta^2}{\gamma_2^2} + 2(1 - \gamma_2)\frac{1}{F_{r_2}^2} \tag{9}$$

by using experimental results. The results are plotted in Fig. 6, and a dependance on β is

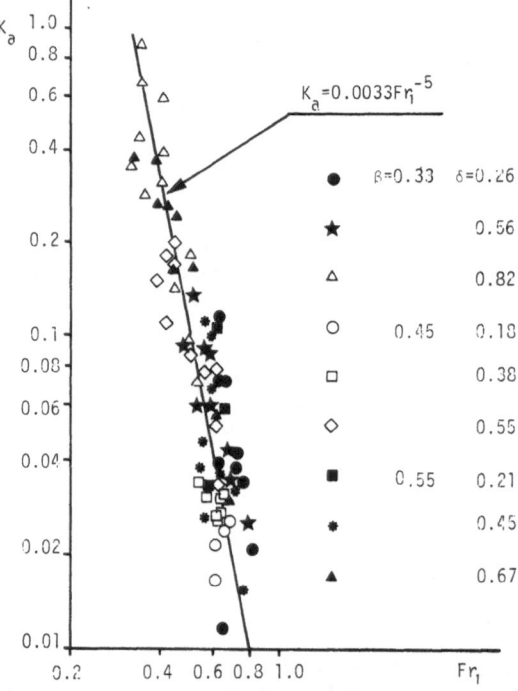

Fig. 5.   Friction coefficient in region II.

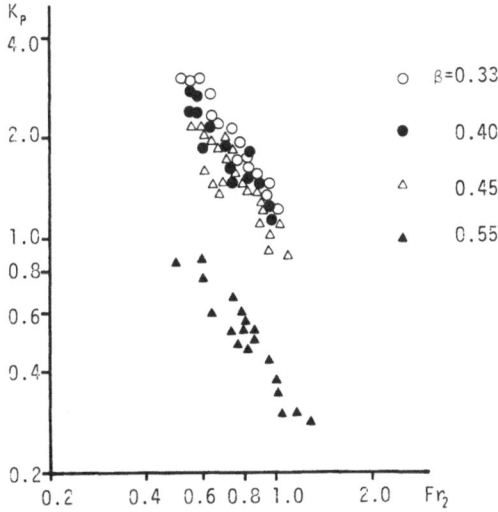

Fig. 6.   Coefficient of expansion in region III with a parameter β.

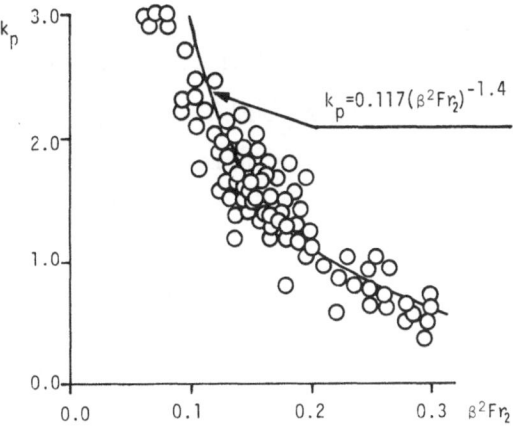

Fig. 7.   Coefficient of expansion in region III as a function of F_{r_2} and β.

seen. Finally, the following formura is given for the coefficient of expansion as plotted in Fig. 7.

$$K_p = 0.117 \left(\beta^2 F_{r_2} \right)^{-1.4} \text{ for } 0.33 < \beta < 0.55 \text{ and } 0.51 < F_{r_2} < 1.30. \qquad (10)$$

3.   Overflow at Low Sea Walls

3.1   Experimental procedure

A channel 7 m long, 0.5 m high, and 0.3 m wide was used in the experiments.

Model sea walls were 10 cm, 15 cm, 20 cm, and 25 cm high, and 2 cm and 5.5 cm thick. Discharges per unit width of the channel varied from 50 cm²/s to 250 cm²/s. Water levels were measured at 1 cm, 10 cm, and 100 cm upstream from the models. Values of H_d, H_s, and L_s, the definitions of which are given in Fig. 8, were also measured.

3.2 Discharge formula in terms of the energy per unit mass

A low sea wall, if overtopped by a steady flow, is identical to the situation at a weir. An example of the comparison of our experimental results with a formula given by Hom-ma (1940) is shown in Fig. 9. The agreement is fairly good. The difference is 25% at the greatest.

However, if the Hom-ma formula or another weir formula is applied to unsteady overflow, a serious unresolved problem remains. In these formulas, the discharge is given in terms of a head measured upstream from the weir at a distance sufficient to eliminate surface contraction. In steady flow, this head can be easily defined while in unsteady flow it is neither defined nor measurable due to the time-dependent motion of the water surface.

In place of this head, the authors suggest that the energy per unit mass, defined as

$$E = \frac{V_A^2}{2g} + H_A \tag{11}$$

should be taken. The discharge per unit width of the channel is related to the energy per unit mass by the following formula with a discharge coefficient

$$q = K\sqrt{g}\,E^{1.5}. \tag{12}$$

After several trials, it was found that the discharge coefficient could be expressed as a function of F_{r_B}. The Froude number F_{r_B} used in Fig. 10 is calculated with a hypothetical water depth, H_B, and velocity, V_B. It is assumed that the water flow concentrates within a layer, the thickness of which is H_B, above the weir crest and that the equation of continuity is given by $q = H_B V_B$. The discharge coefficient is experimentally found to be

$$K = F_{r_B}, \text{ for } 0.35 < F_{r_B} < 0.80. \tag{13}$$

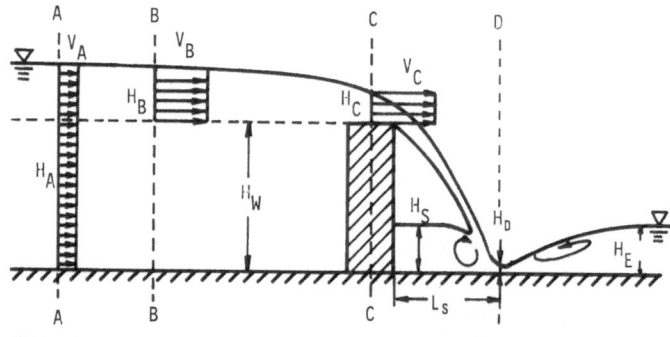

Fig. 8.   Schematic diagram and definition sketch of low sea wall.

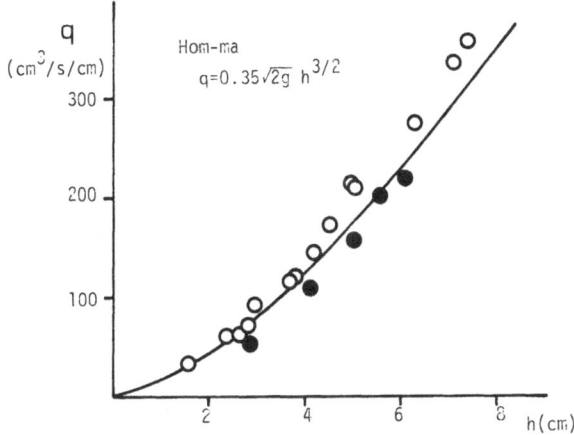

Fig. 9. Comparison between experimental results and Hom-ma's formula.

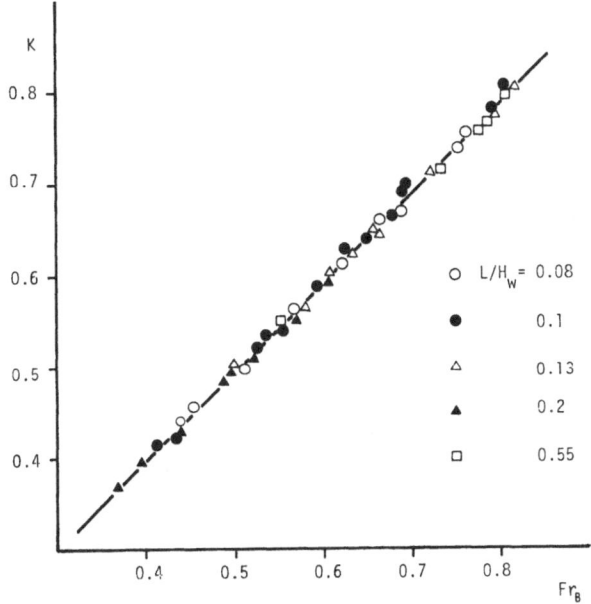

Fig. 10. Discharge coefficient over a weir in terms of energy per unit mass.

3.3 Downstream of a sea wall

If a sea wall is sufficiently high and the water depth after the point of water fall sufficienty shallow, water falls freely, hits the channel bottom and a part of the discharge flows downstream while the rest remains and forms a water cushion near the sea wall. The total head loss is composed of the loss of potential energy due to the fall of water, ΔE_D, and the loss due to the turbulent motion in the water cushion, ΔE_D.

The continuity and energy equations are established between the point where the critical depth appears and the point D where the falling water hits the channel. These are

$$q = H_c V_c = H_D V_D$$
$$\frac{3}{2} H_c + H_w = \frac{V_D^2}{2g} + H_D + \Delta E_D + \Delta E_D' \tag{14}$$

where $\Delta E_D' = H_w + H_c - H_D$, and $\Delta E_D = K_D V_D^2/2g$. The coefficient of loss of energy is calculated with

$$K_D = \left(\frac{H_D}{H_c}\right)^2 - 1. \tag{15}$$

In order to evaluate the water depth at the point of water fall impact, Rand's formulas (1955) are selected from available empirical formulas, because they are formally simple. Rand gave expressions for the water depth, H_D, at the point of water fall, the water depth, H_S, in the water cushion and the horizontal distance, L_s, to the point of water fall. These are given in terms of a dimensionless parameter

$$D_R = \frac{q^2}{gH_w^3} \tag{16}$$

where H_w is the height of the weir. Rand's formulas are

$$\frac{H_D}{H_w} = 0.54 D_R^{0.425}, \frac{H_s}{H_w} = D_R^{0.22}, \frac{L_s}{H_w} = 4.30 D_R^{0.27}. \tag{17}$$

Figure 11 shows the comparison between experimental results and Eq. (17).

4. Numerical Simulation and Comparison with Hydraulic Experiments

4.1 Effect of buildings

For simplicity in the numerical simulation, it was proposed that the effects of buildings and other residential-type structures should be taken into computations through an equivalent roughness. In order to examine the validity and accuracy of the introduction of an equivalent roughness, numerical computations were carried out and the results were compared with the experimental results for unsteady flow.

Equations used in the analysis are the equations for one-dimensional, gradually varied unsteady flow.

$$\frac{\partial A}{\partial t} + \frac{\partial Q}{\partial x} = 0$$

$$\frac{\partial Q}{\partial t} + \frac{\partial}{\partial x}\left(\frac{Q^2}{A}\right) + gA\frac{\partial H}{\partial x} + \frac{gn^2 Q|Q|}{H^{4/3}A} = 0 \tag{18}$$

where Q is the discharge flux. A the sectional area and n the Manning's roughness. The last term in Eq. (18) is equal to gAI_1 where I_1 is the friction slope. We assume that the

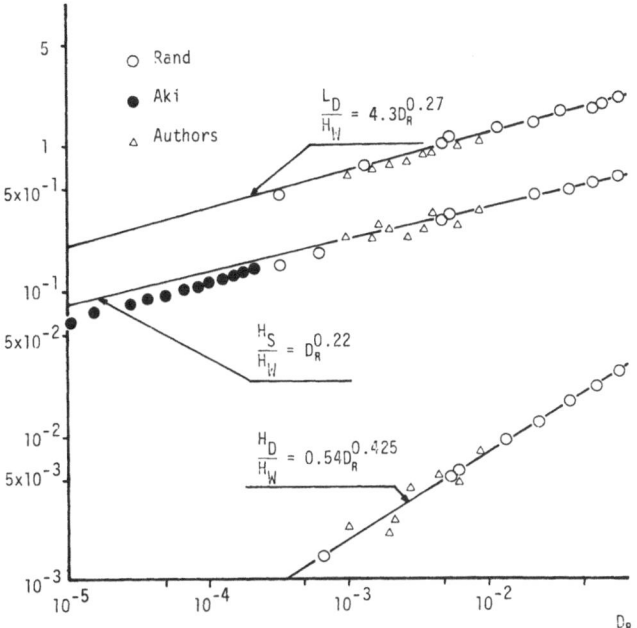

Fig. 11. Comparison with Rand's formula.

instantaneous friction slope in unsteady flow can be approximated by the empirical formulas given in proceeding sections. Then, for example, for region I, I_1 is set equal to $\Delta E_1/l_1$. Therefore, introducing an equivalent roughness n_e, we have

$$gAI_1 = gA\frac{\Delta E_1}{l_1} = \frac{K_e}{2l_1}\frac{Q|Q|}{A} = \frac{gn_e^2 Q|Q|}{H^{4/3}A}. \tag{19}$$

The equivalent roughness is given by

$$n_e^2 = \frac{K_e H^{4/3}}{2gl_1} = \frac{H^{4/3}}{2gl_1}\left(\frac{1}{C_c^2} - 1\right). \tag{20}$$

In a similar way, the equivalent roughness is obtained for regions II and III. Numerical computations can be continued with Eq. (18), by changing the values of n_e from region to region and from time to time.

Hydraulic experiments for unsteady flow were carried out in a channel 100 m long, 0.5 m high, and 1 m wide. Three rows of pillars were installed with $B = 33$ cm, $b = 15$ cm, $m = 18$ cm, and $s = 15$ cm. The still water depth in the experiments was 20 cm. The water surface elevation was measured with resistance-type wave gauges. In the numerical computations, the leap-frog scheme was used. The spatial mesh size was 6 cm.

Figure 12 shows an example of instantaneous wave profiles and velocity distributions obtained in the hydraulic experiments. Figure 13(a) shows the time history of the water surface measured at a point 2 m upstream of the obstacles. This is used as an input wave for the computation.

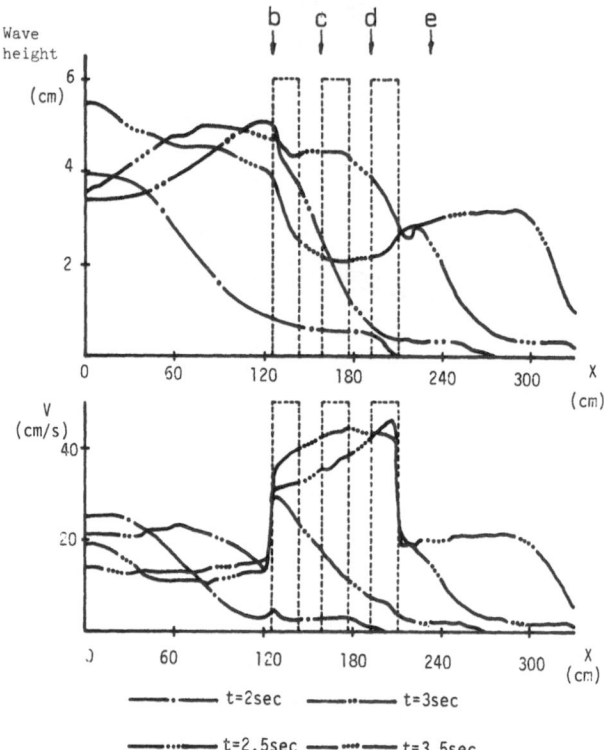

Fig. 12.   Wave profiles and velocity distribution in a hydraulic experiment for unsteady flow.

Figure 13(b) through (e) are comparisons between hydraulic and numerical experiments at several points indicated in Fig. 12. The solid lines are results from the hydraulic experiments and the chain lines are results of the numerical computations for one-dimensional, gradually varied unsteady flow introduced in the present paper.

For comparison, two-dimensional computations were also carried out, in which no dissipation due to the pillars was taken into consideration but no velocity component normal to the surface of the pillars was allowed. Results of the two-dimensional computations with 3 cm spatial mesh size are shown by dotted lines in Fig. 13(b) through (e). The one-dimensional computations give fairly good agreement, with a small reduction in wave height and a slight time-lag of the maximum water level. The two-dimensional computations give poorer results, although a finer spatial mesh was used. The two-dimensional computation required a computation time 60 times longer than one-dimensional case.

4.2   Effect of a low sea wall

Hydraulic experiments for unsteady flow were carried out in the same channel used for those with steady flow. Unsteady flows were generated by sudden openings and closings of a valve installed at the upstream end of the channel. Water surface elevations were measured at points 2 m upstream and just upstream of the sea wall as

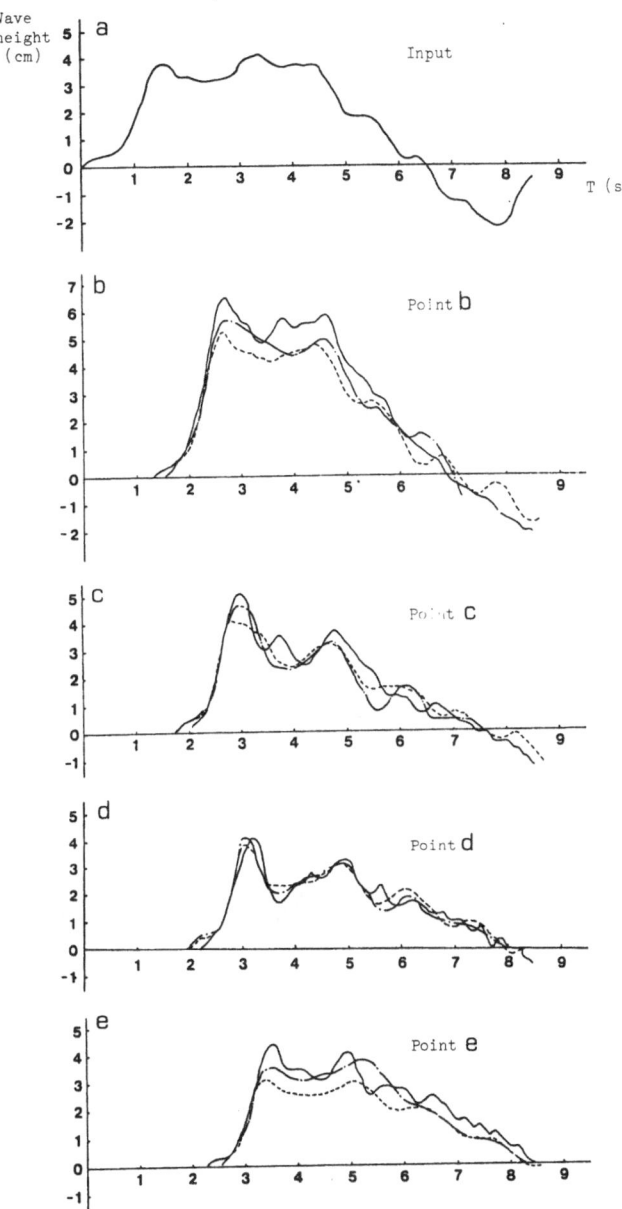

Fig. 13.   Comparison between hydraulic and numerical experiments.

well as at the water cushion downstream. Equation (18) was used in the computation with a modification at the sea wall. In the leap-frog scheme, points where the discharge is computed are different from points where the water surface elevation is computed. Discharge points were arranged at the sea wall. Values obtained at a point upstream of the sea wall by one spatial mesh are used to compute the energy per unit mass, by which the discharge over the sea wall is determined with Eqs. (12) and (13).

In order to continue the computation further downstream, Rand's empirical formulas are used to compute H_D and L_S. At the point of water fall impact, the horizontal position of which is fixed by L_S, the height H_D is combined with discharge q to give the velocity V_D. Then, these values are used to evaluate the head loss, ΔE_D, or the coefficient K_D, which is taken into the computation after being reduced to an equivalent roughness.

Figure 14(a) is the input wave profile obtained at a point 2 m unstream. In Fig. 14(b), wave profiles just upstream of the sea wall are compared. The solid line corresponds to the hydraulic experiment and the chain line represents the result given by the method described above. The dotted line is obtained by the use of Homma's weir

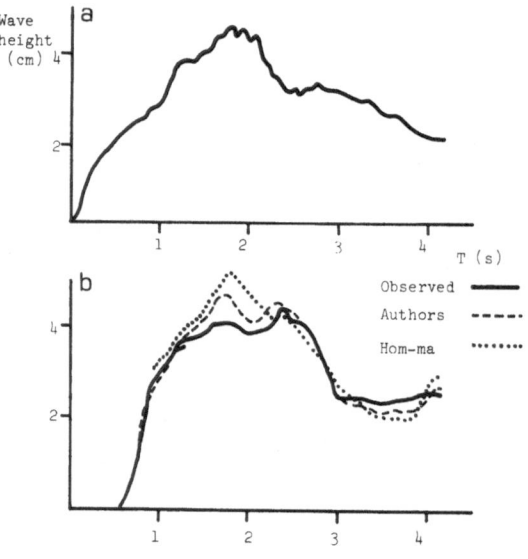

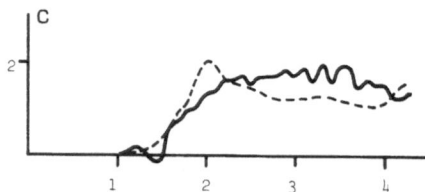

Fig. 14.   Comparison between hydraulic and numerical experiments.

formula instead of Eq. (12). The present method produces a satisfactory result.

Figure 14(c) compares the water surface elevation in the water cushion. The solid line is for the hydraulic experiment and the dotted line shows the present numerical result. Except for small oscillations which are inevitably generated by the water fall, the agreement is satisfactory.

5. Conclusions

Large obstacles such as grouped houses or low sea walls are expected to be effective to some extent in reducing the inundation of tsunamis. A method is proposed and examined to reproduce their effects in numerical simulations. The basis of the method is the use of empirical formulas of the discharge and friction coefficients for steady flow in numerical computation of unsteady flow.

In the case of grouped houses, the houses are classified into three subregions; the region of entry, the intermediate region and the last region. The discharge and friction coefficients of each region are expressed in terms of the Froude numbers or the ratio of contraction. Comparisons with hydraulic experiments for unsteady flow reveal that the coefficients can be easily taken into the numerical computations for unsteady flow through an equivalent roughness to yield good agreement.

In the case of low sea walls, a weir formula, modified by introduction of the energy per unit mass, improves the accuracy of the numerical computations for unsteady flow. Rand's formulas are used to continue the computations further downstream.

In both cases, it is concluded that the present method is applicable to practical problems with satisfactory accuracy.

A part of the research reported herein was supported by a grant from the Ministry of Education.

REFERENCES

D'Aubuisson de Voisins, J. F., *Traité d'hydraulique*, 2d ed., Pitois, Levrant Cie, Paris, 1840.
Fukushima, M., Tsunami inundation at Kushiro Harbour and in Kushiro River, *Tech. Report of the Japan Meteo. Agency*, **8**, 62–68, 1961 (in Japanese).
Hom-ma, M., Discharge coefficient of low dams, *Proc. JSCE*, **6**, **9**, 1940 (in Japanese).
Rand, W., Flow geometry at straight drop spillways, *Proc. ASCE*, **81**, Paper 791, 1955.

Tsunamis—Their Science and Engineering, edited by K. Iida and T. Iwasaki, 527–539.

Land Management Guidelines for Tsunami Hazard Zones*

Jane PREUSS

Urban Regional Research, Seattle, Washington, U.S.A.

(Received November 18, 1981)

The purpose of this project is to develop land Management and Physical Form Guidelines for developments located in Tsunami High Hazard Zones. The research program for the project is organized into two interrelated components based on the dynamics of the tsunami hazard. Part I of the project is concerned with the regional scale of land management. The tsunami issue at the regional scale is related to definition of the hazard zone in terms of locational susceptability. Planning issues at this scale are concerned with the allocation of land uses, with major open space planning and with the location of major connecting transportation routes.

Once an administrative decision has been made to permit development in an identified tsunami hazard zone, the tsunami issues which must be addressed will pertain to the varying aspects of wave forces. Part II of the project therefore develops site planning criteria designed to maximize development potential while minimizing exposure to the hazard. Project scale or site planning consists of the integrated arrangement of three primary elements: buildings and structures; open spaces; and roads. This section develops hazard based performance criteria for each site plan component.

1. Objectives of the Project

The underlying objective of this project is to develop a methodology whereby response to the tsunami hazard is explicitly integrated into the comprehensive planning process. This topic addressed interests of the Problem Focused Research Division of the National Science Foundation relating to siting of projects and development of policies or procedures to integrate technical information on natural hazards into land use plans, urban and coastal zone planning, and site planning. The project which this paper describes focuses on the receiving or land side of tsunami research by developing a two tier approach to land management for previously identified tsunami hazard zones which by the nature of the event are based on a boundary which cannot be defined with certainty. The project utilizes the findings of technical research to develop a range of practical and responsive solutions to land use regulations.

The first phase of the project reviewed land use plans and regulations applicable to coastal hazard areas which had been prepared by the Federal Government, by States and by local communities. The findings of this review indicated that land management

*Funded by NSF Grant Number PFR 782 3884.

occurs at various geographic scales. The substantive issues addressed at each level vary with regard to content as well as site specificity. Both the level of detail and the concerns with respect to the tsunami hazard differ at the various levels of specificity. The most general policies encompass an entire community, subsequent levels of detail are addressed at the regional or coastal segment scale and by the scale of planning which focuses on clusters of land parcels or individual lots. The analysis conducted for this project were organized into two primary sections. Part I develops a methodology which addresses tsunami related issues at the regional or coastal segment scale. Part II develops a site planning methodology applicable at the project scale. The remainder of this paper summarizes the methodology and findings of the project.

2. Part I: Regional Scale Analysis: Methodology

The regional scale of land management is generally concerned with the allocation of land uses, with major open space planning and with the location of major connecting transportation routes. Thus the tsunami issue addressed at the regional scale is related to locational susceptability to the hazard. Objectives of the regional policy analysis are designed to develop a framework for incorporating the tsunami hazard into the comprehensive planning process. Regional plans and land management criteria for coastal hazard areas as utilized by community comprehensive plans in the United States tend to take into consideration a hazard zone definition which includes the projected inundation zone/floodway and probable wave elevations. These projections utilize a method developed for the U.S. Flood Insurance Program based primarily on historic experience (where available), the wave's reaction to off-shore bottom geometry, and coastal configurations. With the exception of the State of Hawaii, land based friction creating factors do not tend to be included in the numerical projections of wave heights.

Since implementation of regional scale plans tend to be carried out by the public sector the focus of Part I was a public policy analysis. The emphasis of Part I was to document the degree of correlation between policies of various governmental agencies which are currently in effect for areas which are susceptible to the tsunami hazard. In order to develop as comprehensive an assessment as possible, plans and regulations were reviewed for three case study communities which differ from each other in terms of government structure, land use pressures and/or response mechanisms to the hazard. The case study communities are: the North Shore on the Island of Kauai in the State of Hawaii; Downtown Hilo in the State of Hawaii; and Downtown Kodiak on the Island of Kodiak in the State of Alaska. Prior to description of the research methodology and findings a brief description of the case study areas is in order.

The North Shore of Kauai, shown in Fig. 1, is a rural area where primary land uses are single family residential, low density resort and agriculture. The North Shore experienced distantly generated tsunami events in 1946 and 1957 which caused considerable damage. Wave height elevations of up to 32 feet above mean high sea level have been projected in conjunction with the U.S. Flood Insurance Program. The North Shore illustrates several common conflicts between defined tsunami hazard related characteristics and land use planning/real estate pressures. For example the

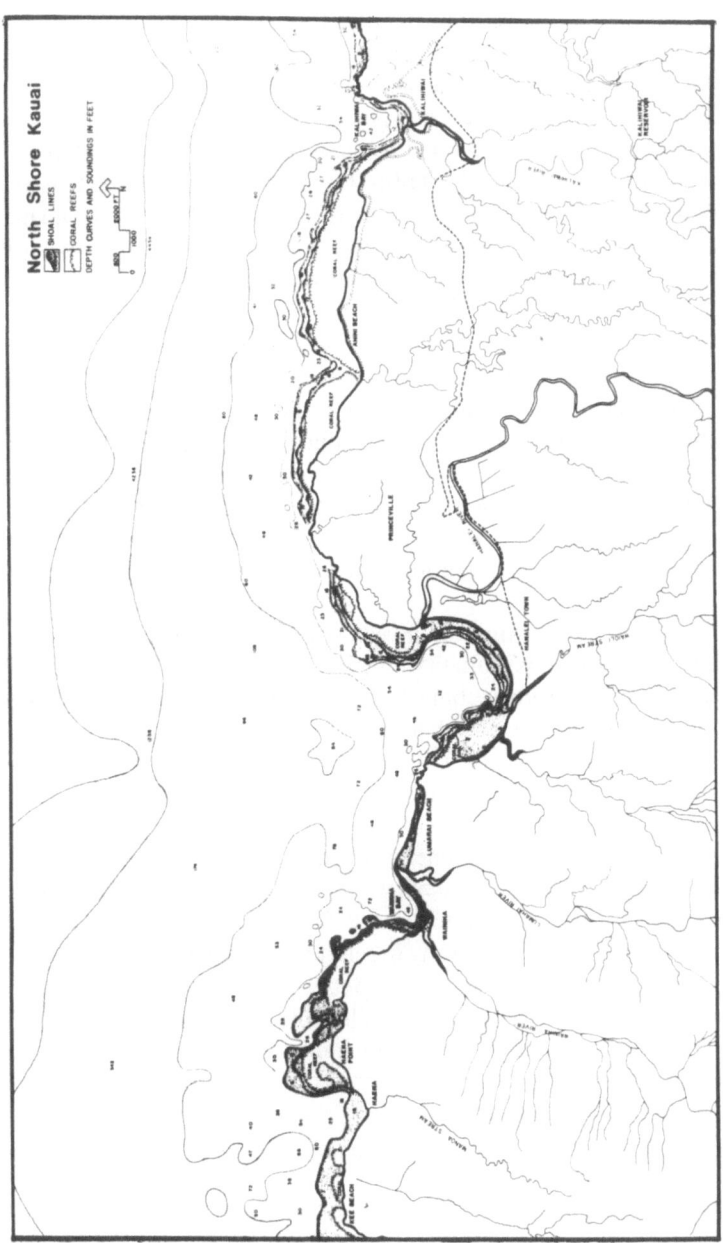

Fig. 1. Case study area: north shore, Island of Kauai, State of Hawaii.

County of Kauai has identified a tsunami hazard zone in the General Plan. Portions of this same area have use designations permitting low density residences with a 35 foot height limit. Compliance with the regulations attendant with the U.S. Flood Insurance Program would however require that the structure be elevated approximately 25 feet above grade. Compliance with these latter regulations results in a building which is economically inefficient and/or in conflict with County zoning.

The City of Hilo is the County Seat and largest town on the Island of Hawaii. In addition the downtown Hilo coastal segment accommodates a large percentage of the island's accessible shoreline. All of the Hilo shoreline experienced extensive devastation in the 1946 and 1960 tsunamis. The area which was damaged in 1960 has for the most part been redeveloped for park purposes. The area damaged in 1946 has been redeveloped for high density resort condominiums and open space. It also accommodates the Island's primary port facilities. Despite or perhaps because of the high demand for the coastal related land uses Hilo, Hawaii exemplifies utilization of the most extensive range of planning measures designed to mitigate tsunami hazards.

Kodiak town on the Island of Kodiak, shown in Fig. 2, is one of the most active fishing ports in the country. The downtown sustained heavy damage as a result of inundation and subsidence caused by the tsunami generated by the 1964 Alaska earthquake. Primary land uses in the downtown area are commercial and industrial/fishing. Kodiak, which is subject to both local and distantly generated tsunamis, has not incorporated consideration of the tsunami hazard into the land use planning process.

During the Regional Scale Analysis investigators first identified the types of policies in each case study which pertained to the hazard zone. It was then possible to identify the degree to which both technical/scientific information and community policies utilized in one plan are correlated with the technical information and community or special purpose objectives reflected in other plans. Base data for the correlation was gathered through an inventory of existing regulations and plans pertaining to Federal coastal programs, State laws in Hawaii and Alaska and county/local programs affecting development in shoreline areas.

Spatially the regional scale as defined by this project encompasses a specified section of coastline which includes a range of coastal configurations. The boundary of the regional sector was determined primarily in relation to use patterns, which in turn tend to be influenced by geographical and topographic factors. Information was collected and organized around the following substantive areas of concern:

Physical conditions: topography and major land forms, natural hazards (tsunami, riverine floods, land slide, etc.).

Land use: existing uses, projected trends.

Special purpose plans: transportation, public improvements and public facilities, open space.

Information for each regional sector was plotted on a series of transparent (acetate) maps at a scale of $1'' = 2,000'$. This map scale is large enough to identify both land use patterns and relationships between hazard delineation and land use. Each transparency represents a) a government plan or program; b) a natural hazard; or c) environmental considerations such as topography. The transparancies are intended to

Fig. 2. Case study area: downtown Kodiak, Kodiak Island, State of Alaska.

be superimposed onto each other in various sequences. When these acetate maps are overlaid onto each other in various combinations they graphically illustrate the degree of spatial correlation and policy coordination which exists between existing regulations and special purpose plans that impact the location and intensity of development in the hazard zone. This method explicitly identifies the existence of geographic relationships and discrepancies between ecological, political and safety related issues relevant to planning for the hazard zone. Figure 3 is a composit of two overlays which correlate the General Plan permitting low density single family use with the flood insurance hazard designation.

Base maps which are prepared for the case study communities indicate coastal configurations, water depths, shoal reefs, and rivers feeding into the area. The issue oriented transparancies include the following:

Natural Hazard Boundaries

Topography

General Plan: Existing and Proposed Land Uses and Open Space

U.S. Flood Insurance Floodway and Wave Elevations (Hawaii Examples)

Local Zoning

Transportation Routes and Public Improvements such as bridges

Civil Defense/Evacuation Routes.

3. Conclusions of Regional Scale Analysis

In all three communities the analysis reveals inconsistencies between a) plans which have been prepared at the local level to reflect special purpose goals such as land use/economic development, open space, transportation or civil defense and b)

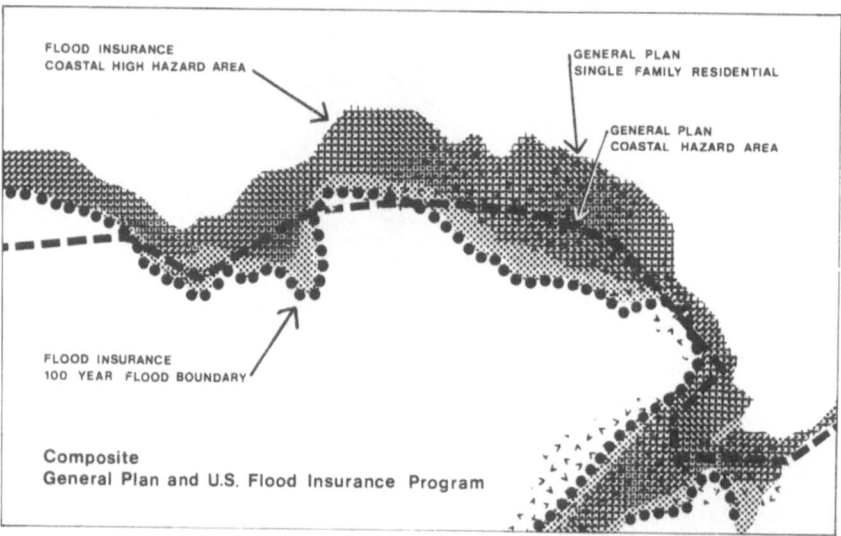

Fig. 3. Composit overlay: general plan and U.S. Flood Insurance Program.

regulations which are adopted by law. For example, major highways tend to be located within the 100 year hazard zone (see Fig. 4). The location of this alignment is not coordinated with evacuation planning. Thus in the event of a disaster neither efficient evacuation or post disaster assistance would be possible. In addition regulations, which have legal standing, often do not conform to the goals which have been identified by special purpose plans. Another example would be construction standards required by the U.S. Flood Insurance Program which would render uses desired by individual communities to be economically infeasible.

Findings of Part I indicate that the tsunami hazard has only begun to be addressed by the area-wide comprehensive planning process. However, although area-wide land use plans generally reflect awareness of the previously identified hazard zones, these plans are usually not adopted by law. Regulations, which on the other hand mandate hazard related restrictions, often do not reflect goals of the community policy plans. Thus regulations which by intent implement community plans are in effect determining land use patterns independently and sometimes contrary to previously adopted and publically sanctioned goals and policies.

There is also a need to coordinate special purpose plans such as for transportation or economic development with hazard planning. For example in all the case study areas evacuation routes are either non-existent or unmaintained. The alignment of these routes should be located outside of the 100 year flood plain in order to take into condideration evacuation and emergency services as well as traffic/routing efficiency. Evacuation/civil defense plans should then be coordinated with transportation planning and land use planning which determines the density of use/number of users feeding into the transportation network. See Fig. 5.

The location of the hazard per se does not seem to have impacted the land use allocation decision making process. For example areas which have historically experienced high wave elevations and even bore formation waves are planned for a wide range of uses including low density single family residential. Often it is difficult if not impossible for a project developer to conform to safety related criteria while developing an economically viable project.

Finally, it was found that the definitions of the hazard zone vary in accordance with the technique which has been used to identify the area of potential hazard. Depending on when they have been prepared various plans in any one community may have utilized more than one definitional technique. For example in Hawaii evacuation routes defined almost 20 years ago are heavily publicized. Thus civil defense warning and public education aimed at protecting human life as opposed to property are not coordinated with land use planning. These boundaries need to be revised to reflect a consistent boundary with other hazard related policies being promulgated for the same areas.

4. Part II: Project Scale: Methodology

The foregoing analysis indicated that land allocation decisions based on susceptibility are primarily a function of tsunami related parameters interacting with off-shore conditions and characteristics of the coast. The project scale addresses characteristics of use pertaining to the location of individual structures. Once an administrative decision has been made to permit development in an identified tsunami hazard zone tsunami effects must be addressed which pertain to the varying aspects of the wave force. Damage to life and property may be due to simple flooding, to the force of waves (especially when a bore is formed) and to horizontal currents associated with the drawdown. Land planning issues responding to interactive wave/land forces (surge, drag and impact) are addressed at the site planning or project scale. Thus site planning criteria must focus on wave force and water volume in such a way that land related factors attributable to roughness/friction created by buildings, trees or the obstructions on land can be organized to minimize the potential for destruction.

Site planning or project scale planning is concerned with the organization and arrangement of three primary elements: structures, open spaces and roads. Implementation of project scale measures is a private sector responsibility where decision making is by individual property owners on a site specific basis. Recommendations at the site specific scale therefore pertain to standards and performance criteria which are to be implemented on a lot by lot basis.

Site planning traditionally follows a process consisting of the following phases:

1) Description of the behavior to be accommodated; specify requirements of the proposed site associated with use.

2) Inventory of physical conditions, which includes identification of opportunities and site limitations such as topography, susceptability to hazards, salient features of the coastal segment (narrow bay, steep cliffs, etc.) soil conditions, etc.

3) Development of a design program or site plan which consists of the conscious arrangement and interrelationship of the three elements: 1) Buildings and Structures, 2) Circulation and Parking, and 3) Open Space and Landscaping.

4) Evaluation of the site plan in terms of: a) Responsiveness to requirements of the proposed use, b) Responsiveness to the physical and environmental characteristics of the site, and c) The expected quantifiable and unquantifiable costs and benefits of the plan.

In order to prepare a responsive site plan it was first necessary to develop tsunami hazard related design oriented objectives and performance criteria for each of the three site plan components. Thus development of these performance criteria, to be integrated into the planning framework, required the addition of analytic step to the traditional site planning analysis. An important aspect of Part II was therefore to prepare a series of objectives and performance standards based on a precise description of the role each site plan element should address with regard to mitigating the tsunami hazard. These criteria explicitly identify the ways in which restrictions imposed by the tsunami hazard could be addressed by each component of the site plan individually and in relation to each other.

The design objectives which have been developed relate to the characteristics or types of wave forces each element will be designed to address. A standard constitutes the performance criteria required to realize the objectives. These standards, or performance criteria correlate safety-related requirements with socio-economic objectives pertaining to use, land use intensity, height limits and visual access to the shoreline from upland areas. Objectives include ways to protect other structures;

ability to withstand surge or drag force; and minimizing impact forces.* Standards include such concerns as: construction type; orientation and location; and relationship to the projected wave and/or to other structures, planting materials, etc. The design objectives and performance standards developed for this project are briefly summarized below according to elements of the site plan.

When reviewing the following discussion it is important to bear in mind that although the design objectives and performance criteria are developed individually for each site plan element the underlying intent of the guidelines is to demonstrate the interdependancy of all the performance criteria. For example standards relating to protective plantings developed in the open space section will effect building standards.

4.1 Site plan element: buildings and structures
4.1.1 Objectives
Design objectives vary in relation to the type of use which is proposed: single family, multi-family, commercial or industrial. Depending on the use requirements, objectives will relate to resisting the wave and or to protecting structures further inland.
4.1.2 Performance standards
Performance standards identify the type of construction which will be appropriate in relation to the wave impact for each type of use. Standards pertain to: a) Building materials, b) Orientation of the building on the site, and c) Location of building on the site.

4.2 Site plan element: open space and landscaping
4.2.1 Objectives
Types of open space to be addressed are major open spaces and linear corridors. Objectives for both types are to create friction; to buffer the built environment from the surge force; and to minimize erosion related impacts attributable to drawdown effects.
4.2.2 Performance standards
Standards are oriented around creation of vegetation massing. Characterstics of vegetation describe the need to balance deep rooted trees for strength with lower lying species of plants with fibrous roots to retard erosion forces.

4.3 Site plan element: roads and streets
4.3.1 Objectives
Objectives are identified for primary and secondary roads. Primary access routes should serve as continuously maintained routes to facilitate evacuation and post disaster assistance. Secondary feeder roads should provide internal or local access and be frequently connected to the primary access route for safety.
4.3.2 Performance standards
Standards pertain to location, alignment and size of the roads. Development of the performance standards is intended to create the design options from which the site

---

*The criteria focus primarily on surge, impact and drag forces. Hydrostatic and buoyancy forces are not specifically included in this project because they are addressed by building design standards.

planner will select discrete elements to incorporate into an individual plan. The next phase of the project was therefore to demonstrate utilization of the performance oriented specifications in the preparation of a series of prototype site plans.

The tsunami sensitive site planning methodology was first applied to a hypothetical coastline exhibiting: 1) a range of uses including single family residential, multi-family residential, resort, commercial and 2) a range of coastal configurations including low lying coast susceptible to bore wave formation, low lying coast without projected bore susceptibility, broad mouthed bay bounded on one side by steep cliffs; narrow coastal area, active port with adjacent commercial district.

For each coastal segment base maps were prepared at a scale of 1″ = 500′. In accordance with a traditional site planning process existing conditions were first defined because they establish the framework within which the subsequent plan must be prepared. Figure 4 illustrates documentation of existing conditions. For purposes of this demonstration project the conditions relating to the tsunami hazard were emphasized. These conditions include ground elevation and slope; projected inundation limit; projected wave elevations; existing sources of friction; and shore configuration.

A series of prototype plans was then prepared for each segment wherein use limitations and use related requirements were specified for the property (single and multi-family residential, commercial, etc.). Plans were then prepared to demonstrate how components of the site plan can be varied in relation to each other while addressing both use requirements and mitigation against surge force, water volume, drawdown and/or impact forces. Figure 5 is a concept drawing of a proposed plan

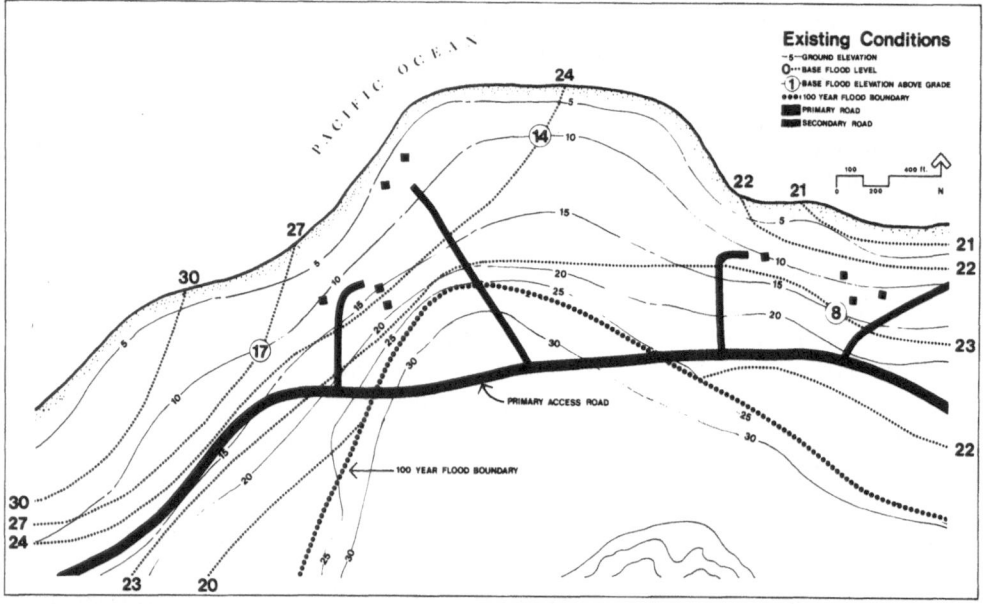

Fig. 4.   Existing conditions analysis.

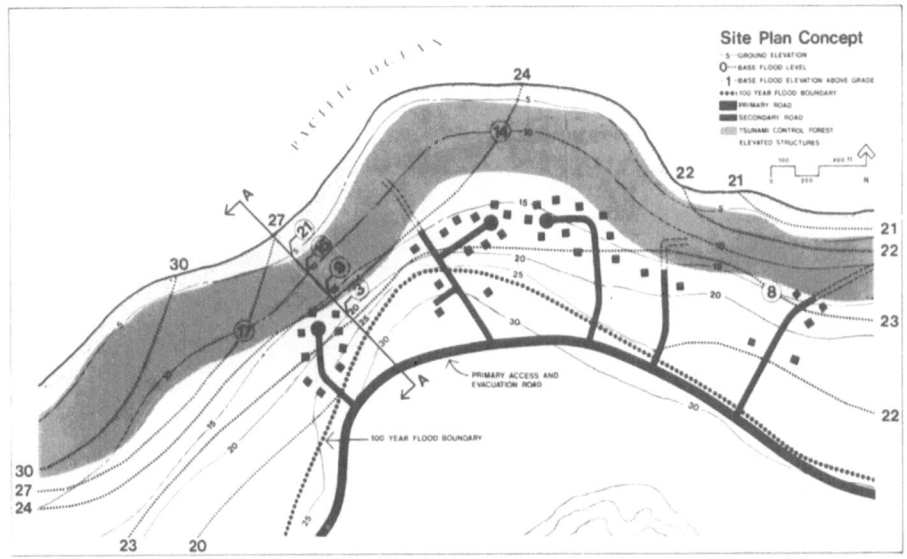

Fig. 5. Site plan concept.

which combines a variety of tsunami oriented measures. Figure 6 is the same site plan concept in section.

These plans illustrate interrelationships and variations in: a) Characteristics of the site (a range of coastal segments with varying topographic configurations. b) Characteristics of the site plan elements in terms of tsunami response. Components of the site plan are combined with differing relationships.

A cost benefit analysis was then conducted for each scheme. The purpose of this analysis was to compare both the relative magnitude of costs and benefits and the incidence of the costs and benefits in terms of implementation responsibility. The various components of the site plan were evaluated from three perspectives:

1) *Hazard Mitigation*: The types of wave force against which protective measures are designed were described in terms of relative level of effectiveness. For example, some types of action may be most effective in addressing impacts attributable to surge force, while others may be designed to protect structures more against drag force.

2) *Socio-Economic Impacts*: Relative magnitude of costs are discussed in terms of the incidence of responsibility to implement the recommended tsunami hazard measures. The relative magnitude of short term initial costs are compared to long term repayment and/or maintenance costs. For example, the implementation of some measures is primarily a public sector responsibility while other measures are a private sector responsibility.

3) *Urban Design Impacts*: Urban design is the organization of the elements of the natural and manmade environment into an ordered framework. The focus of this activity is on the spatial arrangement of activities and the form these activities will assume. Urban design implications are discussed in terms of aesthetics and community

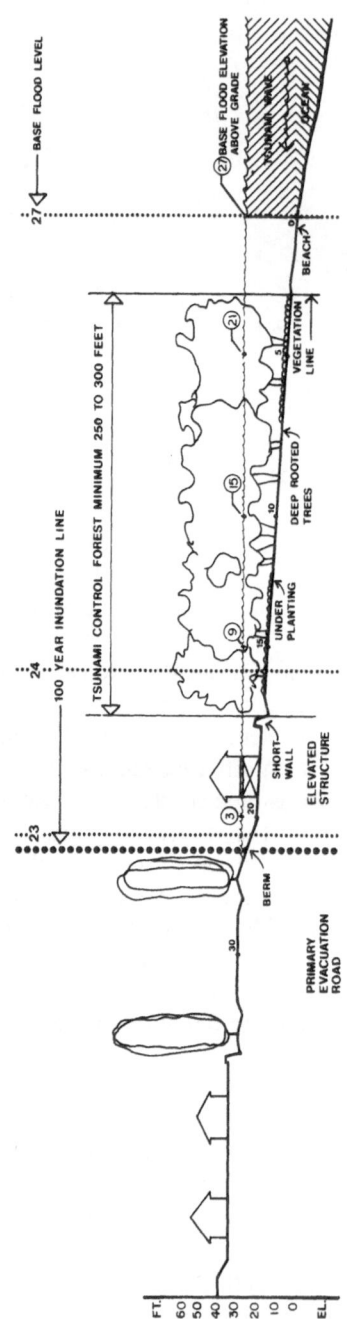

Fig. 6.   Site plan concept: section drawing.

goals pertaining to proposed activity relationships and spatial/form.

In order to demonstrate the applicability of this tsunami hazard oriented approach to project planning a final demonstration site plan was prepared at the even more refined scale of $1'' = 200'$. At this scale a site plan for a multi-family use is developed in greater detail. Using the methodology described above the costs and benefits of this plan can again be evaluated in comparison to a scheme which is not hazard oriented.

5. Conclusions of Project Scale Analysis

Major policy oriented plans reflect public awareness of the hazard. Once a decision has been made to permit development within the hazard zone there are a variety of ways in which to minimize hazard related risks while complying with the socio-economic and aesthetically oriented objectives of the community.

The site planning process must be conducted for each individual project area. Because the characteristics of the tsunami hazard are unique to each site, regulations must permit a flexible process whereby structures may be arranged on the site in response to the demands of the natural environment. The site conditions should dictate the arrangement of uses rather than a traditional zoning code which defines minimum lot size, front and site yard dimensions. The contours of the expanded wave height mitigated by friction factors such as berming and vegatation massing will determine the building location on the site.

If it is not possible for tsunami oriented site planning criteria to guide development of safe, economically viable, aesthetic projects the site is clearly inappropriate for development and should be purchased for use as public open space.

Present guidelines such as those included as part of the U.S. Federal Flood Insurance Program do not comprehensively take into consideration the dynamics of the wave force or interactions of the land with the wave. For example, no specifications are included to minimize drag force. Orientation to the wave is also not considered. These types of guidelines should be made explicit in communities where development is permitted in the hazard zone in order to ensure maximum effective utilization of all protective measures.

I would like to acknowledge the participation on the project of Joseph Morgan, Geographer and Rolf Preuss, Urban Designer.

Symposium Programs

Technical Program

Sunday 24 May
Hotel Sendai Plaza, reception desk (1st floor)
14:00–20:00 Symposium Check-in.

Monday 25 May
Hotel Sendai Plaza, Hall Matsushima-Nishinoma
09:00–10:00 Opening Ceremony

Presiding by Dr. K. Kajiura, Prof. E.R.I. Univ. of Tokyo, Secretary of Organizing Committee, Japan.

Address of Welcome by Dr. T. Iwasaki, Prof. Fac. Engg. Tohoku Univ. Chairman of Organizing Committee, Japan.

Address of Welcome by Mr. S. Yamamoto, Governor of Miyagi Prefecture, Japan.

Opening by Dr. K. Iida, Prof. Fac. Engg. Aichi, Inst. Technology, Chairman of Tsunami Commission, IUGG. Japan.

Remark by Dr. H. Loomis, Prof. Joint Inst. Marine and Atmospheric Research, Univ. Hawaii, U.S.A. Secretary of Tsunami Commission, IUGG.

10:00–10:15 Coffee Break

10:15–12:15 Session A   Tsunami Source and Earthquake and Warning System

Chairman: Dr. W. M. Adams, Professor, Hawaii Institute of Geophysics, University of Hawaii, Honolulu, U.S.A.

Co-chairman: Dr. Z. Suzuki, Professor, Geophysical Institute, Tohoku University, Sendai, Japan.

A–1 Seismic Source Spectrum of Tsunami and Ordinary Earthquake.
M. Takemura and J. Koyama, Tohoku Univ., Japan

A–2 Investigation of Rayleigh Wave Spectra for a Set of Tsunamigenic and Nontsunamigenic Earthquakes.
V. K. Gusiakov, Computing Centre, Novosibirsk, U.S.S.R.

A–3 Use of Long Period Seismic Waves for Fast Evaluation of Tsunami Potential of Large Earthquakes.
H. Kanamori and J. W. Given, Calif. Inst. Tech., U.S.A.

A–4 A New System for Seismic Observations and Tsunami Warning in the Japan Meteorological Agency.

M. Ichikawa and H. Watanabe, Seismological Div., Japan Meteorol. Agency, Japan

A–5 Some Remarks on the Occurence of Tsunamigenic Earthquakes around the Pacific.
K. Iida, Aichi Inst. Tech., Japan

A–6 Mechanism of Tsunami Earthquakes: Evidence from Tsunami Data.
R. P. Comer, M. I. T., U.S.A.

A–7 A New Scale of Tsunami Magnitude, Mt.
Katsu. Abe, Hokkaido Univ., Japan

13:45–15:45 Session B Tsunami Waves and Spectra
Chairman: Dr. S. S. Voyt, Institute of Oceanology, Academy of Sciences of the U.S.S.R., Moscow, U.S.S.R.
Co-chairman: Dr. K. Horikawa, Professor, Department of Civil Engineering University of Tokyo, Japan.

B–1 Verification, Calibration and Quality Assurance for Tsunami Models.
W. M. Adams, Univ. Hawaii, U.S.A.

B–2 Digitization of Tsunamigrams.
S. O. Wigen, Inst. Ocean Sci., Canada

B–3 Paralogism of the Wave in Lituya Bay.
H. Miyoshi, Tokyo Univ. Fisheries. Japan

B–4 An Edge Wave Trapped along a Curved Coast.
Y. Fujinawa, National Res. Center Dis. Prevention, Japan

B–5 Study of Shelf Effect for Tsunami using Spectral Analysis.
Kun. Abe, Nippon Dental Univ. and H. Ishii Tohoku Univ., Japan

B–6 Maximum Entropy Spectral Analysis of Tsunami along the Mexican Coast. 1952–1979
A. J. Sanchez Es. Inv. Ocenogr. and S. F. Farreras, CICESE, Mexico

Hotel Sendai Plaza Hall Matsushima-Higashinoma
13:45–15:45 Session BJ Tsunami Potential Estimation
Chairman: Dr. Li-San Hwang, Tetra Tech, INC., Pasadena, California, U.S.A.
Co-chairman: Dr. H. Watanabe, Nagoya Regional Weather Station, Nagoya, Japan.

BJ–1 Susceptibility of Western Australian Coastline to Tsunami originated South of Indonesia.
H. Allison, CSIRO, Wembley, Australia and S. Nakamura, Kyoto Univ. Japan

BJ–2 Tectonic, Seismic and Tsunami Energies.
T. S. Murty, Inst. Ocean Sci. Canada

BJ–3 Certain Concepts in Numerical Data Analysis Applicable to Tsunami Research.
T. S. Murty, Inst. Ocean Sciences, Canada

BJ–4 Some Statistics relevant to Tsunami: the Coastal Height Statistics and Correlations of Tsunami and Earthquake Parameters.
K. Kajiura, Univ. Tokyo, Japan

BJ–5 Parameters of Tsunami Waves in the Source.
N. R. Mirchina, E. N. Pelinovsky and S. Kh Shavratsky, Inst. of Applied Physics, Acad. Sci. U.S.S.R.

15:45–16:00 Coffee Break

Hotel Sendai Plaza, Hall Matsushima-Nishinoma
16:00–18:00 Session C   Theoretical Arguments on Tsunami Waves
        Chairman: Dr. T. S. Murty, Federal Government of Canada, Fisheries and
        Oceans Institute of Ocean Sciences, Sidney, Canada.
        Co-chairman: Dr. K. Kajiura, Professor, Earthquake Research Institute,
        University of Tokyo, Tokyo, Japan.
        C–1 Focusing and Reflection of a Cylindrical Solitary Wave.
           A. T. Chwang and H. Power, Univ. Iowa, U.S.A.
        C–2 Nonlinear and Dispersive Deformation of Tsunami with Typical Initial
           Profiles on Continental Topographies.
           M. Shibata, INA Ltd. Japan
        C–3 On Some Three-Dimensional Aspects of Tsunami Coastal Effects.
           H. R. Scember and T. Y. Wu, C. I. T., U.S.A.
        C–4 About the Possible Mechanism of Chilean Tsunami in 1960
           S. S. Voyt, A. N. Lebedve and B. I. Sebekin, Inst. Oceanology, U.S.S.R.
        C–5 Tsunami Generation as Finite Depth Cauchy-Poisson Problem or Long
           Wave Problem.
           T. Ichiye, Texas A & M Univ., U.S.A.
        C–6 The Most Dangerous Tsunami Wave Form.
           H. Allison, CSIRO, Australia.

Hotel Sendai Plaza, Hall Keyakinoma, 2nd floor
19:00–21:00 General Reception

Presiding by Dr. K. Horikawa, Prof. Engg. Univ. Tokyo
Address by Dr. T. Iwasaki
Address by Dr. K. Iida, Mr. T. Fukuchi, Dr. B. D. Zetler
Toast by Dr. Y. Kato.

Tuesday 26 May
Hotel Sendai Plaza, Hall Matsushima-Nishinoma
09:00–10:20 Session D   Tsunami Generation and Numerical Simulation of Historical Tsuna-
        mi
        Chairman: Mr. B. D. Zetler, Scripps Inst. of Oceanography, Univ. of Califor-
        nia, La Jolla, U.S.A.
        Co-chairman: Dr. N. Shuto, Professor, Department of Civil Engineering,
        Faculty of Engineering, Tohoku University, Sendai, Japan.
        D–1 A Numerical Model for Tsunami Generation and Propagation.
           Philip L. F. Liu and J. Earickson, Cornell Univ., U.S.A.
        D–2 Numerical Simulation of Historical Tsunami generated off the Tokai
           District, Central Japan.
           I. Aida, Univ. Tokyo, Japan
        D–3 Finite Element Method for Tsunami Wave Propagation in Tokai District.
           K. Iida, Aichi Inst. Tech., T. Suzuki, Chubu Electric Power Co., K.
           Inagaki and K. Hasegawa, Unic Corp., Japan
        D–4 Tsunami Simulation with an Explicit, Variable-grid, Numerical Timestep-
           ping Scheme.

H. G. Loomis, Univ. Hawaii, U.S.A.

10:20–10:35 Coffee Break

10:35–12:15 Session E   Harbor Oscillations by Long Waves and Tsunamis (1)
             Chairman: Dr. F. Raichlen, Professor, W. M. Keck Laboratory, California
             Institute of Technology, Pasadena, California, U.S.A.
             Co-chairman: Dr. Y. Iwagaki, Professor, Department of Civil Engineering,
             Faculty of Engineering, Kyoto University, Kyoto, Japan.
             E–1 Tsunami Response of the Tsugaru Straits.
                     S. Takahashi and I. Yakuwa, Hokkaido Univ. Japan
             E–2 Tsunami intruding into a Bay in a Scope of Numerical Experiment.
                     S. Nakamura, Kyoto Univ., Japan
             E–3 A Hybrid Simulation System developed for Model Tests of Tsunami in a
                  Harbor.
                     T. Iwasaki, Tohoku Univ., Japan
             E–4 Effects of the Continental Shelf on Harbor Resonance.
                     Philip L.-F. Liu, C. I. T., U.S.A.
             E–5 Seiches in Bays Forming a Coupled System.
                     M. Nakano and N. Fujimoto. Tokai Univ., Japan

13:45–15:45 Session F   Tsunami Runup
             Chairman: Dr. T. Y. Wu, California Institute of Technology, Pasadena,
             California, U.S.A.
             Co-chairman: Dr. H. Miyoshi, Professor, Tokyo University of Fisheries,
             Tokyo, Japan.
             F–1 Amplification of Linear Long Waves inside Bays.
                     A. Mano, Tohoku Univ., Japan
             F–2 Numerical Simulation of Tsunami Propagation and Runup.
                     C. Goto and N. Shuto, Tohoku Univ., Japan
             F–3 Tsunami Runup and Backwash on a Dry Bed.
                     K. K. Chu and T. Abe, Asian Inst. Tech., Thailand
             F–4 Research on Shoreline Wave Height and Land Run-up Height of Tsunami
                  on Uniform Sloping Beaches.
                     H. Togashi, Nagasaki Univ., Japan
             F–5 Wave Front Condition and Friction in the Tip Region of the Runup of
                  Tsunami on Dry Bed.
                     H. Matsutomi, Akita Univ., Japan
             F–6 Effects of Large Obstacles on Tsunami Inundation.
                     N. Shuto and C. Goto, Tohoku Univ., Japan

15:45–16:00 Coffee Break

16:00–18:00 Session G   Mitigation of Tsunami Hazards and Socio-economic Effects
             Chairman: Dr. Philip L. F. Liu, Associate Professor, Keck Lab. of Hydraulics
             and Water Resources, California Institute of Technology, Pasadena, Califor-
             nia, U.S.A.
             Co-chairman: Dr. H. Togashi, Professor, Department of Civil Engineering,
             Faculty of Engineering, Nagasaki University, Nagasaki, Japan.

G-1 Types of Tsunami Disasters and Protection Measures in Japan.
     K. Horikawa, Univ. Tokyo and N. Shuto, Tokyo Univ.
G-2 Countermeasures against Tsunami in Fishing Villages of Sanriku Coast in Japan.
     T. Fukuchi and K. Mitsuhashi, Fisheries Agency, Japan
G-3 Design and Construction of Ofunato Tsunami Protection Breakwater.
     T. Matsumoto and Y. Suzuki, Ministry of Transport, Japan
G-4 On the Function of Seawalls and Breakwaters to mitigate Tsunami Hazards.
     T. Iwasaki and A. Mano, Tohoku Univ., Japan
G-5 Land Management Guidelines for Tsunami Hazard Zone.
     J. Preuss, Urban Regional Res., U.S.A.
G-6 Tsunami Impact on Society.
     G. Pararas-Carayannis, ITTC, U.S.A.

Visiting Tsunami Laboratory of Tohoku University
        after lunch on Monday 25 and Tuesday 26.

12:15–13:45 IUGG Tsunami Commission Meeting

Sendai Memorial Hall for Restoration from War Damage
19:30–21:00 Cultural Program
             Koto and Japanese Dancing
             Noh
             Music Concert

Wednesday 27 May
Start from Hotel Sendai Plaza at 08:30. After taking lunch at Kesennuma, arrive at Ofunato at 14:30. Study tour by Motor Coach to Ryori-Minato where a tsunami gate works as a bridge usually and as a gate for an emergency by being rotated in 90°. Check-in Ofunato Grand Hotel at 17:00.

Ofunato Grand Hotel
19:00–21:00 Banquet
Presiding by Dr. H. Togashi, Prof. Engg. Nagasaki Univ.
Address by Mr. Katsuzo Usui, Mayor of Ofunato City
Address by Dr. Iida, Dr. Miyoshi, Dr. T. Wu and Dr. F. Raichlen.

Thursday 28 May
Ofunato Nokyo Kaikan
09:00–10:00 Session H Harbor Oscillations by Long Waves and Tsunamis (2)
             Chairman: Mr. G. C. Dohler, Chairman ITSU, CAN. Hydrographic Service, Ottawa, Canada.
             Co-chairman: Mr. K. Tanimoto, Port and Harbour Research Institute, Ministry of Transport, Yokosuka, Japan.
             H-1 A Numerical Study of the Tsunami Response of a Harbor.
                  P. D. Farrar, USAE Waterways Experiment St. U.S.A.
             H-2 On the Hydraulic Aspects of Tsunami Breakwaters in Japan.
                  K. Tanimoto, Port and Harbour Res. Inst., Japan

H-3   The Excitation of Harbors by Tsunamis.
          F. Raichlen, T. G. Lepelletier and C. K. Tam, C.I.T., U.S.A.

10:00-10:15 Coffee Break

10:15-12:15 Session I   Historical Study of Tsunamis
          Chairman: Dr. Harold G. Loomis, Joint Institute for Marine and Atmospheric
             Research, University of Hawaii, Honolulu, U.S.A.
          Co-chairman: Dr. Y. Nagata, Professor, Geophysical Institute, Faculty of
             Science, University of Tokyo, Tokyo, Japan.
          I-1   Historical Study of Tsunamis at Tofino, Canada.
                   S. O. Wigen, Inst. Ocean Sci., Canada
          I-2   Colombia-Peru Tsunami that observed along the Coast of Japan.
                   T. Hatori, Univ. Tokyo, Japan
          I-3   The Biggest Tsunami in the Sanriku Districts.
                   K. Iida, Aichi Inst. Tech., H. Suzuki, Y. Osawa and H. Miyoshi, Tokyo
                   Univ. Fisheries., Japan
          I-4   The Tsunami and Associated Seiches as recorded on Current Meters and a
                   Pressure Gauge in Otsuchi Bay.
                   N. Shikama, Univ. Tokyo, Japan
          I-5   Historical Study of Tsunami at Miyako, Japan.
                   M. Okada, Japan Meteorological Agency and M. Tada, Miyako
                   Weather Station, Japan
          I-6   Report on the Earthquake and Tsunami of Sep. 20, 1498.
                   Y. Tsuji, National Res. Center Dis. Prevention, Japan

Start from Ofunato Nokyo Kaikan at 13:30. Inspection of the Tsunami trace mark in the
Ofunato City. Visit Okkirai Fishing Port and Tsunami defence works at Urahama, Okkirai. A
tall sea dike in Toni was visited. Bird's-eye view of construction of Kamaishi Tsunami
Breakwater and tsunami defence works in the quay area of the Kamaishi Port were studied.
Check in Hotel Sun Route Kamaishi at 17:30.

Hotel Sun Route Kamaishi, Hall Ho-o-no-ma
19:00-21:00 Closing Ceremony
          Presiding by Dr. N. Shuto, Prof. Engg. Tohoku Univ. Secretary of Executive
          Committee, Japan
          Address by Mr. Tadashi Nakamura, Governor of Iwate Prefecture, Japan
          Address by Mr. Saijiro Hamakawa, Mayor of Kamaishi City, Japan
          Banquet
          Vote of Thanks by Mr. Dohler, Dr. Loomis and Mrs. Zetler
          Thanks on behalf of the Organizing Committee by Dr. T. Iwasaki, Chairman
          of the Organizing Committee
          Closing Remarks by Dr. K. Iida, Chairman of Tsunami Commission, IUGG.

Ladies Program

Monday 25, May
09:00-10:00 Opening Ceremony at Hall Matsushima-Nishinoma in Hotel Sendai Plaza
10:00-10:30 Meeting at Katsura-no-ma in Hotel Sendai Plaza (4th floor).

10:30   Start from the hotel by owner-cabs
11:00–12:00   "Rin-no-ji temple," strolling in a Japanese garden.
12:00–13:30   Restaurant "Azumaya." Lunch invited by Mrs. Kimiko Iwasaki, Mrs. Chairman
                of the Organizing Committee
13:30   Start from "Azumaya."
                Visiting Tohoku University, Bird's-eye View of Sendai City and Suburbs from the
                top of the campus building.
14:40–15:20   Sightseeing "A-O-BA Jyo" the castle of verdue which was built by a feudal lord,
                Masamune Date.
15:40–16:20   Visiting Sendai City Museum
16:30–18:00   Shopping in O-o-machi and Higashi Ichi-Ban-Cho
18:00   Arrive at Hotel Sendai Plaza
19:00–21:00   General Reception

Tuesday 26, May
09:00   Start from Hotel Sendai Plaza by Motor Coach
09:40–10:00   Arrive at Relics of Ruined Temple of Tagajyo which was highly worshipped
                during periods "NARA" (700–780 A.D.) and "HEIAN" (780–1190 A.D.) as a
                temple to pray for peaceful reigning of northern parts by the Government.
10:10–11:00   Visiting the Tohoku Museum of History built in 1974 which displays archeologi-
                cal materials. It is aimed to answer questions on when our ancestors started their
                living and how their lives have been changed. Covered periods are since B.C. 8000
                to B.C. 300 (Jyo-mon), to A.D. 200 (Yayoi) to 700 (Kofun), to 780 (Nara), to 1190
                (Heian), to 1590 (Chusei), to 1868 (Kinsei) to 1940 (Kindai).
11:25–11:40   Visit Remains of "Taga" Castle, which was a local government on Tohoku
                District since 724 to 1126, in which military head-quarters were included. "Ezo"
                was a tribe who lived here as aborigines and has been fighting severely and
                persistently with the central government. Thus building of Taga Castle were burnt
                in battle fire two times. The third destroy was caused by a very large earthquake in
                869 (TEIKAN Great Earthquake) by which a huge tsunami, magnitude 4 was
                said to struck Sendai Plane.
12:00–14:00   Lunch at "Taritsuan" invited by the organizing committee.
14:10–14:30   Bird's-eye View of Matsushima, a scenic bay dotted with hundreds of pine-clad
                islets which is known as one of the "Three landscapes in Japan."
15:10–16:10   Visiting Sendai Museum of Folk Life in History. Interesting exhibition of
                "Tsutsumi Yaki" potteries and "Tsutsumi Ningyo" dolls specially produced in
                Sendai.
19:30–21:00   Cultural Program in Sendai Memorial Hall for Restoration from War Damage

Wednesday 27, May
                Sendai-Kesennuma-Ofunato
19:00–21:00   Banquet at Ofunato Grand Hotel

Thursday 28, May
09:00–12:15   Sightseeing in Ofunato
                (i)   Planting Fishery Center
                       (Nourishing of ear-sells and plaices)
                (ii)   Choanji-temple, a building of Momoyama-era (1600 A.D.)
13:00–17:30   Study Tour

19:00–21:00 Closing Ceremony

Popular Lectures

Tuesday 28 May
Ofunato Nokyo Kaikan
13:30–13:45 Opening Address by Dr. K. Iida, Aichi Inst. Tech., Chairman of Tsunami
                    Commission, IUGG
13:45–14:45 On the 1960 Chilean Earthquake and Tsunami.
                    Dr. Edgar Kausel, University of Chile., Chile
14:45–15:45 Tsunamis, Computors and Mankind.
                    Dr. W. M. Adams, University of Hawaii, U.S.A.

PROGRAM OF POST SYMPOSIUM STUDY TOUR

Friday 29 May
09:00–17:00 Technical visit to Tanohama, Funakoshi, Yamada, Taro and Jyodogahama.
                    Various interesting sites concerning tsunami inundation and defence works were
                    visited.
                    Overnight stay at Miyako.

Saturday 30 May
09:00–17:00 Technical visit to Omoto, another huge defence works against Tsunamis.
                    Sightseeing Ryusendo, a stalactite grotto with a mysteriously calm underground
                    lake, 120 meters deep. A wide plain on a plateau called Hayasaka Heights is
                    peculiar and hard to be found in any part of Japan.
                    Overnight stay in Morioka.

Sunday 31 May
08:30   Leave Morioka
                    Arrive at the Hanamaki Airport.
                    On the way from the Hanamaki Airport to the Sendai Airport, Hiraizumi, the
                    ruined capital of the Fujiwara Dynasty in the 11th century, was visited. The
                    "Golden Pavilion," mummies and treasures of the kings of the dynasty and
                    Motsuji temple with a picturesque garden are in Hiraizumi.
                    Arrive at the Sendai Airport.
                    Tour terminates.

List of Participants

Canada

Mr. Gerhard C. Dohler
Chairman ITSU
CAN. Hydrographic Service
615 Booth STR., Ottawa, KIA 0E6

Dr. Tad S. Murty
Federal Government of Canada
Fisheries and Oceans Institute of
Ocean Sciences
P.O. Box 6000
Sidney, B.C, V8L 4B2

Mr. Sydney O. Wigen
Institute of Ocean Sciences
P.O. Box 6000
Sidney, B.C. V8L 4B2

Mrs. Nancy E. Wigen

Chile

Dr. Edgar Kausel
Dep. of Geophysics
Univ. of Chile
Casilla 2777
Santiago

Denmark

Prof. Helge Lundgren
Institute of Hydrodynamics
and Hydraulics Engineering
Building 115, Techn. Univ.
Denmark DK-2800 Lyngby

France

Mr. Michel Dutzer
CEA
29, 33, Rue de la Fédération
Paris 15^e

Hong Kong

Mr. Kan Kok Chu
73, Taikok Tsui Road
15/F, Hing Wong Bldg.
Kowloon

Japan

Mr. Katsuyuki Abe
Department of Geophysics
Faculty of Science
Hokkaido University
Sapporo, 001

Mr. Kuniaki Abe
Nippon Dental University
Niigata Hamauracho 1-8
Niigata City
Niigata Pref., 951

Mr. Yuhei Adachi
Nippon Tetrapod Co., Ltd.
Shimbashi Fuji Bldg., No. 1-3.
2-chome, Shimbashi, Minato-ku
Tokyo, 105

Prof. Chiaki Agemori
Kochi University
Faculty of Agriculture
B 200, Monobe
Nankoku-shi, 783

Dr. Isamu Aida
Earthquake Research Institute
University of Tokyo
1-1-1 Yayoi-cho Bunkyo-ku
Tokyo 113

Mr. Tadamitsu Aonuma
Tohoku Electric Company Ltd.
7-1, 3-chome, Ichiban-cho
Sendai 980

Dr. Shinichi Arai
  Technical Research Institute
  Hitachi Shipbuilding & Engineering
  Co., Ltd.
  3-22, Sakurajima, 1-chome
  Konohana-ku, Osaka, 554
Mr. Naohiro Fujimoto
  Faculty of Marine
  Science and Technology
  Tokai University
  Kamiyamaguchi 2039-10
  Tokorozawa-shi
  Saitama-ken, 359
Dr. Yukio Fujinawa
  National Research Center for
  Disaster Prevention
  Tennodai 3, Sakura-mura
  Ibaraki-ken, 305
Mr. Tatsuma Fukuchi
  Fishing Port Department, Fisheries
  Agency, Ministry of Agriculture
  Forestry and Fisheries
  1-2-1, Kasumigaseki, Chiyoda-ku
  Tokyo
Mr. Chiaki Goto
  Civil Engineering, Tohoku
  University,
  Aoba, Aramaki
  Sendai, 980
Mr. Naoki Hase
  Co. Ltd., Hase's Geological Survey
  3-5-8, Honcho, Sendai
Mr. Masaru Hashimoto
  C.T.I. Engineering Co., Ltd.
  9th Chuo Bldg., 4-2 Honcho
  Nihonbashi Chuo-ku
  Tokyo
Dr. Tokutaro Hatori
  Earthquake Research Institute
  University of Tokyo
  Yayoi 1-1, Bunkyo-ku
  Tokyo, 113
Mr. Kenkichi Hayashi
  Nippon Koei Co., Ltd.
  4 Kojimachi, 5-chome, Chiyoda-ku

Tokyo, 102
Mr. Toshiyuki Hibiya
  Earthquake Research Institute
  University of Tokyo
  Yayoi, 1-1, Bunkyo-ku
  Tokyo, 113
Dr. Haruo Higuchi
  Ehime University
  3 Bunkyo, Matsuyama, 790
Prof. Mikio Hino
  Dept. of Civil Engineering
  Tokyo Institute of Technology
  O-okayama, Meguro-ku
  Tokyo, 152
Prof. Tomowo Hirasawa
  Obs. Center for Earthquake
  Prediction
  Fac. of Science
  Tohoku University
  Sendai, 980
Prof. Masashi Hom-ma
  Toyo University
  2-8-36, Zoshigaya, Toshima-ku
  Tokyo, 171
Mrs. Toshiko Hom-ma
Prof. Kiyoshi Horikawa
  Department of Civil Engineering
  University of Tokyo
  Bunkyo-ku, Tokyo, 113
Mr. Yoshihisa Hoshino
  Tohoku Regional Construction
  Bureau
  Ministry of Construction
  9-15 Futsukamachi, Sendai
Dr. Masanobu Hosoi
  Nagoya Institute of Technology
  Gokiso-cho Showa-ku
  Nagoya-shi
Prof. Kumizi Iida
  Professor Emeritus
  Nagoya University
  1-39, Higashiyama-motomachi
  Chikusa-ku, Nagoya, 464
Mrs. Fujiko Iida
Mr. Hidetoshi Inagaki

Fishing Port Division
Shizuoka Prefectural Government
9-6, Ohtemachi, Shizuoka City
Shizuoka Prefecture, 420
Mr. Kazuo Inagaki
Unic Corporation
2-8-7 Shinjuku, Shinjuku-ku
Tokyo
Dr. Hajime Ishida
Lab. Coastal Eng.
Dept. of Civil Eng., Kanazawa Univ.
2-40-20, Kodatsuno
Kanazawa, 920
Dr. Hiroshi Ishii
Obs. Center for Earthquake
Prediction, Fac. of Science
Tohoku University
Sendai, 980
Mr. Tadaharu Ishikawa
River Hydraulics Division
Public Works Research Institute
Ministry of Construction
Asahi-1 Toyosato-cho, Tsukuba-
gun, Ibaraki-ken, 300-26
Prof. Yuichi Iwagaki
Department of Civil Engineering
Faculty of Engineering
Kyoto University
Sakyo-ku, Kyoto, 606
Mr. Sin'iti Iwasaki
Earthquake Research Institute
Univ. of Tokyo
Yayoi, 1-1, Bunkyo-ku
Tokyo, 113
Prof. Toshio Iwasaki
Dept. of Civil Engineering
Tohoku University
Sendai, 980
Mrs. Kimiko Iwasaki
Mr. Sumio Kaihatsu
Tohoku Electric Company Ltd.
7-1, 3-chome, Ichiban-cho
Sendai, 980
Dr. Kinjiro Kajiura
Earthquake Research Institute

University of Tokyo
Yayoi, 1-1-1, Bunkyo-ku
Tokyo, 113
Mr. Yoshiro Kano
Tohoku Electric Company Ltd.
7-1, 3-chome, Ichiban-cho
Sendai, 980
Mr. Sinji Kataoka
Ministry of Transport, Bureau of
Ports and Harbours, Disaster
Prevention Division
2-1-3, Kasumigaseki, Chiyoda-ku
Tokyo, 100
Ass. Prof. Sanshiro Kawai
Geophysical Institute
Faculty of Science
Tohoku University
Sendai, 980
Mr. Seiya Kinoshita
River Division
Shizuoka Prefectural Government
9-6, Ohtemachi, Shizuoka City
Shizuoka Prefecture, 420
Mr. Haruo Kitamatsu
Tohoku Electric Company Ltd.
7-1, 3-chome, Ichiban-cho
Sendai, 980
Prof. Hideo Kondo
Muroran Institute of Technology
27-1, Mizumoto-cho, Muroran-shi
Hokkaido, 050
Mr. Kazuo Kondo
Public Works Division
Shizuoka Prefectural Government
9-6, Ohtemachi, Shizuoka City
Shizuoka Prefecture, 420
Dr. Junji Koyama
Geophys. Inst., Tohoku Univ.
Aramaki Aoba
Sendai, 980
Dr. Nicholas C. Kraus
Near Shore Environment Center
1203 Famil-Hongo
1-20-6 Mukogaoka, Bunkyo-ku
Tokyo, 113

Mr. Akibumi Kuriyama
   The President of Maeda Con-
   struction Co. Ltd. Sendai Branch
   11-25, 3-chome, Kokubuncho
   Sendai, 980
Mr. Takao Maebara
   C.T.I. Engineering Co., Ltd.
   9th Chuo Bldg., 4-2 Honcho
   Nihonbashi Chuo-ku
   Tokyo, 103
Dr. Akira Mano
   Dept. of Civil Engineering Fac. of
   Engineering, Tohoku University
   Aza-Aoba, Aramaki
   Sendai, 980
Mr. Kazuaki Masaki
   Aichi Inst. Tech.
   1247 Yachigusa, Yachigusa-machi
   Toyota-shi, 470-03
Mr. Hideo Matsutomi
   Department of Civil Engineering
   Akita University
   1-1, Tegata Gaguen-cho
   Akita-shi, 010
Mr. Koichi Matsuura
   The President of Uminokai
   c/o Goyo Construction Co. Ltd.
   16-20 Futsuka-machi
   Sendai, 980
Mr. Koji Mitsuhashi
   Fishing Port Department, Fisheries
   Agency, Ministry of A.F. and
   Fisheries
   1-2-1, Kasumigaseki, Chiyoda-ku
   Tokyo, 100
Prof. Akira Miura
   College of Industrial Technology
   Nihon University
   1-2-1, Izumi-cho, Narashino-shi
   Chiba-ken, 275
Dr. Motoyasu Miyata
   Geophysical Institute
   University of Tokyo
   Hongo, Bunkyo-ku
   Tokyo, 113

Dr. Hisashi Miyoshi
   Tokyo University of Fisheries
   257-17, Minami Oizumi
   Nerima-ku, Tokyo, 177
Mrs. Y. Miyoshi
Mr. Tatsuya Mochizuki
   Seashore Department
   Bureau of Rivers, Ministry of
   Construction
   1-3, 2-chome, Kasumigaseki,
   Chiyoda-ku, Tokyo, 100
Prof. Yutaka Nagata
   Geophysical Institute, Faculty of
   Science, University of Tokyo
   Hongo, Bunkyo-ku, Tokyo, 113
Mr. Takeo Nakajima
   Earthquake Engineering Division
   Public Works Research Institute
   Asahi 1, Toyosato Machi, Tsukuba
   Gun, Ibaraki Ken, 305
Dr. Shigehisa Nakamura
   Disaster Prevention Research
   Institute Kyoto University
   Uji, Kyoto, 611
Prof. Masito Nakano
   Faculty of Marine Science and
   Technology, Tokai University
   4-8-14 Higashi, Kunitachi-shi
   Tokyo, 186
Mr. Keiji Narushima
   C.T.I. Engineering Co., Ltd.
   9th Chuo Bldg. 4-2 Honcho
   Nihonbashi Chuo-ku
   Tokyo, 604
Mr. Masaru Nishizawa
   Dept. of Civil Engineering
   Fac. of Engineering
   Tohoku University
   Sendai, 980
Prof. Atsushi Numata
   Dept. of Civil Engineering
   Tohoku Institute of Technology
   Aza-Koezi, Nagamachi, Sendai 980
Mrs. Keiko Numata
Prof. Kunihiro Ogihara

Civil Engineering Depart.
Faculty of Engineers
Toyo University
2100, Kujirai, Kawagoe
Saitama, 350

Mr. Masami Okada
Oceanography Division, Japan
Meteorogical Agency
1-3-4 Otemachi Chiyoda-ku
Tokyo, 100

Mr. Akira Saito
Tokai University
1000, Orido, Shimizu, 424

Dr. Tetsuo Sakai
Associate Prof. of Coastal Engineer-
ing Dept. of Civil Eng., Kyoto Univ.
Sakyo-ku, Kyoto, 606

Mr. Toshiro Sasagawa
Tohoku Electric Company Ltd.
7-1, 3-chome, Ichiban-cho, Sendai,
980

Mr. Masayuki Sato
Tohoku Electric Company Ltd.
7-1, 3-chome, Ichiban-cho, Sendai,
980

Dr. Michio Sato
Kagoshima University
1-24-40, Kourimoto
Kagoshima, 890

Prof. Toru Sawaragi
Dept. of Civil Eng. Osaka Un-
iversity,
Yamada-kami, Suita, Osaka, 564

Mr. Masakazu Shibata
INA Civil Eng. Systems Consultants
Ltd.
22-1 Suido-cho, Shinjuku-ku
Tokyo, 162

Mr. Nobuyuki Shikama
Otsuchi Marine Research Center
Ocean Research Institute
University of Tokyo
2-106-1 Akahama, Otsuchi
Iwate, 028-11

Mr. Atsuyuki Shimada

Coastal Hydraulic Section
Hydraulic Department
Civil Engineering Laboratory
Central Research Institute
of Electric Power Industry
1641, Abiko, Abiko-shi
Chiba-ken, 270-11

Mr. Kazutoshi Shimakura
I.N.A. Civil Engineering Consulting
22-1 Suido-cho, Shinjuku-ku
Tokyo, 162

Prof. Nobuo Shuto
Dept. of Civil Engineering
Fac. of Engineering, Tohoku
University
Sendai, 980

Mrs. Noriko Shuto

Mr. Keizo Suda
Engineering & Technical Section
Department, Tohoku Oil Co., Ltd.
1-1 Minato 5-chome
Sendai, 980

Mr. Kyozo Suga
River Hydraulics Division,
Public Works Research Institute
Ministry of Construction
Asahi-1 Toyosato-cho, Tsukubagun
Ibaraki-ken, 300-26

Mr. Keiichi Sugawara
Toa Kensetsu
1-19, Chuo ichome
Sendai City, 980

Mr. Naoyuki Suzuki
Tohoku Electric Company Ltd.
7-1, 3-chome, Ichiban-cho, Sendai,
980

Mr. Takao Suzuki
Chubu Electric Power Company
Inc.
1 Toshin-cho Higashi-ku
Nagoya, 461

Mr. Yuzo Suzuki
Miyako Port Construction Office
The 2nd District Port
Construction Bureau

1-2-5 Takashima, Nishiku
Yokohama, Kanagawa prefecture,
220
Prof. Ziro Suzuki
Geophysical Institute
Tohoku University
Sendai, 980
Ass. Prof. Susumu Takahashi
Faculty of Engineering
Hokkaido University
N-13, W-8 Kitaku
Sapporo, 060
Dr. Masayuki Takemura
Kaiima Institute of Construction
Technology
19-1 Tobitakyu, 2-chome
Chofu-shi, Tokyo, 182
Mr. Ichiro Tanahashi
Urban Planning Department
Building Research Institute
Ministry of Construction
1 Tatehara, Oho-machi
Tsukuba-gun, Ibaraki Prefecture,
300-26
Dr. Hiromichi Tanaka
Dept. of Civil Engineering Fac. of
Engineering, Hachinohe Technical
College
Aza-Uenodaira, Tamonogi
Hachinohe, 031
Mr. Katsutoshi Tanimoto
Port and Harbour Research
Institute
Ministry of Transport
1-1, Nagase 3-Chome, Yokosuka
239
Prof. Yoshiaki Toba
Geophysical Institute,
Faculty of Science,
Tohoku University
Sendai, 980
Prof. Hiroyoshi Togashi
Dept. of Civil Eng., Fac. of Eng.
Nagasaki University
1-14, Bunkyo-machi

Nagasaki City, 852
Mr. Yoshinobu Tsuji
National Research Center for
Disaster, Prevention, Science
and Technology Agency
Nijigahama 9-2, Hiratsuka City
Kanagawa-ken, 254
Ass. Prof. Tadayasu Uehara
Dept. of Civil Engineering, Fac. of
Engineering, Tohoku Gakuin
University
13-1, 1-chome, Chuo, Tagajo-shi,
985
Mrs. Toshiko Uehara
Dr. Hideo Watanabe
Nagoya Local Weather Station
Hiyori-cho 2-18, Chigusa-ku
Nagoya
Mr. Sadahiro Watanabe
Earthquake Preparedness Division
Shizuoka Prefectural Government
9-6, Ohtemachi Shizuoka City
Shizuoka Prefecture, 420
Prof. Isao Yakuwa
Department of Engineering
Faculty of Science
Hokkaido University
North 13. West 8
Sapporo, 060
Mr. Shigeru Yamaki
Department of Geophysics
Faculty of Science
Hokkaido University
West 7, North 10, Kita-ku
Sapporo
Mr. Koichi Yamamoto
River Hydraulics Division
Public Works Research Institute
Ministry of Construction
Asahi-1 Toyosato-cho, Tsukubagun
Ibaraki-ken, 300-26

Peru

Mr. Jorge Del Aguila
Direccion de Hidrografia Y

Navegacion de la Marina
Calle Gamarra 500
Chucuito Callao
Prof. H. Julio Kuroiwa
National University of Engineering
P.O. Box 1301
LIMA-1
Mr. Cesar Vargas
Direccion de Hidrografia Y
Navegacion de la Marina
Calle Gamarra 500
Chucuito Callao
Miss Elvira Vasquez
Port and Terminal Fishery Center
Infrastructure Dept., Ministry of
Fishery
Avenida Mexico, 469-1
La Victoria, Lima

Thailand

Dr. Tetsuo Abe
Asian Institute of Technology
c/o Water Resources Engg.
Division, AIT,
P.O. Box 2754, Bangkok

U.S.A.

Prof. William M. Adams
University of Hawaii
Hawaii Institute of Geophysics
2525 Correa Road Honolulu
Hawaii, 96822
Mrs. N. N. Adams
Dr. Fred E. Camfield
U.S. Army, Corps of Engineers
Coastal Engineering Research
Center, Kingman Building Fort
Belvuir Virginia, 22060
Prof. Allen T. Chwang
Institute of Hydraulic Research
University of Iowa
Iowa City, Iowa, 52242
Mr.Robert P. Comer
Department of Earth and Planetary
Science

Massachusetts Institute of
Technology
54-521, 77 Massachusetts Avenue,
Cambridge, Massachusetts, 02139
Prof. Salvador Farreras
Cicese Oceanography Dept.
P.O. Box 4844
San Ysidro, California, 92073
Ph. D. Li-San Hwang
Tetra Tech, INC.
630 N. Rosemead Boulevard
Pasadena, California, 91107
Prof. Takashi Ichiye
Department of Oceanography
Texas A & M University
College Station, Texas, 77843
Associate Prof. Philip L.-F. Liu
Cornell University
California Institute of Technology
W. M. Keck Laboratory of
Hydraulics
California Institute of Technology
Pasadena, California, 91125
Dr. Harold G. Loomis
Joint Institute for Marine
and Atmospheric Research
University of Hawaii
2525 Correa Road
Honolulu, Hawaii, 96822
Prof. Hiroo Kanamori
Seismological Laboratory, 252-21
California Institute of Technology
Pasadena, California, 91125
Dr. George Pararas-Carayannis
International Tsunami Information
Center (UNESCO/IOC)
P.O. Box 50027
Honolulu, Hawaii, 96850
Mrs. Irene Pararas-Carayannis
Mrs. Jane Preuss
Urban Regional Research
1426 Fifth Avenue
Suite 308
Seattle, Washington, 98115
Prof. Fredric Raichlen

W.M. Keck Lab. of
Hydraulics & Water Resources
California Institute of Technology
Pasadena, CA 91125
Mrs. J. Raichlen
Prof. Antonio Sanchez
    Escuela Superior de
    Ciencias Marinas (UABC)
    P.O. Box 4844
    San Ysidro, California, 92073
Prof. Theodore Y. Wu
    California Institute of Technology
    104-44, California Institute of Tech-
    nology

Pasadena, CA 91125
Mr. Bernard D. Zetler
    Scripps Inst. of Oceanography
    I.G.P.P. Code A-025
    University of California
    San Diego
    La Jolla, CA 92093
Mrs. H. E. Zetler

U.S.S.R.

Prof. Serguei Voyt
    Institute of Oceanology Academic of
    Sciences of the USSR
    Moscow

Author Index

557

Subject Index